Circuit Analysis for Analog, Mixed-Signal and RF Designers

IEEE Press
445 Hoes Lane
Piscataway, NJ 08854

Circuit Analysis for Analog, Mixed-Signal and RF Designers

Shanthi Pavan

NT Alexander Institute-Chair Professor, Indian Institute of Technology Madras, Chennai, India

Gabor C. Temes

Oregon State University, Corvallis, OR, USA; and University of California, Los Angeles, CA, USA

IEEE Press
WILEY

Published by John Wiley & Sons, Inc., Hoboken, New Jersey.

For general information on our other products and services or for technical support, please contact our Customer Care Department within the United States at (800) 762-2974, outside the United States at (317) 572-3993 or fax (317) 572-4002.

Wiley also publishes its books in a variety of electronic formats. Some content that appears in print may not be available in electronic formats. For more information about Wiley products, visit our web site at www.wiley.com.

Library of Congress Cataloging-in-Publication Data:

Names: Pavan, Shanthi, 1973- author | Temes, Gabor C., 1929- author
Title: Circuit analysis for analog, mixed-signal and RF designers / Shanthi Pavan, NT Alexander Institute Chair Professor, Indian Institute of Technology, Madras, Gabor C. Temes, Oregon State University, Corvallis, OR, USA; and University of California Los Angeles, Los Angeles, CA, USA.
Description: Piscataway, NJ : IEEE Press | Includes bibliographical references and index.
Identifiers: LCCN 2026019223 | ISBN 9781394275007 hardback | ISBN 9781394275021 adobe pdf | ISBN 9781394275014 epub
Subjects: LCSH: Electric circuit analysis | Mixed signal circuits | Analog integrated circuits | Radio frequency integrated circuits
Classification: LCC TK454 .P38 2026
LC record available at https://lccn.loc.gov/2026019223

Cover Design: Wiley
Cover Image: Shanthi Pavan

Contents

	About the Authors	xi
	Preface	xiii
1	**Introduction and Basic Concepts**	1
1.1	Classification of Circuits	2
1.1.1	Linear and Nonlinear Circuits	2
1.1.2	Time Invariant and Time-Varying Circuits	3
1.1.3	Passive and Active Circuits	3
1.2	Circuit Components	5
1.3	Kirchhoff's Laws and the Incidence Matrix	7
1.4	Tellegen's Theorem	10
1.5	The Incremental Network and Small-Signal Analysis	14
1.6	Solving Linear Networks Using Modified Nodal Analysis	16
1.6.1	Networks with Multiple Ideal Opamps	24
1.6.2	Sinusoidal Steady-State: Capacitors and Inductors	28
1.7	Bilinear Form of a Transfer Function, and the Extra-Element Theorem	31
1.8	Transposed MNA Stamps	36
1.9	Reciprocity and Inter-Reciprocity	39
1.10	Why are Reciprocity and Inter-Reciprocity Useful?	44
1.10.1	Transfer Functions in a Biquadratic Section	47
1.11	Transfer-Function Sensitivity	49
	Problems	54
	References	59
2	**Noise in Linear Time-Invariant Networks**	61
2.1	Noise in Network Components	61
2.1.1	Noise Models for Other Devices	63
2.1.2	Relationship Between Mean-Square Noise and Power Spectral Density	64
2.1.3	Single- and Double-Sided Noise Spectral Density	64
2.1.4	Noise Processed by a Linear Time-Invariant System	65

2.2	Multiple Noise Sources	69
2.3	Equivalent Noise of a One-Port	69
2.3.1	Noise in RLC Networks and the Nyquist Noise Theorem	71
2.4	Total Integrated Noise in RLC Networks	72
2.4.1	Bode's Noise Theorem	77
2.5	Equivalent Noise Sources in a Two-Port Network	79
2.5.1	A Source-Independent Noise Model for the Two-Port	84
2.6	Noise with Impedance Scaling and Differential Operation	86
2.6.1	Impedance Scaling	86
2.6.2	Differential Operation	88
2.7	The Noise Factor of a Two-Port	89
2.7.1	SNR and the "Sweet Spot" in R_s	90
2.8	Example: Noise-Factor Calculations	91
	Problems	96
	References	102
3	**Weak Nonlinearities in Circuits**	105
3.1	Second-Order Intermodulation Distortion	107
3.2	IIP_2 of Cascaded Stages	112
3.3	Differential Operation	112
3.4	Third-Order Nonlinearity	113
3.4.1	Third-Order Intermodulation Distortion	114
3.4.2	IIP_3 of Cascaded Stages	120
3.5	DC Transfer-Curve Derivatives: A Quick Nonlinearity Diagnostic	120
3.6	The Method of Current Injection	122
3.6.1	A Weakly Nonlinear Negative Feedback Loop	131
3.6.2	AC Coupling with a Nonlinear Capacitor	134
3.6.3	Distortion in a Common-Collector Amplifier	137
3.6.4	Case Study: Distortion in an Active-RC Biquad	138
3.6.5	Intermodulation Distortion in a Tow-Thomas Biquad due to OTA Weak Nonlinearities	140
	Problems	149
	References	153
4	**Introduction to Distributed Circuits**	155
4.1	The Transmission Line	156
4.1.1	Analysis of the Uniformly Distributed Lossless Transmission Line	157
4.2	Transmission-Line Circuit Analysis	160
4.2.1	Impedance Transformation Due to a Lossless Line	168
4.3	The Uniformly Distributed RC Line	171
4.3.1	Modeling Pitfalls	173

4.4 Scattering Parameters and the Scattering Matrix 175
4.4.1 Deriving the Scattering Matrix from the Admittance Matrix 181
4.5 Some Properties of Scattering Matrices 183
4.5.1 The Scattering Matrix of a Lossless Network 183
4.5.2 The Scattering Matrix of a Reciprocal Network 184
4.6 Measuring S-Parameters: The Vector Network Analyzer 185
Problems 188
References 193

5 Basics of Linear Time-Varying Circuits and Systems 195
5.1 Motivation 195
5.2 Linearity and Time (In)variance 201
5.3 Linear Time-Varying Systems 203
5.4 Linear Periodically Time-Varying (LPTV) Systems 208
5.5 Calculating Harmonic Transfer Functions 215
5.5.1 MNA Stamps of Periodically Time-Varying Controlled Sources 222
5.5.2 Determining $H_{-l}(j2\pi l f_s)$ and $H_l(-j2\pi l f_s)$ 230
5.6 Impedance and Admittance in LPTV Circuits 234
5.7 Thevenin- and Norton-Equivalent Circuits 236
5.8 The N-Path Principle 238
Problems 248
References 254

6 LPTV Circuits and Systems: Advanced Topics 257
6.1 Reciprocity and Inter-Reciprocity 257
6.1.1 The Frequency-Reversal Theorem 262
6.1.2 Implications of the Frequency-Reversal Theorem: Driving Point Impedance of an LPTV Network and Its Adjoint 269
6.1.3 Implications of the Frequency-Reversal Theorem: Two-Port Parameters 269
6.1.4 The Adjoint Network with Periodically-Varying Controlled Sources 270
6.1.5 Multiple Inputs 272
6.1.6 Signal-Flow Graphs 275
6.2 A Chopped Amplifier as an LPTV System 277
6.3 Time Domain Implications of Reciprocity and Inter-Reciprocity 283
6.4 LPTV Systems with Sampled Outputs 287
6.4.1 The Switched-RC Kernel 292
6.4.2 Integrator with Sampled Feedback 295
6.4.3 Offset Sampling 297
6.5 LPTV Systems Driven by a Class of Modulated Signals 299
6.5.1 Example: LPTV Smoothing Filter 303

6.6	Noise in LPTV Networks	304
6.6.1	Cascaded LPTV Systems	307
6.6.2	Average Noise Spectral Density in RLC Networks with Periodically Operated Switches	310
6.6.3	The Switched-RC Kernel	312
6.6.4	First-Order Switched-Capacitor Filter	314
6.6.5	Charge-Rotating Switched-Capacitor Filter	316
6.6.6	Total Integrated Noise in RLC Networks with Periodically Operated Switches	318
6.7	Application Examples	320
6.7.1	Example 1	320
6.7.2	Example 2	321
6.7.3	Example 3	322
	Problems	325
	References	328
7	**Continuous-Time Filters**	331
7.1	Magnitude-Approximation Principles	334
7.2	The Maximally Flat (Butterworth) Approximation	337
7.2.1	Standard Form of a Second-Order Section	340
7.3	The Chebyshev Approximation	344
7.3.1	Even-Order Chebyshev Filters	351
7.4	Butterworth Versus Chebyshev Filters	351
7.5	Realization of Filters	353
7.5.1	Realizing a Second-Order Section	356
7.5.2	The Rauch Filter	362
7.6	Noise in Opamp-RC Sections	366
7.7	Dynamic-Range Scaling	373
7.8	Fully-Differential Signal Processing, Integrators and Filters	377
7.8.1	Fully-Differential Integrators and Filters	381
7.9	Effect of Opamp Finite Gain-Bandwidth Product	383
7.9.1	Addressing Finite Opamp Bandwidth	391
7.10	Transconductance-Capacitance Integrators and Filters	394
7.10.1	The First-Order Gm-C Filter	394
7.10.2	The Gm-C Biquad	395
	Problems	401
	References	406
8	**Switched-Capacitor Circuits and Filters**	409
8.1	Switched-Capacitor Filters	416
8.1.1	The Switched-Capacitor Integrator	417
8.1.2	Stray-Insensitive Integration	420

8.2 Switched-Capacitor Filter Stages 424
8.2.1 Second-Order Lowpass Filters 427
8.3 Switched-Capacitor Amplifiers 434
8.4 Nonideal Effects in Switched-Capacitor Circuits 437
8.4.1 Single-Supply Operation 438
8.4.2 Opamps Versus OTAs 439
8.4.3 Switches 440
8.4.4 Capacitor Nonidealities 444
8.4.5 Operational-Amplifier Nonidealities 445
8.5 Noise in Switched-Capacitor Circuits 450
8.5.1 The Noise Penalty Due to Sampling 450
8.5.2 Noise in Passive Switched-Capacitor Circuits 451
8.5.3 Noise in a Switched-Capacitor Integrator 452
8.6 The History of Switched-Capacitor Filters 456
Problems 458
References 465

Index 467

About the Authors

Shanthi Pavan received the B. Tech. degree in Electronics and Communication Engineering from the Indian Institute of Technology Madras in 1995, followed by the M.S. and Sc.D. degrees in Electrical Engineering from Columbia University, New York, in 1997 and 1999, respectively. His early professional work included high-speed analog filters and data converters at Texas Instruments, Warren, New Jersey, and microwave integrated circuits for data communications at Bigbear Networks, Sunnyvale, California.

He joined the faculty of the Department of Electrical Engineering at IIT Madras in 2002, where he currently holds the NT Alexander Institute-Chair Professorship. His research interests span high-speed analog circuit design, mixed-signal and RF integrated circuits, delta-sigma data converters, and signal-processing techniques for analog front ends.

Prof. Pavan has served in multiple editorial and professional leadership roles, including the Vice-President of Publications of the IEEE Solid-State Circuits Society, Editor-in-Chief of the *IEEE Transactions on Circuits and Systems I: Regular Papers*, and as Dean (Research) at the Indian Institute of Technology, Madras. He has also served as a Distinguished Lecturer for both the IEEE Solid-State Circuits Society and the IEEE Circuits and Systems Society.

His recognitions include the J.C. Bose Fellowship, the Shanti Swarup Bhatnagar Award in Engineering Sciences, the IEEE Circuits and Systems Society Darlington Best Paper Award, the Swarnajayanti Fellowship, and several institute-level teaching and research awards. He is a Fellow of the IEEE, the Indian National Academy of Engineering, and the Indian National Science Academy. He is co-author of *Understanding Delta-Sigma Data Converters (2nd ed.)*, which received the Wiley-IEEE Press Professional Book Award in 2020.

Gabor C. Temes received his undergraduate degrees from the Technical University of Budapest and Eötvös University, Budapest, Hungary, in 1952 and 1955, respectively. He received the Ph.D. degree in Electrical Engineering from the University of Ottawa, ON, Canada, in 1961, and an honorary doctorate from the Technical University of Budapest, Budapest, Hungary, in 1991.

He has held academic positions at the Technical University of Budapest, Stanford University, and the University of California, Los Angeles. He also worked in industry at Bell-Northern Research, Ottawa, Canada, as well as at Ampex Corporation. He is currently a professor in the School of Electrical Engineering and Computer Science at Oregon State University and a Distinguished Professor Emeritus of UCLA. His recent research has focused on CMOS analog integrated circuits and data converters.

He served as Editor of the *IEEE Transactions on Circuits and Systems (CAS)* and as Vice President of the IEEE Circuits and Systems Society. He has received numerous IEEE awards, including the CAS Darlington Award, the Andrew Chi Prize Award of the Instrumentation and Measurement Society, the Education Award, and the Technical Achievement Award of the IEEE CAS Society. He also received the IEEE Graduate Teaching Award. He was the recipient of the IEEE Gustav Robert Kirchhoff Award and the Semiconductor Industry Association–SRC University Researcher Award. He is a member of the U.S. National Academy of Engineering and the European Academy of Sciences and Arts, and he is a Fellow of the U.S. National Academy of Inventors.

Preface

Analog and mixed-signal integrated circuits are the foundation of modern electronics, with circuit analysis serving as the bedrock of their design. While many texts cover circuit theory and synthesis, they often fall into two extremes: undergraduate books focusing on basic linear analysis (phasors, theorems, and transforms), or classic graduate texts from the 1960s and 70s that dive into deep mathematics often disconnected from modern design needs.

As designers, we identified the tools required by the aspiring analog, mixed-signal, and RF engineer. We believe a designer's toolkit must include:

a. Noise analysis: Rigorous calculation and interpretation of noise in electronic circuits.

b. Weak nonlinearities: Techniques for analyzing weakly nonlinear behavior in electronic circuits.

c. Distributed networks: A fundamental understanding of circuits involving ideal transmission lines.

d. Linear periodically time-varying (LPTV) systems: The analysis of circuits that change over time, such as mixers and samplers.

To our knowledge, no single text covers these essential topics from the ground up. While each subject is vast, a designer primarily needs a "working knowledge"—the ability to apply theory and interpret results from simulations or measurements. These topics are often squeezed into design courses that are already dense with architecture and topology, leaving little room for fundamental analysis.

This book addresses that reality. Based on over a decade of graduate-level instruction at the Indian Institute of Technology Madras, the University of California at Los Angeles, and Oregon State University, the material is designed to be taken alongside analog design electives. In our experience, this concurrent approach allows students to immediately apply theoretical analysis to practical circuit architectures, leading to a deeper appreciation and mastery of the craft.

Organization of the Book

Chapter 1 reviews the preliminaries, and forms the foundation for the rest of this book. The student has likely seen several portions of this material in earlier classes, in which case this is a refresher. We begin with the modified nodal analysis (MNA) formulation for systematically writing circuit equations in a linear time-invariant network. We use this formulation to derive the extra-element theorem. It turns out that the very important concepts of reciprocity and inter-reciprocity, which form a central theme through much of this book, are natural consequences of interpreting the MNA equations of the network. We apply these principles to the sensitivity analysis of networks.

Analog and mixed-signal circuits are fundamentally limited by noise. Chapter 2 introduces the student to techniques for analysis of noise in electronic circuits. Building on the foundations laid in Chapter 1, we derive the Nyquist and Bode's noise theorems, useful to quantify noise in RLC networks. We also derive the notion of equivalent noise voltage and current of two ports, which naturally leads to the concept of noise factor, which is essential for link-budget calculations.

In Chapter 3, we delve into what happens to circuit performance when the input signal strength is too large. Network elements, even those that are supposed to be linear, are weakly nonlinear in practice. Consequently, errors which depend on the input in a nonlinear fashion, called distortion, manifest at the circuit output. These errors are problematic, and often degrade performance of the system. In this chapter, we introduce the student to various kinds of nonlinear errors, such as harmonic and intermodulation distortion. We then develop the method of current injection, which is a simple and systematic way of analyzing the effect of weak nonlinearities in electrical networks. We apply this technique to several practical circuits like amplifiers and filters.

When circuit dimensions become comparable to the signal wavelength, lumped-element assumptions fail. Chapter 4 introduces the student to analysis of circuits with ideal transmission lines. One might wonder if this is useful in practice, since circuit dimensions are very small in an IC environment. Though the circuits might themselves be very small, they often need to interface to other ICs, or to measurement equipment. These interfaces can have dimensions that are a significant fraction of a wavelength, or even several wavelengths long, and need to be treated as distributed elements. We then motivate the need for scattering parameters as the standard parameter set to characterize networks at high frequencies. We conclude this chapter by introducing the student to the operating principle behind the vector network analyzer (VNA).

The next two chapters are intended to introduce the reader to linear periodically time-varying (LPTV) circuits and systems. Given the ubiquitous use of time variance in RF and mixed-signal circuits, a working knowledge and understanding of LPTV systems is an essential part of a designer's training. Yet, we are not aware of any textbook that builds these foundations. Consequently, designers usually pick these skills up "on the job," and tend to rely extensively on simulations. While there is no substitute for CAD tools, an inadequate understanding of what happens under the hood in a simulator is problematic as it results in designers "flying blind." Chapter 5 introduces the basics of linear time-varying and periodically time-varying circuits, and the notion of harmonic transfer functions. We then describe the MNA formulation of circuit equations for LPTV circuits, which forms the basis for more advanced material in Chapter 6. At every step, we draw parallels to the (more familiar) analysis of their time-invariant counterparts. Finally, we develop the N-path principle, which is a technique that is extensively used in applications that range from time-interleaved data converters to tunable RF filters.

If reciprocity and inter-reciprocity were useful tools in the analysis of linear time-invariant networks, they are even more so in LPTV circuits and systems. Chapter 6 discusses this important aspect of LPTV circuits. The discussion builds on the MNA formulation of the circuit equations of Chapter 5, and derives the transfer-function and frequency-reversal theorems. In many practical applications, the output of an LPTV network is sampled. Examples include sample-and-hold circuits, delta-sigma data converters, and switched-capacitor filters. We derive the properties of such systems, which are facilitated by our study of inter-reciprocity. Next, we discuss noise in LPTV circuits. It turns out that the noise in RLC networks with periodically operating switches can be determined using analogs of the Nyquist and Bode noise theorems.

Our philosophy is that the basic ideas and principles behind the analysis of LPTV circuits have strong parallels to those used in LTI circuit analysis, which students are familiar with. Though the discussions are necessarily heavy with notation, we attempt to give the "intuition behind the math" and make the field less intimidating to a novice. Aspects relating to circuit simulation are sprinkled at appropriate junctures throughout the text.

Chapters 7 and 8 are the "application chapters"; they cover commonly used subsystems in analog and mixed-signal designs. The first of these chapters, namely Chapter 7, focuses on continuous-time filters. We introduce the reader to the principles of magnitude approximation, covering Butterworth and Chebyshev low-pass filters. The chapter details the physical realization of these filters using opamp-RC, Rauch, and G_m-C topologies. Aspects central to filter design practice, namely dynamic-range scaling, differential operation, and analysis and compensation of finite opamp bandwidth are covered.

The last chapter of the book is dedicated to switched-capacitor circuits and filters. Switched-capacitor (SC) technology, which allows for precise filter and amplifier realization in CMOS processes, is a very important technology used in many analog subsystems like filters and data converters. The aim of this chapter, which assumes no prior background in this area, is to build and analyze basic switched-capacitor circuits from the ground up. After deriving topologies for basic building blocks like integrators and filters, we focus on non-ideal effects such as charge injection, and noise.

This puts the student on solid ground when pursuing advanced analog and mixed-signal design courses.

At IIT Madras, the audience for this 14-week course with four lectures per week (called Advanced Electrical Networks [AEN]) consists of senior undergraduate and graduate students interested in circuits who are concurrently taking other analog design electives. In our experience, the circuit-design courses give context for the topics covered in the semester-long AEN class, and are better appreciated by the students as they see the practical relevance of the topics.

The book has benefited from the extensive feedback of several students, and colleagues from academia and industry. In particular, we would like to acknowledge Jaival Shah and Danyal Shamsi (IIT Madras), Sashank Krishnamurthy (Intel Corporation), John Khoury (Silicon Laboratories Inc.), Stef van Zanten and Eric Klumperink (University of Twente), and Ashwin Kumar (IIT Kanpur). The National Programme on Technology Enhanced Learning (NPTEL), an initiative of the Government of India, facilitated the recording and hosting of live lectures covering the subject matter discussed in this book. They are available as two massive open online courses (MOOCs), namely Circuit Analysis for Analog Designers (https://nptel.ac.in/courses/117106148) and Introduction to Time-Varying Electrical Networks (https://nptel.ac.in/courses/108106174).

Given the constraints of space and time, several worthy topics had to be given short shrift, and some eliminated altogether. Nevertheless, we hope that the book will be useful to students, and engineers for self study.

Shanthi Pavan
Chennai, India

Gabor C. Temes
Corvallis, USA

1

Introduction and Basic Concepts

Understanding how electrical circuits behave is fundamental to all of integrated circuit (IC) design. Whether we are designing amplifiers, filters, or data converters, every system can ultimately be reduced to a network of interconnected elements that obey a few simple physical laws. This chapter develops the methods used to analyze such networks systematically and efficiently. The material discussed here feeds into the rest of the chapters of this book in one way or another.

We begin by revisiting the essential circuit laws—Ohm's law, Kirchhoff's current law (KCL), and Kirchhoff's voltage law (KVL)—and use them to relate currents and voltages in electrical networks [1]. These laws form the basis for setting up equations that describe circuit behavior. We give the physical intuition behind the Tellegen's theorem, a result that seems very intriguing at first [2].

We then introduce the concept of nodal analysis, which provides a structured way of formulating circuit equations in linear networks. From there, we build up to modified nodal analysis (MNA) [3], a powerful and general technique that can be applied to any linear circuit. MNA forms the computational foundation of circuit simulators such as SPICE [4] and provides a unified framework for hand and computer-based analysis [5] alike.

As circuits become more complex, elements such as controlled sources must be incorporated. We will see how these components can be represented through compact "MNA stamps" that can be directly inserted into circuit equations. Using this approach, we can efficiently analyze circuits that include both passive and active devices without changing the overall formulation. We use the MNA approach to conclude that any transfer function in a linear network can be related to a component value in a bilinear form. This leads to the so-called extra-element theorem [6].

It turns out that the MNA formulation of network equations naturally leads to the concept of reciprocity and the adjoint (inter-reciprocal) network. The adjoint network is a *very* valuable computational and analytical tool that forms the basis for advanced topics such as noise and sensitivity computation [7]. The concepts of reciprocity and inter-reciprocity [8] are a continuous theme throughout this book.

1.1 Classification of Circuits

1.1.1 Linear and Nonlinear Circuits

An electric circuit is linear if its response (with all initial conditions being zero) is proportional to the excitation. For such a circuit, superposition is applicable, i.e., the overall response is the sum of the responses due to the individual sources. Mathematically, if the circuit's response to $x_1(t)$ and $x_2(t)$ are $y_1(t)$ and $y_2(t)$ respectively, the response to $\alpha x_1(t) + \beta x_2(t)$ is $\alpha y_1(t) + \beta y_2(t)$ for all constants α and β (Figure 1.1).

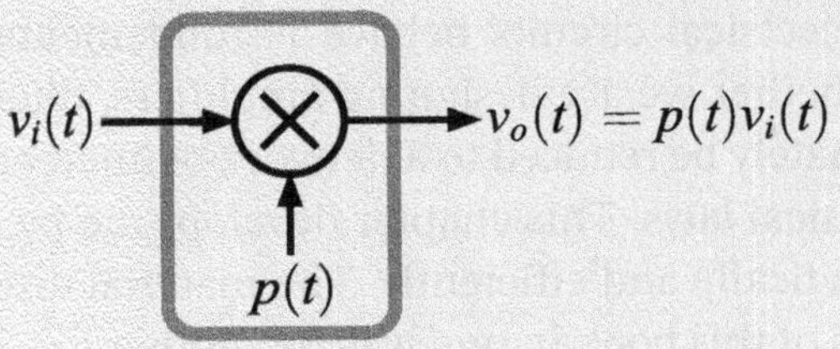

Figure 1.1 Is this a linear system?

The system above has an input/output relationship given by $v_o(t) = p(t)v_i(t)$. Is it a linear system?

Input	Output
$v_a(t)$	$p(t)v_a(t)$
$v_b(t)$	$p(t)v_b(t)$
$\alpha v_a(t) + \beta v_b(t)$	$p(t)(\alpha v_a(t) + \beta v_b(t))$
	$= \alpha p(t)v_a(t) + \beta p(t)v_b(t)$

From the input–output table above, we see that superposition is valid. The system, therefore, is linear.

Circuits where superposition does not hold are nonlinear. In practice, electronic circuits may behave approximately linearly, but only within a range of signals (Figure 1.2).

The system above has an output given by

$$v_o(t) = \{A(\alpha v_1^3(t) + v_2(t))\}^{1/3}\beta,$$

indicating that it is nonlinear in general. However, *in the special case that* $v_2 = 0$, $v_o(t) = (\alpha A)^{1/3}\beta v_1(t)$, leading us to conclude that the system behaves linearly in spite of signals being processed in a nonlinear fashion inside. Such systems are called externally linear, internally nonlinear systems.

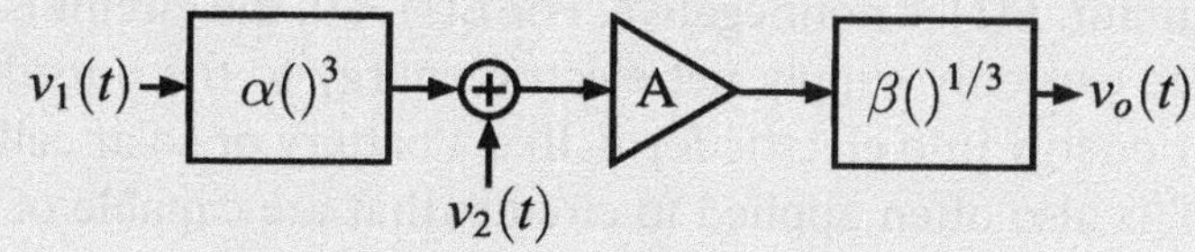

Figure 1.2 Is this system linear or nonlinear?

1.1.2 Time Invariant and Time-Varying Circuits

A circuit is time invariant if its response, when initially relaxed, to a given excitation is the same regardless of the time instant at which the source is connected. It is time-varying otherwise. Time (in)variance and (non)linearity are different aspects of a circuit that do not have anything to do with each other. The reader is encouraged to think of examples of linear and time-invariant, nonlinear and time-invariant, linear and time-varying, and nonlinear and time-varying systems.

1.1.3 Passive and Active Circuits

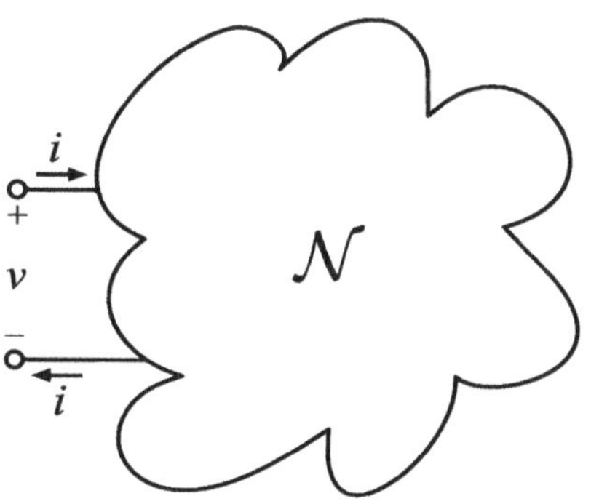

Figure 1.3 Associated directions of voltage and current at a network port. $\mathcal{N}$ is a symbol used to denote a network.

A network $\mathcal{N}$ is passive if it cannot provide more energy to the outside world than what was stored in it. Assume that $\mathcal{N}$ is connected to the outside world through two terminals (Figure 1.3). Note that the reference directions of the voltage and currents are such that the reference current flows from the positive to the negative terminal in the circuit. These are associated directions and will be assumed throughout this work. The power flowing into the network from the outside at time t is then $P(t) = v(t)i(t)$, and the energy delivered into it during the time period from 0 to t is

$$E(t) = \int_0^t v(t)i(t)\,dt. \tag{1.1}$$

For a passive circuit, $E(t)$ is nonnegative. For $E(t) < 0$, the circuit is active. Strictly speaking, a device can only supply net electric energy to the outside if it converts chemical or solar energy into electric form, like a battery or solar cell. However, the adjective "active" is also often applied to circuits that are capable of converting the power from a fixed voltage or current source into a magnified signal, like a voltage amplifier. For a circuit with several terminal pairs, the total energy delivered at all terminal pairs determines the active or passive property.

Terminal pairs where the current entering at one terminal is the same as the current leaving at the other are called ports. Thus, the structure in Figure 1.3 is a one-port circuit. Circuits may be *lossless* or *lossy*. The energy stored in a lossless circuit equals the energy supplied to it from the outside. For a lossy circuit, the stored energy is less than the energy input from outside. Thus, in a lossy network, the electromagnetic energy is transformed into some other form of energy, such as heat or light. A circuit is *memoryless* if its response at a given time depends only on its excitation at the same time. This will happen if there are no energy-storing elements in the circuit. Otherwise, the circuit is called *memoried*, or *dynamic*.

Energy Efficiency of Charging a Capacitor

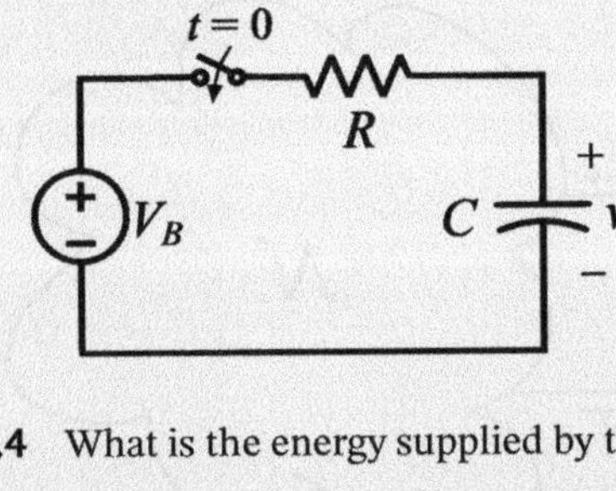

Figure 1.4 What is the energy supplied by the battery?

Figure 1.4 shows a simple circuit that illustrates some of the concepts discussed. A battery with a voltage V_B is switched at $t = 0$ to the series-connected R and C. (R may be the on-resistance of the switch.) The capacitor is initially uncharged. Let us determine the energy stored in the capacitor after the charging is completed, and the energy delivered by the battery. The final energy stored in C is $(1/2)CV_B^2$. The current flow during charging is $i = Cdv/dt$, so the total energy E_B supplied by the battery at the end of the charging process is

$$E_B(\infty) = \int_0^\infty V_B C \frac{dv}{dt} dt = CV_B^2. \tag{1.2}$$

This is twice the energy stored in C. Thus, half of the energy provided by the battery is lost in the charging process, dissipated in the switch resistance R. This energy is given by

$$E_R(\infty) = \int_0^{\infty} Ri^2(t)dt = \frac{1}{2}CV_B^2, \tag{1.3}$$

as expected. It is interesting to note that the energy efficiency of the charging process is 50%, independent of the resistance of the switch. However, the speed of the charging process is increased when R is lowered.

1.2 Circuit Components

The circuit components discussed next include passive elements such as R, L, C, as well as active elements such as controlled sources and operational amplifiers. It will be assumed that all elements are ideal, i.e., linear and time-invariant (LTI). Then, their defining voltage–current relations are the following:

$$\text{Resistor } R : v = Ri \quad \text{and} \quad i = Gv, \text{ where } G = 1/R. \tag{1.4}$$

The equation above is Ohm's law. Here, R is the resistance and G the conductance of the resistor. Both have positive values. The power dissipated in the resistor is given by $P = vi = Ri^2$. P is clearly nonnegative, and therefore so is its time integral, the energy E. Hence, the resistor is a passive element.

$$\text{Capacitor } C : i = C\frac{dv}{dt} \quad \text{and} \quad v = \frac{1}{C}\int_0^t i(x)dx + v(0). \tag{1.5}$$

Here, $C > 0$ is the capacitance of the element. Finding the energy entering a capacitor using the time integral of the power vi shows that its value at time t is $E(t) = (1/2)Cv^2(t) > 0$. Thus, the capacitor is also a passive element. From (1.5), we see that there cannot be a step change in the voltage across a capacitor unless an impulse current flows through it. Equivalently, if the voltage across a capacitor experiences a step change, one must conclude that an impulse current has flown through it.

$$\text{Inductor } L : v = L\frac{di}{dt} \quad \text{and} \quad i = \frac{1}{L}\int_0^t v(x)dx + i(0). \tag{1.6}$$

Here, $L > 0$ is the inductance of the element. Integrating the power vi gives $E(t) = (1/2)Li^2(t) > 0$ for the energy that entered the inductor by time t. Thus, the inductor is also a passive element. From (1.6), we see that there cannot be a step-change in the current through an inductor unless an impulse voltage is applied across it. This also means that if the current through an inductor experiences a step change, an impulse voltage must have occurred across its terminals.

The energy entering the resistor leaves the circuit in the form of radiation of heat or an electromagnetic wave and is lost for the circuit. However, the energy entering the capacitor is stored in the electric field between the plates; the energy of the inductor is stored in the magnetic field within and around the device. The stored energies can be redeemed by the circuit.

The voltages and currents in (1.4)–(1.6) are physical quantities. They are generally functions of time. As the equations indicate, the relations describing the operation of the circuit in which they are present are integro-differential ones. They can be reduced to algebraic equations, and thus their solution greatly simplified, by using the Laplace transformation. Denoting the transformed variables by capital letters, this changes the voltage–current relation of the capacitor into $I = sCV$ and $V = I/sC$. Similarly, for an inductor, the relations become $V = sLI$ and $I = V/sL$. For the resistor, Ohm's law remains unchanged between V and I.

The ratio $Z = V/I$ between the Laplace-transformed quantities is called the impedance, and its reciprocal $Y = I/V$ is called the admittance. The Laplace-transformed variables are closely related to the phasors used in sinusoidal steady-state analysis. The equations used for Laplace-transformed variables in terms of the variable s remain valid for phasors if s is replaced by $j\omega$. The admittance of a capacitor is $j\omega C$, which is zero at DC. This means that the average current flowing through a capacitor in a circuit "that works" (in the sense that none of the voltages or currents blow up to ∞) is zero. Likewise, since the DC impedance of an inductor is zero, the average voltage across it must be zero.

The active components often encountered include the independent and dependent (controlled) sources. The voltage across an independent voltage source is defined by the designer, and it is independent of the current drawn from it. Similarly, the current

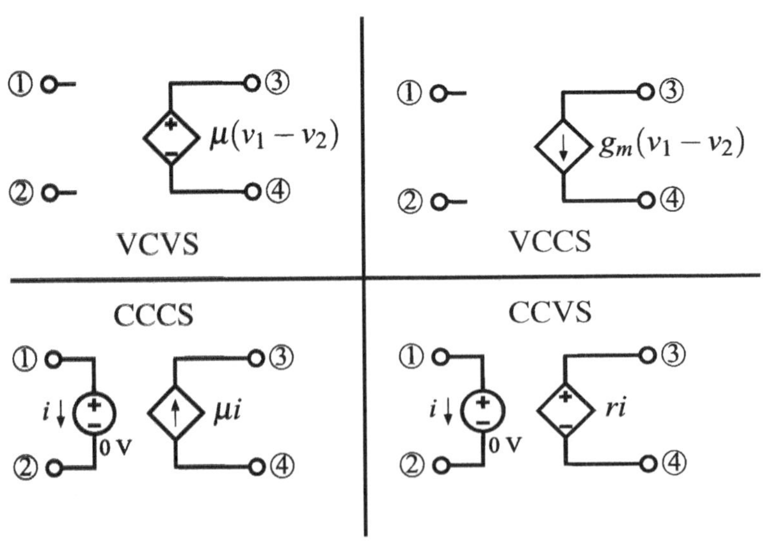

Figure 1.5 The four linear dependent (controlled) sources.

provided by an independent current source is a given quantity and is independent of the voltage across it.

A dependent voltage source provides a voltage v_j in branch j which is a function of another voltage v_k (VCVS) or current i_k (CCVS) in the circuit. Usually, the function is linear, so the voltage is given by $v_j = \mu v_k$ or by $v_j = ri_k$. Similarly, a dependent current source has a value determined by the voltage (VCCS) or current (CCCS) in another branch. Usually, it is a linear dependence, so that the current is given by $i_j = gv_k$ or $i_j = \mu i_k$. Figure 1.5 illustrates the symbols used for independent and dependent sources.

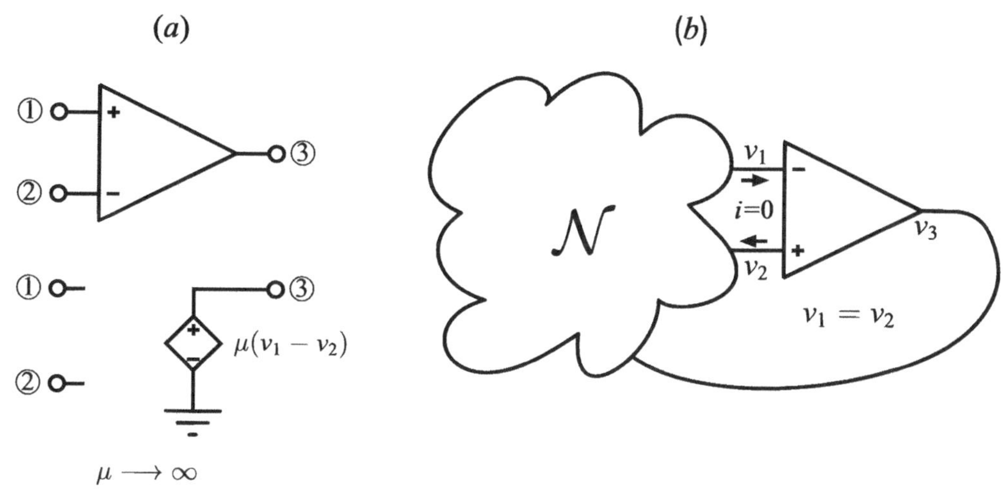

Figure 1.6 (a) Model for an ideal opamp. (b) The input terminals form a virtual short if the opamp has DC negative feedback around it.

An important concept in analog circuit design is that of the operational amplifier (Figure 1.6a), often called the workhorse of analog engineering. It can be thought of as a voltage-controlled voltage source with an output voltage $v_3 = \mu(v_1 - v_2)$, where μ is ideally infinite. When an opamp is embedded in a circuit as shown in Figure 1.6b, *and there is DC negative feedback around it*, v_1 and v_2 are equal. Since the opamp is a VCVS, $i = 0$. Its input terminals, therefore, are said to form a *virtual* short.[1]

1.3 Kirchhoff's Laws and the Incidence Matrix

The analysis of circuits is based on Kirchhoff's current law (KCL) and voltage law (KVL). The KCL states that the sum of all currents leaving any node at any time is zero. This is because charges ideally cannot accumulate at a node. (In a real circuit, parasitic capacitances are present at all nodes. Their currents must be included to make

[1]When there is a *real* short circuit between two nodes, the voltage between them is zero, and the current can be arbitrary. In a virtual short, the voltage between the nodes is zero, and the current flowing between the nodes *is also zero*.

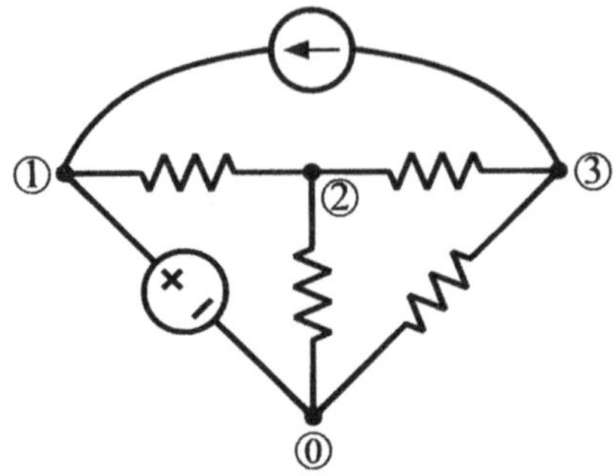

Figure 1.7 Example network with four nodes and six branches.

this statement valid.) The KVL states that the sum of all branch voltages in a loop is zero. Kirchhoff's laws are illustrated for a simple circuit in Figure 1.7. The "skeleton" of the circuit, called its *graph*, is shown in Figure 1.8. The junctions between elements are the nodes, and the elements themselves are abstracted out as branches. Our example circuit consists of four nodes and six branches. The nodes and branches are numbered. The voltage of the jth node is denoted by v_j. The voltage across and current through the kth branch are denoted by e_k and i_k, respectively. As discussed earlier (see Figure 1.3), the direction of current in a branch is chosen so that $e_k(t)i_k(t)$ represents the instantaneous power dissipated in the branch.

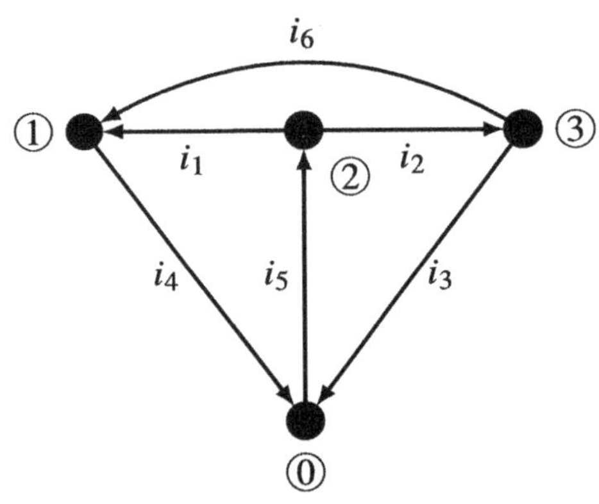

Figure 1.8 Graph of the network of Figure 1.7.

The aim of circuit analysis is to "solve" the network, which means that we need to determine the node voltages, branch voltages, and branch currents. Since currents only flow in response to potential differences, any one of the nodes can be chosen as the datum or reference with respect to which the potential of all other nodes is measured. This is numbered as node ⓪ and is often referred to as the *ground node*. In a network with n nodes, therefore, there are $(n - 1)$ node voltages to be found. The network is solved by using Kirchhoff's current law (KCL), Kirchhoff's voltage law (KVL), and the branch relationships, i.e., the relationships between e_k and i_k.

KCL for our example circuit leads to

$$
\begin{aligned}
-i_1 + i_4 - i_6 &= 0 \\
i_1 + i_2 - i_5 &= 0 \\
-i_2 + i_3 + i_6 &= 0 \\
-i_3 - i_4 + i_5 &= 0.
\end{aligned}
$$

It is efficient at this point to introduce vector and matrix notations. We define the incidence matrix **A** of the graph of the circuit. Its element in row j and column k is

$$
\begin{aligned}
a_{jk} &= +1, && \text{if branch } k \text{ leaves node } j \\
a_{jk} &= -1, && \text{if branch } k \text{ enters node } j \\
a_{jk} &= \ \ 0, && \text{otherwise.}
\end{aligned}
$$

The incidence matrix for the graph of Figure 1.8 is shown below. Each row corresponds to a node, and there is a column for each branch. **A** contains all the information there is in the graph. For instance, examining the third column indicates that the branch b_3 originates at node ③ and terminates at node ⓪. The second row of **A** tells us that branches b_1 and b_2 leave node ② while b_5 arrives into the node. Notice that there are precisely two nonzero entries in every column and that the sum of every column is zero. This means that one row of **A** can be removed without losing any information.

$$
\begin{array}{c}
\\ ① \\ ② \\ ③ \\ ⓪
\end{array}
\begin{array}{c}
\begin{matrix} b_1 & b_2 & b_3 & b_4 & b_5 & b_6 \end{matrix} \\
\begin{bmatrix}
-1 & 0 & 0 & 1 & 0 & -1 \\
1 & 1 & 0 & 0 & -1 & 0 \\
0 & -1 & 1 & 0 & 0 & 1 \\
0 & 0 & -1 & -1 & 1 & 0
\end{bmatrix}
\end{array}.
$$

Removing the (redundant) row corresponding to the ground node, we have the so-called *reduced* incidence matrix denoted by $\mathbf{A}_r$ given below.

$$
\begin{array}{c}
\\ ① \\ ② \\ ③
\end{array}
\begin{array}{c}
\begin{matrix} b_1 & b_2 & b_3 & b_4 & b_5 & b_6 \end{matrix} \\
\begin{bmatrix}
-1 & 0 & 0 & 1 & 0 & -1 \\
1 & 1 & 0 & 0 & -1 & 0 \\
0 & -1 & 1 & 0 & 0 & 1
\end{bmatrix}
\end{array}.
$$

$\mathbf{A}_r$ is an $(n-1) \times b$ matrix, where n and b denote the number of nodes and branches in the circuit, respectively. The branch currents can be collected into a column vector $\mathbf{i} = \begin{bmatrix} i_1 & i_2 & \dots & i_b \end{bmatrix}^T$. Then, all the KCL equations can be written in the simple form

$$
\mathbf{A}_r \mathbf{i} = \mathbf{0}. \tag{1.7}
$$

where $\mathbf{0}$ is the column vector of $(n-1)$ zeros. The (reduced) incidence matrices only depend on the network's graph, and say nothing of the components constituting the branches.

Kirchhoff's voltage law (KVL) can simply be stated as follows. The branch voltage is the difference between its node voltages. This is equivalent to saying that the sum of branch voltages in a loop is zero. Since the potentials of all other nodes are measured with respect to that of node ⓪, the latter's voltage is assumed to be 0. KVL for our circuit example gives the equations

$$\begin{aligned} e_1 &= v_2 - v_1 \\ e_2 &= v_2 - v_3 \\ e_3 &= v_3 \\ e_4 &= v_1 \\ e_5 &= -v_2 \\ e_6 &= v_3 - v_1. \end{aligned}$$

Similarly, collecting the branch and node voltages into column vectors $\mathbf{e}$ and $\mathbf{v}$, respectively, the KVL equations can be written in the concise form

$$\mathbf{e} = \mathbf{A}_r^T \mathbf{v}. \tag{1.8}$$

The equations (1.7) and (1.8) above hold for all networks—linear, nonlinear, time invariant or time varying.

1.4 Tellegen's Theorem

The sum total of the instantaneous power dissipated in all branches of the network is given by

$$P_{\text{tot}} = \sum_k e_k i_k = \mathbf{e}^T \mathbf{i}. \tag{1.9}$$

Using (1.8) and (1.7), we see that

$$P_{\text{tot}} = \mathbf{v}^T \underbrace{\mathbf{A}_r \mathbf{i}}_{=\mathbf{0}\,(\text{KCL})} = \mathbf{v}^T \mathbf{0} = 0. \tag{1.10}$$

This indicates that the sum of instantaneous powers in all branches of a network is zero; the power dissipation in a branch is due to power generated in other branches of the circuit. It is important to note that this is satisfied at *every instant* of time—not in some average sense.[2] This is one form of the famous *Tellegen's Theorem*. The reader might wonder what all the fuss is about; net power cannot be generated or dissipated

---

[2]Evidently, Nature (unlike our economists) takes Polonius' advice of neither being a borrower nor lender (of power) seriously.

in a network after all! However, there is more to this than meets the eye, as we see below.

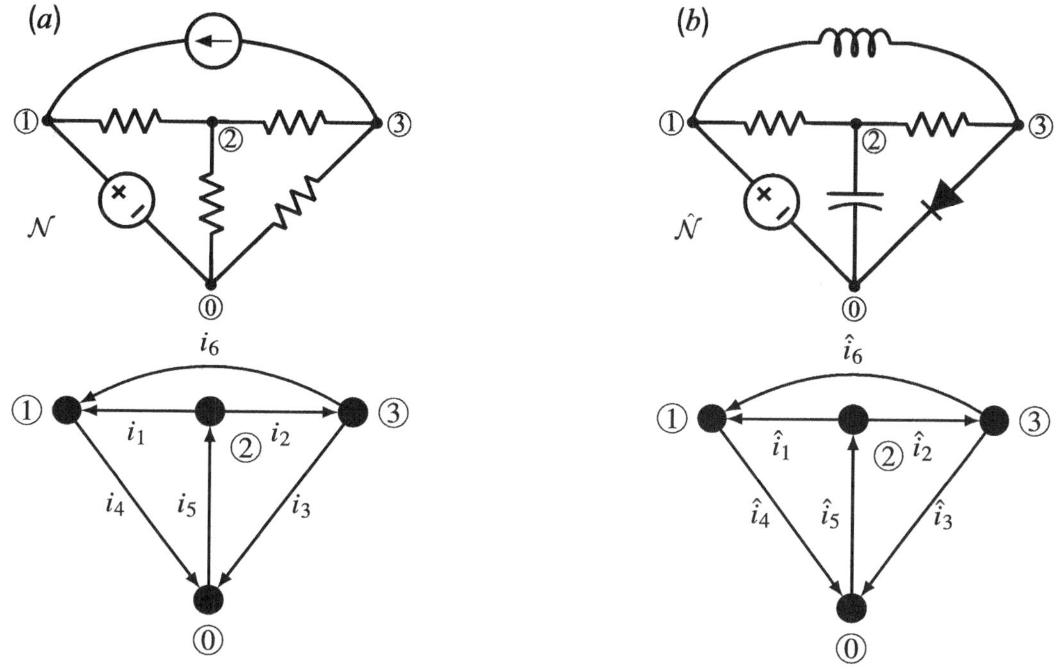

Figure 1.9 Two networks with the same graph. The branch voltages and currents of the networks are denoted by e_k, i_k and $\hat{e}_k$, $\hat{i}_k$ respectively.

Consider the networks $\mathcal{N}$ and $\hat{\mathcal{N}}$ shown in Figure 1.9a,b respectively. They are evidently different. Assuming the resistors obey Ohm's law, $\mathcal{N}$ is linear and memoryless, while $\hat{\mathcal{N}}$ is nonlinear and has memory. However, their graphs are identical, and the branches and nodes are numbered the same way. The branch voltages and currents of $\mathcal{N}$ are denoted by e_k, and i_k and those in $\hat{\mathcal{N}}$ are denoted by $\hat{e}_k$, and $\hat{i}_k$ respectively.

Let us compute the quantity $\sum_k e_k\hat{i}_k$, which has the dimensions of power. To be clear, what we are doing is computing the product of a branch voltage in $\mathcal{N}$ and the current in the corresponding branch in $\hat{\mathcal{N}}$ and summing this "power-like" quantity over all branches. In vector notation, this is $\mathbf{e}^T\hat{\mathbf{i}}$. By KVL, $\mathbf{e}^T = \mathbf{v}^T\mathbf{A}_r$. We thus have the surprising result

$$\sum_k e_k\hat{i}_k = \mathbf{e}^T\hat{\mathbf{i}} = \mathbf{v}^T\mathbf{A}_r\hat{\mathbf{i}} = \mathbf{v}^T\mathbf{0} = 0. \tag{1.11}$$

In a similar fashion, it is easy to see that

$$\sum_k \hat{e}_k\hat{i}_k = \sum_k e_k\hat{i}_k = \sum_k \hat{e}_k i_k = 0 \tag{1.12}$$

where the summation runs across all the branches in the network. What throws one off about this result is that it appears that $\mathcal{N}$ and $\hat{\mathcal{N}}$ must somehow know what is happening in the other network for the result to be true. Clearly, however, there is nothing in common between the networks other than that their graphs (and therefore their incidence matrices) are the same. It seems difficult to wrap our heads around the fact we can make any statement at all about $\sum_k e_k\hat{i}_k$ given that the networks are completely independent, let alone the sum being zero! Fortunately, as we see below, this result is not as intriguing as it seems at first.

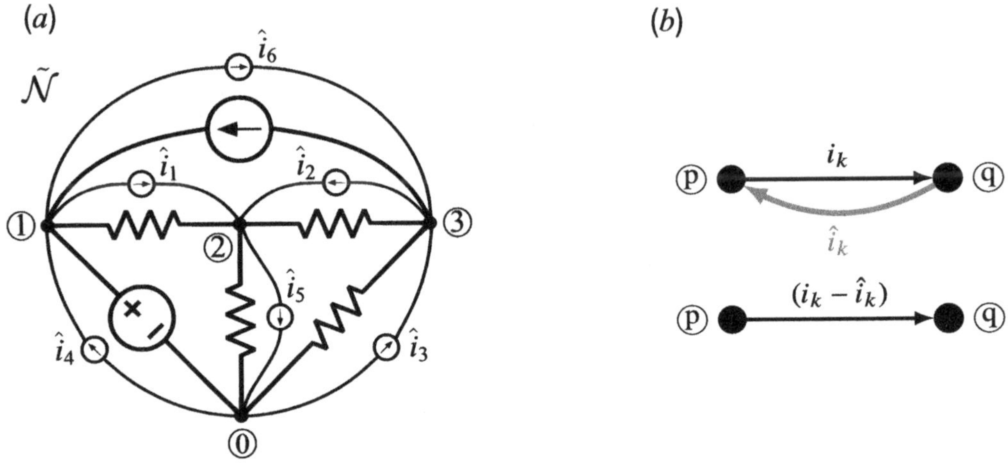

Figure 1.10 (a) $\tilde{\mathcal{N}}$ is formed by adding currents sources in parallel with the branches of $\mathcal{N}$. (b) Every branch in $\mathcal{N}$ has in (anti)parallel with it a current source whose value is the same as that of the corresponding branch in $\hat{\mathcal{N}}$. The composite branch carries a current $(i_k - \hat{i}_k)$.

To intuitively understand the result, consider the new network $\tilde{\mathcal{N}}$ shown in Figure 1.10a. It is formed from $\mathcal{N}$ using the following artifice. In parallel with every branch of $\mathcal{N}$, we add a current source whose value is identical to the current flowing in the corresponding branch of $\hat{\mathcal{N}}$, but in the opposite direction. Since the branch currents $\hat{i}_k$ satisfy KCL, it follows that the net current injected into each node is zero. In general, the kth branch has in parallel with it a current source $\hat{i}_k$ connected in the opposite direction as shown in Figure 1.10b. This composite branch can be thought of as a single branch with a current $(i_k - \hat{i}_k)$.

Since the currents injected into every node of $\mathcal{N}$ are zero, it follows that the node voltages $v_1, \dots, v_3$ in $\tilde{\mathcal{N}}$ in Figure 1.10a are identical to those in $\mathcal{N}$ in Figure 1.9a.

The graph of $\tilde{\mathcal{N}}$ and a simplified version are given in Figure 1.11. Using Tellegen's theorem on the network of Figure 1.11b, we have

$$\sum_k e_k(i_k - \hat{i}_k) = \sum_k e_k i_k - \sum_k e_k\hat{i}_k = 0. \tag{1.13}$$

This is a just a statement of power conservation applied to $\tilde{\mathcal{N}}$. Since $\sum_k e_k i_k = 0$ (by power conservation in $\mathcal{N}$), it is evident that

$$\sum_k e_k\hat{i}_k = 0. \tag{1.14}$$

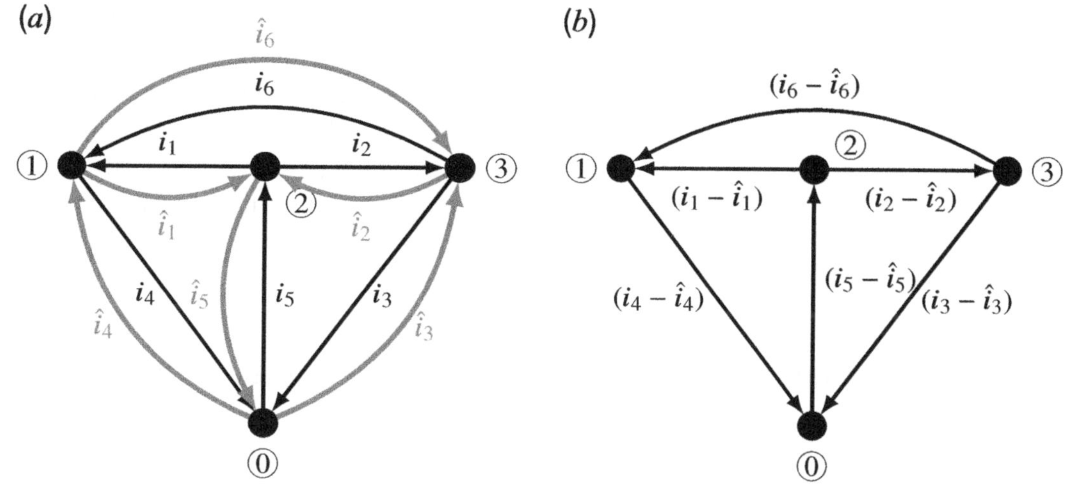

Figure 1.11 Graph of $\tilde{\mathcal{N}}$, and its simplified equivalent.

We started with $\mathcal{N}$ and added current sources to arrive at $\tilde{\mathcal{N}}$. Had we started with $\hat{\mathcal{N}}$ instead, we would conclude that

$$\sum_k \hat{e}_k i_k = 0. \tag{1.15}$$

To summarize, though Tellegen's theorem when applied to different networks looks intriguing at first, it can be intuitively understood.

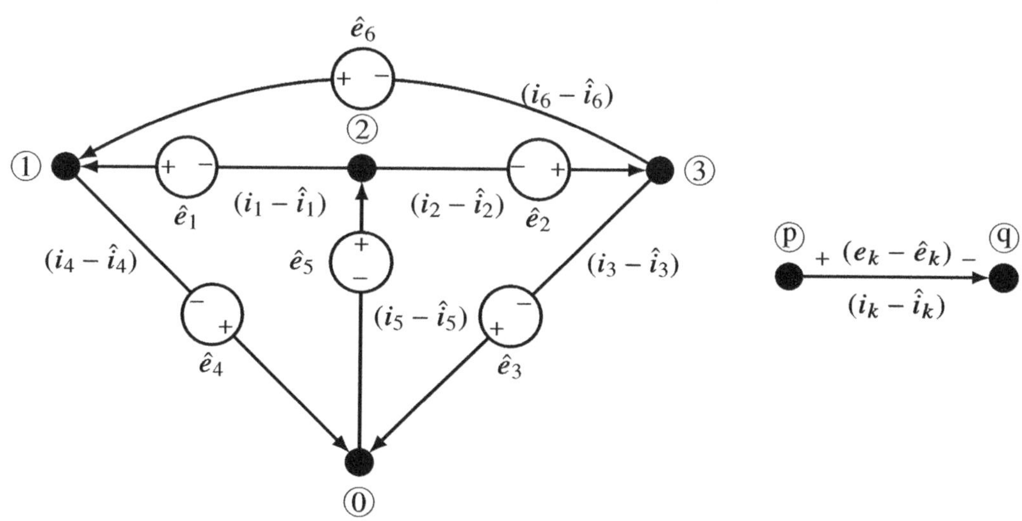

Figure 1.12 Inserting voltage sources in every branch of $\tilde{\mathcal{N}}$, in a direction as to oppose the branch voltage. The voltage source is chosen to be that of the corresponding branch in $\hat{\mathcal{N}}$.

We now describe a modification of the graph of Figure 1.11b that is useful in understanding the incremental behavior of nonlinear networks. As discussed earlier, the branch voltages in Figure 1.11b are identical to those in $\mathcal{N}$ since the net current we injected into every node when we formed $\tilde{\mathcal{N}}$ was zero. Taking this a step further, we insert in series with the kth branch of the graph of Figure 1.9b a voltage source identical to that in the corresponding branch of $\hat{\mathcal{N}}$, namely $\hat{e}_k$. The voltage source is inserted in the sense that it decreases the branch voltage as shown in Figure 1.12. Going around every loop, we see that the net voltage we have introduced into the loop is zero, since the $\hat{e}_k$'s satisfy KVL. As a consequence, the branch currents cannot change; they remain $(i_k - \hat{i}_k)$.

1.5 The Incremental Network and Small-Signal Analysis

What does all this mean in practice, and why is this construct useful? What we have concluded from the previous section is the following. Suppose that two networks, $\mathcal{N}$ and $\hat{\mathcal{N}}$, have the same graph and their branch voltages/currents are denoted by e_k/i_k and $\hat{e}_k/\hat{i}_k$, respectively. A new network with the same graph but with branch voltages/currents $(e_k - \hat{e}_k)/(i_k - \hat{i}_k)$ satisfy KVL and KCL, respectively. It turns out, as we see below, that this is the basis for small-signal (also known as incremental) circuit analysis.

Consider the two networks shown in Figure 1.13a,b. They are identical except that the sources in $\hat{\mathcal{N}}$ are increased by ΔV_a and ΔI_b. Note that the increments *need not be* small. $g_1(\cdot)$ and $g_2(\cdot)$ denote nonlinear conductances so that the branch current i_1 is related to the branch voltage according to $i_1 = g_1(e_1)$.

The "difference" network is shown in Figure 1.13c. The branch voltages and currents in this network are Δe_k and Δi_k, which are used to denote $(\hat{e}_k - e_k)$ and $(\hat{i}_k - i_k)$ respectively. We see that independent sources are the increments in the voltage and current excitations, namely ΔV_a and ΔI_b. The branch relations in the network of Figure 1.13c relate Δe_k to Δi_k. For instance, the first branch relation is derived as follows.

$$\begin{aligned} i_1 &= g_1(e_1) && \text{from } \mathcal{N} \\ i_1 + \Delta i_1 &= g_1(e_1 + \Delta e_1) && \text{from } \hat{\mathcal{N}} \\ \Delta i_1 &= g_1(e_1 + \Delta e_1) - g_1(e_1). \end{aligned} \tag{1.16}$$

Using the Taylor expansion for $g_1(\cdot)$ around e_1, we have

$$\Delta i_1 = \left.\frac{dg_1(x)}{dx}\right|_{x=e_1} \Delta e_1 + \frac{1}{2}\left.\frac{d^2 g_1(x)}{dx^2}\right|_{x=e_1} \Delta e_1^2 + \cdots \tag{1.17}$$

If the ΔV_a and ΔI_b qualify as small-signal excitations so that the higher-order terms in the Taylor series are negligible, the first branch can be replaced by a *linear* conductance whose value is given by $dg_1(x)/dx$ evaluated at $x = e_1$. e_1 is the so-called "operating point" of the nonlinear element $g_1(\cdot)$, and the difference network becomes linear. It is commonly referred to as the incremental circuit, where the component values depend on the operating point as shown in Figure 1.13d.

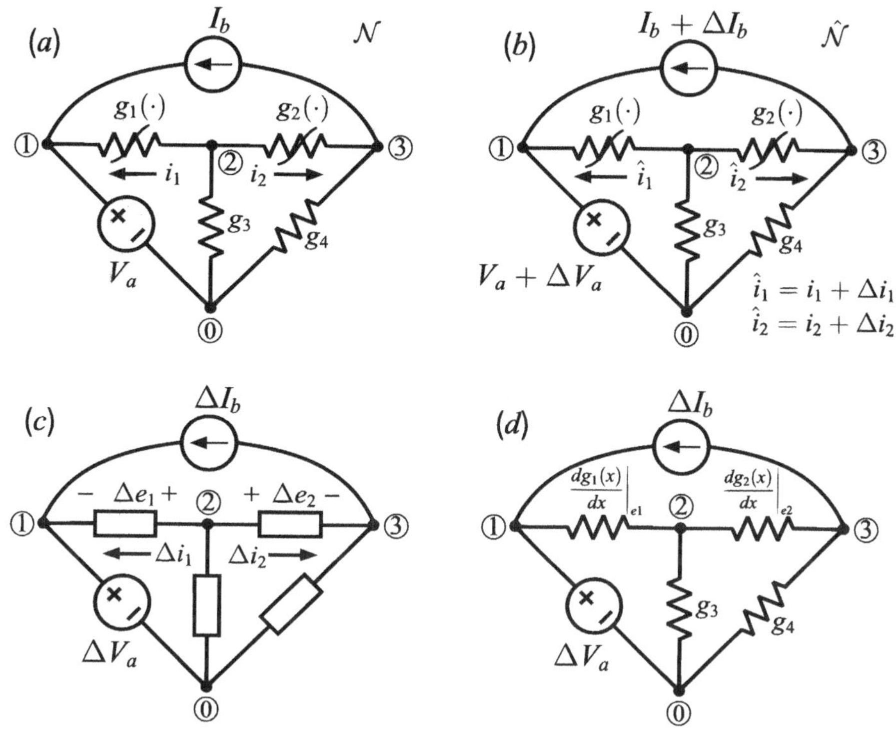

Figure 1.13 (a) Original network, consisting of nonlinear conductances $g_1(\cdot)$ and $g_2(\cdot)$. (b) Network in (a), but the exciting sources changed. The changes need not necessarily be small. (c) The changes in branch currents and voltages satisfy KCL/KVL. The elements in the network are nonlinear. (d) In the special case where the changes in branch voltages can be considered small, the nonlinear elements can be replaced by their incremental equivalents.

To summarize, to determine the changes in branch voltages and currents in a network $\mathcal{N}$ with nonlinear conductances due to a small change in the excitation sources, we proceed as follows.

a. Determine the branch voltages and currents in $\mathcal{N}$. Since the relations that govern the branch currents and voltages are nonlinear in general, it is often not possible to solve the set of KCL/KVL equations analytically. In such a case, the solution must be found numerically. The branch voltages/currents constitute the operating points of the corresponding conductances.

b. Create a linear network with the same graph as $\mathcal{N}$. This is the *incremental* network. The sources that excite this network are the *changes* in the exciting sources. The component values are linear conductances whose values are the slopes of the nonlinear conductances determined at the corresponding operating point. The branch voltages and currents in the incremental network correspond to changes in the branch voltages/currents in $\mathcal{N}$ due to a change in the exciting sources.

c. Solve the incremental network. This can be done analytically, since the network is linear. This will yield all the Δe_k and Δi_k. The branch voltages and currents in

$\mathcal{N}$ due to the changed excitations are simply given by $(e_k + \Delta e_k)$ and $(i_k + \Delta i_k)$ respectively.

1.6 Solving Linear Networks Using Modified Nodal Analysis

In this section, we develop the modified nodal analysis (MNA) method of writing down circuit equations. For simplicity, we consider memoryless networks. The same concepts hold for networks containing inductors and capacitors operating in the sinusoidal steady state. We start with the basic example of Figure 1.14, which consists of conductances and independent current sources. How do we "solve" the network? By "solve", we mean—determining all the node voltages and branch currents in the network.

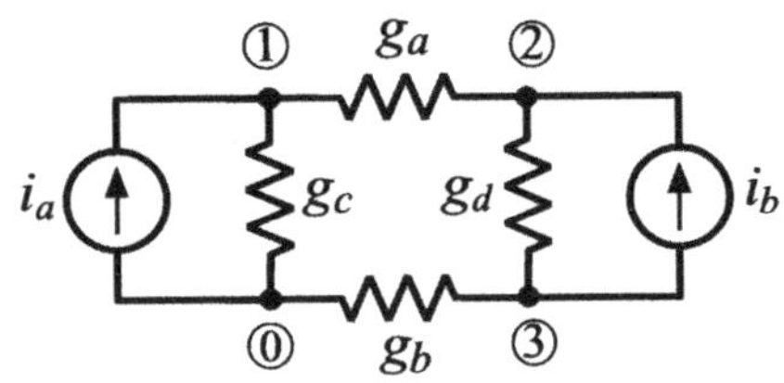

Figure 1.14 Example network to illustrate the nodal analysis.

a. Choose an arbitrary node as the ground node, with respect to which all other node potentials are measured. In the example of Figure 1.14, the node ⓪ is chosen as the ground node.

b. The unknowns to be evaluated are the node voltages. In our example network, they are v_1, v_2, and v_3.

c. Write Kirchhoff's current law (KCL) at every node other than the ground node. For our example network, we have the following.

$$
\begin{aligned}
g_a(v_1 - v_2) + g_c v_1 &= i_a \\
g_a(v_2 - v_1) + g_d(v_2 - v_3) &= i_b \\
g_b v_3 + g_d(v_3 - v_2) &= -i_b
\end{aligned}
$$

The set of equations above is expressed in matrix form as follows. Let us take a look at the anatomy of the matrices—this will perhaps help us write the matrices right away without having to go through the effort of writing KCL at each node, collecting terms, etc.

$$
\begin{array}{l}
\\
\text{KCL at ①} \\
\text{KCL at ②} \\
\text{KCL at ③}
\end{array}
\begin{array}{c}
\begin{array}{ccc} \quad ① \quad & \quad ② \quad & \quad ③ \quad \end{array} \\
\begin{bmatrix}
g_a + g_c & -g_a & 0 \\
-g_a & g_a + g_d & -g_d \\
0 & -g_d & g_b + g_d
\end{bmatrix}
\end{array}
\begin{bmatrix} v_1 \\ v_2 \\ v_3 \end{bmatrix}
=
\begin{bmatrix} i_a \\ i_b \\ -i_b \end{bmatrix}
\begin{array}{l}
\leftarrow \text{Indep. source} \\
\leftarrow \text{Indep. source} \\
\leftarrow \text{Indep. source}
\end{array}
\tag{1.18}
$$

The unknowns are the node voltages v_1, v_2, and v_3. They form the column vector of unknowns on the left-hand side. In general, if the number of nodes is n, there will be $(n-1)$ unknowns. The square matrix on the left-hand side is 3×3: this makes sense, as there are three unknowns and three equations. Each row of the matrix arises from writing KCL at a node. Each column of the matrix corresponds to a node potential. For instance, the entry in the second row and first column, namely $-g_a$, indicates that $-g_a$ is the coefficient of v_1 in the KCL equation written for node ②. The right-hand side is a column vector with known entries, which are the independent currents *entering* the nodes.

d. Careful observation of the square matrix on the left-hand side of (1.18) reveals that each element in the network has a unique pattern of entries associated with it. For instance, the conductance g_a connected between nodes ① and ② appears in rows 1 and 2, and columns 1 and 2. This makes sense—after all, the current through g_a must appear in the KCL equations written at nodes ① and ②. Furthermore, the current through g_a depends on v_1 and v_2; g_a must thereby appear in columns 1 and 2. This specific pattern is called the *stamp* of the conductance. The stamp of a conductance connected between nodes ⓟ and ⓠ is given in Figure 1.15. It is a square matrix whose dimension is equal to the number of unknowns. Its only nonzero entries are shown in the figure. In similar fashion, we see that a current source connected between nodes ⓟ and ⓠ is given in Figure 1.16. It is a column matrix with height equal to the number of unknowns. Its only nonzero entries are shown in the figure.

e. The unknown, namely the node-voltage vector, is determined by inverting the conductance matrix and multiplying it with the excitation vector.

$$\begin{array}{c} \\ p \\ q \end{array}\begin{array}{c} \begin{array}{cc} p & q \end{array} \\ \begin{bmatrix} g & -g \\ -g & g \end{bmatrix} \end{array}$$

Figure 1.15 Conductance connected between the pth and qth nodes, and its stamp.

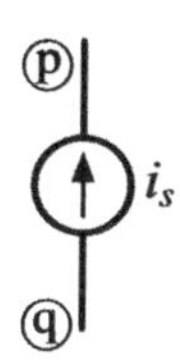

$$\begin{array}{c} p \\ q \end{array}\begin{bmatrix} i_s \\ -i_s \end{bmatrix}$$

Figure 1.16 Current source connected between the pth and qth nodes, and its stamp.

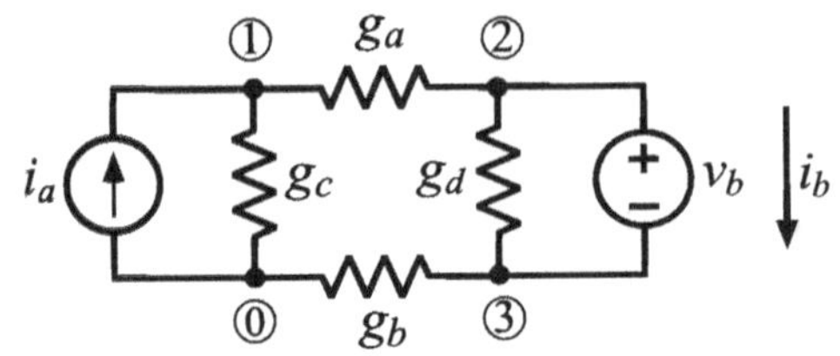

Figure 1.17 Example network containing a voltage-source excitation.

Consider now a network that contains a voltage-source excitation, such as the one shown in Figure 1.17. The voltage source poses a problem for nodal analysis, since the current that flows through is independent of the voltage. This is unlike a conductance, where the current depends on the voltage across it. Evidently, our procedure for solving the network must be modified. To address the problem, we denote the current through the voltage source v_b by i_b and treat it as an unknown. An additional unknown will need an additional equation. It is simply the constraint imposed by v_b, which is

$$v_2 - v_3 = v_b. \tag{1.19}$$

In matrix form, the equations are given by

$$\underbrace{\begin{bmatrix} g_a + g_c & -g_a & 0 & 0 \\ -g_a & g_a + g_d & -g_d & 1 \\ 0 & -g_d & g_d + g_b & -1 \\ 0 & 1 & -1 & 0 \end{bmatrix}}_{\mathbf{A}} \underbrace{\begin{bmatrix} v_1 \\ v_2 \\ v_3 \\ i_b \end{bmatrix}}_{\mathbf{X}} = \underbrace{\begin{bmatrix} i_a \\ 0 \\ 0 \\ v_b \end{bmatrix}}_{\mathbf{E}} \tag{1.20}$$

A few observations are in order. The shaded entries are due to the voltage source. The network is captured in the *augmented nodal matrix*, denoted by **A**. Its order has increased by 1 when compared to the network in Figure 1.14, since the number of unknowns is now 4. The 1 in the last column of row 2, and −1 in the last column of row 3 make explicit that i_b is being drawn from node ② and pushed into node ③. i_b is added to the column of unknowns, denoted by **X**. v_b is added as a new row to **E**, which is the vector of independent *excitation sources* on the right-hand side. **X** is seen to be given by $\mathbf{X} = \mathbf{A}^{-1}\mathbf{E}$. The way of writing network equations is nodal analysis, except that it has been *modified* to be able to support voltage sources. It is therefore called *modified nodal analysis* (MNA), and the stamps of the circuit elements are all referred to as MNA stamps. Figure 1.18 shows the MNA stamp of a voltage source connected between the pth and q^{th} node.

To summarize, when a voltage source is encountered, a new row and column are added to the augmented nodal matrix. The extra column has 1 and −1 in the *p*th and *q*th positions, respectively. The new row is the transpose of the extra column. The

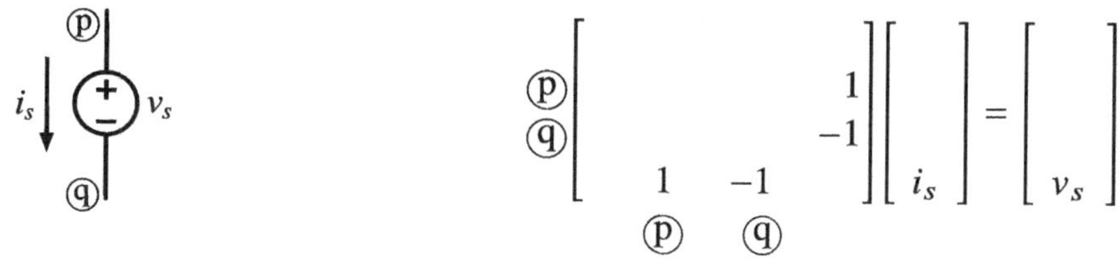

Figure 1.18 Voltage source connected between the pth and qth nodes, and its MNA stamp.

current through the voltage source is appended to the column of unknowns. The value of the voltage source is appended to the vector of independent sources.

Let us now move to the MNA stamps of controlled sources. We first consider the voltage-controlled current source (VCCS), using the network of Figure 1.19 for illustration.

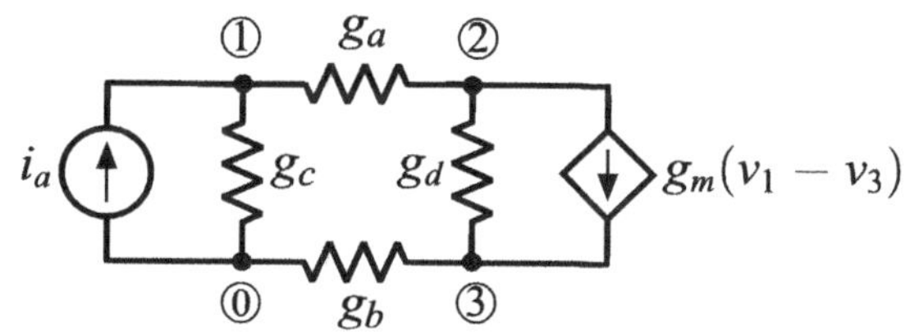

Figure 1.19 Example network containing a VCCS.

Applying KCL at each node yields three equations, which in matrix form are written as follows. The terms shaded in gray are due to the VCCS. Since the current source is connected between nodes ② and ③, the g_m terms appear in the second and third rows. Furthermore, since the current is proportional to the potential difference between ① and ③, g_m appears in the first and third columns. Since no new unknowns are introduced, the three node potentials need to be determined. Figure 1.20 shows a VCCS connected between arbitrary nodes and the corresponding MNA stamp.

$$\begin{bmatrix} g_a + g_c & -g_a & 0 \\ -g_a + g_m & g_a + g_d & -g_d - g_m \\ -g_m & -g_d & g_b + g_d + g_m \end{bmatrix} \begin{bmatrix} v_1 \\ v_2 \\ v_3 \end{bmatrix} = \begin{bmatrix} i_a \\ 0 \\ 0 \end{bmatrix}. \tag{1.21}$$

Figure 1.20 A voltage-controlled current source and its MNA stamp.

Let us now derive the stamp of a voltage-controlled voltage source (VCVS). We use the network of Figure 1.21 for illustration. The four node voltages are unknowns, as is the current i_x through the voltage source.

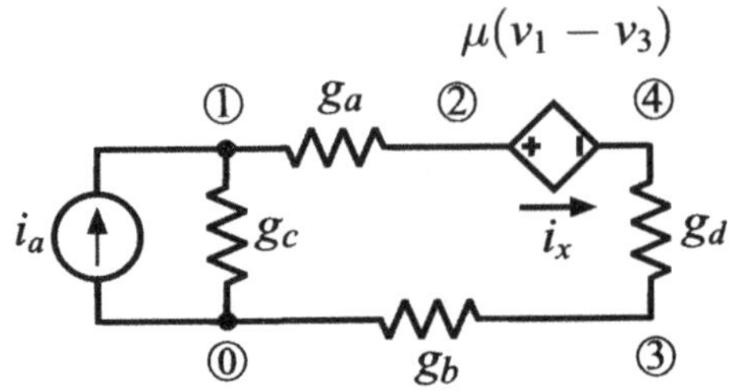

Figure 1.21 Example network containing a VCVS.

Applying KCL at each node yields four equations. In addition, the VCVS yields the additional equation

$$v_2 - v_4 - \mu v_1 + \mu v_3 = 0. \tag{1.22}$$

In matrix form, the five equations are expressed as follows. The terms shaded in gray are due to the VCVS. Since i_x flows from node ② to ④, 1 and −1 appear in the last column of the second and fourth rows respectively.

$$\begin{bmatrix} g_a + g_c & -g_a & 0 & 0 & 0 \\ -g_a & g_a & 0 & 0 & 1 \\ 0 & 0 & g_b + g_d & -g_d & 0 \\ 0 & 0 & -g_d & g_d & -1 \\ -\mu & 1 & \mu & -1 & 0 \end{bmatrix} \begin{bmatrix} v_1 \\ v_2 \\ v_3 \\ v_4 \\ i_x \end{bmatrix} = \begin{bmatrix} i_a \\ 0 \\ 0 \\ 0 \\ 0 \end{bmatrix}. \tag{1.23}$$

Figure 1.22 shows a VCVS connected between arbitrary nodes, and the corresponding MNA stamp.

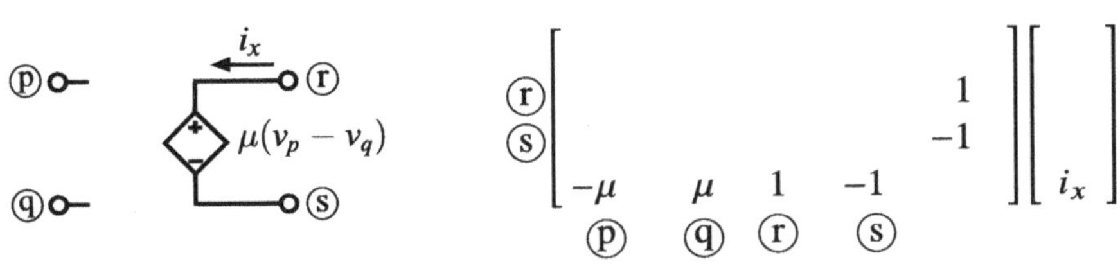

Figure 1.22 A VCVS and its MNA stamp.

Next, we derive the stamp of a current-controlled current source (CCCS). We use the network for Figure 1.23 for illustration. The four node voltages are unknowns, as is the current i_x through the zero-voltage source. Applying KCL at each node yields four equations. In addition, the zero-voltage source implies that

$$v_1 - v_4 = 0. \tag{1.24}$$

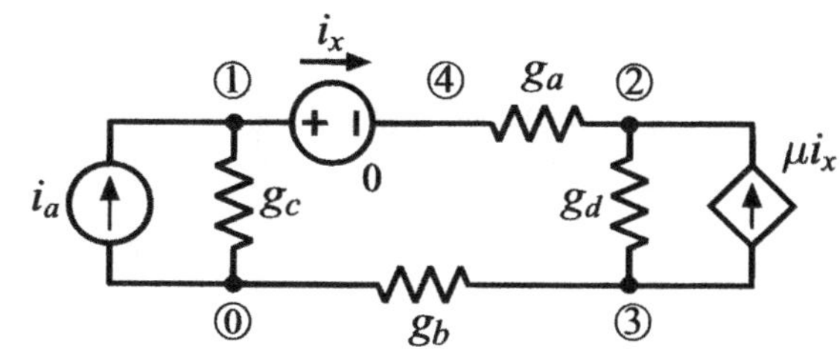

Figure 1.23 Example network containing a CCCS.

In matrix form, the five equations are expressed as follows. The terms shaded in gray are due to the CCCS. Since i_x flows from node ① to ④, 1 and −1 appear in the last column of the first and fourth rows, respectively.

$$\begin{bmatrix} g_c & 0 & 0 & 0 & 1 \\ 0 & g_a + g_d & -g_d & -g_a & -\mu \\ 0 & -g_d & g_b + g_d & 0 & \mu \\ 0 & -g_a & 0 & g_a & -1 \\ 1 & 0 & 0 & -1 & 0 \end{bmatrix} \begin{bmatrix} v_1 \\ v_2 \\ v_3 \\ v_4 \\ i_x \end{bmatrix} = \begin{bmatrix} i_s \\ 0 \\ 0 \\ 0 \\ 0 \end{bmatrix}. \tag{1.25}$$

Figure 1.24 shows a CCCS connected between arbitrary nodes, and the corresponding MNA stamp.

The last controlled source remaining is the current-controlled voltage source (CCVS). We derive its MNA stamp using the network of Figure 1.25 for illustration. The unknowns are the four node voltages and the currents i_x and i_y through the zero- and controlled-voltage sources, respectively. Applying KCL at each node yields four equations. There are *two* additional unknowns; two more equations are needed. They are the constraints imposed by the voltage sources and given by

$$v_1 - v_4 = 0 \tag{1.26}$$

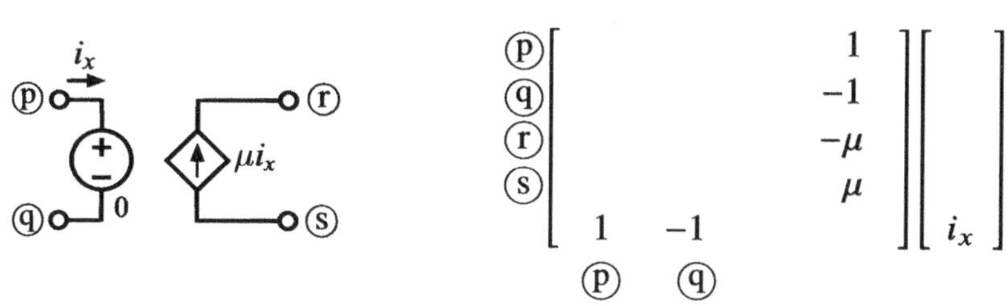

Figure 1.24 A CCCS, and its MNA stamp.

and

$$v_2 - v_3 = Ri_x. \tag{1.27}$$

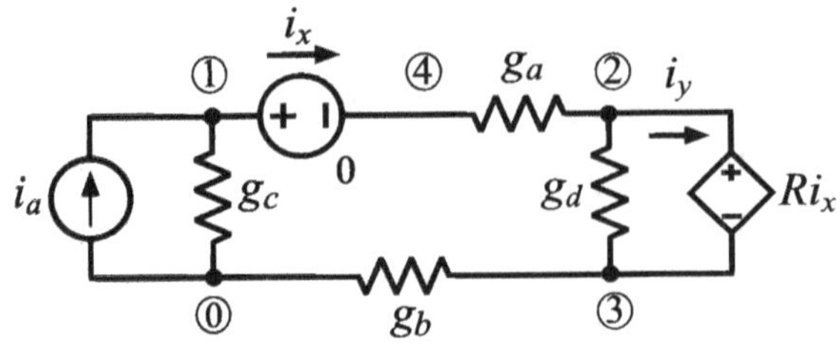

Figure 1.25 Example network containing a CCVS.

In matrix form, the six equations are expressed as shown below. The terms shaded in gray are due to the CCVS. Since i_x flows from node ① to ④, 1 and −1 appear in the penultimate column of the first and fourth rows, respectively. As i_y flows from node ② to ③, 1 and −1 appear in the last column of the second and third rows, respectively.

$$\begin{bmatrix} g_c & 0 & 0 & 0 & 1 & 0 \\ 0 & g_a + g_d & -g_d & -g_a & 0 & 1 \\ 0 & -g_d & g_b + g_d & 0 & 0 & -1 \\ 0 & -g_a & 0 & g_a & -1 & 0 \\ 1 & 0 & 0 & -1 & 0 & 0 \\ 0 & 1 & -1 & 0 & -R & 0 \end{bmatrix} \begin{bmatrix} v_1 \\ v_2 \\ v_3 \\ v_4 \\ i_x \\ i_y \end{bmatrix} = \begin{bmatrix} i_a \\ 0 \\ 0 \\ 0 \\ 0 \\ 0 \end{bmatrix} \tag{1.28}$$

Figure 1.26 shows a CCVS connected between arbitrary nodes and the corresponding MNA stamp.

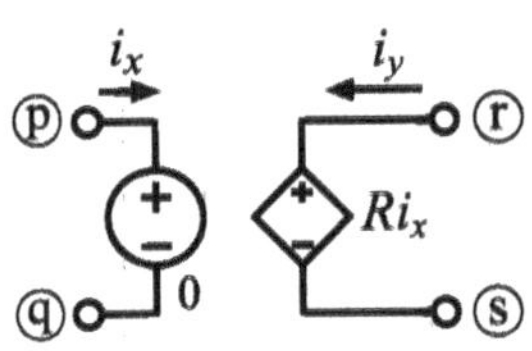

$$\begin{array}{c} ⓟ \\ ⓠ \\ ⓡ \\ ⓢ \\ \\ \\ \end{array} \begin{bmatrix} & & & & 1 & \\ & & & & -1 & \\ & & & & & 1 \\ & & & & & -1 \\ 1 & -1 & & & 0 & 0 \\ & & 1 & -1 & -R & 0 \end{bmatrix} \begin{bmatrix} \\ \\ \\ \\ i_x \\ i_y \end{bmatrix} \begin{bmatrix} \\ \\ \\ \\ \\ \\ \end{bmatrix}$$

$$ⓟ \quad ⓠ \quad ⓡ \quad ⓢ$$

Figure 1.26 A CCVS and its MNA stamp.

The ideal opamp is a commonly used network element. *If an opamp has DC negative feedback around it,* its two input terminals form a virtual short. This means that they have the same potential, and no current flows into these terminals. Let us derive the MNA stamp of an opamp, using the network of Figure 1.27 for illustration. The unknowns are the five node voltages, the currents i_x and i_z flowing into the opamps, and i_y flowing into the input source v_s. Applying KCL at each node yields five equations. Three more equations are needed, as there are as many additional unknowns. The equations are

$$v_1 = v_s$$
$$v_1 = v_3$$
$$v_4 = 0.$$

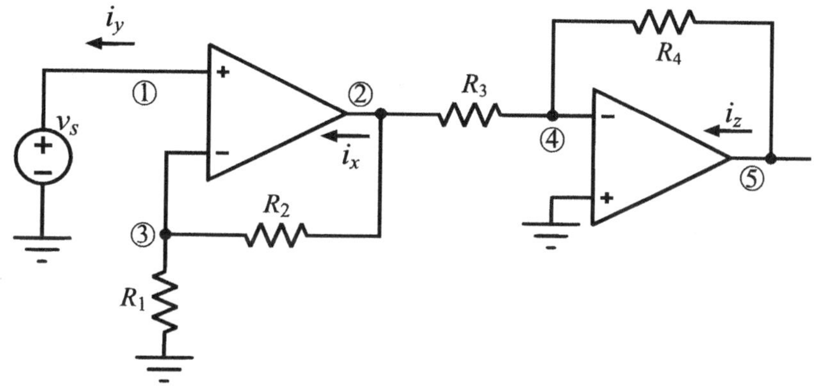

Figure 1.27 Example network containing ideal opamps.

The first of these is imposed by the voltage source, while the other two are the result of the opamps' input terminals forming a virtual short. In matrix form, the eight equations are expressed as shown below. The terms shaded in gray are due to the opamps. Since i_x flows out of ②, 1 appears in the penultimate column of the second row. Likewise, as i_z flows out of node ⑤, 1 appears in the last column of the fifth row.

$$\begin{bmatrix} 0 & 0 & 0 & 0 & 0 & 1 & 0 & 0 \\ 0 & g_2+g_3 & -g_2 & -g_3 & 0 & 0 & 1 & 0 \\ 0 & -g_2 & g_1+g_2 & 0 & 0 & 0 & 0 & 0 \\ 0 & -g_3 & 0 & g_3+g_4 & -g_4 & 0 & 0 & 0 \\ 0 & 0 & 0 & -g_4 & g_4 & 0 & 0 & 1 \\ 1 & 0 & 0 & 0 & 0 & 0 & 0 & 0 \\ 1 & 0 & -1 & 0 & 0 & 0 & 0 & 0 \\ 0 & 0 & 0 & 1 & 0 & 0 & 0 & 0 \end{bmatrix} \begin{bmatrix} v_1 \\ v_2 \\ v_3 \\ v_4 \\ v_5 \\ i_y \\ i_x \\ i_z \end{bmatrix} = \begin{bmatrix} 0 \\ 0 \\ 0 \\ 0 \\ 0 \\ v_s \\ 0 \\ 0 \end{bmatrix}. \quad (1.29)$$

Figure 1.28 gives the MNA stamp of an opamp connected between arbitrary nodes. It is important to bear in mind that there must be DC negative feedback around the opamp for the virtual-short concept to hold.

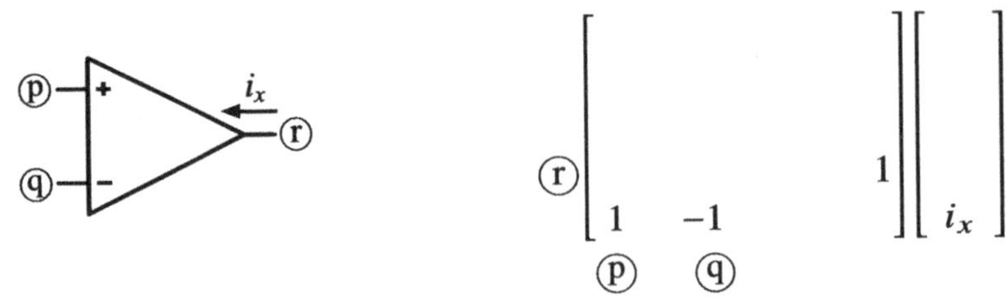

Figure 1.28 An ideal opamp and its MNA stamp.

1.6.1 Networks with Multiple Ideal Opamps

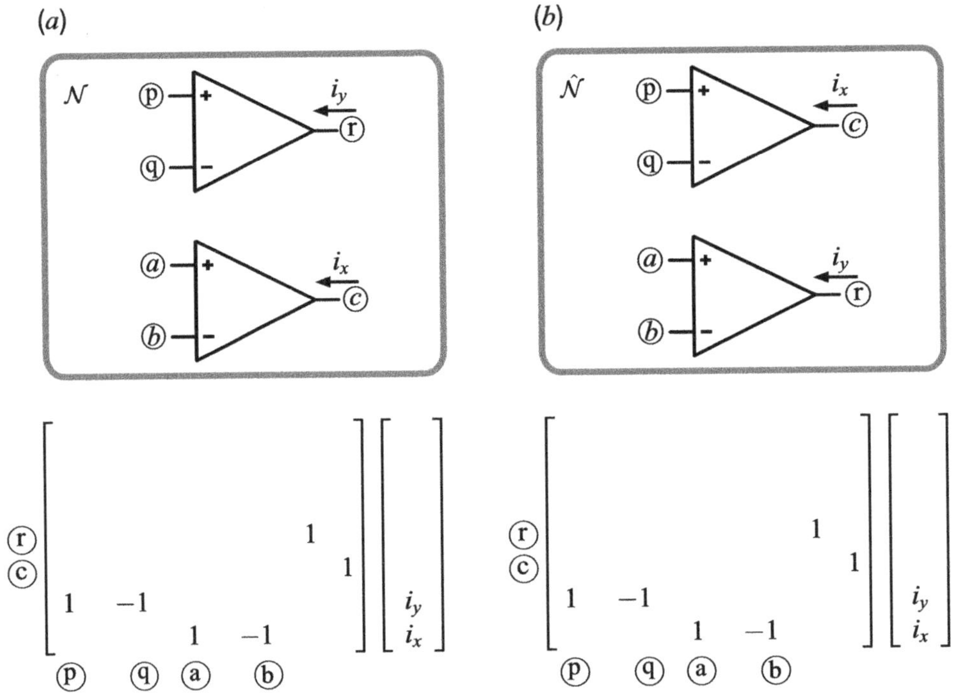

Figure 1.29 (a) Network $\mathcal{N}$ containing two ideal opamps. (b) Network $\hat{\mathcal{N}}$ with the outputs of the opamps interchanged. All node voltages and branch currents remain unchanged since the MNA equations are identical.

We now make a useful observation in circuits that contain multiple ideal opamps. For starters, consider a network $\mathcal{N}$ with two ideal opamps connected to nodes ⓟ, ⓠ, ⓡ and ⓐ, ⓑ, ⓒ as shown in Figure 1.29a. The relevant portion of the MNA matrices, i.e., that pertaining to the two opamps is shown below $\mathcal{N}$. Let us now try the reverse—to determine the connections of the opamps from the MNA matrices. From the last two rows, it is evident that $v_p = v_q$ and $v_a = v_b$, indicating that ⓟ–ⓠ and ⓐ–ⓑ form virtual shorts. What it is not clear, however, is how rows ⓡ and ⓒ should be assigned to the opamps. Interchanging the roles of ⓡ and ⓒ, as seen in $\hat{\mathcal{N}}$ (Figure 1.29b) would result in *the same* MNA equations! The node voltages and branch currents in $\mathcal{N}$ and $\hat{\mathcal{N}}$ are thus seen to be identical, though the circuits might appear very different. In a circuit with m ideal opamps, there are therefore $m!$ seemingly different variations that basically are the same circuit.[3] The "trick" of interchanging opamp outputs, therefore, is useful to generate alternative circuits that attempt to realize the same functionality—some of which may be better than others when practical opamp nonidealities are considered. One should note that the signs of the opamps in each of the alternative circuits have to be reevaluated to ensure that there is indeed DC negative feedback around them.

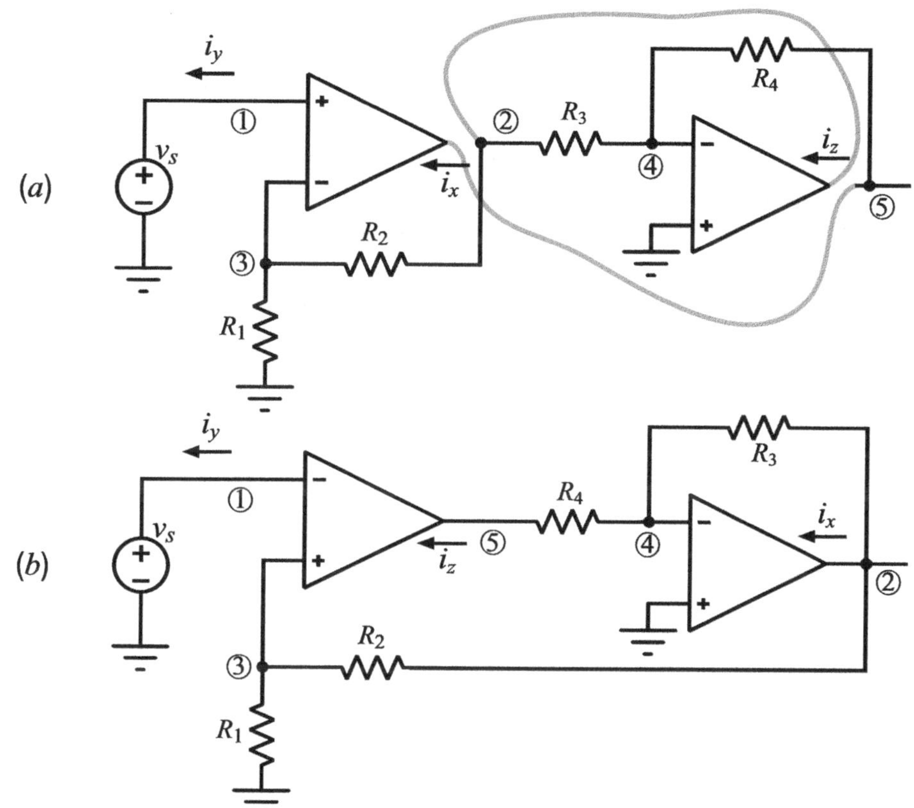

Figure 1.30 (a) Circuit of Figure 1.27, but with the output terminals of the opamps interchanged. The output is v_5. (b) Redrawing the circuit.

We illustrate with an example. Referring to Figure 1.27, let v_5 denote the output of the circuit. The voltage gain is given by

$$\frac{v_5}{v_S} = -\left(1 + \frac{R_2}{R_1}\right)\frac{R_4}{R_3}. \tag{1.30}$$

Figure 1.30a shows the circuit of Figure 1.27, but with the outputs of the opamps interchanged. Part (b) of the figure shows a redrawn version. The output is still v_5.

[3]The ! in $m!$ not only denotes the factorial but also excitement at this result!

Note that the signs of the first opamp have to be appropriately chosen so that both opamps have dc negative feedback around them. Furthermore, the output is the potential of node ⑤. Straightforward analysis shows that

$$\frac{v_2}{v_s} = \left(1 + \frac{R_2}{R_1}\right). \tag{1.31}$$

As a consequence,

$$\frac{v_5}{v_s} = -\left(1 + \frac{R_2}{R_1}\right)\frac{R_4}{R_3}. \tag{1.32}$$

The gain is indeed the same in both circuits, even though they look very different; the amplifier of Figure 1.27 is a cascade of two stages, while that of Figure 1.30b uses overall feedback. However, as our analysis shows, these two very different-looking circuits are actually the same *if the opamps are ideal*. The reader should verify that the MNA equations for the circuit of Figure 1.30b are identical to those for the network of Figure 1.27. It must therefore follow that the node voltages and branch currents in the circuits must therefore be the same. We reiterate that this only holds when the opamps are ideal—with real opamps, however, the performance of the two circuits will not be the same—one could perform better than the other.

To summarize, a systematic way of generating the MNA equations of a network is the following.

a. The augmented nodal matrix **A** is square with a dimension that equals the number of node voltages plus the number of currents through voltage sources. **A** is initialized with all zero entries.

b. Going element by element in the network, we add to **A** the MNA stamp of the corresponding element. It is thus seen that **A** is simply the sum of the MNA stamps of all the elements in the network.

c. The unknown vector **X** is determined using $\mathbf{X} = \mathbf{A}^{-1}\mathbf{E}$, where **E** is the source (or excitation) vector.

Figure 1.31a–e tabulates the MNA stamps of conductance and the four controlled sources for ready reference.

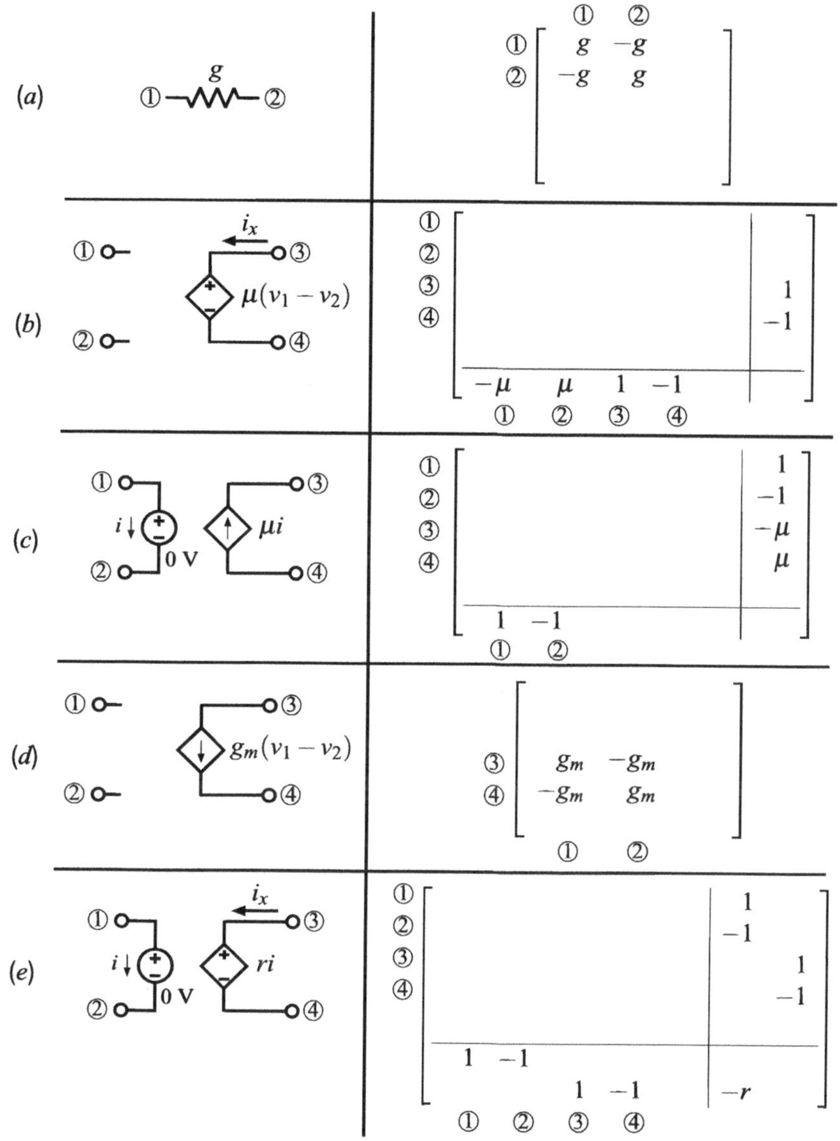

Figure 1.31 Summary of MNA stamps.

1.6.2 Sinusoidal Steady-State: Capacitors and Inductors

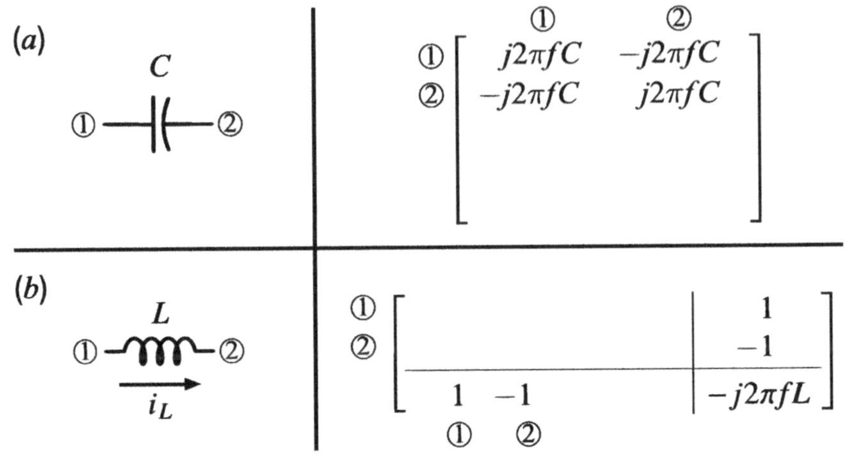

Figure 1.32 The MNA stamp of (a) a capacitor and (b) an inductor. Sinusoidal steady-state operation at a frequency f is assumed.

The discussion so far focused on memoryless networks. It is straightforward to see that the principles hold for networks operating in the sinusoidal steady state. The unknowns and excitations are now phasors. When capacitors and inductors are included in the network, their (frequency-dependent) MNA stamps are necessary. It is easy to see that the MNA stamp of the capacitor is a symmetric matrix as shown in Figure 1.32a. The stamp is seen to be similar to that of a conductance, except that the entries correspond to the frequency-dependent admittance of the capacitor. It is tempting to use the same approach for the inductor. This, however, is problematic since the entries in the stamp would then be of the form $\pm 1/(j2\pi fL)$ which go to infinity at DC. The solution to this problem is to introduce the inductor current as a new unknown I_L and add the equation $V_1 - V_2 - j2\pi fLI_L = 0$ to the set of nodal equations. In the nodal matrix, this corresponds to an additional row and column as shown in Figure 1.32b.

The same approach can be used to write the MNA equations when transfer functions need to be computed; $j2\pi f$ in the discussion above is replaced by the Laplace variable s.

The reader might wonder *why* all the discussion about writing MNA equations is relevant in the first place, when it seems more suited for a book devoted to describing the operation of circuit simulation programs. It turns out that understanding the MNA formulation is valuable in multiple ways, such as the following.

- As we will see going forward, the MNA formulation allows one to systematically derive reciprocal and inter-reciprocal networks, which are immensely useful in their own right.
- During the design of analog and mixed-signal subsystems and systems, it is often necessary to be able to simulate circuit blocks in the time domain. In many

cases, the level of detail needed is more than just the "top-level" transfer function. The MNA formulation allows a designer to create state-space models for circuit blocks, which can in turn be used for system-level simulations in tools like MATLAB.

Example: Descriptor State-Space Modeling of a Second-Order Filter

Suppose that the second-order filter structure shown in Figure 1.33 needs to be simulated in the time domain as part of a larger system. The opamp is modeled as a VCCS with a transconductance given by g_{ota}. Furthermore, assume that we are interested in the voltage waveform at *all* nodes. The reader is no doubt familiar with the state-space description of a linear system, namely

$$\dot{\mathbf{x}} = \mathbf{A}\mathbf{x} + \mathbf{B}\mathbf{u} \quad , \quad \mathbf{y} = \mathbf{C}\mathbf{x} + \mathbf{D}\mathbf{u}. \tag{1.33}$$

$\mathbf{x}$ is the vector of state variables, $\mathbf{A}$, $\mathbf{B}$, $\mathbf{C}$, $\mathbf{D}$ are the state matrices, and $\mathbf{y}$ and $\mathbf{u}$ denote the output and input vectors, respectively.

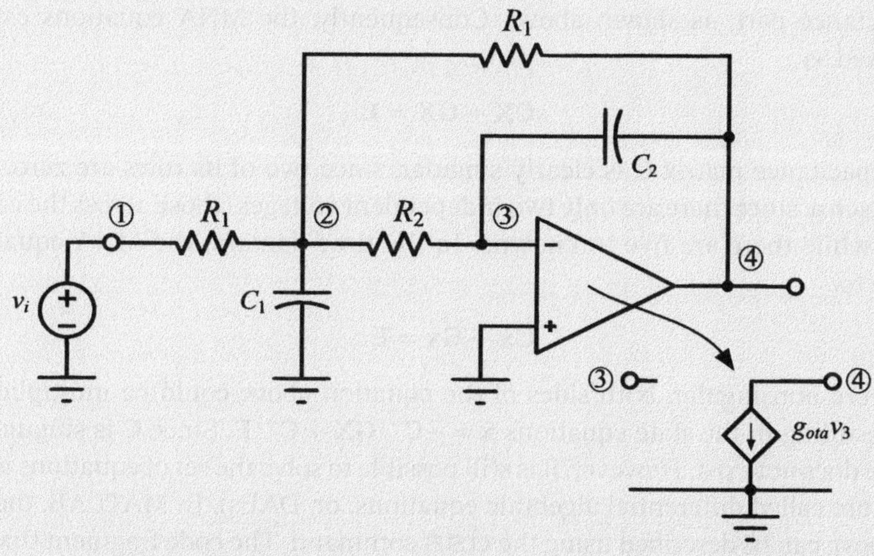

Figure 1.33 Modeling an analog sub-block for time-domain simulations.

When applied to the filter of Figure 1.33, the "standard" state-space description is problematic due to the following. The states are the capacitor voltages v_2 and $(v_3 - v_4)$. All other node voltages and voltage-source currents (a total of five in number) are linearly related to the state variables. Writing out the state matrices in terms of the node voltages is cumbersome.

(Continued)

(Continued)

The problem above is avoided by the use of the descriptor-state space formulation, which allows us to use the MNA equations directly. The mechanics of the process is first described, following which we delve into practical details. The MNA equations, written in terms of the Laplace-transformed unknowns in the network (namely the node voltages and input-source current), are expressed as follows.

$$s\underbrace{\begin{bmatrix} 0 & 0 & 0 & 0 & 0 \\ 0 & C_1 & 0 & 0 & 0 \\ 0 & 0 & C_2 & -C_2 & 0 \\ 0 & 0 & -C_2 & C_2 & 0 \\ 0 & 0 & 0 & 0 & 0 \end{bmatrix}}_{\mathbf{C}}\underbrace{\begin{bmatrix} V_1 \\ V_2 \\ V_3 \\ V_4 \\ I_x \end{bmatrix}}_{\mathbf{X}} + \underbrace{\begin{bmatrix} G_1 & -G_1 & 0 & 0 & 1 \\ -G_1 & 2G_1+G_2 & -G_2 & -G_1 & 0 \\ 0 & -G_2 & G_2 & 0 & 0 \\ 0 & -G_1 & g_{ota} & G_1 & 0 \\ 1 & 0 & 0 & 0 & 0 \end{bmatrix}}_{\mathbf{G}}\underbrace{\begin{bmatrix} V_1 \\ V_2 \\ V_3 \\ V_4 \\ I_x \end{bmatrix}}_{\mathbf{X}} = \underbrace{\begin{bmatrix} 0 \\ 0 \\ 0 \\ 0 \\ V_i \end{bmatrix}}_{\mathbf{E}}$$

The "capacitor" terms in the augmented nodal matrix are separated from the conductance part, as shown above. Consequently, the MNA equations can be expressed as

$$s\mathbf{CX} + \mathbf{GX} = \mathbf{E}.$$

The capacitance matrix $\mathbf{C}$ is clearly singular, since two of its rows are zero. This makes sense since there are only two independent voltages (those across the capacitors), while there are five unknowns. In the time domain, the MNA equations reduce to

$$\mathbf{C\dot{x}} + \mathbf{Gx} = \mathbf{E}.$$

If $\mathbf{C}$ were nonsingular, both sides of the equation above could be multiplied by $\mathbf{C}^{-1}$, resulting in the state equations $\dot{\mathbf{x}} = -\mathbf{C}^{-1}\mathbf{GX} + \mathbf{C}^{-1}\mathbf{E}$. Since $\mathbf{C}$ is singular, its inverse does not exist. However, it is still possible to solve the set of equations above (these are called differential algebraic equations, or, DAEs). In MATLAB, the system above can be described using the `dss` command. The code fragment that can be used to plot all the unknowns due to a step change in v_i is given below.

Listing 1.1 Descriptor State-Space Systems

```
% MATLAB code
%
% C\dot{x} = -Gx + e;
% A is the matrix that multiplies x.
% B = e/v_i = [0 0 0 0 1]';
%
A = -G;
B = [0 0 0 0 1]';
```

```
D = 0;
sys_biquad = dss(A,B,eye(5),D,C);
% Defines biquad in descriptor state-space form
% The outputs are all the node voltages, and current
    through the input source.
[x, t] = step(sys_biquad,10); % computes step response
    for 10 seconds
```

1.7 Bilinear Form of a Transfer Function, and the Extra-Element Theorem

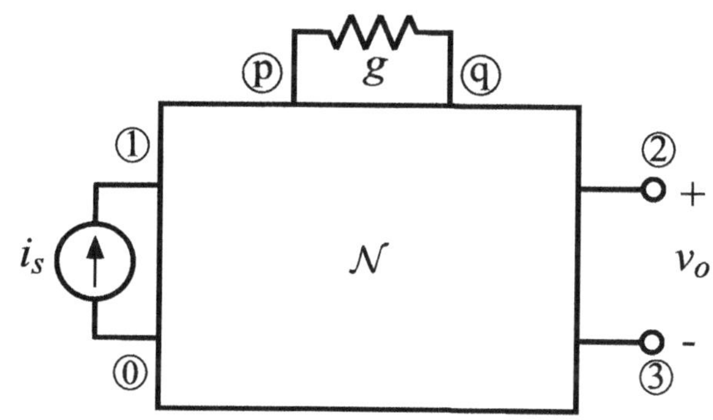

Figure 1.34 How does v_o/i_s depend on g?

Consider a network $\mathcal{N}$, as shown in Figure 1.34. How does the transimpedance v_o/i_s depend on a conductance g connected between nodes Ⓟ and Ⓠ? Let us use the "MNA way" to answer this question. Writing the nodal equations, we have

$$\begin{array}{c} \\ \textcircled{p} \\ \textcircled{q} \\ \\ \end{array} \begin{bmatrix} & \textcircled{p} & \textcircled{q} & \\ & & & \\ & g & -g & \\ & -g & g & \\ & & & \end{bmatrix} \begin{bmatrix} v_1 \\ v_2 \\ v_3 \\ \vdots \end{bmatrix} = \begin{bmatrix} i_s \\ \\ \\ \\ \end{bmatrix}. \tag{1.34}$$

The empty spaces in the nodal matrix are entries that do not have g in them. Furthermore, the entries marked g and $-g$ could have additional terms in them. We express the equations as

$$\mathbf{A}(g)\mathbf{V} = \mathbf{E}, \tag{1.35}$$

to reflect the fact that the nodal matrix **A** depends on g. **V** is the vector of (unknown) node voltages, and **E** is the vector of excitations. The empty rows in **E** are zero. In the discussion that follows, we use $\mathbf{A}_k$ to denote a matrix where the kth column of **A** is replaced by $\mathbf{E}/i_s$. As an example, $\mathbf{A}_1$ is given by

$$\mathbf{A}_1(g) = \begin{array}{c} \\ \textcircled{1} \\ \textcircled{p} \\ \textcircled{q} \\ \\ \end{array} \begin{array}{c} \begin{array}{ccc} \textcircled{1} & \textcircled{p} & \textcircled{q} \end{array} \\ \left[\begin{array}{cccc} 1 & & & \\ 0 & & g & -g \\ 0 & & -g & g \\ 0 & & & \end{array}\right] \end{array}. \tag{1.36}$$

From Figure 1.34, we see that $v_o = v_2 - v_3$. Using Cramer's rule, we have

$$\frac{v_o}{i_s} = \frac{\det(\mathbf{A}_2(g)) - \det(\mathbf{A}_3(g))}{\det(\mathbf{A}(g))}, \tag{1.37}$$

where $\det(\mathbf{A})$ denotes the determinant of **A**. How does $\det(\mathbf{A}(g))$ depend on g? To answer this question, we perform a series of row and column operations on $\mathbf{A}(g)$ as follows.

$$\begin{array}{c} \begin{array}{cc} \textcircled{p} & \textcircled{q} \end{array} \\ \begin{array}{c} \textcircled{p} \\ \textcircled{q} \end{array} \left[\begin{array}{cc} g & -g \\ -g & g \end{array}\right] \end{array} \longrightarrow \begin{array}{c} \begin{array}{cc} \textcircled{p} & \textcircled{q} \end{array} \\ \begin{array}{c} \textcircled{p} \\ \textcircled{q}+\textcircled{p} \end{array} \left[\begin{array}{cc} g & -g \\ & \end{array}\right] \end{array} \longrightarrow \begin{array}{c} \begin{array}{cc} \textcircled{p} & \textcircled{q}+\textcircled{p} \end{array} \\ \begin{array}{c} \textcircled{p} \\ \textcircled{q} \end{array} \left[\begin{array}{cc} g & \\ & \end{array}\right] \end{array}$$

$$\longrightarrow \begin{array}{c} \begin{array}{cc} \textcircled{p} & \textcircled{q} \end{array} \\ \begin{array}{c} \textcircled{p} \\ \textcircled{q} \end{array} \left[\begin{array}{cc} g & \\ & \end{array}\right] \end{array}$$

We first add the pth row of $\mathbf{A}(g)$ to the qth row. As a consequence, the qth row will no longer have g in it. Next, we add the pth column of $\mathbf{A}(g)$ to the qth column. The resulting matrix will contain g only it its pth row and pth column. By inspection, it is clear that the determinant of this matrix must be of the form $k_1 g + k_2$. In other words, the highest power of g must be unity. Since this matrix is derived from $\mathbf{A}(g)$ through row and column additions, its determinant is identical to $\det(\mathbf{A}(g))$. We thus see that

$$\det(\mathbf{A}(g)) = k_1 g + k_2. \tag{1.38}$$

Using similar arguments, it is straightforward to see that

$$\det(\mathbf{A}_2(g)) - \det(\mathbf{A}_3(g)) = k_3 g + k_4. \tag{1.39}$$

From the equations above, we conclude that the transfer function v_o/i_s must be of the form

$$\boxed{\frac{v_o}{i_s} = \frac{k_3 g + k_4}{k_1 g + k_2}.} \tag{1.40}$$

It is thus seen that the transfer function is a *bilinear form*, i.e., a ratio of first-order polynomials in g. While we derived this for a current excitation and voltage output, it is straightforward to see that this is valid for *any* transfer function. It turns out, as we see below, that (1.40) lends itself to an alternative (and sometimes useful) interpretation. It can be recast as

$$\frac{v_o}{i_s} = \frac{k_4}{k_2}\frac{1 + \frac{k_3}{k_4}g}{1 + \frac{k_1}{k_2}g} = \left(\frac{v_o}{i_s}\right)_{oc} \frac{1 + (z_n/R)}{1 + (z_d/R)}, \tag{1.41}$$

where $R = 1/g$.

a. k_4/k_2 is the transfer function of $\mathcal{N}$ when $g = 0$, i.e., when the conductance is replaced by an open circuit. It is therefore denoted by $(v_o/i_s)_{oc}$.

b. z_n can be interpreted as follows. From (1.41), we see that choosing R to be $-z_n$ results in $v_o/i_s = 0$. In other words, if the conductance g is chosen so that the output v_o is zero (i.e., the output has been nulled) when $\mathcal{N}$ is excited by the current i_s, then that value of conductance is given by $-1/z_n$. Rather than vary g to obtain an output null, one could instead inject a current i_x between nodes ⓟ and ⓠ as shown in Figure 1.35, and vary it so that the output v_o is zero. Let the resulting voltage that is developed across the current source be v_x. It is clear that $v_x/i_x = z_n$.

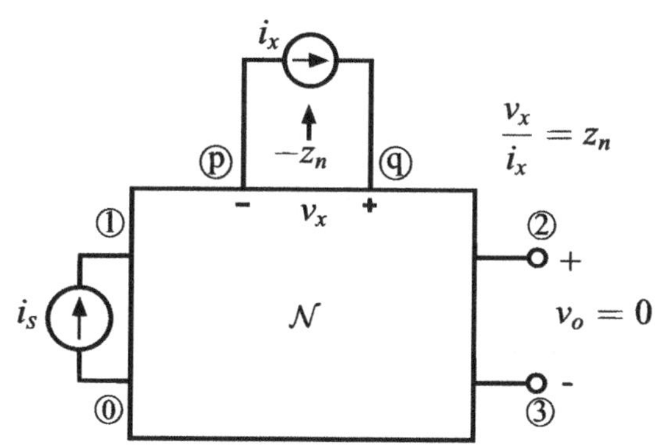

Figure 1.35 If R is replaced by $-z_n$, v_o will be zero. i_x is varied to obtain a null output, and $v_x/i_x = z_n$.

c. As seen from (1.40), z_d is given by k_1/k_2. Recall that

$$\det(\mathbf{A}(g)) = k_1 g + k_2. \tag{1.42}$$

This means that k_1 and k_2 can be obtained as follows.

$$k_1 = \left.\frac{d}{dg}\det(\mathbf{A}(g))\right|_{g=0}, \qquad k_2 = \det(\mathbf{A}(0)). \tag{1.43}$$

Recall that the derivative of the determinant of a matrix with respect to a variable x can be expressed as

$$\frac{d}{dx}\det\begin{bmatrix} a(x) & b(x) & \cdots \\ c(x) & d(x) & \cdots \\ \cdots & \cdots & \cdots \end{bmatrix} = \det\begin{bmatrix} a'(x) & b(x) & \cdots \\ c'(x) & d(x) & \cdots \\ \cdots & \cdots & \cdots \end{bmatrix} + \det\begin{bmatrix} a(x) & b'(x) & \cdots \\ c(x) & d'(x) & \cdots \\ \cdots & \cdots & \cdots \end{bmatrix} + \cdots$$

where the prime denotes the derivative of the function with respect to its argument. Using this, we see that z_d can be expressed as follows:

$$z_d = \frac{k_1}{k_2} = \frac{1}{\det(\mathbf{A}(0))} \left.\frac{d(\det(\mathbf{A}(g)))}{dg}\right|_{g=0}$$

$$= \frac{\det\left\{ \begin{matrix} & \textcircled{p} & \textcircled{q} \\ & \begin{bmatrix} 0 & \\ 1 & \\ -1 & \\ 0 & \end{bmatrix} \\ \begin{matrix} \\ \textcircled{p} \\ \textcircled{q} \\ \\ \end{matrix} \end{matrix} \right\} - \det\left\{ \begin{matrix} & \textcircled{p} & \textcircled{q} \\ & \begin{bmatrix} & 0 \\ & 1 \\ & -1 \\ & 0 \end{bmatrix} \\ \begin{matrix} \\ \textcircled{p} \\ \textcircled{q} \\ \\ \end{matrix} \end{matrix} \right\}}{\det(\mathbf{A}(0))} = \frac{v_p - v_q}{1\,\mathrm{A}}.$$

In the expression above, the numerator is the difference between two determinants. These determinants are $\det(\mathbf{A}(0))$ with the pth and qth columns replaced by column vectors that contain 1 and -1 in the pth and qth rows, respectively. This makes sense, as only the pth and qth columns of $\mathbf{A}(g)$ contain g. It is clear that z_d can be interpreted as the difference between the potentials of nodes ⓟ and ⓠ when a 1A current source is connected between these nodes as shown in Figure 1.36. Note that the input excitation is disabled, i.e., $i_s = 0$.

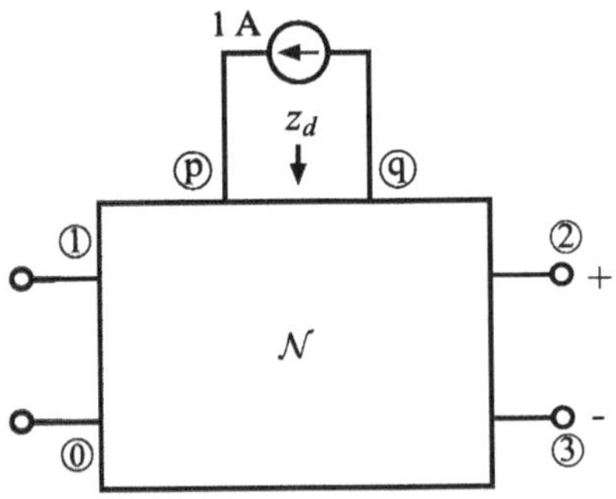

Figure 1.36 z_d is the impedance looking into nodes p–q when $i_s = 0$.

To summarize:

a. The transfer function of a network $\mathcal{N}$ is a bilinear function of its component values. Denoting the impedance of a component by z, any transfer function T can be expressed as

$$T = T_{\infty}\left(\frac{1+\frac{z_n}{z}}{1+\frac{z_d}{z}}\right). \tag{1.44}$$

T can be a voltage gain, current gain, (trans)impedance, or (trans)admittance. T_{∞} denotes the value of T when z is infinite (1.44), which quantifies the effect of introducing z into $\mathcal{N}$ on T_{∞} by multiplying with a correction factor, is called the **extra-element theorem.**

b. z_n is found by replacing z by a current source i_x whose value is such that the output is nulled and determining the ratio of the voltage developed across it to i_x. Since the network has two excitations and the output is zero, this is referred to as *null double injection.*

c. z_d is the impedance looking into $\mathcal{N}$ when the input excitation is set to 0; i.e., the input voltage (current) is replaced by a short (open) circuit as seen in Figure 1.37d.

Example: Using the Extra-Element Theorem

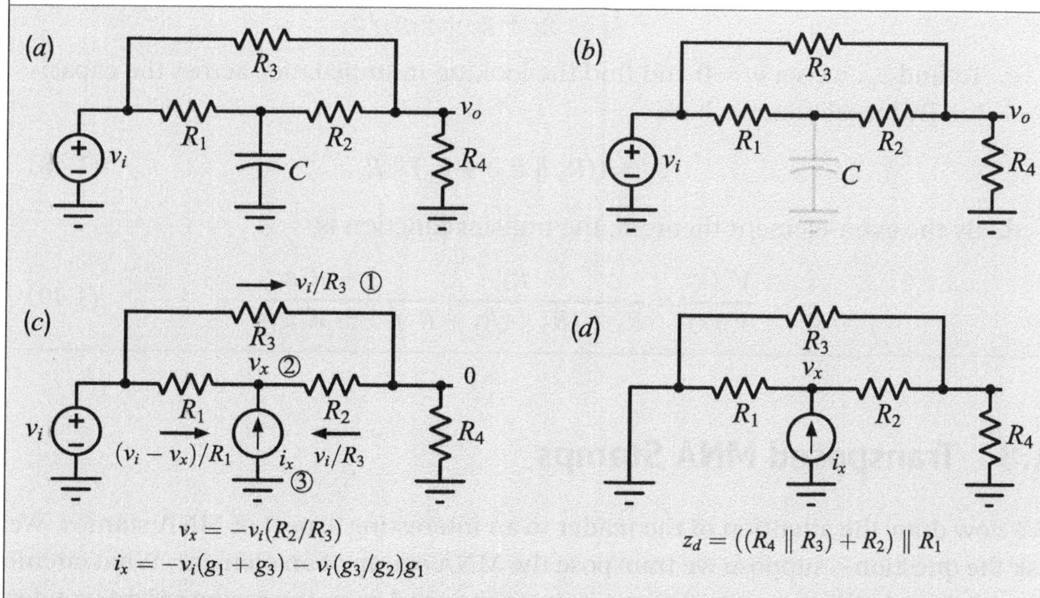

Figure 1.37 (a) $\mathcal{N}$: we treat C as the extra element. (b) Finding the transfer function without C. (c) Finding z_n. (d) Finding z_d.

(Continued)

(Continued)

We illustrate the application of the extra-element theorem using the network of Figure 1.37a. Suppose we wish to find $V_o(s)/V_i(s)$. We treat the capacitor C as the "extra element", and proceed as follows.

a. We set $C = 0$ to find the transfer function $\left.\frac{V_o(s)}{V_i(s)}\right|_{oc}$. The resulting network is shown in Figure 1.37b. It is seen that

$$\left.\frac{V_o(s)}{V_i(s)}\right|_{oc} = \frac{R_4}{R_4 + (R_3 \parallel (R_1 + R_2))}. \tag{1.45}$$

b. To find z_n, we replace the capacitor with a current source i_x with the aim of nulling the output. The resulting circuit is shown in Figure 1.37c. We observe the following.

① Since $v_o = 0$, the current through R_3 is v_i/R_3.
② This current flows through R_2, resulting in $v_x = -v_i(R_2/R_3)$.
③ Applying KCL, we see that

$$-i_x = v_i g_3 + (v_i - v_x)g_1 = v_i(g_1 + g_3 + g_1 g_3/g_2). \tag{1.46}$$

z_n is thus given by

$$z_n = \frac{v_x}{i_x} = \frac{g_3/g_2}{g_1 + g_3 + g_1 g_3/g_2}. \tag{1.47}$$

c. To find z_d, we set $v_i = 0$ and find the looking-in impedance across the capacitor. By inspection, we have

$$z_d = ((R_4 \parallel R_3) + R_2) \parallel R_1. \tag{1.48}$$

d. By the extra-element theorem, the transfer function is

$$\frac{V_o(s)}{V_i(s)} = \frac{R_4}{R_4 + (R_3 \parallel (R_1 + R_2))} \frac{1 + sCz_n}{1 + sCz_d}. \tag{1.49}$$

1.8 Transposed MNA Stamps

We now draw the attention of the reader to an interesting aspect of MNA stamps. We ask the question—suppose we transpose the MNA stamp of an element. What circuit element would the transposed stamp correspond to? Again, the reader might wonder why this discussion is useful. First, this exercise might be seen as one that deepens our understanding of MNA stamps. More importantly, it turns out that it aids the seamless derivation of reciprocal and inter-reciprocal networks.

We begin with the simplest possible element, namely a conductance. Since the MNA stamp of a conductance is a symmetric matrix, the element corresponding to its transpose is the conductance itself. The MNA stamps of a capacitor and an inductor are also symmetric. It thus follows that the elements that correspond to the transpose of their MNA stamps are the elements themselves.

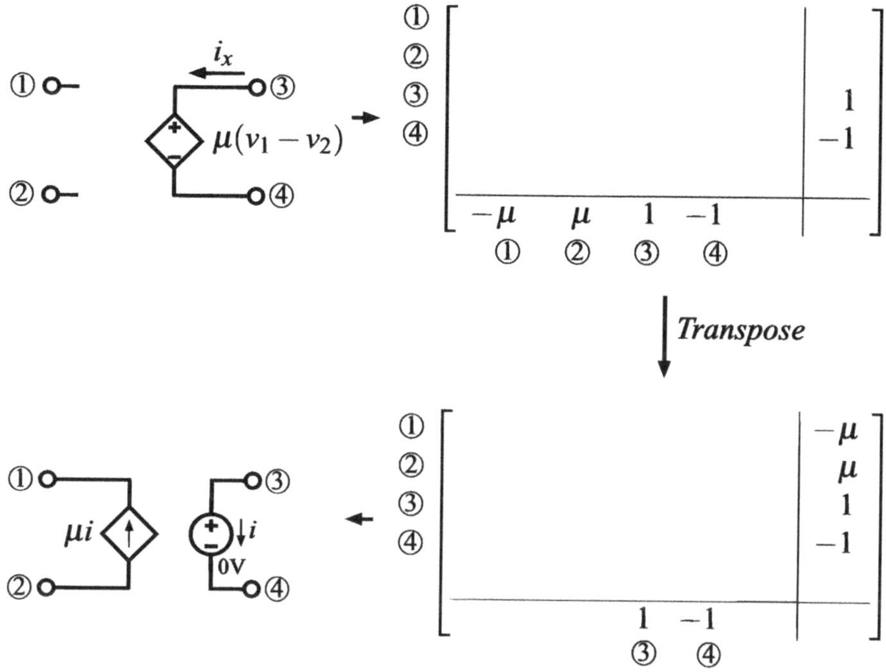

Figure 1.38 A VCVS and its MNA stamp. The transposed MNA stamp corresponds to a CCCS with its controlling and controlled ports interchanged.

The situation is different with controlled sources, as we explain with the help of an example (Figure 1.38). The figure shows a voltage-controlled voltage source and its MNA stamp. The transposed MNA stamp can be immediately recognized as that of a current-controlled current source. However, the controlling and controlled ports are reversed. A sanity check is the following. When $\mu = 0$, the VCVS is an open circuit (i.e., conductance of zero) between its input terminals and a short circuit (infinite conductance) between the output terminals. As we have seen earlier, a network of conductances remains unchanged when its MNA matrix is transposed. It must thus follow that with $\mu = 0$, the element corresponding to the transpose of the MNA stamp of the VCVS must also reduce to an open circuit between ① and ②, and a short circuit between nodes ③ and ④, which is indeed the case, as seen from Figure 1.38.

Practical Tip: Current Upwards or Downwards?

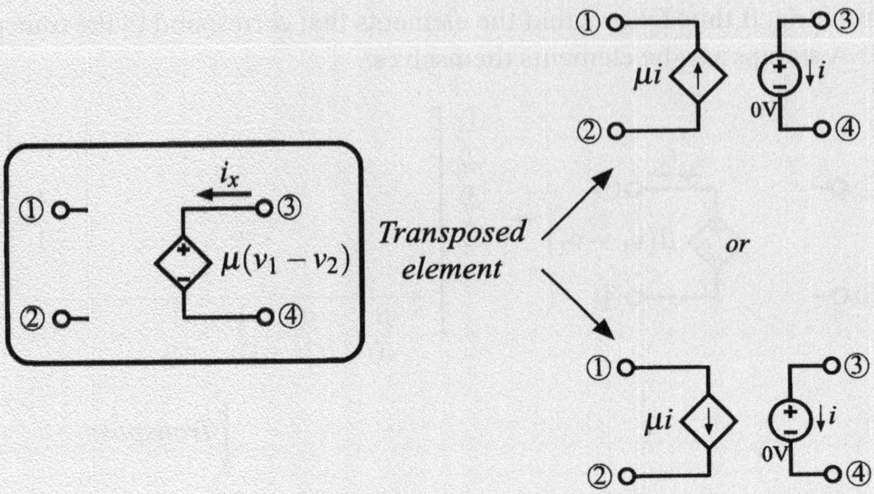

Figure 1.39 Does the μi in the transposed element go up or down?

A common source of confusion is the orientation of the CCCS. A simple way to mark the correct direction, without having to write out the MNA stamp, transpose it, and recognize the orientation, is the following. Assume that μ is positive. Applying a positive voltage at the input port of the VCVS yields a positive voltage at the output port (Figure 1.39). The same must happen in the CCCS. Attempting to apply a positive voltage to the controlling port of the CCCS causes current to flow downwards in Figure 1.38. That should cause an increase in the voltage of the controlled port, just like in the VCVS. This means that the current μi should be chosen to flow *upwards.*

Figure 1.40 shows the four controlled sources and the elements that correspond to the transpose of their MNA stamps (which are also referred to as the inter-reciprocal elements). In all cases, note that when the controlling parameter is set to zero, the port impedances in the original and the inter-reciprocal elements remain the same.

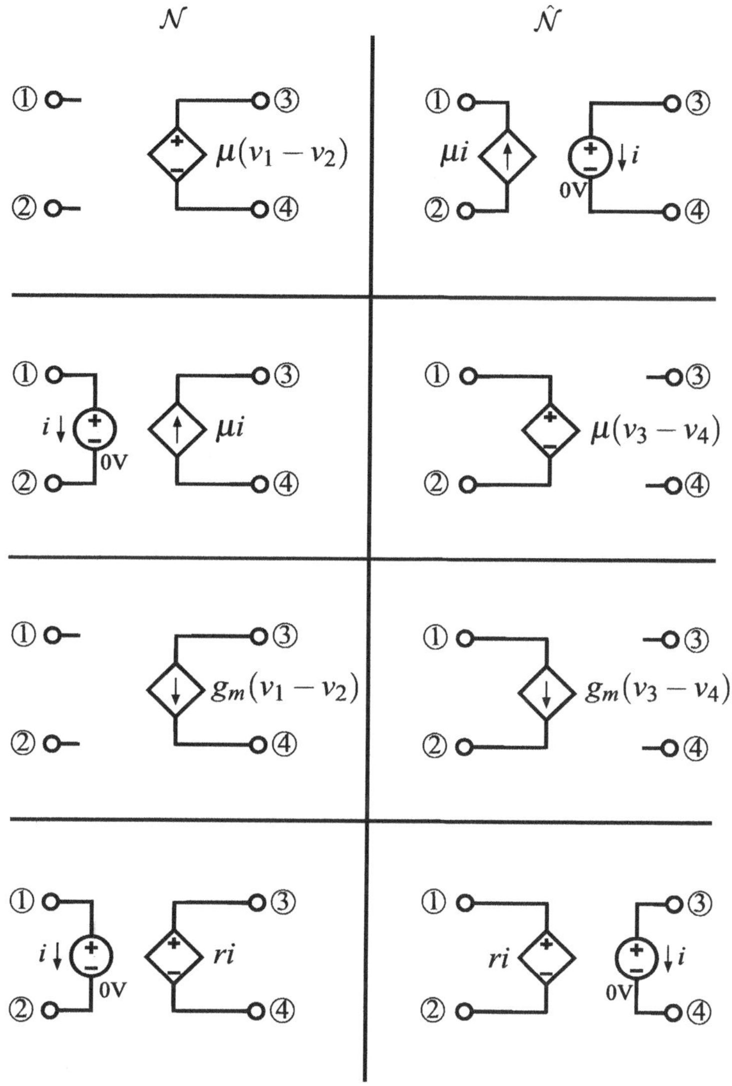

Figure 1.40 Inter-reciprocal elements corresponding to the four controlled sources.

1.9 Reciprocity and Inter-Reciprocity

In this section, we describe the property of inter-reciprocity using the background given in the previous section. We illustrate with an example. For simplicity, the discussions below use memoryless networks. Extension to networks with memory operating in the sinusoidal steady state is straightforward.

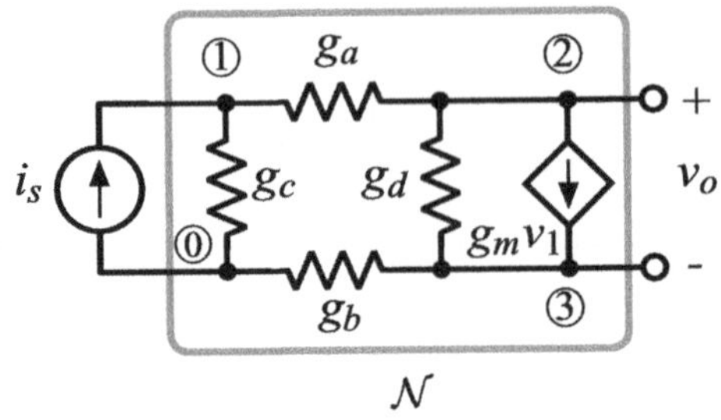

Figure 1.41 Example network used to illustrate reciprocity and inter-reciprocity.

Consider the network $\mathcal{N}$ of Figure 1.41. The network equations are seen to be

$$\underbrace{\begin{bmatrix} g_a + g_c & -g_a & 0 \\ -g_a + g_m & g_a + g_d & -g_d \\ -g_m & -g_d & g_b + g_d \end{bmatrix}}_{\mathbf{A}} \underbrace{\begin{bmatrix} v_1 \\ v_2 \\ v_3 \end{bmatrix}}_{\mathbf{X}} = \underbrace{\begin{bmatrix} i_s \\ 0 \\ 0 \end{bmatrix}}_{\mathbf{E}}. \tag{1.50}$$

The output voltage is given by $v_o = (v_2 - v_3)$, and can be written in matrix form as

$$v_o = \begin{bmatrix} 0 & 1 - 1 \end{bmatrix} \mathbf{X}. \tag{1.51}$$

Using (1.50), we see that v_o can be expressed as

$$v_o = \underbrace{\begin{bmatrix} 0 & 1 & -1 \end{bmatrix}}_{\textit{Measurement}} \mathbf{A}^{-1} \underbrace{\begin{bmatrix} i_s \\ 0 \\ 0 \end{bmatrix}}_{\textit{Excitation}}. \tag{1.52}$$

The equation above can be interpreted as follows. The excitation vector $\mathbf{E}$ is processed by the network represented by the MNA matrix $\mathbf{A}$ to result in a node-voltage vector $\mathbf{X} = \mathbf{A}^{-1}\mathbf{E}$. The output voltage v_o is obtained by subjecting $\mathbf{X}$ to the *measurement* matrix, given by $\begin{bmatrix} 0 & 1 - 1 \end{bmatrix}$ in the example above.

Since v_o is a scalar, (1.52) can be rewritten by transposing its right-hand side, and moving the scalar i_s as shown below.

$$v_o = \underbrace{\begin{bmatrix} 1 & 0 & 0 \end{bmatrix}}_{\text{Measurement}} (\mathbf{A}^T)^{-1} \underbrace{\begin{bmatrix} 0 \\ i_s \\ -i_s \end{bmatrix}}_{\text{Excitation}}. \tag{1.53}$$

The equation above can be *interpreted* as follows. The excitation vector $\begin{bmatrix} 0 & i_s - i_s \end{bmatrix}^T$ is processed by a network whose MNA matrix is $\mathbf{A}^T$. The resulting node-voltage vector is subjected to the measurement vector $\begin{bmatrix}1 & 0 & 0\end{bmatrix}$. The output that results is v_o. A few remarks are in order.

a. The exciting current is placed between nodes ② and ③. These nodes represented the output port in $\mathcal{N}$.

b. Recall that **A** was obtained by summing up the MNA stamps of the constituent network elements in $\mathcal{N}$. The network $\hat{\mathcal{N}}$ whose MNA matrix is $\mathbf{A}^T$ is thus seen to be comprised of elements whose MNA stamps are the transpose of the corresponding elements in $\mathcal{N}$. This suggests that if $\mathcal{N}$ is given, $\hat{\mathcal{N}}$ can be obtained by appropriate element-wise replacement. A conductance in $\mathcal{N}$ remains unchanged since the MNA stamp of a conductance is symmetric. Controlled sources are replaced according to the rules given in Figure 1.40. It is straightforward to see that the MNA stamp of the element on the right in Figure 1.40 is the transpose of that of the element on the left. $\hat{\mathcal{N}}$ is called the inter-reciprocal network. It is also referred to as the adjoint network.

c. The output voltage is measured between ① and ground. Recall that the exciting current source was connected between these nodes in $\mathcal{N}$.

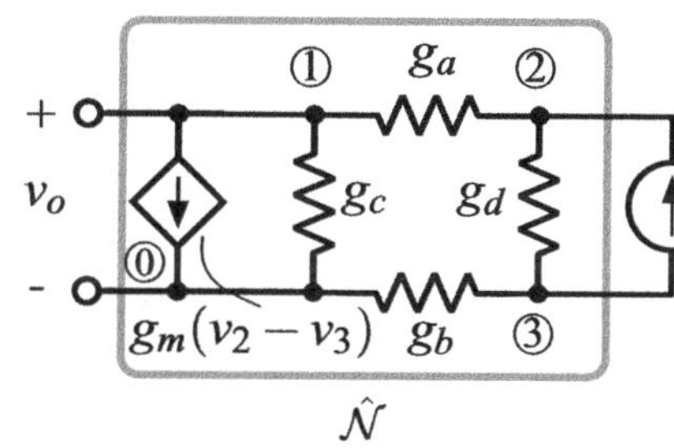

Figure 1.42 Exciting the inter-reciprocal network corresponding to the network of Figure 1.41 at the "output" port yields the same voltage v_o at the "input" port.

The interpretation of (1.53) as described above leads to the network shown in Figure 1.42. From the discussion above, we see that inter-reciprocity follows quite naturally from the way one interprets the relationship between v_o and i_s. In the special case of a network without any dependent sources, $\mathcal{N}$ and $\hat{\mathcal{N}}$ are identical, which leads to the more familiar concept of reciprocity.

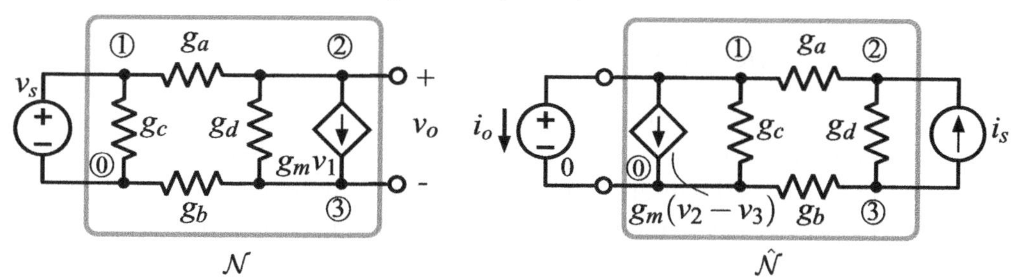

Figure 1.43 $\mathcal{N}$ is excited with a voltage, and the output is a voltage. The inter-reciprocal network is excited with a current, and the output is also a current.

Let us now try this out with a voltage-source excitation. Consider the network $\mathcal{N}$ of Figure 1.43. The MNA equations, written in matrix form, are given by

$$\underbrace{\begin{bmatrix} g_a + g_c & -g_a & 0 & 1 \\ -g_a + g_m & g_a + g_d & -g_d & 0 \\ -g_m & -g_d & g_b + g_d & 0 \\ 1 & 0 & 0 & 0 \end{bmatrix}}_{\mathbf{A}} \underbrace{\begin{bmatrix} v_1 \\ v_2 \\ v_3 \\ i_s \end{bmatrix}}_{\mathbf{X}} = \underbrace{\begin{bmatrix} 0 \\ 0 \\ 0 \\ v_s \end{bmatrix}}_{\mathbf{E}}. \tag{1.54}$$

The output voltage is given by $v_o = (v_2 - v_3)$, and can be written in matrix form as

$$v_o = \begin{bmatrix} 0 & 1 - 1 & 0 \end{bmatrix} \mathbf{X}. \tag{1.55}$$

Using (1.54), we see that v_o/v_s can be expressed as

$$\frac{v_o}{v_s} = \underbrace{\begin{bmatrix} 0 & 1 & -1 & 0 \end{bmatrix}}_{\text{Measurement}} \mathbf{A}^{-1} \underbrace{\begin{bmatrix} 0 \\ 0 \\ 0 \\ 1 \end{bmatrix}}_{\text{Excitation}} = \underbrace{\begin{bmatrix} 0 & 0 & 0 & 1 \end{bmatrix}}_{\text{Measurement}} (\mathbf{A}^T)^{-1} \underbrace{\begin{bmatrix} 0 \\ 1 \\ -1 \\ 0 \end{bmatrix}}_{\text{Excitation}} = \frac{i_o}{i_s}. \tag{1.56}$$

The equation above can be *interpreted* as follows. The excitation vector $\begin{bmatrix} 0 & i_s - i_s & 0 \end{bmatrix}^T$ is processed by a network whose MNA matrix is $\mathbf{A}^T$. The resulting vector of unknowns is subjected to the measurement vector $\begin{bmatrix} 0 & 0 & 0 & 1 \end{bmatrix}$. Recall that the first three rows of the excitation vector represent currents. The first three columns of the measurement row matrix correspond to node voltages, while the last column corresponds to the current flowing through the voltage source connected between node ① and ground. This interpretation is captured in the network $\hat{\mathcal{N}}$ in Figure 1.43. The network itself is derived from $\mathcal{N}$ using element-by-element replacement rules that we have already discussed. $\hat{\mathcal{N}}$ is excited by a current source between nodes ② and ③, and by a zero-volt source between nodes ① and ⓪. The "measurement" is the current flowing through the voltage source. By inter-reciprocity, v_o/v_s in $\mathcal{N}$ is identical to i_o/i_s in $\hat{\mathcal{N}}$.

Practical Point: Which of the Following Is Correct?

The figure above shows a common source of confusion with respect to reciprocal and inter-reciprocal networks. The network $\mathcal{N}$ with a voltage-source excitation is shown on the left. $\hat{\mathcal{N}}$ is the inter-reciprocal network. Which of the two pictures on the right (b or c) correctly demonstrates the concept of inter-reciprocity?

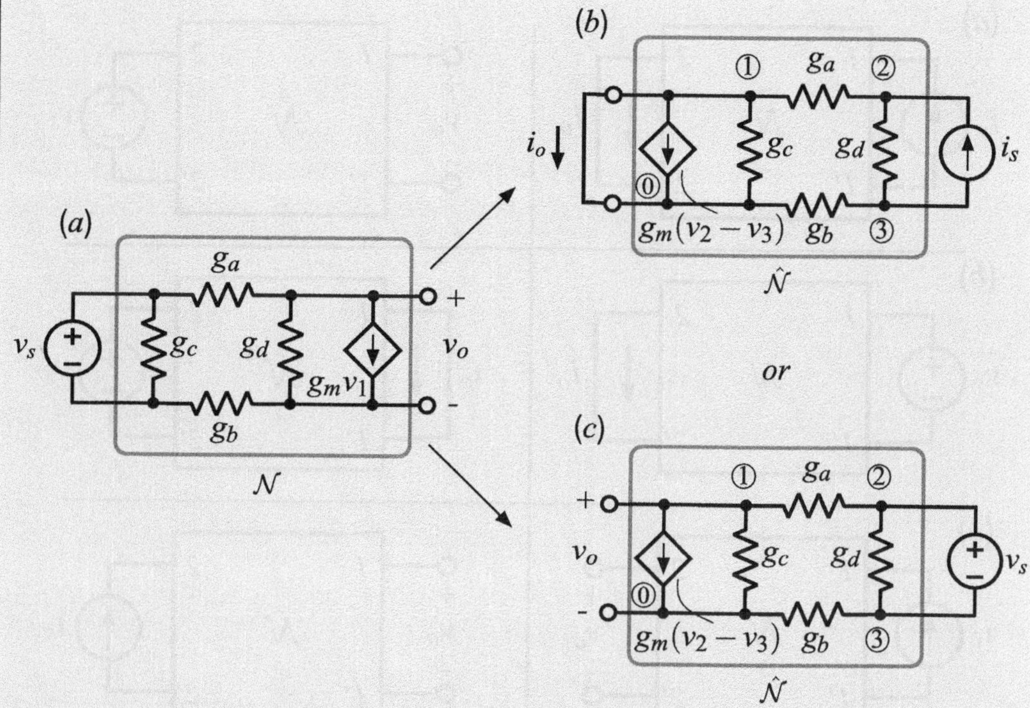

Figure 1.44 A common source of confusion.

The key point to note is that when the independent sources in $\mathcal{N}$ and $\hat{\mathcal{N}}$ are "de-energized", meaning that the voltage source is replaced by a short circuit and a current source by an open circuit, the port impedances of $\mathcal{N}$ and $\hat{\mathcal{N}}$ should be the same. For instance, the ports of the left- and right-hand sides of $\mathcal{N}$ are short and open circuits, respectively when $v_s = 0$. The same holds for the network of Figure 1.44b, but not for that of Figure 1.44c. Therefore, v_o/v_s for $\mathcal{N}$ and i_o/i_s for $\hat{\mathcal{N}}$ are identical. However, v_o/v_s in Figure 1.44c is *not* identical to v_o/v_s in Figure 1.44a.

The example of Figure 1.41 considered a network with an excitation that was a current, while the output was a voltage. It is straightforward to see (using the approach described above) that a voltage input and/or current output also behave in a similar manner, as shown in Figure 1.45. An observation is in order here. Notice that in Figure 1.45a–d, setting the excitation sources to zero makes the terminating impedances at ports 1–1′ and 2–2′ identical in $\mathcal{N}$ and $\hat{\mathcal{N}}$.

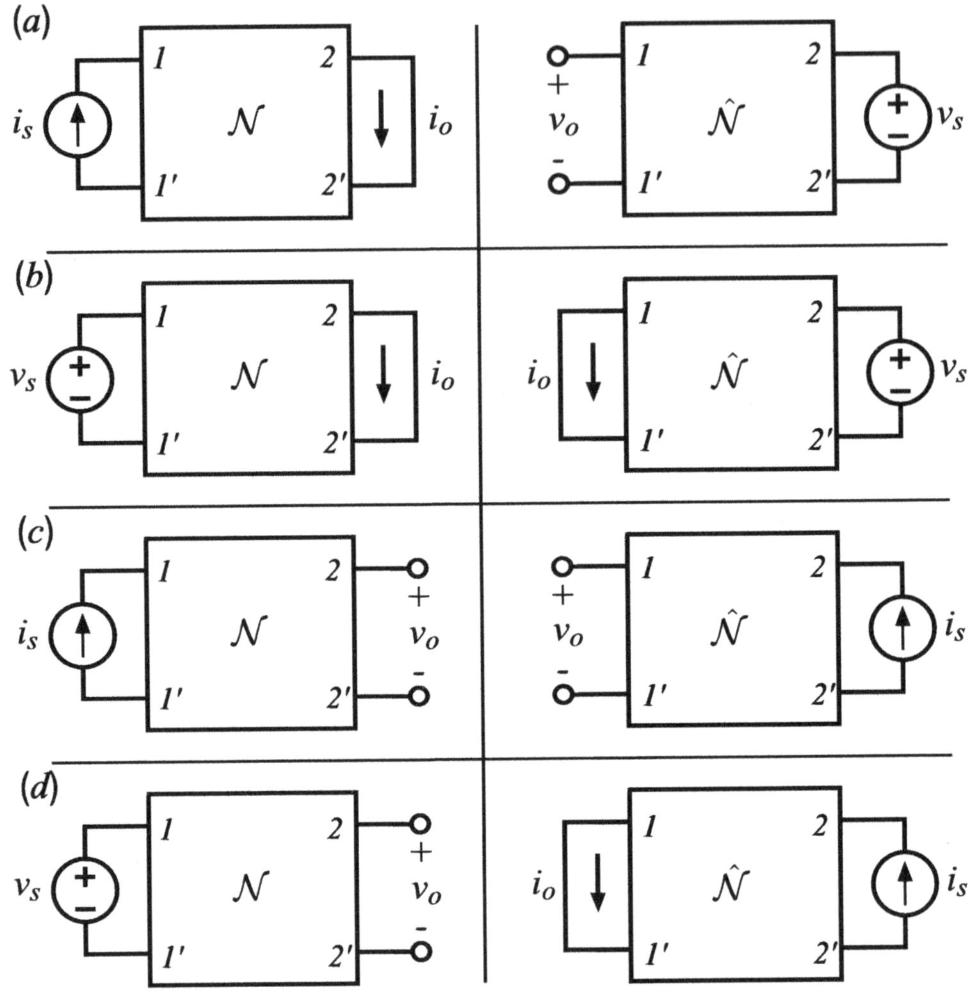

Figure 1.45 The ratio of the response to the excitation in $\mathcal{N}$ is identical to that in $\hat{\mathcal{N}}$. $\hat{\mathcal{N}}$ is derived by replacing every element in $\mathcal{N}$ with one whose MNA stamp is the transpose of that in $\mathcal{N}$.

1.10 Why are Reciprocity and Inter-Reciprocity Useful?

While the concepts described in the previous section are interesting and the end results intriguing, the immediate question that comes to mind is if all of this is anything more than a mathematical curiosity. It turns out, as we will see below, that reciprocity and inter-reciprocity are very relevant in practice. Consider the problem of determining the output of a network when excited by multiple inputs, as shown in Figure 1.46a. Such a situation occurs during noise analysis. For example, evaluating the output noise of a standard fully differential two-stage Miller-compensated opamp would need us to determine about a dozen transfer functions. One would naturally

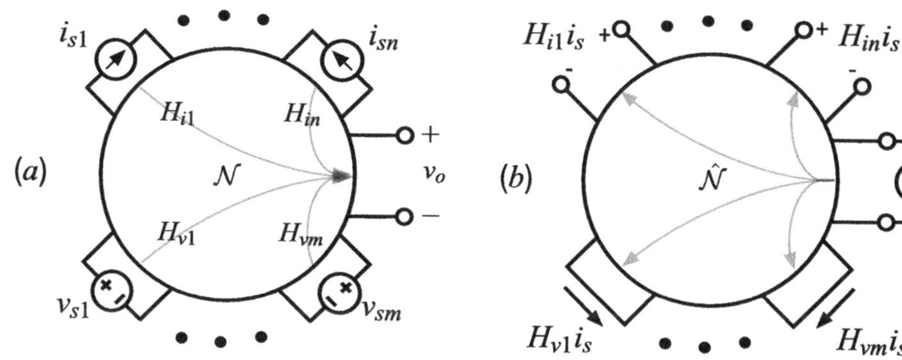

Figure 1.46 (a) The problem of determining the transfer functions from many inputs to a single output. Applying superposition would need the network to be evaluated $(m+n)$ times. (b) Using the inter-reciprocal (or adjoint) network yields all the $(m+n)$ transfer functions in one shot.

be tempted to solve this problem using superposition—i.e., by analyzing the network by considering one source at a time. The output is seen to be

$$v_o = \sum_{k=1}^{m} H_{vk} v_{sk} + \sum_{l=1}^{n} H_{il} i_{sl}. \tag{1.57}$$

For our example in Figure 1.46a, this would result in a total of $(m + n)$ evaluations of the network, with each evaluation yielding the transfer function from one source to the output. This is tedious and computationally inefficient.

The problem can be solved in a *much more* efficient manner by using the concept of inter-reciprocity, as shown in Figure 1.46b. We form the inter-reciprocal network using the rules described in Section 1.9. We excite the output port of $\hat{\mathcal{N}}$ with a current source and de-energize all the independent sources. Consequently, the voltage sources are replaced by short circuits, while the current sources are opened. We evaluate the network *only once*. By the properties of the inter-reciprocal network, the currents flowing in the short-circuited ports (corresponding to the exciting voltage sources in $\mathcal{N}$) are given by $H_{vk}i_s$. Similarly, the voltages across the open-circuited ports (which correspond to the exciting current sources in $\mathcal{N}$) are given by $H_{il}i_s$. Once the H_{vk} and H_{il} are determined, they can be used in (1.57) to determine v_o. The labor saved with respect to evaluating the desired $(m + n)$ transfer functions (naively) using superposition is evident. In a practical circuit simulator, the overhead in determining the adjoint is negligible, since all that one needs to do is to transpose the MNA matrix. This is easily accomplished in an algorithm that finds the matrix inverse, since all that is needed in a computer implementation is an interchange of the row and column addresses when fetching the corresponding matrix elements from memory. Not surprisingly, the adjoint (inter-reciprocity) network is routinely used in circuit simulators for the purposes of noise analysis.

Example: R-2R DAC

Figure 1.47a shows an R-2R DAC. The four current sources are identical and have strength I. Depending on the bits $b_0, \dots, b_3$ the currents are steered either into the network of resistors or into ground. The opamp is ideal. To find the DAC output v_o, one could use superposition to find the transfer resistance from each current source to the output.

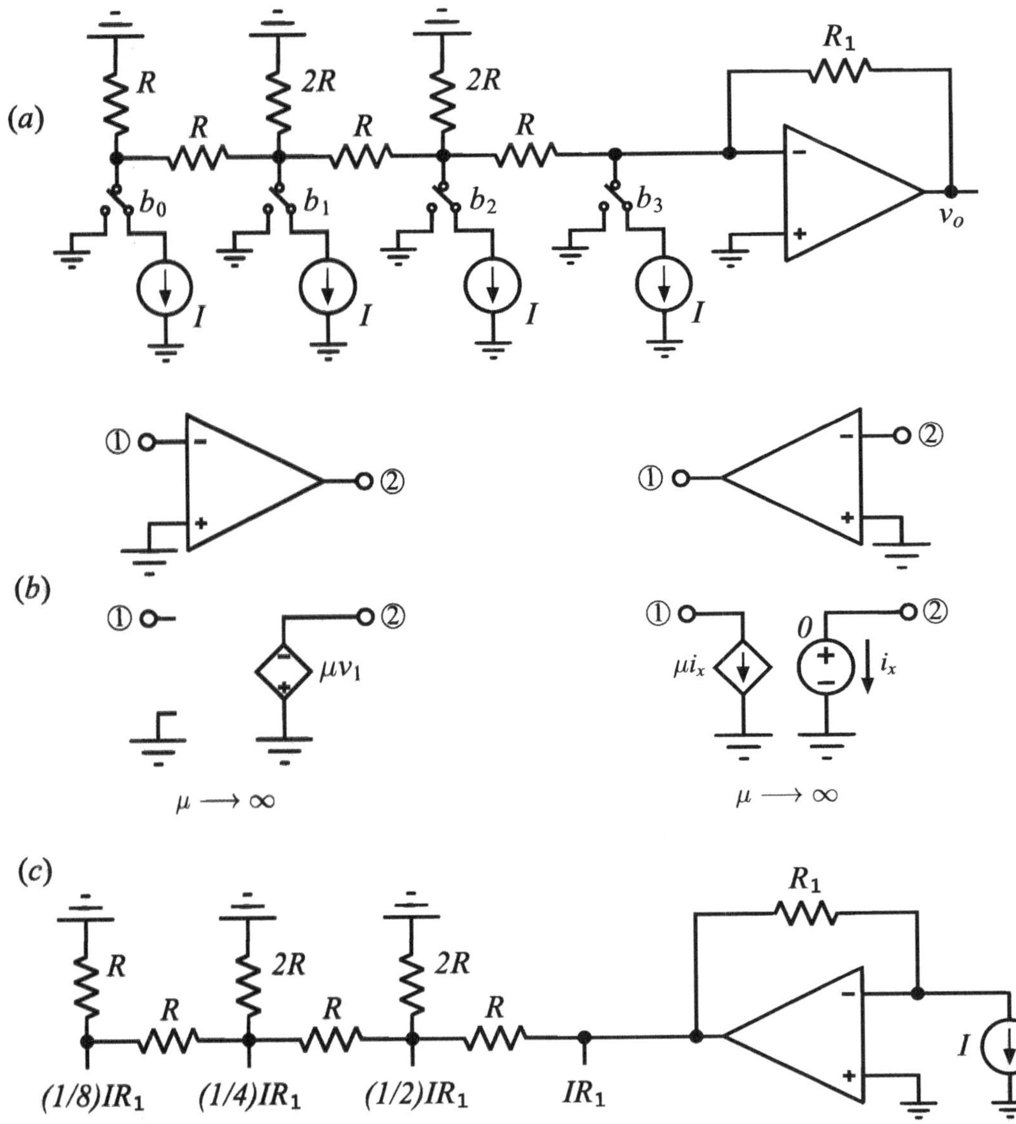

Figure 1.47 (a) An R-2R DAC. Transfer resistances from the four current sources to the output voltage v_o need to be found. (b) The opamp interpreted as a voltage-controlled voltage source. The transpose element corresponding to the opamp. $i_x \to 0$ as $\mu \to \infty$. (c) The inter-reciprocal (or adjoint) network can be used to obtain all the transfer functions in one shot.

A more efficient way is to use the inter-reciprocal network. To draw the adjoint, we need to replace every element with its inter-reciprocal cousin. The resistors, therefore, remain as they are in the original network. What do we do with the opamp? Remember that the opamp in the original network can be thought of as a VCVS with infinite gain, as shown in Figure 1.47b.[4] In the adjoint (Figure 1.47b), therefore, it must be replaced by a CCCS with infinite current gain, but with the controlling and controlled ports reversed. Again, thanks to negative feedback and infinite gain $i_x = 0$. We thus see that not only is the node potential ② zero, but also $i_x = 0$. There is therefore a virtual short between ② and ground. Consequently, the CCCS with infinite gain is also an ideal opamp. To draw the adjoint of the network of Figure 1.47a, therefore, the opamp needs to be "flipped" as shown in Figure 1.47c. By inter-reciprocity, the transfer resistances from the current sources to v_o in the original network can be obtained by exciting the "output" port of the inter-reciprocal network with a current I and determining the voltages developed across the current sources. It is seen (by inspection) that the transfer resistances are R_1, $(1/2)R_1, \dots, (1/8)R_1$. The DAC output is thus given by

$$v_o = IR_1\left(b_3 + \frac{1}{2}b_2 + \frac{1}{4}b_1 + \frac{1}{8}b_0\right). \tag{1.58}$$

1.10.1 Transfer Functions in a Biquadratic Section

Consider the network of Figure 1.48a. The opamps are ideal. The input source is v_s, and there are internal voltage sources $v_1, \dots, v_4$.[5] It is required to find the transfer functions from the internal sources to the output.

Straightforward analysis shows that

$$\frac{V_o(s)}{V_s(s)} = \frac{1}{D(s)} = \frac{1}{\frac{s^2}{\omega_o^2} + \frac{s}{Q\omega_o} + 1}$$

where $\omega_o^2 = 1/(R_1R_2C_1C_2)$ and $Q = \sqrt{C_2/C_1}R_3/\sqrt{R_1R_2}$. Using superposition to find the required transfer functions is too cumbersome. The inter-reciprocal network, shown in Figure 1.48b, comes in handy. The opamps are flipped, and the VCVS with a gain of -1 is replaced by a CCCS with the same gain but with the controlling and controlled ports flipped. The adjoint network is excited by a current source i_s. The transfer functions from i_s to $i_1, \dots, i_4$ in the adjoint are the same as those from $v_1, \dots, v_4$ to v_o in the original network. Furthermore, with respect to the adjoint, we note that:

a. The voltages across the resistors labeled R_1 are identical. Consequently, $i_1 = i_3$. Furthermore, $I_1(s)/I_s(s) = H(s)$.

[4] Negative feedback and infinite gain force the virtual short between nodes ① and ground.
[5] As we will see in Chapter 7, these represent noise sources.

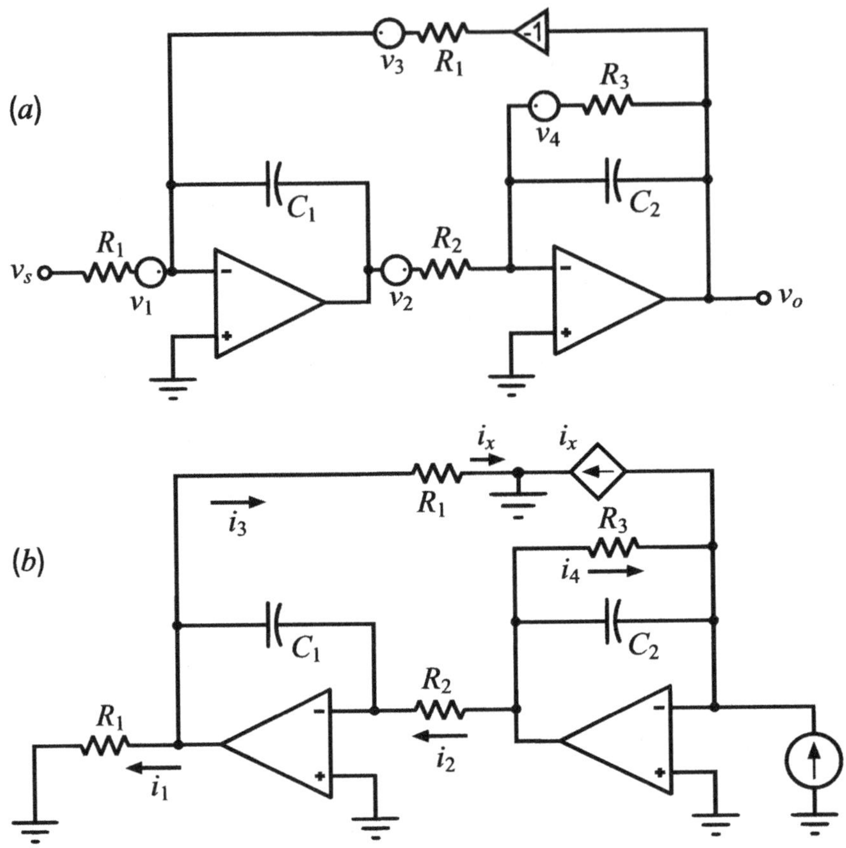

Figure 1.48 (a) A biquadratic section with several internal voltage sources. (b) The inter-reciprocal (or adjoint) network can be used to obtain all the transfer functions in one shot.

b. The voltages across the resistors labeled R_2 and R_3 are identical. As a result, $i_2R_2 = i_4R_3$.

c. $I_2(s) = -sC_1R_1I_1(s)$.

From the observations above, and using the properties of the inter-reciprocal network, we conclude that

$$\frac{V_o(s)}{V_1(s)} = \frac{I_1(s)}{I_s(s)} = H(s)$$

$$\frac{V_o(s)}{V_2(s)} = \frac{I_2(s)}{I_s(s)} = -sC_1R_1H(s)$$

$$\frac{V_o(s)}{V_3(s)} = \frac{I_3(s)}{I_s(s)} = H(s)$$

$$\frac{V_o(s)}{V_4(s)} = \frac{I_4(s)}{I_s(s)} = -sC_1R_1(R_2/R_3)H(s).$$

1.11 Transfer-Function Sensitivity

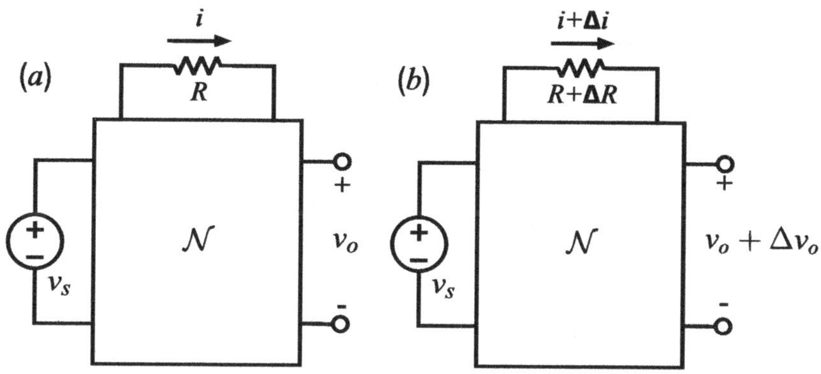

Figure 1.49 (a) Original network. (b) Network with a resistor changed from R to $R + \Delta R$.

It is often desired to find the change in the transfer function of a network due to small changes in the values of its elements. It turns out that the inter-reciprocal (or adjoint) network concept is very useful in such computations. To develop this, we first consider the change in the transfer function of a network $\mathcal{N}$ due to the change in a single resistor R. In the discussion that follows, we assume networks without memory, but the results also hold for Laplace-transformed quantities. The original network, shown in Figure 1.49a, is excited by v_s and results in an output $v_o = Hv_s$. Let i denote the current flowing through the resistor. Suppose R changes to $(R + \Delta R)$ as in Figure 1.49b. The resulting output is $(v_o + \Delta v_o)$ and the current is modified to $(i + \Delta i)$. The resistor $(R + \Delta R)$ can be split into a resistor R in series with ΔR.

Since the current flowing through ΔR is $(i + \Delta i)$, it can be replaced by a voltage source $\Delta v_1 = i\Delta R$ (neglecting the second-order term $\Delta i \Delta R$) as shown in Figure 1.50a.

Consider the network in Figure 1.50a, where a voltage source v_x is inserted in series with R. We denote the transfer function from v_s to v_1 by H_{s1}, while that from v_x to v_o is denoted by H_{1o}. Using this, we see that the output in Figure 1.50b can be expressed as

$$v_o + \Delta v_o = Hv_s + \underbrace{\frac{H_{s1}v_s}{R}}_{i} \Delta R H_{1o}. \tag{1.59}$$

The change in the transfer function H due to the change in the resistor R is thus seen to be

$$\Delta H = \frac{\Delta v_o}{v_s} = H_{s1}H_{1o}\frac{\Delta R}{R}. \tag{1.60}$$

The *sensitivity* of the transfer function H to the resistor R, formally defined as

$$S_R^H = \frac{\Delta H/H}{\Delta R/R}, \tag{1.61}$$

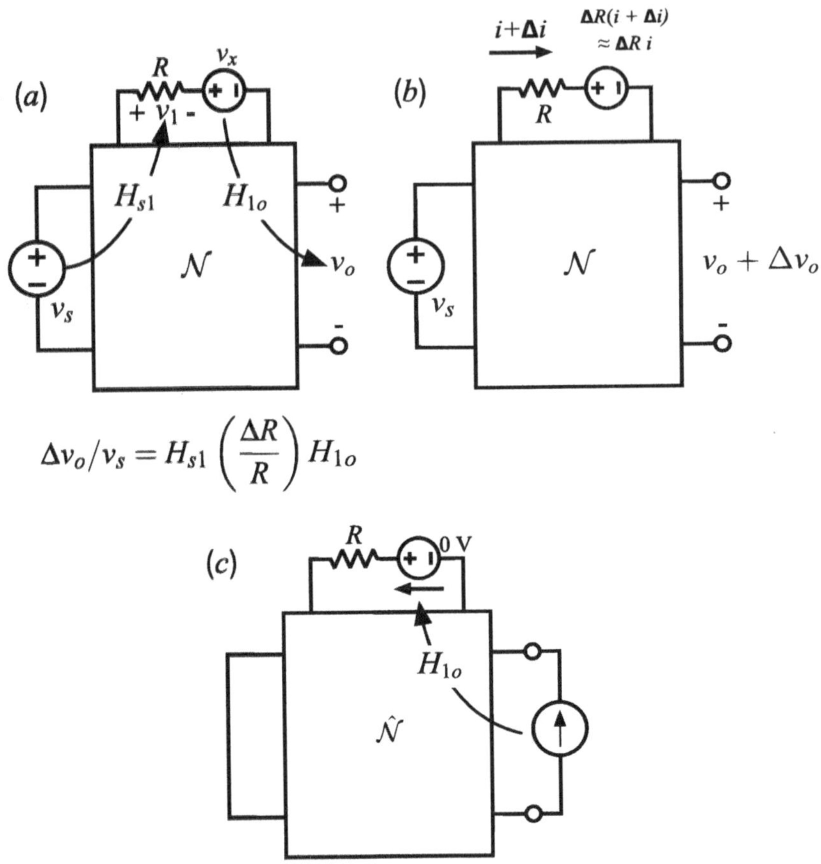

Figure 1.50 (a) Interpreting the effect of ΔR as a voltage source $I\Delta R$ in series with R. (b) Defining the transfer functions H_{s1} and H_{1o}. (c) Determining H_{1o} using the adjoint network.

is seen to be $S_R^H = H_{s1}H_{1o}/H$. How does one find H_{1o}? One way is to evaluate $\mathcal{N}$ again, now with v_x as the excitation. An alternative method is to form the adjoint network $\hat{\mathcal{N}}$, excite its output port with a current and determine the current flowing through v_x, as shown in Figure 1.50c. Regardless of the method chosen, we see that evaluation of the sensitivity to a single parameter needs the solution of *two* networks. This seems reasonable—to find the change in H due to a change in R, one would need to find H for the two different values of R and subtract the transfer functions. Furthermore, it seems that using the adjoint has no particular advantage in the sense that the analysis is no easier than the simple-minded method.

Now, what if one needed to determine the sensitivity to many (say m) parameters? The discussion above would have us believe that one would need to evaluate the network twice for each parameter change, thereby requiring a total of $2m$ analyses. Using the adjoint network concept, however, multi-parameter sensitivity can be computed by solving the network just *twice*, irrespective of the number of parameters. The idea is illustrated using Figure 1.51. Part (a) of the figure shows the network $\mathcal{N}$. Suppose

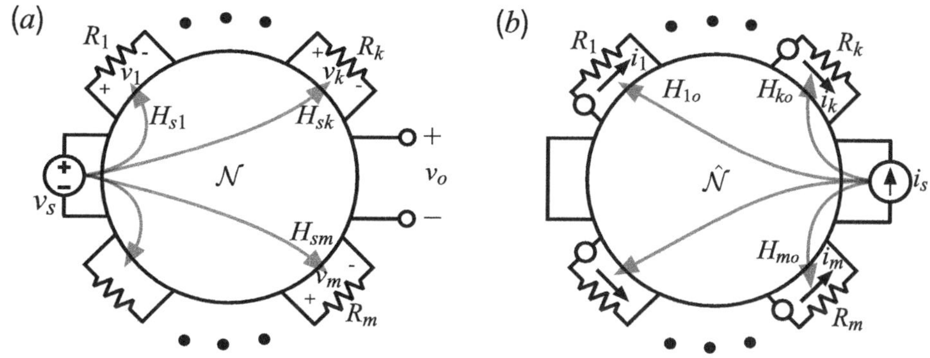

Figure 1.51 (a) Original network, the solution of which yields all the H_{sk}. (b) Solving the adjoint network yields all the H_{ko} in one shot.

that we need to determine the change in the transfer function from v_s to v_o due to small changes in the values of the resistors $R_1, \ldots, R_m$. From the discussion in the single-parameter case, we see that

$$\Delta H = \frac{\Delta v_o}{v_s} = \sum_{k=1}^{m} H_{sk} H_{ko} \frac{\Delta R_k}{R_k}. \tag{1.62}$$

The H_{sk} represent the transfer functions from v_s to the voltages across the resistors R_k. Evaluating these transfer functions would need one evaluation of $\mathcal{N}$. The H_{ko} represent the transfer functions from voltage sources inserted in series with each of the resistors to the output. To evaluate these one at a time would need us to solve $\mathcal{N}$ m times. An efficient way of accomplishing this is to use the adjoint network $\hat{\mathcal{N}}$ as shown in Figure 1.51b. The input source is replaced by a short circuit. The output port is excited by a current i_s, and the currents through the voltage sources are determined to yield the H_{ko}. This needs $\hat{\mathcal{N}}$ to be evaluated only once. The H_{sk} and the H_{ko} found in the two steps above can then be used in (1.62) to yield ΔH.

The discussion above assumed that the resistors changed by small amounts; the extension of this analysis to controlled sources is straightforward and is illustrated for a VCVS in Figure 1.52. Part (a) of the figure shows a VCVS with gain μ. It is part of a larger network $\mathcal{N}$. To find the change in v_o due to a change in μ, the transfer function H_{s1} from v_s to the controlled voltage as well as H_{1o} (which is that from v_x to v_o) have to be found. The latter is determined from the adjoint network as shown in Figure 1.52b.

To summarize, therefore, the change in the transfer function due to m parameters can be determined by solving *two* networks once each—the original network $\mathcal{N}$ and its adjoint $\hat{\mathcal{N}}$.

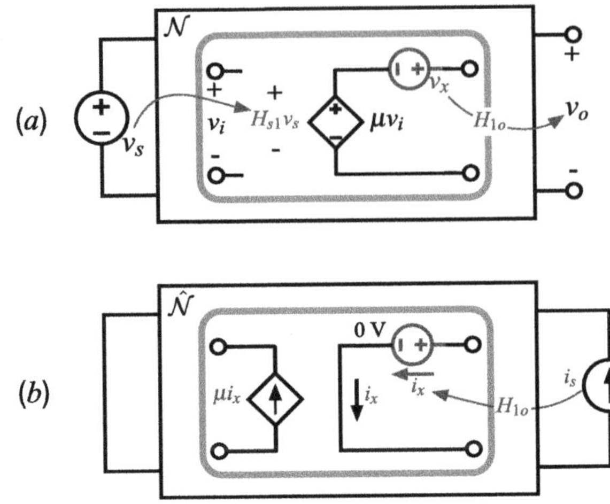

$$\frac{\Delta v_o}{v_s} = H_{s1} H_{1o} \Delta\mu$$

Figure 1.52 (a) Original network. H_{s1} and H_{1o} need to be determined. (b) Determining H_{1o} using the adjoint network.

Example: Effect of Component Variations on a Series-Feedback Amplifier

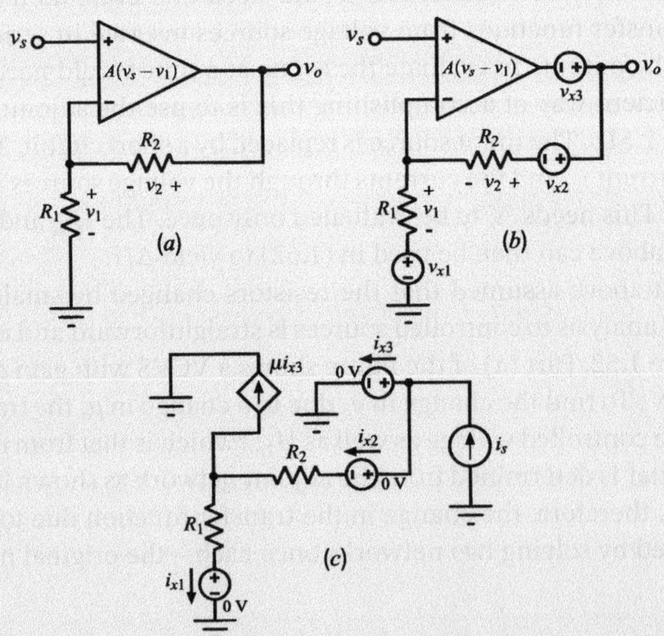

Figure 1.53 (a) Original network. (b) H_{s1} and H_{1o} need to be determined. (c) Determining H_{1o} using the adjoint network.

Consider the series-feedback amplifier shown in Figure 1.53a. It is desired to find the change in v_o due to changes in A, R_1, and R_2. The first step is to determine the H_{sk} from v_o, v_1, and v_2. Denoting

$$LG \equiv A\frac{R_1}{R_1 + R_2} \tag{1.63}$$

straightforward analysis yields

$$\begin{aligned} v_1 &= v_s\frac{LG}{(1+LG)} \\ v_2 &= v_s\frac{LG}{(1+LG)}\frac{R_2}{R_1} \\ v_o &= v_s\frac{LG}{(1+LG)}\left(1+\frac{R_2}{R_1}\right). \end{aligned} \tag{1.64}$$

The next step is to determine the H_{ko}. For this, voltage sources v_{x1}, v_{x2} and v_{x3} are inserted in series with R_1, R_2 and the output of the VCVS respectively, as shown in Figure 1.53b. To determine all the H_{ko} in one shot, the adjoint network is constructed. The result is shown in Figure 1.53c. The "output" port is excited by a current i_s, and the currents through the 0 V voltage sources are determined. Analysis of this network yields

$$\begin{aligned} i_{x1} &= -i_s\frac{LG}{(1+LG)}\frac{R_2}{R_1} \\ i_{x2} &= i_s\frac{LG}{(1+LG)} \\ i_{x3} &= i_s\frac{1}{(1+LG)}. \end{aligned} \tag{1.65}$$

Using the equations above along with (1.62), we obtain

$$\begin{aligned} \frac{\Delta v_o}{v_s} &= \frac{1}{(1+LG)^2}\Delta A - \frac{LG^2}{(1+LG)^2}\frac{R_2}{R_1^2}\Delta R_1 \\ &\quad + \frac{LG^2}{(1+LG)^2}\frac{\Delta R_2}{R_1}. \end{aligned} \tag{1.66}$$

The equation above is easily verified by differentiating the expression for the gain of the amplifier of Figure 1.53a. As $A \to \infty$, (1.66) reduces to

$$\frac{\Delta v_o}{v_s} \to -\frac{R_2}{R_1^2}\Delta R_1 + \frac{\Delta R_2}{R_1} \tag{1.67}$$

which makes intuitive sense.

Chapter Summary

- **Kirchhoff's Current and Voltage Laws, and Tellegen's Theorem:** Circuit theory rests on Ohm's law, Kirchhoff's Current Law (KCL), and Kirchhoff's Voltage Law (KVL), which together provide the framework for analyzing any linear electrical network. Tellegen's theorem is a consequence of KCL/KVL.
- **Modified Nodal Analysis (MNA) Formulation of Circuit Equations:** This is an extension of nodal analysis that accommodates components whose current cannot be determined from the branch voltage (e.g., voltage sources, inductors). Using MNA "stamps" of components, the circuit equations can be written by inspection. The MNA formulation is not only the cornerstone of circuit simulators but also immensely useful in understanding reciprocity and inter-reciprocity.
- **Bilinear Form of a Transfer Function and the Extra-Element Theorem:** Any transfer function in a linear network is a bilinear form with respect to a component value. This observation is not only useful in noise analysis (see Chapter 2) but also facilitates the derivation of the extra-element theorem.
- **Reciprocity, Inter-reciprocity, and Adjoint Networks:** These concepts are immensely useful when one is required to determine the transfer functions from multiple inputs to a single output. Such a scenario is routinely encountered in noise and sensitivity analysis. The MNA formulation of circuit equations is a way of understanding (inter)reciprocity from the ground up.

Problems

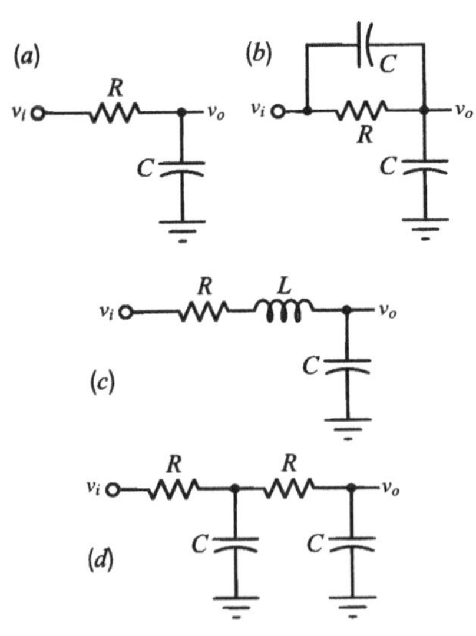

Figure 1.54

Problem 1. *Figure 1.54 shows four networks. v_i is a unit step.*

- *For each network, determine $v_o(0+)$, $v_o(\infty)$, and $\frac{dv_o(t)}{dt}$ at $t = 0+$.*
- *For the networks in Figure 1.54c,d determine $\frac{d^2v_o(t)}{dt^2}$ at $t = 0+$.*

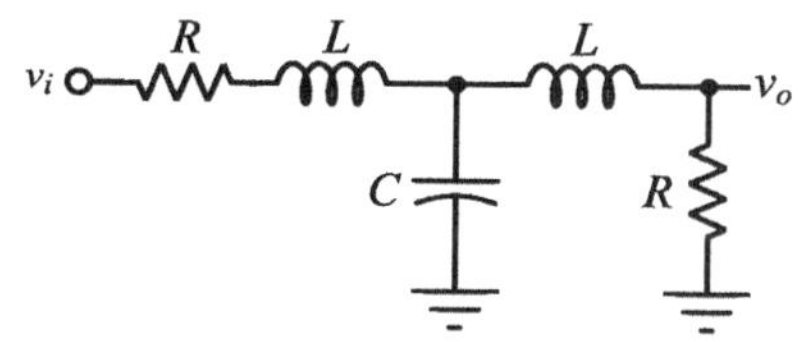

Figure 1.55

Problem 2. *In Figure 1.55, v_i is a unit step. Determine $\frac{d^3v_o(t)}{dt^3}$ at $t = 0+$.*

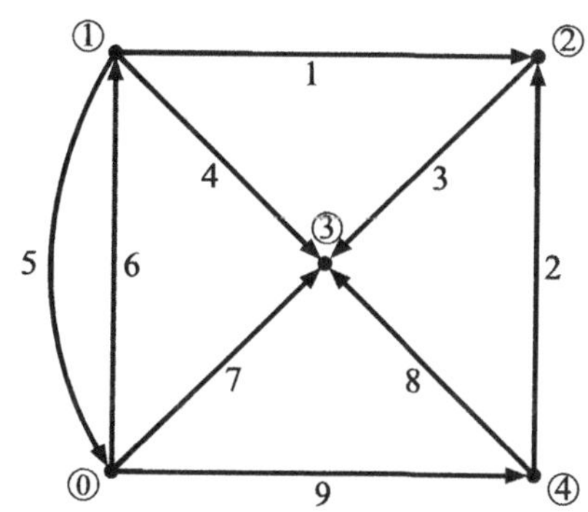

Figure 1.56

Problem 3. *Figure 1.56 shows the directed graph of a network, with the nodes and branches marked. Choose node 0 to be the reference (datum) and determine the incidence matrix of this graph.*

Problem 4. *Consider two topologically identical networks $\mathcal{N}$ and $\hat{\mathcal{N}}$. We denote the branch currents and branch voltages in these networks (as a function of time) by $v_k(t)$, $i_k(t)$ and $\hat{v}_k(t)$, $\hat{i}_k(t)$ respectively. Further, let t_1 and t_2 be two different time instants. Determine*

a. $\sum_k v_k(t_1)i_k(t_2)$

b. $\sum_k v_k(t_1)\hat{i}_k(t_2)$

c. $\sum_k \hat{v}_k(t_1)\frac{di_k}{dt}\Big|_{t=t_2}$

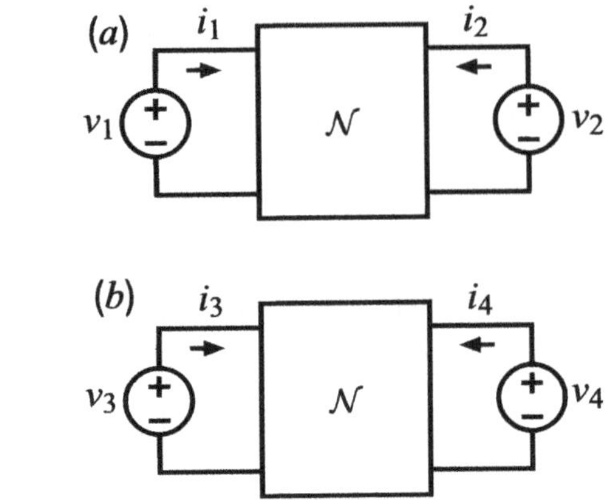

Figure 1.57

Problem 5. *$\mathcal{N}$ consists of resistors only. We perform the following two experiments indicated in Figure 1.57:*

Experiment 1. Drive $\mathcal{N}$ with two voltage sources v_1 and v_2 and measure the resulting port currents i_1 and i_2 respectively.

Experiment 2. Drive $\mathcal{N}$ with two voltage sources v_3 and v_4 and measure the resulting port currents i_3 and i_4 respectively.

Prove that the two sets of measurements are related as follows:

$$v_1 i_3 + v_2 i_4 = v_3 i_1 + v_4 i_2$$

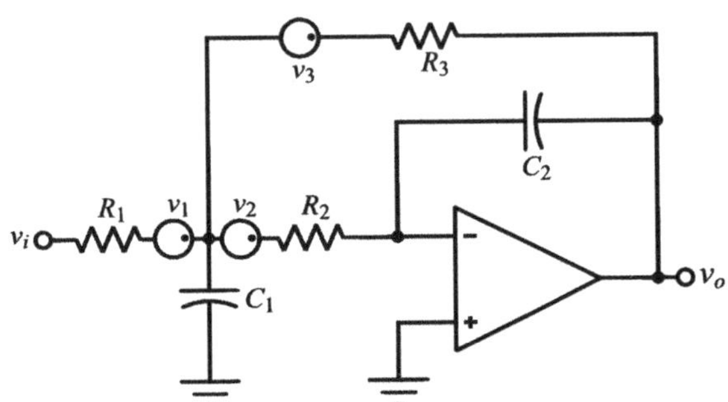

Figure 1.58

Problem 6. *In the network of Figure 1.58, the opamp is ideal. Determine the transfer functions from v_i, v_1, v_2, v_3 to v_o. The positive terminals of the voltage sources are marked with a dot.*

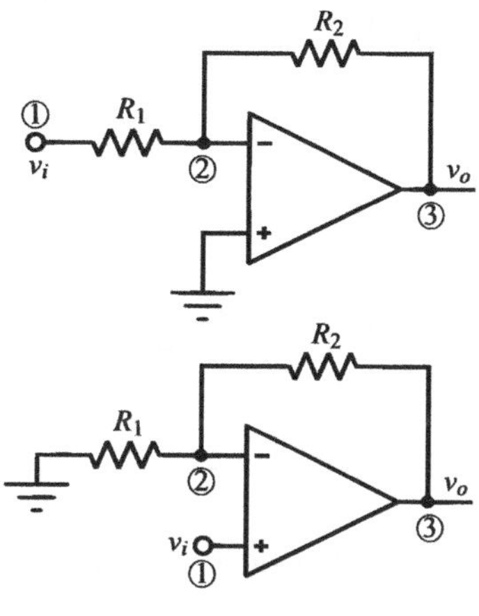

Figure 1.59

Problem 7. *In the networks of Figure 1.59, the opamps are ideal. Write the MNA equations for these networks. The nodes are numbered.*

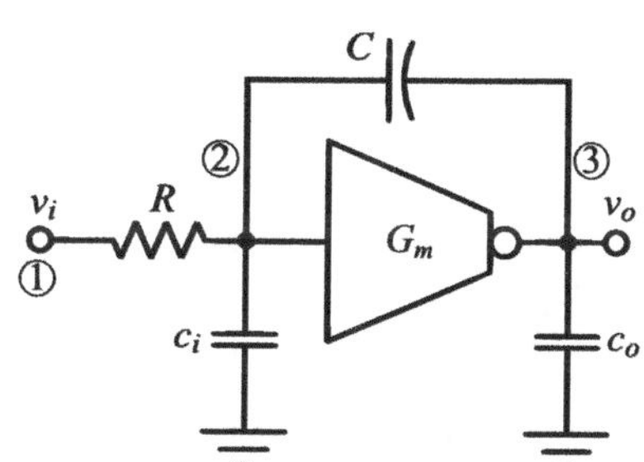

Figure 1.60

Problem 8. *Write the MNA equations for the active-RC integrator of Figure 1.60. The nodes are numbered. What is the input–output transfer function as* $G_m \to \infty$*?*

Problem 9. *Two integrators of the kind shown in Figure 1.60 are cascaded. Write the MNA equations for the cascade. You might want to use the results of the previous problem to reduce the effort involved.*

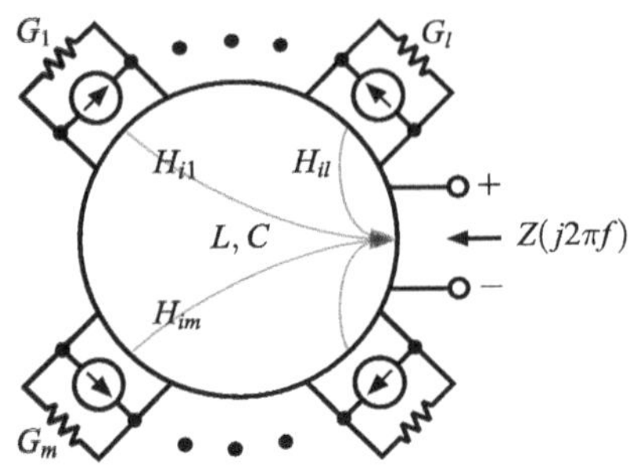

Figure 1.61

Problem 10. *Figure 1.61 shows a network that consists of only resistors, inductors and capacitors. All the resistors (m in total) are pulled out of the network as shown and current sources placed in parallel with them. Denote the transfer function from the lth current source to the output by $H_{il}(j2\pi f)$. If $Z(j2\pi f)$ denotes the impedance looking into the network (with all current sources turned off), show that*

$$\sum_{l=1}^{m} G_l |H_{il}(j2\pi f)|^2 = Re\{Z(j2\pi f)\},$$

where G_l denotes the lth conductance.

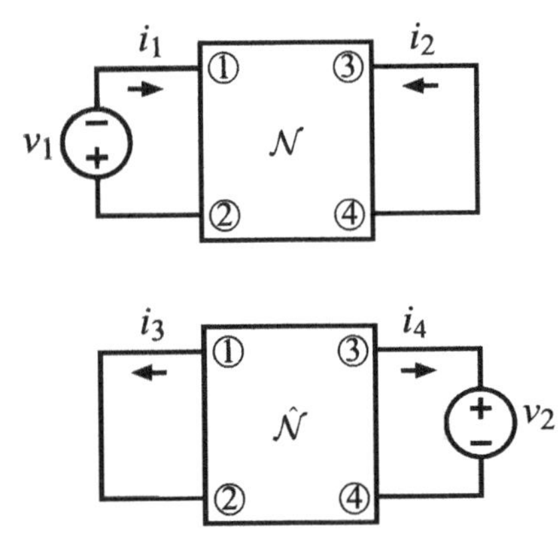

Figure 1.62

Problem 11. *In Figure 1.62, $\hat{\mathcal{N}}$ is the adjoint network corresponding to $\mathcal{N}$. Determine*

$$\frac{v_1 i_3}{v_2 i_2} + 2.$$

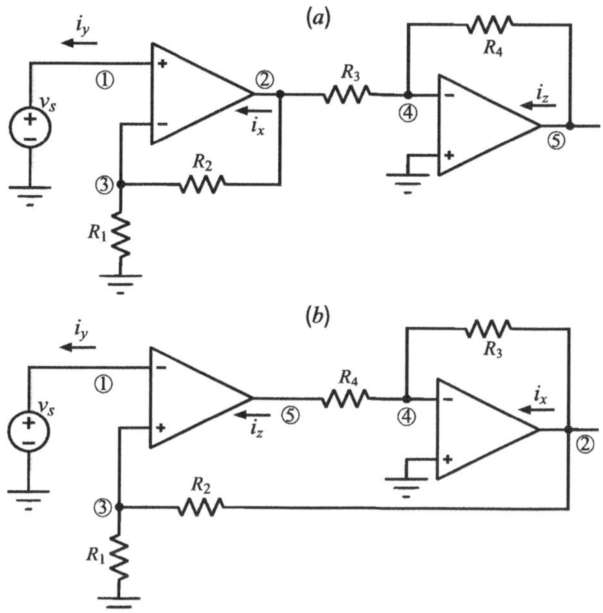

Figure 1.63

Problem 12. *As we have seen, the gain from v_s to v_5 is identical in both the circuits of Figure 1.63 if the opamps are ideal. Calculate the gain of the circuits if the gain of the opamps is not infinite, but a value $A \gg 1$. Repeat the exercise if the opamp's transfer function is modeled as $A(s) = \frac{2\pi f_u}{s}$.*

References

1 L. O. Chua, C. A. Desoer, and E. S. Kuh, *Linear and Nonlinear Circuits*. New York: McGraw-Hill, 1987.

2 B. D. H. Tellegen, "A general network theorem, with applications," *Philips Research Reports*, vol. 7, pp. 259–269, 1952.

3 C.-W. Ho, A. E. Ruehli, and P. A. Brennan, "The modified nodal approach to network analysis," *IEEE Transactions on Circuits and Systems*, vol. 22, no. 6, pp. 504–509, 1975.

4 L. W. Nagel, "Spice2: A computer program to simulate semiconductor circuits," Electronics Research Laboratory, University of California, Berkeley, Tech. Rep. ERL-M520, 1975.

5 J. Vlach and K. Singhal, *Computer Methods for Circuit Analysis and Design*, 2nd. Van Nostrand Reinhold, 1993.

6 R. D. Middlebrook, "Null double injection and the extra element theorem," *IEEE Transactions on Education*, vol. 32, no. 3, pp. 167–180, 1989.

7 S. W. Director and R. A. Rohrer, "The generalized adjoint network and network sensitivities," *IEEE Transactions on Circuit Theory*, vol. 16, no. 3, pp. 318–323, 1969.

8 S. Pavan and G. C. Temes, "Reciprocity and inter-reciprocity: A tutorial—Part 1: Linear time-invariant networks," *IEEE Transactions on Circuits and Systems I: Regular Papers*, vol. 70, no. 9, pp. 3413–3421, 2023.

2

Noise in Linear Time-Invariant Networks

Noise sets the lower limit on the smallest signal that a system can process while still producing an intelligible output and therefore fixes the effective resolution, dynamic range, and sensitivity of every analog and mixed-signal block we design or use. This chapter develops what we need to understand and analyze how noise propagates through linear time-invariant (LTI) circuits. We focus on input-/output-referred models, power spectral densities, and how noise is processed by transfer functions. The emphasis is on engineering intuition; the aim is to give the circuit designer the ability to make reliable noise calculations.

As one can imagine, the study of noise is a vast subject in itself, and entire books describing one aspect of noise or the other have been written [1, 2]. The treatment here is aimed at circuit and system work: it is sufficient for budgeting, specification, topology comparison, and architecture-level trade-offs. We do *not* delve into the underlying physical mechanisms (microscopic origins of thermal, shot, or flicker noise) beyond what is needed to parameterize models. Readers seeking either: (i) the detailed device-physics origins of noise or (ii) a rigorous stochastic-process foundation for the results used here should consult advanced books on the subject.

We first describe noise models for commonly used components. We then study how noise propagates in networks and how to determine output noise. Next, we analyze how the effect of multiple noise sources in a network can be abstracted if one is only interested in the port behavior. This naturally leads to a metric, namely the noise factor, that quantifies the degradation in SNR experienced by a signal as it is processed by a network.

2.1 Noise in Network Components

Suppose a resistor of value R in thermal equilibrium with its surroundings (at an absolute temperature T) is connected to an ideal bandpass filter with center frequency f_o, bandwidth Δf, and unity passband gain as shown in Figure 2.1. The input impedance of the filter is infinite. *It turns out* that a random waveform $v_o(t)$ is observed at the filter's output as shown in the figure. Furthermore, the following hold:

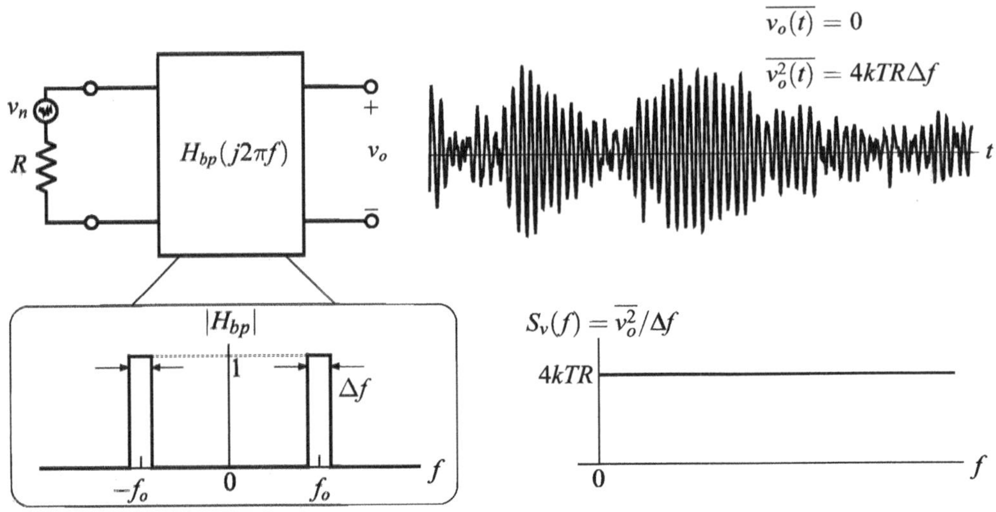

Figure 2.1 A resistor placed at the input of a bandpass filter with center frequency f_o and bandwidth Δf. The output of the filter is $v_o(t)$ and has a mean-square value of $4kTR\Delta f$.

a. The average value of $v_o(t)$ is zero, irrespective of R, f_o, or Δf. As Δf becomes very small, $v_o(t)$ largely looks like a sinusoid of frequency f_o with a slowly varying envelope and phase as shown in Figure 2.1.

b. The mean-square value of $v_o(t)$ is given by $4kTR\Delta f$, where k is the Boltzmann's constant (1.38×10^{-23}J/K).[1] Interestingly, $\overline{v_o^2}$ is *independent* of the bandpass filter's center frequency f_o.

c. If a histogram of the values taken on by $v_o(t)$ were plotted, one would obtain a Gaussian distribution with zero mean and a variance $4kTR\Delta f$.

d. Evidently, the random voltage waveform at the output of the bandpass filter is a consequence of the agitation of charge carriers in the resistor. This is modeled as a noise voltage source v_n in series with the resistor as shown in Figure 2.1.

e. Suppose the same experiment was repeated with another resistor of value R. We denote the waveform at the output of the bandpass filter by $\hat{v}_o(t)$. The mean and variance of $\hat{v}_o(t)$ are zero and $4kTR\Delta f$; however, $\hat{v}_o(t)$ and $v_o(t)$ are independent. Consequently, $\overline{v_o(t)\hat{v}_o(t)} = 0$.

f. Suppose the limit $\overline{v_o^2}/\Delta f$ as $\Delta f \to 0$ was plotted as a function of the bandpass filter's center frequency. One would obtain a flat line of value $4kTR$ as shown in Figure 2.1. This is termed the spectral density of the noise voltage v_n and is given by $4kTR$ V^2/Hz [3, 4].

$$S_v(f) \equiv \lim_{\Delta f \to 0} \frac{\overline{v_o^2}}{\Delta f} = 4\,kTR \qquad \text{(Noise spectral density).} \qquad (2.1)$$

[1]The reader should convince him/herself that the units of $4kTR\Delta f$ are indeed V^2.

Since the noise spectral density remains constant with frequency, it is termed "white noise," in line with "white light," which consists of all colors (i.e., frequencies) in equal measure. Since the origin of resistor noise is the *thermal* agitation of charge carriers, it is termed "thermal noise."

Intuitively speaking, the random output voltage associated with a resistor can be thought of as the sum of sinusoids of all frequencies, each with a slowly varying envelope and phase. The bandpass filter in Figure 2.1 picks out a narrow set of frequencies (i.e., in the range $f_o - \Delta f/2$ to $f_o + \Delta f/2$) while suppressing the rest. The mean-square value of the waveform at the output of the bandpass filter is given by $S_v(f) \cdot \Delta f = 4kTR\Delta f$.

Figure 2.1 also gives us a way of interpreting the spectral density of a noise source, which is the following. Process the noise waveform through a bandpass filter with a center frequency f and bandwidth Δf. Denote the output of the bandpass filter by $v_o(t)$. The spectral density $S(f)$ is given by the limiting value of $v_o^2(t)/\Delta f$ as $\Delta f \to 0$.

2.1.1 Noise Models for Other Devices

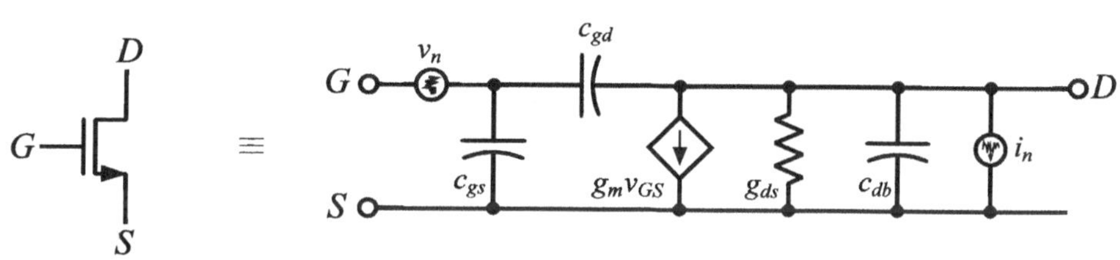

Figure 2.2 Noise model for a MOS transistor operating in saturation, valid for low frequencies. The substrate is assumed to be connected to the source. v_n models flicker noise, while i_n models thermal noise.

Lossless inductors and capacitors do not add noise. The noise model of a nonlinear device depends on the operating point, just like the small-signal equivalent circuit of the device. For instance, when a MOS transistor operates in the triode (linear) region, the channel behaves like a resistor r_{ds}, resulting in a drain-current noise with PSD $S_i = 4kT/r_{\text{ds}}$.

When operating in the saturation region, the following noise model is widely used at low frequencies. A white noise current i_n is added between the source and drain as shown in Figure 2.2. Its spectral density is given by [5, 6].

$$S_i(f) = \frac{8}{3}\eta k T g_m \tag{2.2}$$

where g_m denotes the transconductance of the device, and η is a correction factor. The transistor is also associated with $1/f$ or *flicker* noise, modeled as a noise voltage v_n in series with the gate, as indicated in Figure 2.2. Its spectral density is given by

$$S_v(f) = \frac{K_f}{WLC_{\text{ox}}f}. \tag{2.3}$$

Here, K_f is a process- and operating-point-dependent parameter, W and L denote the width and length of the transistor, respectively, and C_{ox} is the oxide capacitance per unit area.

2.1.2 Relationship Between Mean-Square Noise and Power Spectral Density

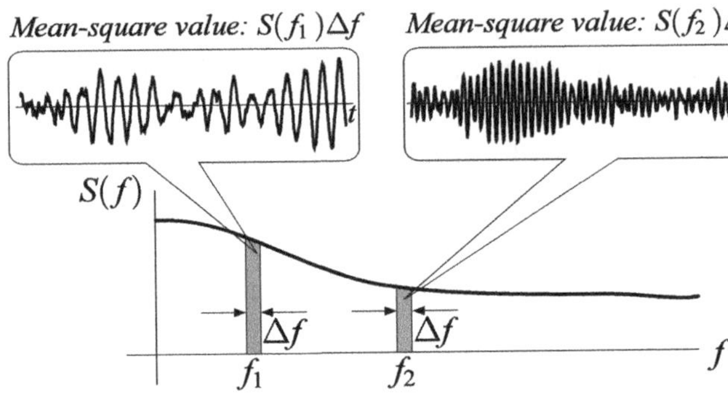

Figure 2.3 Time-domain interpretation of noise with spectral density $S(f)$.

Consider a noise source with a spectral density $S(f)$, as shown in Figure 2.3. What is the mean-square value of the noise waveform? Remember that each chunk of width Δf centered around f can be thought of as a sinusoid with a mean-square value $S(f)\Delta f$. The figure shows the time-domain waveform corresponding to the spectra around f_1 and f_2. As expected, they look like sinusoids with frequencies f_1 and f_2 with randomly varying envelope and phase (the smaller the Δf, the slower these variations will be). The noise waveform is a sum of such sinusoids, each with a different frequency. The mean-square value of a sum of sinusoids with differing frequencies is simply the sum of their mean-square values. Consequently, it is easy to see that

$$\overline{v_o^2} = \int_0^{\infty} S(f)\, df. \tag{2.4}$$

2.1.3 Single- and Double-Sided Noise Spectral Density

Recall that we interpreted noise spectral density $S(f)$ as the mean-square value at the output of a narrowband filter with a center-frequency f normalized to the bandwidth Δf, excited by the noise source. Physically, the filter's center frequency f must be positive. Thus, $S(f)$ is nonzero only for $f \geq 0$ and zero for negative frequencies. This is what is called the single-sided noise spectral density and is commonly used in circuit work. For instance, resistor noise has a *single-sided* noise-spectral density of $4kTR$.

Another way of bookkeeping the noise power is possible. This is called the *double-sided* noise spectral density and recognizes that the magnitude response of the band-pass filter is even and therefore has a passband of width Δf around $-f$. The noise power spectral density is assumed to be an even function of frequency, having the same value at negative frequencies. The double-sided noise spectral density, therefore, is one half its single-sided cousin but extends to negative frequencies. For instance, the voltage noise produced by a resistor would have the double-sided spectral density of $2kTR$. To determine the mean-square noise, the double-sided density has to be integrated over the frequency range $-\infty$ to ∞. We will generally use the single-sided density unless otherwise mentioned.

2.1.4 Noise Processed by a Linear Time-Invariant System

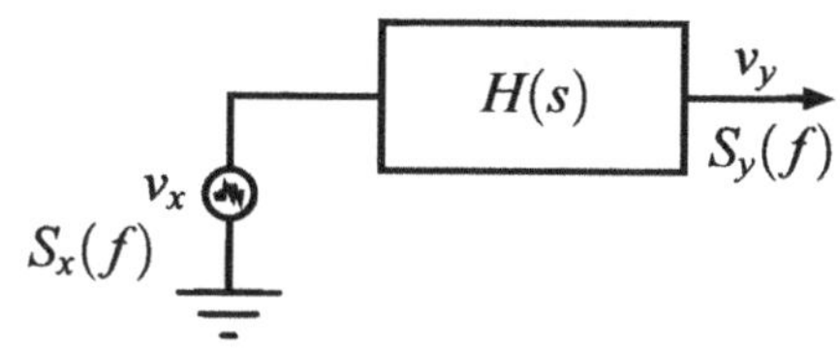

Figure 2.4 Noise processed by a linear time-invariant system with transfer function $H(s)$.

Noise is often processed by a linear system, as shown in Figure 2.4. Suppose the input (single-sided) noise spectral density is known to be $S_x(f)$. The transfer function of the system is denoted by $H(s)$. What is the noise spectral density at the output?

The input can be thought of as a sum of sinusoids with slowly varying envelope and phase. The power (i.e., mean-square value) of the content at frequency f in a bandwidth Δf is given by $S_x(f)\Delta f$. The power gain experienced by this sinusoid as it is processed by $H(s)$ is $|H(j2\pi f)|^2$. Consequently,

$$S_y(f) = |H(j2\pi f)|^2 S_x(f), \tag{2.5}$$

and

$$\overline{v_y^2} = \int_0^\infty S_y(f)\, df = \int_0^\infty |H(j2\pi f)|^2 S_x(f)\, df. \tag{2.6}$$

A special case of the result above is when $H(s)$ is a constant, denoted by α. In that case, $S_y(f) = |\alpha|^2 S_x(f)$.

Example: Equivalent Noise Model for a Resistor

We are already familiar with a model for the noise of the resistor; it consists of a voltage source v_n in series with the resistor as shown in Figure 2.5a. The spectral density of v_n is $4kTR$. The network can be transformed into the Norton equivalent, with a current source i_n in shunt with the resistor as in Figure 2.5b. What is the spectral density of the noise current? Clearly, $i_n = v_n/R$. i_n can be thought of as processing v_n by a "gain" $(1/R)$. Consequently,

$$S_i(f) = \left(\frac{1}{R}\right)^2 S_v(f) = 4kT\frac{1}{R}\ \mathrm{A}^2/\mathrm{Hz}. \tag{2.7}$$

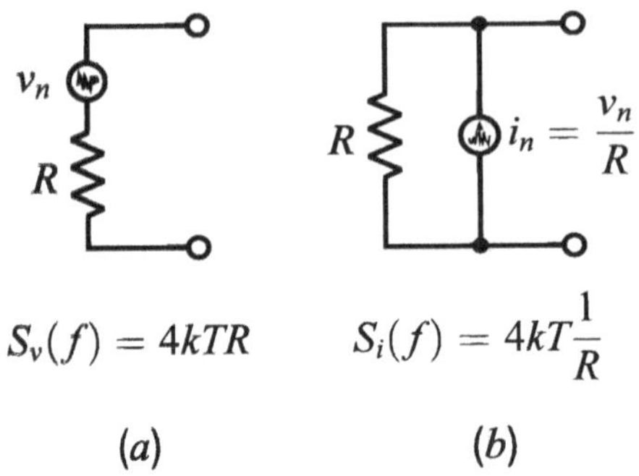

Figure 2.5 (a) Model for noise in a resistor of value R. (b) Equivalent noise model with a current source in parallel with the resistor.

Example: Output Noise of an RC Lowpass Filter

Figure 2.6a shows a first-order RC lowpass filter. The resistor adds thermal noise, which is represented by a voltage source v_n in series with it. Its single-sided noise spectral density (NSD) is $4kTR$ V^2/Hz. The squared magnitude response of the transfer function from v_n to the output is

$$|H(j2\pi f)|^2 = \frac{1}{1 + (2\pi fRC)^2}. \tag{2.8}$$

The output noise voltage spectral density is thus given by

$$S_v(f) = \frac{4kTR}{1 + 4\pi^2 R^2 C^2 f^2}. \tag{2.9}$$

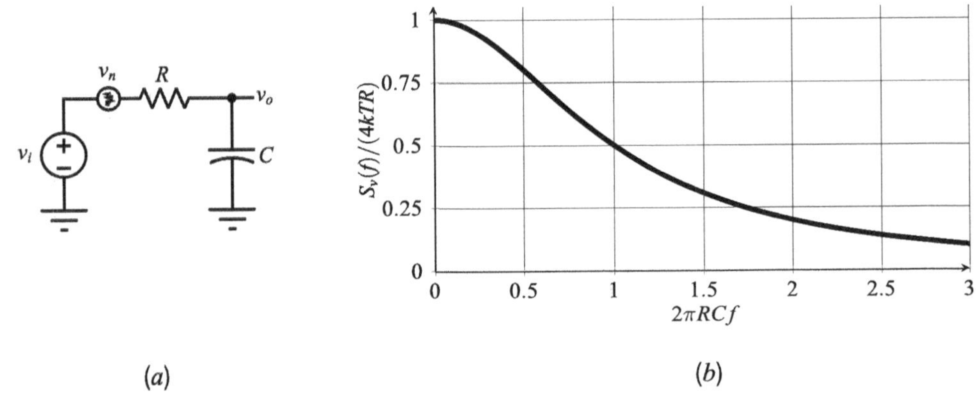

Figure 2.6 (a) The first-order RC lowpass filter. (b) Normalized output noise voltage spectral density as a function of normalized frequency.

As seen in Figure 2.6b, the NSD is $4kTR$ at dc and falls off at high frequencies. This makes sense due to the following. The capacitor is an open circuit at DC, and the noise of the resistor comes "straight through." High-frequency components of the noise are attenuated by the lowpass filter. The half-power frequency, which is the frequency at which the NSD is half its peak value, is $1/(2\pi RC)$. It is instructive and useful to find the total integrated noise across the capacitor, which is given by

$$\overline{v_o^2} = \int_0^\infty S_v(f)\, df.$$

We proceed as follows. Denoting

$$u = 2\pi f RC \quad \text{so that} \quad df = \frac{du}{2\pi RC},$$

the integral becomes

$$\overline{v_o^2} = \frac{2kT}{\pi C} \underbrace{\int_0^\infty \frac{1}{1+u^2}\, du}_{=\pi/2} = \frac{kT}{C}.$$

Interestingly, the mean square noise is independent of the value of the resistor even though it is the resistor that generates noise! What is the intuition behind this seemingly puzzling result? It is the following. Suppose R is doubled. The spectral density of the noise generated by the resistor increases by 2×; however, the bandwidth of the filter decreases by the same factor. Evidently, the increased noise of the resistor is compensated by the reduced bandwidth to result in an output mean-square noise that is independent of R.

Noise in a Track-and-Hold Circuit

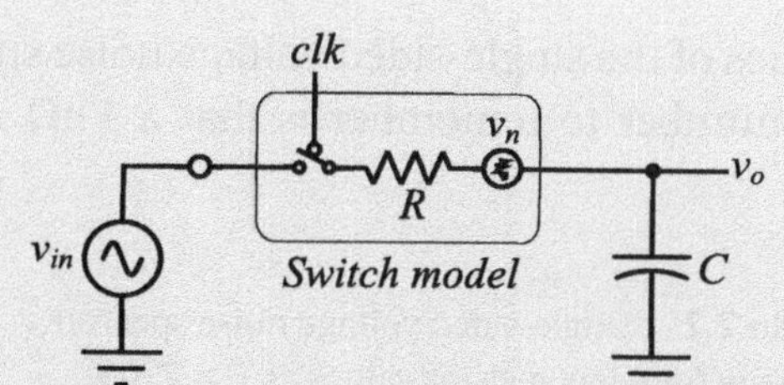

Figure 2.7 Simplified model of a track-and-hold circuit.

It turns out that the result above is very relevant in practice. Figure 2.7 shows the simplified model of a track-and-hold circuit, which forms the front end of many types of analog-to-digital converters. It consists of a switch and a capacitor of value C. The former is realized using a MOS transistor. It has a non-zero on-resistance denoted by R, which is associated with a noise voltage v_n whose single-sided voltage spectral density is $4kTR$.

(Continued)

(Continued)

When the clock signal (*clk*) is low, the switch is closed, and $v_o(t)$ tracks $v_{in}(t)$. For good tracking, the switch resistance must be chosen so that RC is much smaller than the tracking interval. v_o consists of two components: the first is the input voltage, and the second is the noise due to v_n, which has a mean-square value kT/C. When the switch opens, the noise is "trapped" on the capacitor. We thus see that kT/C noise limits the SNR of the track-and-hold circuit, and thereby that of the analog-to-digital converter that processes this sampled voltage. To increase the SNR by a factor of 2 (=3 dB), therefore, C has to be doubled.

Table 2.1 gives the RMS noise for a few values of C. Suppose the input is sinusoidal with an amplitude of 1.4 V. The RMS value of the sampled signal component is 1 V. For $C = 1$ pF, the SNR of the sampled sequence is limited to about 84 dB.

Table 2.1 RMS noise as a function of capacitance at $T = 290$ K.

Capacitance (C)	RMS noise ($\sqrt{kT/C}$)
1 fF	2008 μV
10 fF	632.7 μV
100 fF	200.8 μV
1 pF	63.3 μV
10 pF	20.1 μV
100 pF	6.33 μV

Table 2.2 gives the values of the single-sided voltage noise spectral density for a few resistor values. A good number to remember is that a 1 kΩ resistor has an NSD of 4 nV/$\sqrt{\text{Hz}}$.

Table 2.2 Single-sided voltage noise spectral density for a few resistor values at $T = 290$ K.

Resistance (R)	NSD
50 Ω	0.89 nV/$\sqrt{\text{Hz}}$
100 Ω	1.27 nV/$\sqrt{\text{Hz}}$
1000 Ω	4 nV/$\sqrt{\text{Hz}}$

2.2 Multiple Noise Sources

The next question that arises is what happens when there are multiple noise sources in a network. We begin with the simple resistor divider shown in Figure 2.8.

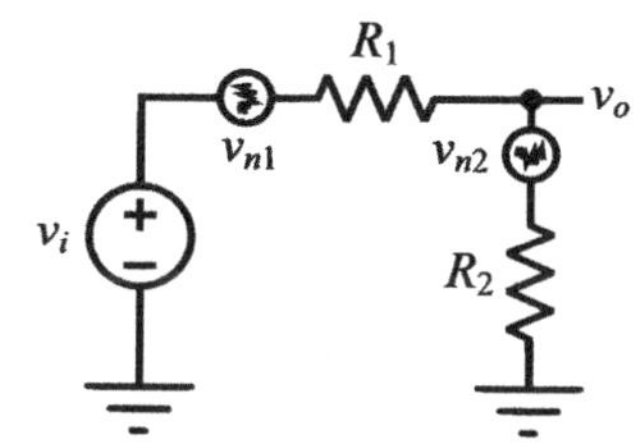

Figure 2.8 A simple resistor-divider network.

The noise of the resistors, modeled by v_{n1} and v_{n2} are independent. How does the output depend on v_{n1} and v_{n2}? If $v_i = 0$, it is evident that

$$v_o = \frac{R_2}{R_1 + R_2} v_{n1} + \frac{R_1}{R_1 + R_2} v_{n2}. \tag{2.10}$$

As the noise sources are zero mean and independent, their contributions to the noise power at the output can simply be added. The output voltage NSD is thus given by

$$S_o(f) = 4kTR_1\left(\frac{R_2}{R_1 + R_2}\right)^2 + 4kTR_2\left(\frac{R_1}{R_1 + R_2}\right)^2 = 4kT\frac{R_1R_2}{R_1 + R_2}. \tag{2.11}$$

In most cases in circuit work, it turns out that noise sources are independent. This means that the output noise spectral densities due to the individual sources can simply be added. This suggests the following way to determine the output NSD in an arbitrary network with independent noise sources.

a. Determine the transfer function from each source to the desired output. For instance, let $H_l(j2\pi f)$ denote the transfer function from the lth noise source to the output. Furthermore, the NSD of the noise source is denoted by $S_l(f)$.
b. Calculate the noise spectral density at the output due to the lth noise source as $S_l(f)|H_l(j2\pi f)|^2$. Do this for all the noise sources.
c. Determine the output NSD as

$$S_o(f) = \sum_l S_l(f)|H_l(j2\pi f)|^2. \tag{2.12}$$

2.3 Equivalent Noise of a One-Port

Consider the noisy one-port network shown in Figure 2.9a. The noise currents and voltages are marked as $i_{s1}, \ldots, i_{sn}$ and $v_{s1}, \ldots, v_{sm}$ respectively. Suppose we are only interested in its terminal behavior, i.e., with respect to the port. As far as the port

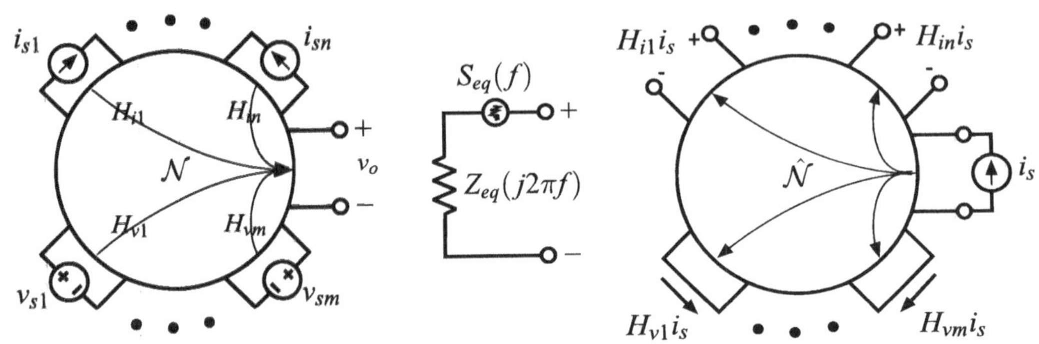

(a) (b) (c)

Figure 2.9 (a) A noisy one-port with $(m + n)$ internal noise sources. (b) Equivalent network modeling the port behavior. (c) Using the inter-reciprocal (or adjoint) network yields all the $(m + n)$ transfer functions in one shot.

impedance is concerned, it is simply the Thevenin impedance. In series with it is a noise voltage source, whose noise spectral density can be found using 2.12.

Determining the $H_l(j2\pi f)$ is straightforward. However, a "complicated" network is likely to have many noise sources. Evaluating the transfer function from each source to the output using superposition is tedious and computationally expensive since it involves solving the network $(m + n)$ times. Fortunately, we can avoid this by exploiting the concept of inter-reciprocity that we learned in Chapter 1. Shown in Figure 2.9c is the adjoint network $\hat{\mathcal{N}}$, formed using the rules described in Chapter 1. We excite the output port of $\hat{\mathcal{N}}$ with a current source and de-energize all the independent sources. Consequently, the voltage sources are replaced by short circuits, while the current sources are opened. We evaluate the network *only once*. By inter-reciprocity, the currents flowing in the short-circuited ports (corresponding to the exciting voltage sources in $\mathcal{N}$) are given by $H_{vk}i_s$. Similarly, the voltages across the open-circuited ports (which correspond to the exciting current sources in $\mathcal{N}$) are given by $H_{il}i_s$. Once the H_{vk} and H_{il} are determined, they can be used in (2.12) to determine $S_o(f)$. The labor saved with respect to evaluating the desired $(m + n)$ transfer functions (naively) using superposition is evident.

In a practical circuit simulator, the overhead in determining the adjoint is negligible, since all that one needs to do is to transpose the MNA matrix. This is easily accomplished in an algorithm that finds the matrix inverse, since all that is needed in a computer implementation is an interchange of the row and column addresses when fetching the corresponding matrix elements from memory. Not surprisingly, the adjoint (inter-reciprocal) network is routinely used in circuit simulators for the purposes of noise analysis. The adjoint and inter-reciprocity, therefore, are not mere mathematical curiosities; they are valuable tools for analysis and computation.

The one-port equivalent of $\mathcal{N}$ of Figure 2.9a is the network of Figure 2.9b. In the special (and useful) case of $\mathcal{N}$ being an RLC network, $S_{eq}(f)$ takes a special form as we see next.

2.3.1 Noise in RLC Networks and the Nyquist Noise Theorem

It is often necessary to determine the voltage noise spectral density at the output of an RLC network. It turns out, as we see below, that the driving point impedance of the network is related to the noise spectral density. This relationship, called the Nyquist noise theorem, is a useful result that exploits reciprocity and passivity.

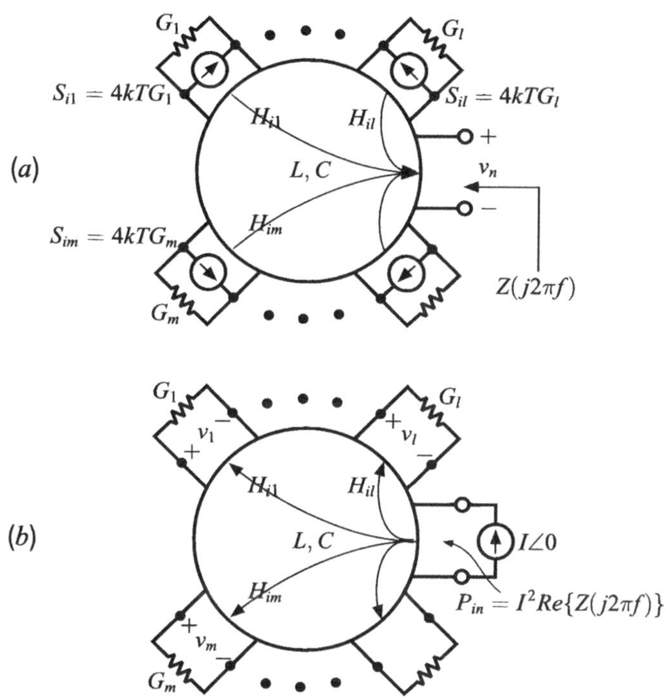

Figure 2.10 (a) Noise due to an RLC network—the impedance looking into the output port is denoted by $Z(j2\pi f)$. (b) Using reciprocity and power conservation to relate $Z(j2\pi f)$ to the H_{im}.

Consider the RLC one port of Figure 2.10a. Its driving-point impedance is denoted by $Z(j2\pi f)$. The conductances are all shown explicitly. The noise of a conductance G_l is modeled by a current source in parallel with it. The single-sided spectral density of the noise current is given by $4kTG_l$ A^2/Hz. It is desired to find the voltage noise spectral density $S_v(f)$ at the output of the one port. Denoting the transfer function from the noise current of the mth conductance to the output voltage by $H_{im}(j2\pi f)$, and since the noise sources are independent, we have

$$S_v(f) = 4kT \sum_{m=1}^{M} G_m |H_{im}(j2\pi f)|^2. \tag{2.13}$$

Suppose a sinusoidal current at a frequency f was injected across the output port as shown in Figure 2.10b, where all the noise sources are set to zero. Using reciprocity, we see that the H_m is also the transfer function that relates the current I injected at the port to the voltage across G_m. The power dissipated in the network is seen to be $P_{in} = I^2\text{Re}\{Z(j2\pi f)\}$. Since all of this has to be dissipated in the conductances, we have

$$P_{in} = I^2\text{Re}\{Z(j2\pi f)\} = I^2 \sum_{m=1}^{M} G_m|H_{im}(j2\pi f)|^2. \tag{2.14}$$

Using this with (2.13), we obtain

$$\boxed{S_v(f) = 4kT\text{Re}\{Z(j2\pi f)\}.} \tag{2.15}$$

Arguments similar to the ones used above can be used to determine the short-circuit current noise spectral density of the RLC network. It turns out that the result is $S_i(f) = 4kT\text{Re}\{Y(j2\pi f)\}$.

2.4 Total Integrated Noise in RLC Networks

A circuit designer is often required to evaluate the total integrated noise in electrical networks with linear passive components. As we have already seen, the most common network in which the total noise is evaluated is the humble RC network shown in Figure 2.11a.

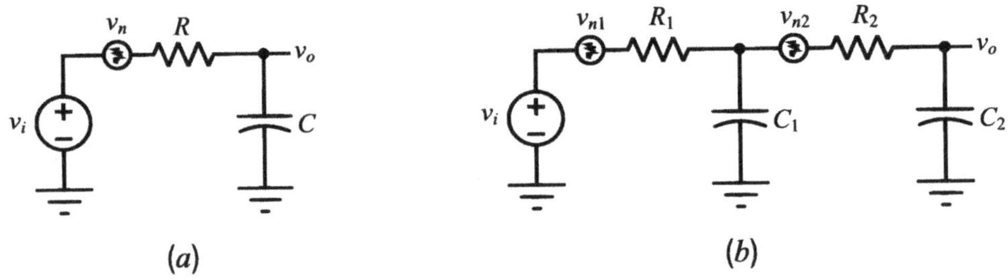

Figure 2.11 (a) First and (b) second-order RC networks.

The resistor has a voltage noise spectral density given by $4kTR$ V^2/Hz. To evaluate the mean square noise across the capacitor, we first evaluate the transfer function $H_1(j2\pi f)$ from v_n to v_o. The total noise is then given by

$$\overline{v_o^2} = 4kT \int_0^{\infty} R|H_1(j2\pi f)|^2 df = \frac{kT}{C}. \tag{2.16}$$

For a more "complicated" network, the procedure for determining the total noise remains the same. For instance, to determine $\overline{v_o^2}$ in the network of Figure 2.11b, we evaluate the transfer functions from v_{n1} and v_{n2} to v_o, and denote them by $H_1(j2\pi f)$ and $H_2(j2\pi f)$, respectively.

$$\overline{v_o^2} = 4kT \int_0^{\infty} (R_1|H_1(j2\pi f)|^2 + R_2|H_2(j2\pi f)|^2)df. \tag{2.17}$$

Since $H_1(j2\pi f)$ and $H_2(j2\pi f)$ are second-order transfer functions, evaluation of the integral above is complicated and challenging. Interestingly, the mean-square noise voltage happens to be kT/C_2. From the discussion above, we see that determining the total integrated noise in a high-order network using integrals of the form in (2.17) is bound to bury us in a mountain of algebra. Fortunately, it turns out that one can (again) exploit reciprocity and passivity to greatly simplify the evaluation of total integrated noise, as shown below [7].

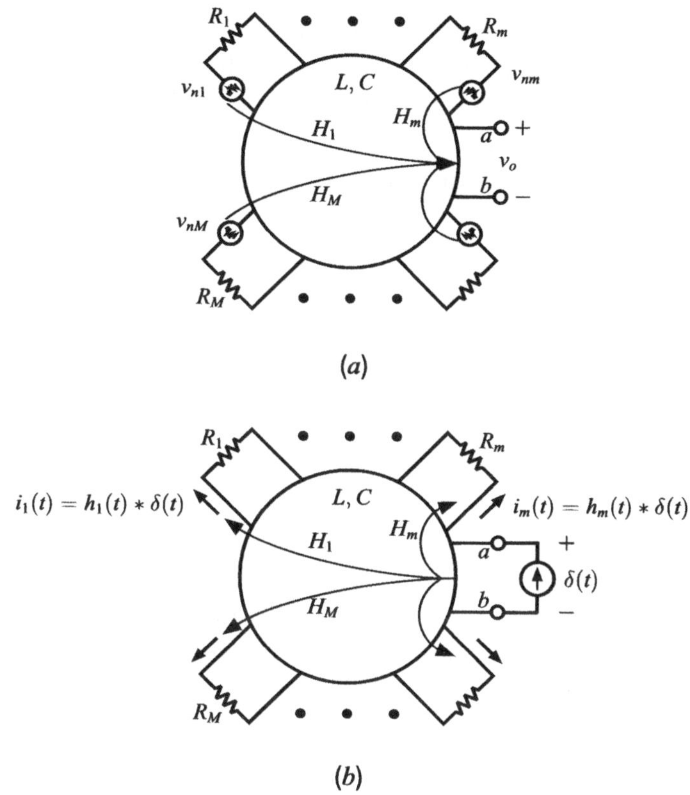

Figure 2.12 (a) Original network with noise sources $v_1, \ldots, v_n$. The voltage transfer function from v_l to v_o is denoted by $H_l(j2\pi f)$. (b) The impulse responses $h_l(t)$ corresponding to $H_l(j2\pi f)$ can be obtained by exciting the output with a current impulse and observing the currents in R_l.

Consider the generic linear passive RLC network of Figure 2.12a. There are M resistors in the network, and the thermal noise generated by each of them is modeled by a voltage source in series. We denote the transfer function from the mth voltage source to the output v_o by $H_m(j2\pi f)$. Since the noise sources are independent, the total mean square noise at the output is given by

$$\overline{v_o^2} = 4kT \sum_{m=1}^{M} \int_0^{\infty} R_m |H_m(j2\pi f)|^2 df. \tag{2.18}$$

By Parseval's theorem,

$$\int_{-\infty}^{\infty} |H_m(j2\pi f)|^2 df = 2\int_0^{\infty} |H_m(j2\pi f)|^2 df = \int_0^{\infty} |h_m(t)|^2 dt,$$

where $h_m(t)$ is the impulse response corresponding to $H_m(j2\pi f)$. Thus, $\overline{v_o^2}$ is

$$\overline{v_o^2} = 2kT \sum_{m=1}^{M} \int_0^{\infty} R_m h_m^2(t) dt. \tag{2.19}$$

The $h_m(t)$ can be found by using reciprocity, as shown in Figure 2.12b. The output port is excited by a current impulse. The current through the mth resistor R_m is given by $i_m(t) = h_m(t) * \delta(t)$, where $*$ denotes convolution. $i_m(t)$ is a current and *not* the impulse response $h_m(t)$ (which has units of 1/second). To obtain $h_m(t)$, the current $i_m(t)$ in R_m should be divided by one coulomb. The division by one coulomb is necessary for dimensional correctness (see the gray box below for a more detailed discussion).

Dimensions of Impulse Response

The frequency response $H(j2\pi f)$ of a linear network that relates an input and output having the same dimensions (say, voltage) is dimensionless. The impulse response $h(t)$ corresponding to $H(j2\pi f)$ is related to the latter according to

$$h(t) = \int_{-\infty}^{\infty} H(j2\pi f) e^{j2\pi f t}\, df. \tag{2.20}$$

Thus, $h(t)$ must have dimensions of s^{-1}.
Referring to Figure 2.12, the reciprocity theorem states that the impulse response $h_1(t)$ relating $v_{n1}(t)$ to $v_o(t)$ in part (a) of the figure is the same as the impulse response relating the current injected into port *a-b* to $i_1(t)$ in Figure 2.12b. Thus, if the exciting current in Figure 2.12b is a current impulse, $i_1(t) = h_1(t) \times \delta(t)$ which is

$$\underbrace{i_1(t)}_{\text{amperes}} = \int_{-\infty}^{\infty} \underbrace{h_1(\tau)}_{s^{-1}} \underbrace{\delta(t-\tau)}_{\text{amperes}} \underbrace{d\tau}_{s}\ . \tag{2.21}$$

To obtain $h_1(t)$ from $i_1(t)$, therefore, one needs to divide $i_1(t)$ (which has dimensions of amperes) by 1 ampere · second = 1 coulomb.

A subtle, but important, point is the following. $i_1(t)$ has the *same shape* as $h_1(t)$ since it is obtained by convolving $h_1(t)$ with an impulse. However, $i_1(t)$ and $h_1(t)$ *do not have the same dimensions*. The former has units of current, while the latter has units of s^{-1}.

We now examine the energy delivered and dissipated when the network of Figure 2.12b is excited by a current impulse. Let E_0 denote the energy delivered by the impulse current to the network. The subscript 0 indicates that all of this energy has to be delivered at $t = 0$, since the input current $\delta(t)$ is zero for $t \neq 0$. Further, let E_∞ denote the energy that remains in the network of Figure 2.12b at $t = \infty$.

The difference $(E_0 - E_\infty)$ must have been dissipated in the resistors. Energy conservation, therefore, dictates that

$$E_0 - E_\infty = \sum_m \int_0^\infty R_m i_m^2(t)\, dt. \tag{2.22}$$

Using $i_m(t) = h_m(t) \cdot 1$ coulomb in the above, along with (2.19), we have

$$\boxed{\overline{v_o^2} = \frac{2kT}{(1\,\text{coulomb})^2}\,(E_0 - E_\infty)\,.} \tag{2.23}$$

The equation above is seen to greatly simplify the evaluation of the total integrated noise in RLC time-invariant networks. To a circuit designer, the immense value of this theorem lies in the fact that the total integrated noise of a complicated passive network can be obtained by inspection, as we show in several examples that follow.

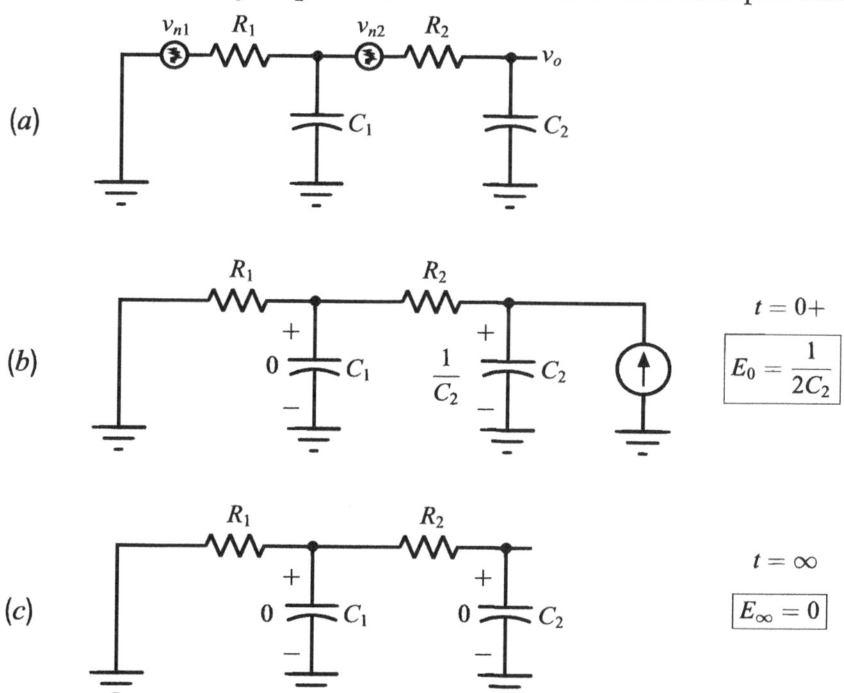

Figure 2.13 (a) RC network under consideration. The mean-square noise across C_2 is of interest. (b) Determining E_0. (c) Finding $E\infty$.

Consider the network of Figure 2.13a. We wish to determine the mean-square noise $\overline{v_o^2}$. To find E_0, we inject an impulse current into C_2, as shown in Figure 2.13b. At $t = 0+$, it is easily seen that the voltage across C_2 is 1 coulomb/C_2, while that across C_1 remains zero. E_0 is thus seen to be 1 coulomb^2/$(2C_2)$. At $t = \infty$, both capacitors are discharged, resulting in $E_\infty = 0$. Consequently,

$$\overline{v_o^2} = 2kT\frac{E_0 - E_\infty}{1\,\text{coulomb}^2} = \frac{kT}{C_2}. \tag{2.24}$$

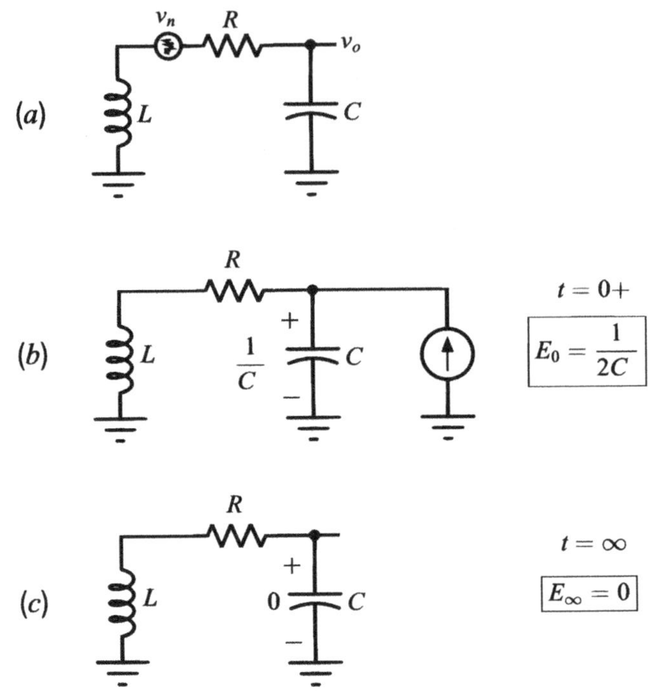

Figure 2.14 (a) RC network under consideration. The mean-square noise across C is of interest. (b) Determining E_0. (c) Finding $E\infty$.

Another example network is shown in Figure 2.14a. It is desired to find the mean-square noise across C. To find E_0, we inject an impulse current into C, as shown in Figure 2.14b. At $t = 0+$, the voltage across C is 1 coulomb/C, and the inductor current remains zero. E_0 is thus seen to be 1 coulomb^2/(2C). At $t = \infty$, the capacitor voltage and inductor current are both zero, resulting in $E_\infty = 0$. The mean-square noise is thus given by

$$\overline{v_o^2} = 2kT\frac{E_0 - E_\infty}{1\ \text{coulomb}^2} = \frac{kT}{C}. \tag{2.25}$$

Lest the reader conclude that E_∞ is always zero, we consider the example network of Figure 2.15a. We are interested in determining the mean-square noise across C_2. To that end, we inject an impulse current $\delta(t)$ amperes into the output port. At $t = 0+$, the voltage across C_2 is 1 coulomb/C_2, and the energy stored in the network is $E_0 = 1\ \text{coulomb}^2/(2C_2)$, as shown in Figure 2.15b. At $t = \infty$, the currents through all the resistors are zero, and all capacitors have the same voltage, given by $v_\infty = 1\ \text{coulomb}/(C_0 + C_1 + C_2)$. The energy stored in the network is seen to be

$$E_\infty = \frac{1}{2}\underbrace{(Q_0 + Q_1 + Q_2)}_{1\ \text{coulomb}} v_\infty = \frac{1\ \text{coulomb}^2}{2(C_0 + C_1 + C_2)}. \tag{2.26}$$

Using (2.23), the mean square noise is

$$\overline{v_o^2} = kT\left(\frac{1}{C_2} - \frac{1}{C_0 + C_1 + C_2}\right) = \frac{kT(C_0 + C_1)}{C_2(C_0 + C_1 + C_2)}.$$

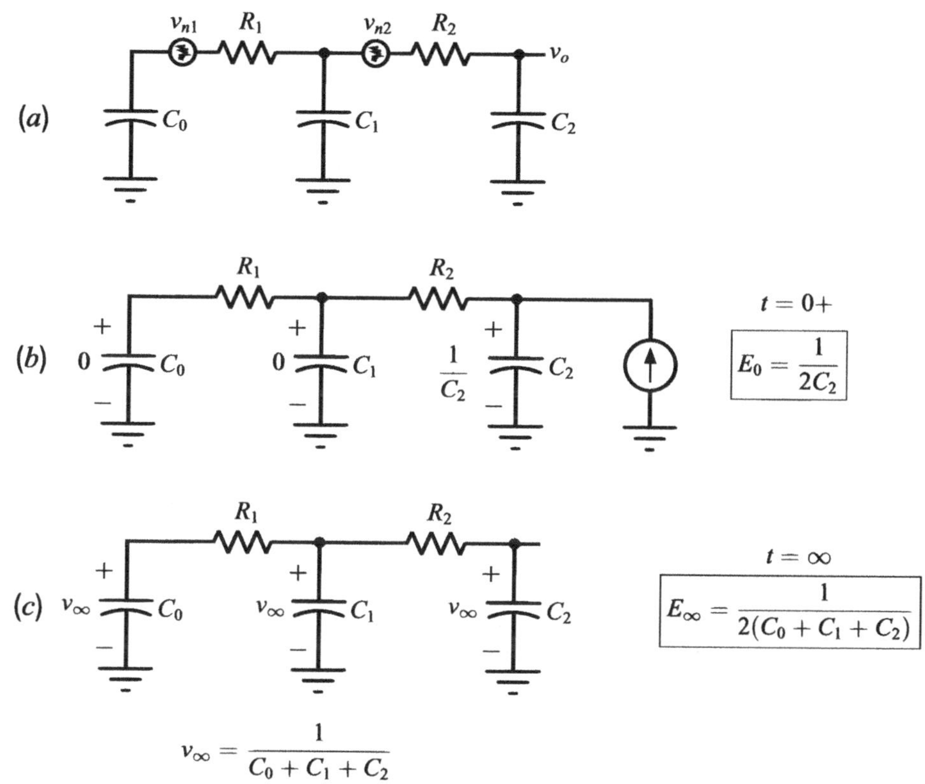

Figure 2.15 (a) RC network under consideration. The mean-square noise across C_2 is of interest. (b) Determining E_0. (c) Finding $E\infty$.

2.4.1 Bode's Noise Theorem

The discussion above used time-domain arguments to determine the mean-square noise. A frequency domain approach is also possible, as we see next. This is called the *Bode Noise Theorem.*

Consider the passive RLC network of Figure 2.16a. The driving point impedance of the network is denoted by $Z(s)$, where s is the Laplace variable. As we have already seen, Nyquist's noise theorem states that the noise spectral density $S_{vo}(f)$ is given by $4kT\ \mathrm{Re}(Z(j2\pi f))$. Thus,

$$\overline{v_o^2} = 4kT \int_0^\infty \mathrm{Re}(Z(j2\pi f))\, df. \tag{2.27}$$

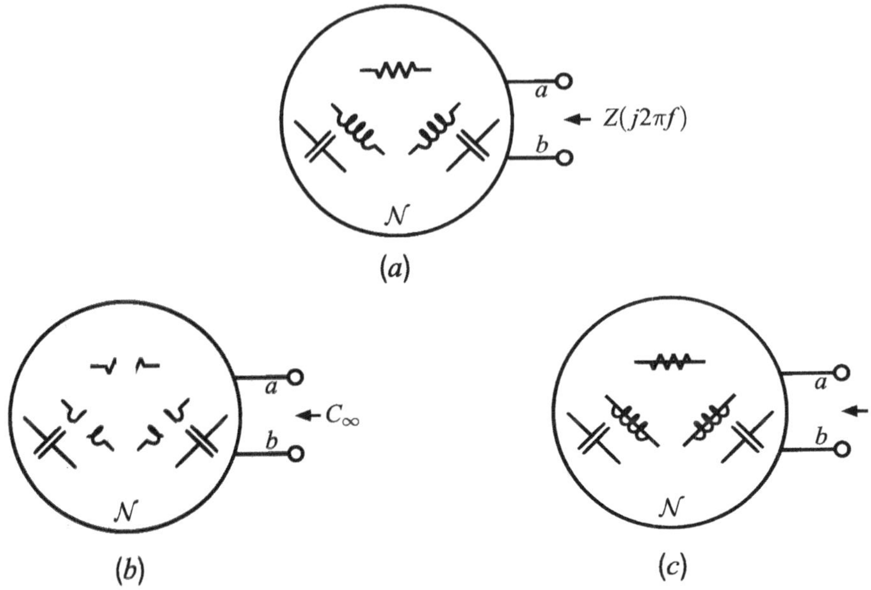

Figure 2.16 (a) The double-sided voltage noise spectral density of a passive impedance $Z(s)$ is given by $2kT\text{Re}(Z(j2\pi f))$. (b, c) Physical interpretation of C_∞ and C_0: C_∞ is the looking-in capacitance of the network when all resistors and inductors are opened, and C_0 is the capacitance looking in when all the resistors and inductors are shorted.

Using the properties of the driving-point impedance of a passive RLC network and contour integration, Bode showed that

$$\int_{-\infty}^{\infty} \text{Re}(Z(j2\pi f))\, df = \frac{1}{2}\left(\frac{1}{C_\infty} - \frac{1}{C_0}\right), \tag{2.28}$$

where

$$\frac{1}{C_\infty} = \lim_{s\to\infty} sZ(s), \tag{2.29}$$

and

$$\frac{1}{C_0} = \lim_{s\to 0} sZ(s). \tag{2.30}$$

Physically, C_∞ can be interpreted as the capacitance looking into the network when all the resistors and inductors have been replaced by open circuits, as shown in Figure 2.16b. Similarly, C_0 is the capacitance looking into the network when all the resistors and inductors have been replaced by short circuits, as in Figure 2.16c. Given a network, it is straightforward to determine C_∞ and C_0 by inspection. Using Nyquist and Bode's results, the mean-square noise v_o^2 is thus seen to be

$$\overline{v_o^2} = kT\left(\frac{1}{C_\infty} - \frac{1}{C_0}\right). \tag{2.31}$$

The equation above is called the Bode Noise Theorem.

We now show that the result (2.23) reduces to Bode's result (2.31). As discussed earlier, E_0 is the energy delivered by the impulse current to the network of Figure 2.16b. The Laplace transform of the voltage developed across the terminals of the impulse current $\delta(t)$ is given by $V(s) = Z(s)$ ampere · second, where $Z(s)$ is the driving-point impedance of the network. The ampere · second is the physical unit of the Laplace transform of the unit current impulse. The voltage at $t = 0+$, obtained using the initial value theorem, is $v(0+) = \lim_{s\to\infty} sZ(s) \cdot 1$ coulomb. Thus,

$$E_0 = \frac{1}{2}v(0+)\underbrace{\int_{-\infty}^{\infty} \delta(t)\,dt}_{1\text{ coulomb}} = \frac{1}{2}\underbrace{\lim_{s\to\infty} sZ(s)}_{1/C_\infty} \cdot 1\text{ coulomb}^2. \tag{2.32}$$

To evaluate E_∞ for the network of Figure 2.16c, we proceed as follows. In steady-state ($t \to \infty$), all currents in the resistors and inductors are zero, and they can be replaced by short circuits. What we are left with then is a network of capacitors. Two possibilities arise with regard to this capacitive network.

a. There is a DC path from node a to node b. In this case, voltage v_∞ is 0. E_∞ is thus zero.
b. There is no DC path from node a to node b. The energy stored in the network is given by $E_\infty = (1/2)Q_a v_\infty$, where Q_a is the total charge on all capacitor plates connected to node a. Since there is no DC path from a to b, it follows that Q_a must equal the charge injected by the impulse current, namely 1 coulomb. Thus, $E_\infty = (1/2)v_\infty \cdot 1$ coulomb.

In either case, E_∞ is seen to be given by $(1/2)v_\infty \cdot 1$ coulomb. From the final value theorem, $v_\infty = \lim_{s\to 0} sZ(s) \cdot 1$ coulomb. From the discussion above, it thus follows that

$$E_\infty = \frac{1}{2}v(\infty) \cdot 1\text{ coulomb} = \frac{1}{2}\underbrace{\lim_{s\to 0} sZ(s)}_{1/C_0} \cdot 1\text{ coulomb}^2. \tag{2.33}$$

Bode's Noise Theorem results from using (2.32) and (2.33) in (2.23).

2.5 Equivalent Noise Sources in a Two-Port Network

In most practical situations a two-port contains numerous passive elements and active devices, each contributing its own noise. For small-signal analysis, we first find the operating point, linearize the nonlinear elements about that point, and *then* attach the corresponding noise sources to obtain a purely linear RLC network augmented by controlled sources and independent noise voltages/currents. For example, the small-signal model of a MOS transistor (with transconductance g_m and output resistance r_o) is accompanied by a channel noise current with single-sided spectral density.

$$S_i(f) = \frac{8}{3}\eta\, kT\, g_m \quad [\mathrm{A}^2/\mathrm{Hz}],$$

where η is a process/geometry-dependent nonideality factor. By contrast, a resistor of value R is modeled with the familiar voltage noise spectrum $4kTR$ across its terminals. In the resulting linear network, the output response contains both the processed input signal and the contributions of these internal noise sources propagated through the network's transfer functions.

From a user's perspective, a complicated internal schematic can be abstracted by port characteristics. The internal details matter to the designer; however, what is useful to a user is a concise representation that captures both *signal* transfer and *noise* added by the two-port. The signal transfer is modeled by any convenient two-port description (e.g., *y*-, *z*-, *h*-, or g-parameters). What about noise?

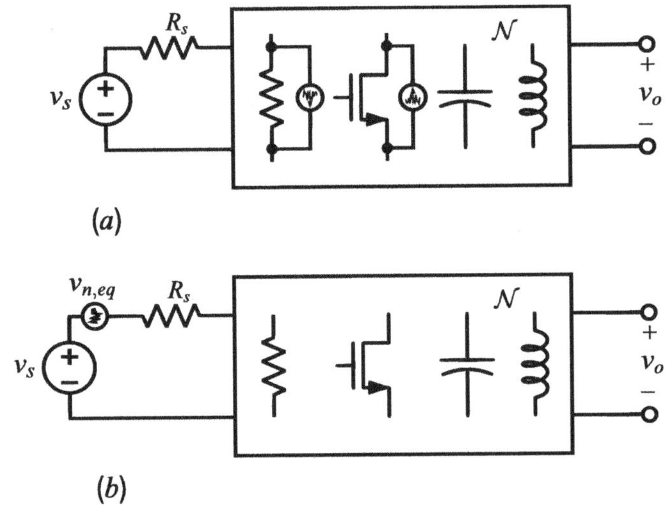

Figure 2.17 (a) A two-port network with multiple noise sources, driven by a source with resistance R_S. (b) Replacing the effect of all the noise sources by an equivalent noise voltage $v_{n,eq}$ at the input.

An effective way to summarize a complicated set of internal noise sources is to *refer* their effect to the input as an *equivalent* noise source whose propagation to the output matches that of the real, noisy networks. Figure 2.17 shows the basic idea. Part (a) shows a network driven by an input source with a resistance R_s. The desired output is the voltage v_o. There are multiple noise sources inside the network, causing the output to be noisy. It turns out, as we see below, that the effect of all the internal noise sources can be modeled by a single *equivalent* noise voltage source in series with the input, namely $v_{n,eq}$, as shown in Figure 2.17b, where the noise sources in the network are turned off. In other words, we replace the many internal sources by a single input-side source such that the output noise spectral density remains unchanged.

To determine the spectral density of the equivalent input-referred source, we proceed as follows. We assume that all the noise sources are voltages—this does not lead to a loss of generality. Furthermore, it is instructive to first consider the noise sources as being deterministic.

a. Determine the transfer function from each noise source to the output in Figure 2.17a. Denote these by $H_l(j2\pi f)$ for $l = 1, 2, 3, \ldots.$ If we denote the transforms of the noise sources (initially assumed to be deterministic) by $V_l(j2\pi f)$, the output is seen to be

$$V_o(j2\pi f) = \sum_l H_l(j2\pi f)V_l(j2\pi f). \tag{2.34}$$

b. Determine the transfer function from the equivalent noise source, namely $v_{n,eq}$ in Figure 2.17b, to the output. Denote this by $H_{eq}(j2\pi f)$.

$$V_o(j2\pi f) = H_{eq}(j2\pi f)V_{eq}(j2\pi f). \tag{2.35}$$

c. If the equivalent voltage source is to model the effect of the myriad sources inside the network, it is evident that

$$H_{eq}(j2\pi f)V_{eq}(j2\pi f) = \sum_l H_l(j2\pi f)V_l(j2\pi f). \tag{2.36}$$

$V_{eq}(j2\pi f)$ is thus seen to be

$$V_{eq}(j2\pi f) = \sum_l \frac{H_l(j2\pi f)}{H_{eq}(j2\pi f)} V_l(j2\pi f). \tag{2.37}$$

Assuming the noise sources are independent, the spectral density of the equivalent input referred noise voltage is given by

$$S_{eq}(f) = \sum_l \left| \frac{H_l(j2\pi f)}{H_{eq}(j2\pi f)} \right|^2 S_l(f), \tag{2.38}$$

where the S_l denote the spectral densities of the internal noise sources.

Example: Resistive Attenuator

Consider the resistor divider of Figure 2.18a. We first consider v_{n1} and v_{n2} to be deterministic. The noise at the output, obtained by setting $v_s = 0$ is given by

$$v_o = \frac{R_2}{R_1 + R_2 + R_s} v_{n1} + \frac{R_1 + R_s}{R_1 + R_2 + R_s} v_{n2}. \tag{2.39}$$

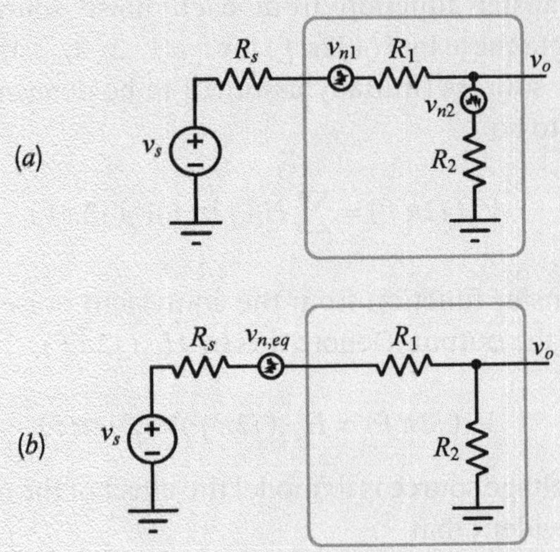

Figure 2.18 (a) Resistive attenuator. (b) Equivalent input-referred voltage noise.

The transfer function from v_s to v_o, namely H_{eq}, is seen to be $R_2/(R_1 + R_2 + R_s)$. Consequently, the equivalent input-referred voltage $v_{n,eq}$ can be obtained from

$$\frac{R_2}{R_1 + R_2 + R_s} v_{n,eq} = \frac{R_2}{R_1 + R_2 + R_s} v_{n1} + \frac{R_1 + R_s}{R_1 + R_2 + R_s} v_{n2}, \tag{2.40}$$

resulting in

$$v_{n,eq} = v_{n1} + \frac{R_1 + R_s}{R_2} v_{n2}. \tag{2.41}$$

The NSD of the equivalent source, namely S_{eq} is given by

$$S_{eq} = 4kTR_1 + 4kT\frac{(R_1 + R_s)^2}{R_2^2} R_2. \tag{2.42}$$

It is clear from Figure 2.17 that $H_{eq}(j2\pi f)$ and the $H_l(j2\pi f)$ depend on the source resistance R_s. This makes sense since R_s is "part of the network." Consequently, $S_{eq}(j2\pi f)$ depends on the "external" component R_s. This is somewhat of an anticlimax, since we expected that the equivalent noise source to be a function of the network, independent of the external world. Fortunately, however, it turns out that a source-resistance independent representation is possible. To see this, it is necessary to examine the dependence of $V_{eq}(j2\pi f)$ on R_s. The key network property that allows us to infer this is the bilinear dependence of any transfer function in a linear network on a network parameter. We derived this in Chapter 1; we give a summary below.

Suppose we pick any one element parameter p (e.g., the source resistance R_s) and treat all others as fixed; then any transfer function $H(p)$ is a *bilinear* (linear fractional) function of p:

$$H(p) = \frac{A\,p + B}{C\,p + D}$$

for some constants $A,\ B,\ C,\ D$ that depend only on the rest of the network and $j2\pi f$. In particular, no transfer function can contain p^2 or higher powers in either numerator or denominator when only a single element is varied. Furthermore, when multiple transfer functions are considered, it turns out that the denominators remain unchanged. For instance, all the H_l can be expressed in the form

$$H_l(p) = \frac{A_l p + B_l}{Cp + D}.$$

The numerator polynomials depend on the source, but the denominators remain unchanged. In the network of Figure 2.18, for example, the denominators of all the transfer functions are $(R_1 + R_2 + R_s)$, while the numerators of the transfer functions from v_{n1} and v_{n2} to v_o are R_2 and $(R_1 + R_s)$, respectively. The example network of Figure 2.18 is memoryless. In general, the constants $A,\ B,\ C,\ D$ depend on frequency (through reactive elements), but the *form* $H(p) = (Ap + B)/(Cp + D)$ with respect to any single element remains valid for all linear time-invariant networks.

An interesting observation regarding the input-to-output transfer function H_{eq} is in order. It has the bilinear form

$$H_{\text{eq}}(j2\pi f) = \frac{A_{\text{eq}}\,R_s + B_{\text{eq}}}{C\,R_s + D}.$$

In the limit as $R_s \to \infty$, the expression above approaches A_{eq}/C. Physically, it is apparent that as $R_s \to \infty$, the input source is effectively disconnected from the network, so the output should vanish; therefore, A_{eq} *must be* 0. $H_{\text{eq}}(j2\pi f)$ is thus of the form

$$H_{\text{eq}}(j2\pi f) = \frac{B_{\text{eq}}}{C\,R_s + D}.$$

Revisiting (2.37), and expressing each of the transfer functions in their bilinear forms, we obtain

$$\begin{aligned}H_{\text{eq}}(j2\pi f)\,V_{\text{eq}}(j2\pi f) &= \frac{B_{eq}}{CR_s + D}V_{\text{eq}}(j2\pi f)\\ &= \sum_l H_l(j2\pi f)V_l(j2\pi f) = \sum_l \frac{A_l R_s + B_l}{CR_s + D}\,V_l(j2\pi f)\end{aligned}$$

from which we have

$$V_{\mathrm{eq}}(j2\pi f) = \sum_l \left(\frac{A_l}{B_{eq}}R_s + \frac{B_l}{B_{eq}}\right) V_l(j2\pi f)$$

$$= \underbrace{\sum_l \frac{B_l}{B_{\mathrm{eq}}} V_l(j2\pi f)}_{\equiv V_a(j2\pi f)} + \underbrace{\left(\sum_l \frac{A_l}{B_{\mathrm{eq}}} V_l(j2\pi f)\right)}_{\equiv I_a(j2\pi f)} R_s.$$

Thus the *equivalent input-referred voltage* decomposes as

$$\boxed{V_{\mathrm{eq}}(j2\pi f) = V_a(j2\pi f) + I_a(j2\pi f)\, R_s.}$$

where V_a (voltage source in series) and I_a are *network-only* linear combinations of the internal sources. Why does the R_s term make intuitive sense? As R_s becomes very large, the input port becomes weakly coupled, so the signal gain H_{eq} falls; to reproduce the same output noise, the input-referred voltage must grow with R_s. This is precisely the role of the $I_a(j2\pi f)R_s$ term. A few observations are in order.

a. Recall that A_{eq}, A_l, B_{eq} and B_l are independent of R_s. They depend *only* on the network $\mathcal{N}$. Consequently, $V_a(j2\pi f)$ and $I_a(j2\pi f)$ are inherent properties of $\mathcal{N}$.

b. Since the noise sources are independent, the spectral densities of $I_a(j2\pi f)$ and $V_a(j2\pi f)$ are given by

$$S_i(f) = \sum_l \left|\frac{A_l(j2\pi f)}{B_{eq}(j2\pi f)}\right|^2 S_l(f)$$

$$S_v(f) = \sum_l \left|\frac{B_l(j2\pi f)}{B_{eq}(j2\pi f)}\right|^2 S_l(f).$$

c. $V_a(j2\pi f)$ and $I_a(j2\pi f)$ are *different* linear combinations of the same noise sources $V_l(j2\pi f)$. It thus follows that they are *correlated.*

2.5.1 A Source-Independent Noise Model for the Two-Port

It would be wonderful if we could specify the noisy two-port with an *intrinsic* input noise model that does not depend on R_s. This is indeed possible by moving sources and invoking Norton–Thévenin equivalences, as shown in Figure 2.19a,b. Since v_a and i_a are independent of R_s, they are properties of the network without reference to the source as shown in Figure 2.19c. In other words, the effect of the many noise sources in $\mathcal{N}$ can be modeled by the equivalent sources v_a and i_a. The spectral densities of these sources are, in general, dependent on frequency. Both v_a and i_a are properties of the network alone (functions of the internal sources and topology), not of R_s.

Two limiting cases highlight the roles of v_a and i_a.

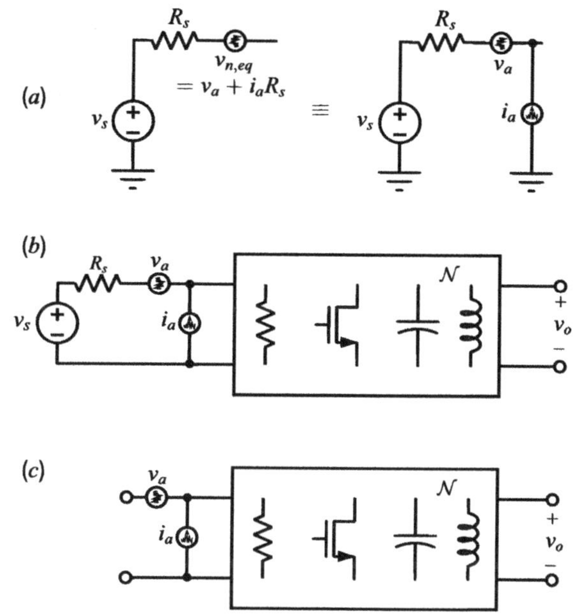

Figure 2.19 (a) Decomposing the equivalent noise voltage into a voltage and current source. (b) Equivalent noise sources of the network. (c) Since v_a and i_a do not depend on R_s, they are inherent properties of the network.

- **Ideal voltage drive** ($R_s \to 0$): the current-source term $i_a R_s$ is negligible; the noise is dominated by v_a.
- **Ideal current drive** ($R_s \to \infty$): the series voltage source is inconsequential; the input noise is dominated by i_a.

Input-Referred Noise Sources in a Resistive Attenuator

Consider again the resistor divider of Figure 2.20a. We saw earlier that the input-referred equivalent noise source $v_{n,eq}$ is given by

$$v_{n,eq} = v_{n1} + \frac{R_1 + R_s}{R_2} v_{n2}. \tag{2.43}$$

Consequently,

$$v_a = v_{n1} + \frac{R_1}{R_2} v_{n2} \quad , \quad i_a = \frac{1}{R_2} v_{n2}. \tag{2.44}$$

The NSD of the equivalent input-referred noise voltage and current sources, namely S_v and S_i are given by

$$S_v = 4kTR_1 + 4kT\frac{R_1^2}{R_2} \quad , \quad S_i = 4kT\frac{1}{R_2}. \tag{2.45}$$

It is evident that v_a and i_a are correlated, since v_a is in part formed by v_{n2}.

(Continued)

(Continued)

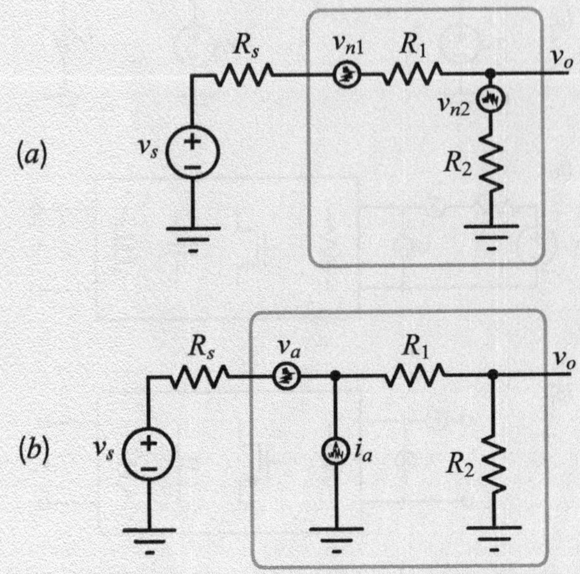

$$v_a = v_{n1} + \frac{R_1}{R_2} v_{n2}$$

$$i_a = \frac{v_{n2}}{R_2}$$

Figure 2.20 (a) Resistive attenuator. (b) Equivalent input-referred sources.

2.6 Noise with Impedance Scaling and Differential Operation

2.6.1 Impedance Scaling

We now discuss two important aspects that relate noise to power dissipation. Consider the linear network $\mathcal{N}$, shown in Figure 2.21a. 1 and 1′ denote the output terminals. Apart from inductors, resistors, and capacitors, it includes linear controlled sources, realized using active devices like transistors. The components in $\mathcal{N}$ produce noise, resulting in an open-circuit noise voltage spectral density denoted by $S_{v1}(f)$. Suppose the outputs of two identical instances of $\mathcal{N}$, driven by the same input v_s were shorted as shown in Figure 2.21a. The equivalent representation of the output is given in Figure 2.21b. The output noise of each instance has the same noise spectral density; however, the noise voltages are independent. Consequently, the Thevenin equivalent is that shown in Figure 2.21c. The signal component of the Thevenin voltage is what one would obtain with a single instance of $\mathcal{N}$. The Thevenin impedance reduces by a

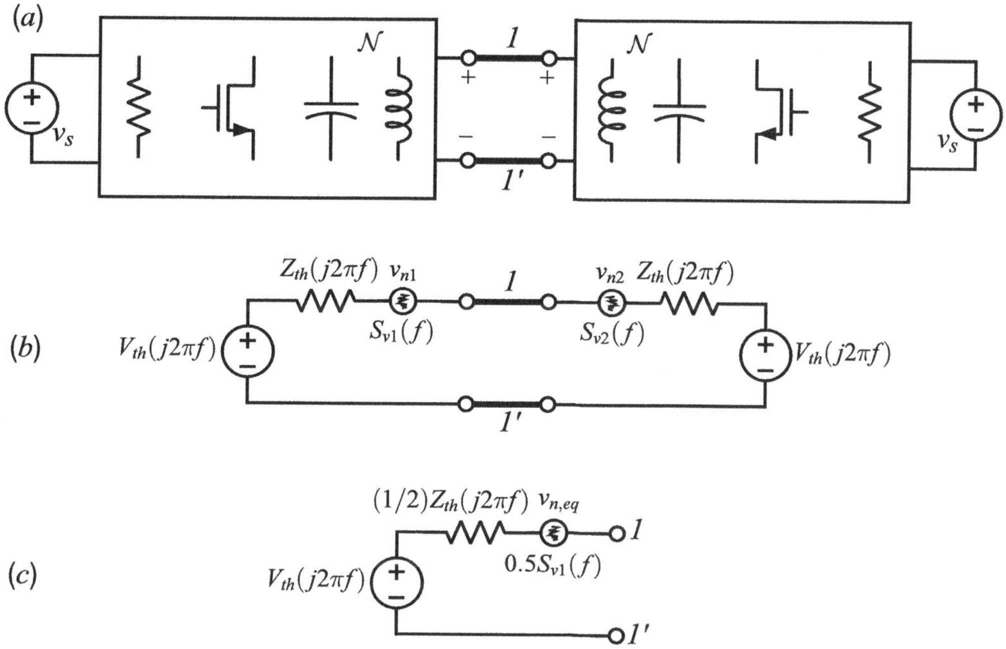

Figure 2.21 (a) The outputs of two identical instances of a linear network $\mathcal{N}$, driven by the same input are connected together. (b) Equivalent representation of the output. The noise produced by each instance is independent of the other, but v_{n1} and v_{n2} have the same noise spectral density. (c) The simplified equivalent of (b). The NSD is half that of the individual networks. The SNR is improved by a factor of 2(=3 dB).

factor of 2, since two networks are connected in parallel. The noise component, which is $(1/2)(v_{n1} + v_{n2})$ has a spectral density $(1/4)(S_{v1}(f) + S_{v2}(f)) = S_{v1}(f)/2$. The SNR has thus increased by a factor of 2 (i.e., 3 dB). This increase in SNR does not come for free; two instances of $\mathcal{N}$ will dissipate twice as much power. This is a fundamental trade-off; it costs twice the power to improve SNR by a factor of 2. In general, shorting the outputs of n networks will cause the SNR to increase by a factor of n at the cost of n times the power dissipation. The Thevenin impedance will be reduced by a factor of n.

It is not necessary to make multiple copies of $\mathcal{N}$ and short their outputs to achieve the results described above. One could instead increase all (trans)conductances in $\mathcal{N}$ by the same factor. This is illustrated in Figure 2.22 for $n = 2$. Conceptually, the corresponding nodes in two copies of $\mathcal{N}$ are connected together. If $\mathcal{N}$ was noiseless, the output would be unaffected, since the voltages at identical nodes in both copies of $\mathcal{N}$ are the same. In effect, this is equivalent to doubling (trans)conductance in every branch of $\mathcal{N}$. It is straightforward to see that the output noise voltage spectral density reduces by a factor of 2. In general, increasing *all* (trans)conductances in a network by a factor of n reduces noise power by a factor of n at the expense of power dissipation that increases by the same factor.

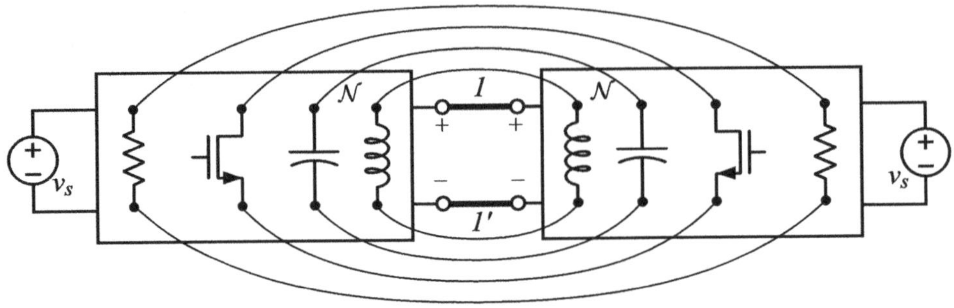

Figure 2.22 Impedance (down) scaling by a factor of 2.

2.6.2 Differential Operation

Figure 2.23a shows the principle behind an extensively used circuit technique called differential operation. Two identical networks $\mathcal{N}$ are excited by equal and opposite inputs, namely v_s and $-v_s$. The output of interest is the *difference* between the outputs of these networks, as shown in the figure. The port-equivalent representations of the networks are given in Figure 2.23b. By linearity, the Thevenin voltages are of

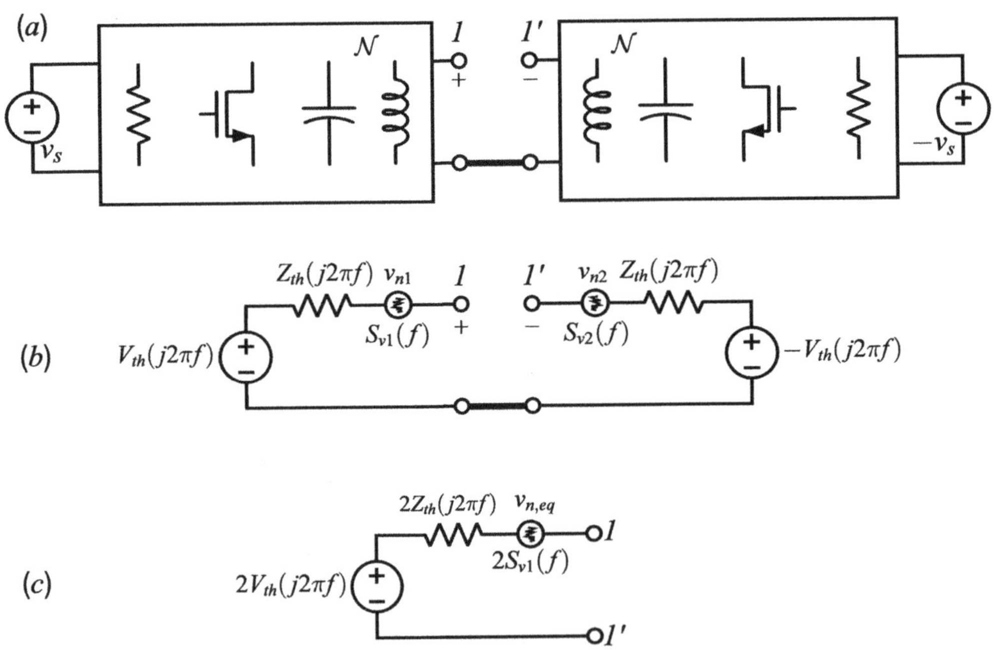

Figure 2.23 (a) The outputs of two identical instances of a linear network $\mathcal{N}$, driven by equal and opposite inputs. (b) Equivalent representation of the output. The noise produced by each instance is independent of the other, but v_{n1} and v_{n2} have the same noise spectral density. (c) The simplified equivalent of (b). The NSD is twice that of the individual networks. The signal power is four times higher, however, resulting in an SNR improvement by a factor of 2(=3 dB).

opposite signs. The noise sources v_{n1} and v_{n2} have the same spectral density but are independent. The Thevenin equivalent of the output seen between terminals 1 and 1′ is shown in Figure 2.23c. The open-circuit voltage and Thevenin impedance are $2V_{th}(j2\pi f)$ and $2Z_{th}(j2\pi f)$, respectively. Since the signal amplitude has doubled, its power has increased by a factor of 4. The noise spectral density, however, has only increased by a factor of 2 since v_{n1} and v_{n2} are independent. Consequently, the SNR increases by a factor of 2. However, this increase is not free, as it results in twice the power dissipation.

2.7 The Noise Factor of a Two-Port

As a signal is processed by a two-port network, the signal-to-noise ratio is degraded, as explained using Figure 2.24. Part (a) of the figure shows the source, along with its resistance R_s which adds noise with a spectral density $4kTR_s$. At a particular frequency f, the SNR is given by

$$\mathrm{SNR}_i = \frac{S(f)}{4kTR_s},$$

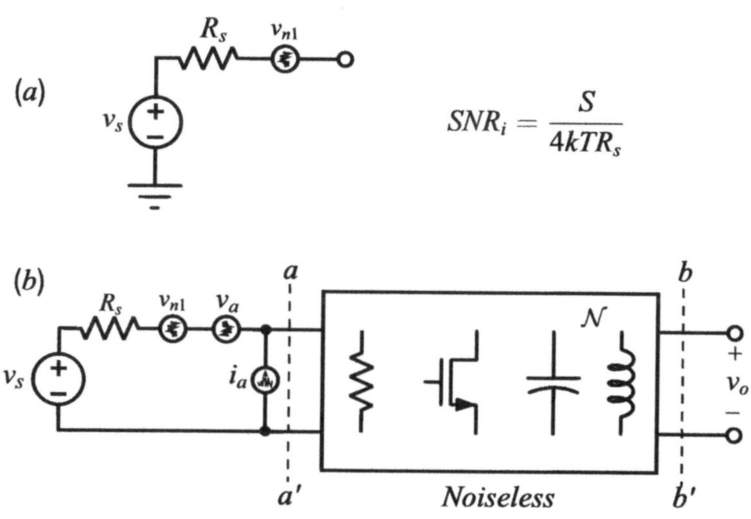

Figure 2.24 (a) The SNR of the source at a frequency f is given by $S(f)/4kTR_s$. (b) The SNR at planes a–a’ and b–b’ is the same, and is lower than that of the source.

where $S(f)$ denotes the spectral density of the input signal. Let us examine what happens when the source is connected to the two-port, as in Figure 2.24b. The noise sources in the network are represented by their equivalent input-referred voltage and current sources. To quantify the degradation of the SNR due to the network, we proceed as follows. We first recognize that the SNR of the voltage at a–a′ and at the output b–b′ is the same, since $\mathcal{N}$ is noiseless (the effect of all internal noise in $\mathcal{N}$ is accounted for by v_a and i_a). The SNR at a–a′ (and b–b′) is obtained by examining the Thevenin equivalent of the circuit to the left of a–a′. The signal power (at a certain frequency f

and a bandwidth Δf) is still $S(f)\Delta f$, while the noise power is $4kTR_s\Delta f + \overline{(v_a + i_aR_s)^2}$ where v_a and i_a denote the noise quantities in a bandwidth Δf. The SNR at the output of $\mathcal{N}$ is thus given by

$$\text{SNR}_o = \frac{S(f)\Delta f}{4kTR_s\Delta f + \overline{(v_a + i_aR_s)^2}}.$$

The degradation of the SNR due to processing by the two-port, called the *noise factor*, is seen to be

$$F = \frac{\text{SNR}_i}{\text{SNR}_o} = 1 + \frac{\overline{(v_a + i_aR_s)^2}}{4kTR_s\Delta f}. \tag{2.46}$$

$F \geq 1$ and is a function of the operating frequency, since v_a and i_a (and their correlation) are frequency dependent. The noise figure (NF) is the noise factor expressed in dB and is given by $\text{NF} = 10\log_{10}(F)$ dB. The noise figure of the two-port would be 0 dB if it were noiseless.

Another way of interpreting the noise factor is the following. We see that (2.46) can be expressed as

$$F = \frac{\{4kTR_s\Delta f + \overline{(v_a + i_aR_s)^2}\}|H(j2\pi f)|^2}{4kTR_s\Delta f|H(j2\pi f)|^2} \equiv \frac{\overline{v_{n1}^2}}{\overline{v_{n2}^2}},$$

where $H(j2\pi f)$ denotes the voltage gain from a–a' to v_o in Figure 2.24b. $\overline{v_{n2}^2}$ is what a *noiseless* version of the network would produce at its output when driven by the real source (same signal transfer as the actual network), while $\overline{v_{n1}^2}$ is what one actually measures with the two-port's internal noise sources turned on. A 3 dB noise figure corresponds to $F = 2$, i.e., the network adds as much output noise as comes from the source after being processed by the network [8].

2.7.1 SNR and the "Sweet Spot" in R_s

Referring to Figure 2.24b, a question that arises is whether there is a "special" value of R_s that maximizes the SNR at the output of the network.[2] To see that there could be an optimum R_s that achieves the best output SNR, we note the following.

a. In Figure 2.24b, the SNR at a–a' is identical to that at the network's output, namely b–b'.

b. To maximize output SNR, therefore, we should minimize $\overline{(v_a + i_aR_s)^2}/4kTR_s\Delta f$.

c. Making R_s very small will emphasize the noise due to the v_a term, whereas making R_s very large makes i_aR_s dominate; thus, an intermediate "sweet spot" in R_s can minimize the total input-referred noise for the chosen frequency. Moreover, the precise optimum depends on the correlation between v_a and i_a in addition to their spectral densities.

[2]It is important to note that achieving the peak SNR *does not* mean that other metrics, like the voltage gain, are maximized.

d. Since v_a and i_a are correlated, the optimum source impedance is in general a frequency-dependent complex quantity.

e. *If* v_a and i_a are uncorrelated and their spectral densities are white, it is straightforward to see that the optimum source resistance is given by $R_{s,opt} = \sqrt{S_{va}/S_{ia}}$.

2.8 Example: Noise-Factor Calculations

This section develops the noise factor (noise figure) through a few simple examples. We keep to low-frequency (memoryless) models so that gains are real numbers; the frequency-dependent case follows by replacing gains with transfer functions.

Example 1: A "Network" That Is Just a Resistor

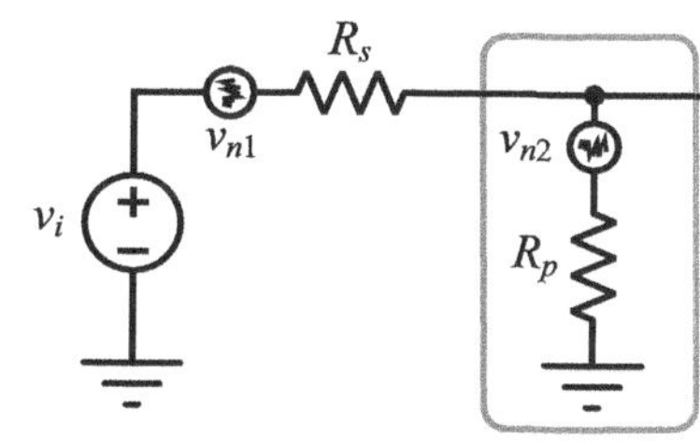

Figure 2.25 Determining the noise factor of a resistive attenuator.

Consider a source resistance R_s driving a network that is merely a shunt resistor R_p (an attenuator), as shown in Figure 2.25. The output is taken across R_p. Let us compute the noise factor

$$F = \frac{\text{(output noise with noisy network)}}{\text{(output noise with noiseless network)}},$$

at a frequency f.

Noisy network

Both R_s and R_p are noisy. The output voltage-noise spectral density is simply the thermal noise of the parallel combination:

$$S_o^{\text{noisy}} = 4kT\frac{R_s R_p}{R_s + R_p}.$$

Noiseless network

We suppress the noise of R_p; only R_s contributes. The transfer function from v_{n1} to the output is $R_p/(R_s + R_p)$, so

$$S_o^{\text{noiseless}} = 4kTR_s\left(\frac{R_p}{R_s + R_p}\right)^2.$$

Noise factor and checks

Taking the ratio and simplifying, we have

$$\boxed{F = 1 + \frac{R_s}{R_p}.}$$

Sanity checks: $R_p \to \infty$ gives $F \to 1$ (no SNR loss). If $R_p = R_s$, then $F = 2$ (NF = 3 dB): the absolute output noise drops by 2×, but the signal is attenuated by 2× in amplitude (4× in power), so SNR degrades by 2×. The noise figure quantifies *SNR loss*, not the absolute noise floor.

What is the optimum R_s that results in the maximum SNR at the output of the network? It is straightforward to see that the equivalent input noise voltage of the "network" is zero. Consequently, $S_{va} = 0$. The optimum R_s, which is $\sqrt{S_{va}/S_{ia}}$, is zero. This, of course, makes intuitive sense.

Example 2: Low-Frequency MOS Common-Source Amplifier

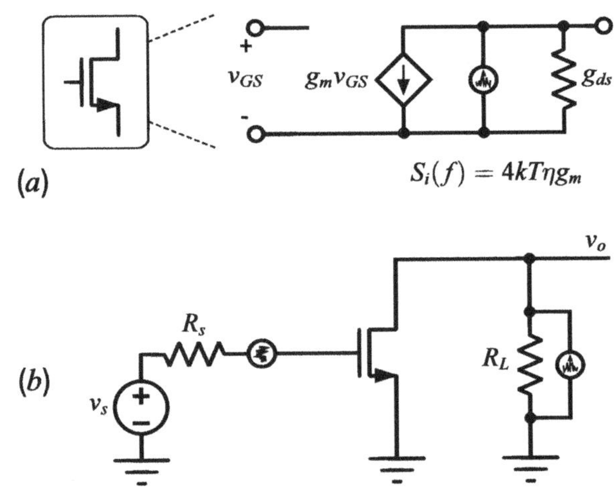

Figure 2.26 (a) MOSFET model (neglecting flicker noise). (b) A common-source amplifier.

Let us determine the noise factor of a common-source amplifier. The model of the MOS transistor at low frequencies is shown in Figure 2.26a. The transistor's noise is modeled by a current source between the drain and source with one-sided density $S_i = (8/3)kT\,\eta\, g_m$. The common-source amplifier is shown in Figure 2.26b. The input source has a resistance R_s, and drives the gate. The load at the drain is a resistor R_L, which contributes a shunt current noise with spectral density $4kT/R_L$. We assume $g_{ds} \to 0$ for simplicity.

Noise due to R_s:

Assuming that the transistor and load are noiseless, only R_s contributes noise. The small-signal gain from v_s to v_o is $-g_m R_L$; hence

$$S_{vo,1} = 4kTR_s\,(g_m R_L)^2.$$

Noise due to the transistor and R_L:

Additional output noise arises from the two shunt *current* sources at the drain, each converted to a voltage by R_L:

$$S_{vo,2} = \left((8/3)kT\eta g_m + 4kTG_L\right)R_L^2.$$

The total output noise spectral density is thus given by

$$S_{vo} = 4kTR_s(g_m R_L)^2 \; + \; \left((8/3)kT\eta g_m + 4kTG_L\right)R_L^2.$$

The noise factor is the ratio of the total output noise to that due to R_s alone and is given by

$$\boxed{F = \frac{S_{vo}}{S_{vo,1}} = 1 + \frac{2\eta}{3g_m R_s} + \frac{1}{g_m^2 R_s R_L}.}$$

From the expression, we see that in a design with a high gain (i.e., $g_m R_L \gg 1$), the contribution of the load resistor is much smaller than that due to the transistor. This makes sense due to the following. A drain-current disturbance must be "divided by g_m^2" to appear as an equivalent gate-voltage disturbance; increasing $g_m R_L$ reduces the *relative* impact of R_L's noise compared to transistor noise.

Measuring the Noise Factor

We saw that the noise factor of a network at a frequency f can be interpreted as the ratio of its output noise spectral density at f to that which would be obtained if the internal noise sources were turned off. While this is a straightforward approach for *analysis/simulation*, it is not possible to "turn off" the internal noise sources in measurements. How then does one measure the noise factor in practice?

The principle [9] behind noise factor measurement is shown in Figure 2.27. The source and $\mathcal{N}$ are at a temperature T_1, and the resulting noise spectral density at the output of the two-port at the frequency f, denoted by $S_1(f)$, is measured as shown in Figure 2.27a. One way to identify the contribution of $\mathcal{N}$ to the output noise is to turn the internal noise sources off; evidently that is not practical. What is possible, however, is to make the source resistance "noisier." A way (injury-prone, perhaps) of doing this is to heat the resistor R_s. A better way is to use an active device whose resistance is R_s when operated at the temperature T_1, but which has

(Continued)

(Continued)

a noise spectral density $4kT_2R_s$ where $T_2 \gg T_1$, as shown in Figure 2.27b. This is the origin of the term "hot" measurement. Note that $\mathcal{N}$ still operates at T_1. The resulting noise spectral density at the output of the two-port is S_2. The ratio S_2/S_1 is called the Y-factor, and is denoted by Y (not to be confused with admittance).

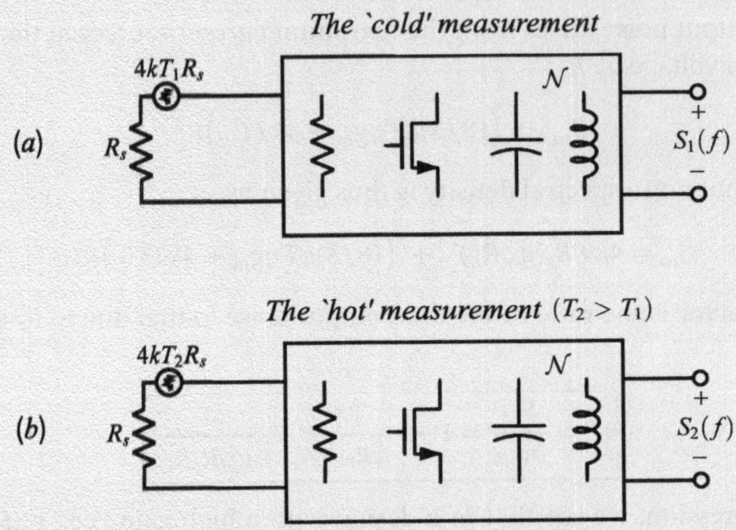

Figure 2.27 The principle behind the hot-cold measurement of noise factor. (a) The "cold" measurement, where the source and $\mathcal{N}$ are at temperature T_1. (b) The "hot" measurement, where $\mathcal{N}$ is at T_1, but the source resistance is at $T_2 > T_1$.

Let N_s and N_{dut} denote the contributions of the source and network to the output noise spectral density in the cold measurement, respectively. We have

$$S_1 = N_s + N_{\text{dut}}$$
$$S_2 = N_s\left(\frac{T_2}{T_1}\right) + N_{\text{dut}}$$

Recall that the noise factor F at T_1 is, by definition, S_1/N_s. Using $Y = S_2/S_1$ with the equations above, we obtain

$$F = \frac{T_2 - T_1}{T_1(Y-1)}. \tag{2.47}$$

As a sanity check, suppose $Y = 1$. This means that $S_2 = S_1$, indicating that making the source noisier has had no effect on the output noise. This can only happen when $N_{\text{dut}}/N_s \to \infty$, meaning that $F \to \infty$.

On the other hand, suppose the measured $Y = T_2/T_1$, which means that $S_2 = S_1(T_2/T_1)$. This indicates that the output noise has increased by the same factor as the source noise. This is only possible if $N_{\text{dut}} = 0$, implying that $F = 1$. This is indeed borne out by (2.47).

Summary

- **Component noise** Thermal noise of a resistor R:

$$S_v(f) = 4kTR \quad \text{and} \quad S_i(f) = 4kT/R.$$

  These are single-sided densities. (Other device noises can be represented with parameterized PSDs as needed.)
- **Noise through LTI networks.** If $Y(s) = H(s)X(s)$, then $S_y(f) = |H(j2\pi f)|^2 S_x(f)$. For multiple independent sources $\{n_l\}$ with transfers $\{H_l(s)\}$,

$$S_o(f) = \sum_l |H_l(j2\pi f)|^2 S_l(f),$$

  and the corresponding mean-square at the output is $\overline{v_o^2} = \int_0^\infty S_o(f)\, df$.
- **Nyquist noise theorem.** For an RLC one-port with impedance $Z(j2\pi f)$,

$$\boxed{S_v(f) = 4kT \,\text{Re}\{Z(j2\pi f)\}.}$$

  This result follows from power balance/reciprocity and unifies the resistor noise formula with that applicable to general RLC networks.
- **Total integrated noise via energy.** In RLC networks, the total mean-square voltage noise across a port can be written as an energy difference associated with a unit-charge experiment (provided the mean-square noise is finite):

$$\overline{v_o^2} = \frac{2kT}{(1\ \text{coulomb})^2}\left(E_0 - E_\infty\right),$$

  which reduces to $\overline{v_o^2} = kT/C$ for the first-order RC network measured across C.
- **Equivalent noise sources.** The effect of multiple noise sources internal to a two port can be modeled by equivalent input-referred noise voltage and current sources.
- **SNR, noise factor, and measurement.** For a two-port, the noise factor F quantifies the degradation in the SNR due to the network. The noise figure is the noise factor expressed in dB, i.e., $\text{NF} = 10 \log_{10}(F)$ dB. The noise factor can be measured using what is called the "hot-cold" method.

Problems

The opamps in the problems that follow are ideal. However, unless otherwise mentioned, they are associated with a noise voltage at their inputs denoted by v_a (the corresponding spectral density is denoted by $S_a(f)$) (Figure 2.28).

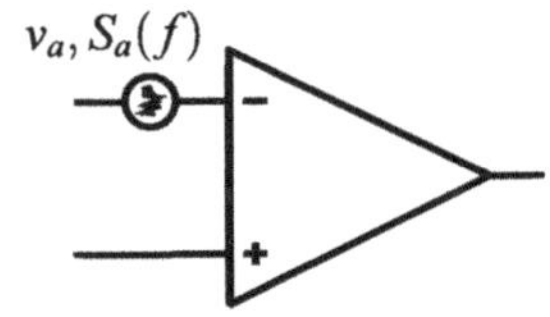

Figure 2.28 Noise model for an opamp, which is otherwise ideal.

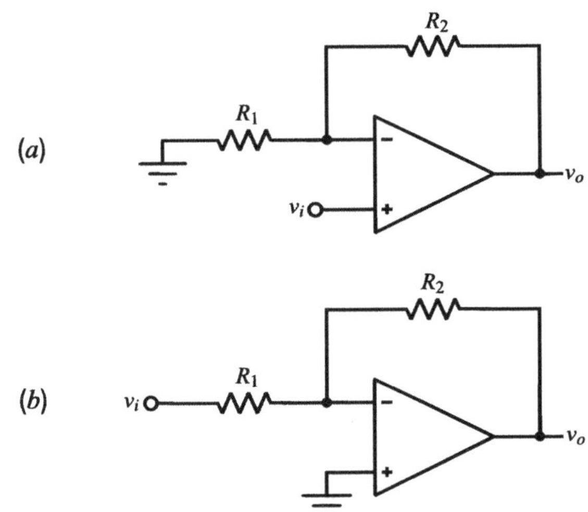

Figure 2.29

Problem 1. *Determine the output noise voltage spectral density and that of the equivalent input-referred noise voltage and current for the two amplifiers in Figure 2.29.*

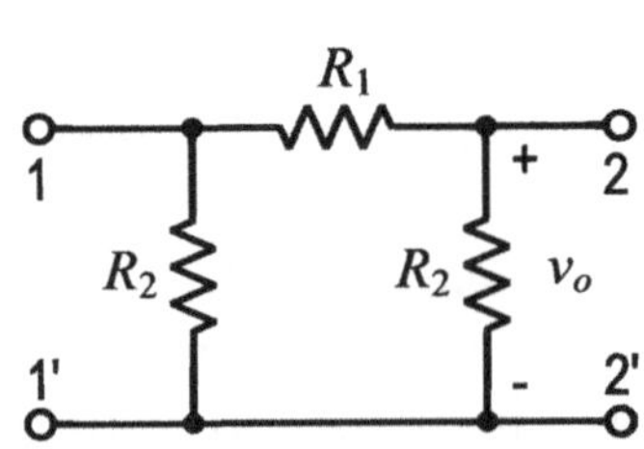

Figure 2.30

Problem 2. *Determine the spectral densities of the equivalent input-referred equivalent noise voltage and current of the passive two-port network in Figure 2.30.*

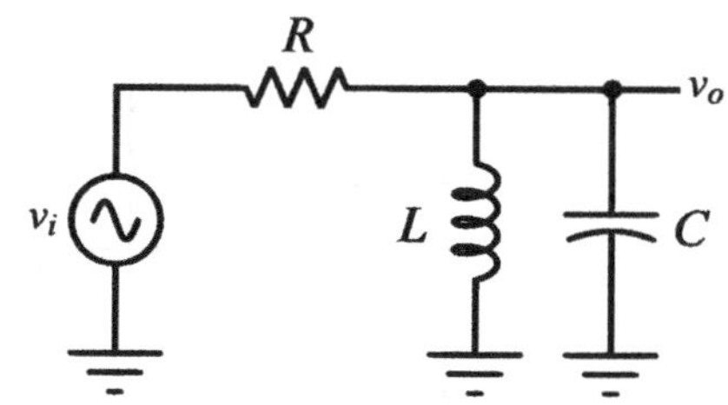

Figure 2.31

Problem 3. *The network of Figure 2.31 shows a second-order bandpass filter whose transfer function is of the form*

$$H(s) = \frac{G\frac{s}{\omega_o}}{1 + \frac{s}{\omega_0 Q} + \frac{s^2}{\omega_0^2}} \tag{2.48}$$

Determine G, ω_0, Q *in terms of circuit parameters. Determine the output noise-voltage spectral density and the total integrated output noise.*

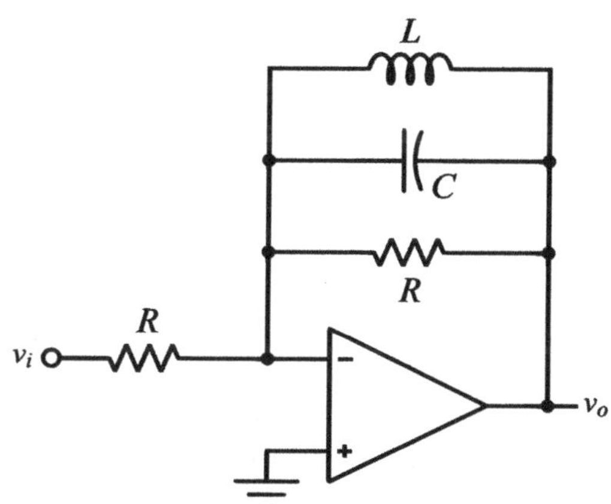

Figure 2.32

Problem 4. *Repeat the previous problem for the circuit shown in Figure 2.32. Assume that the opamp is ideal and noiseless.*

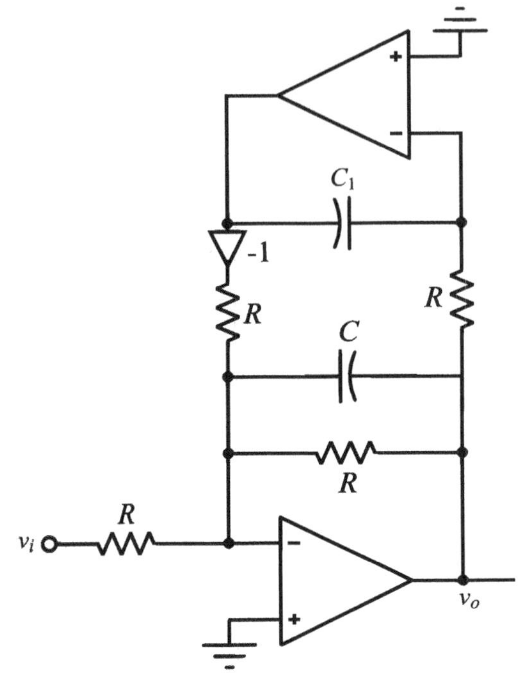

Figure 2.33

Problem 5. *Physical inductors tend to be bulky and can only be integrated when they are very small (a few tens of nH). This led to the invention of electronic ways of emulating the behavior of inductors, and one popular approach is shown in Figure 2.33.*

Determine how C_1 must be chosen so that the circuit achieves the same transfer function as that realized by the circuit of Figure 2.32. Determine the mean-square noise at the output, and compare it with that of the previous problem. What do you conclude? (Assume that the opamps are noiseless.)

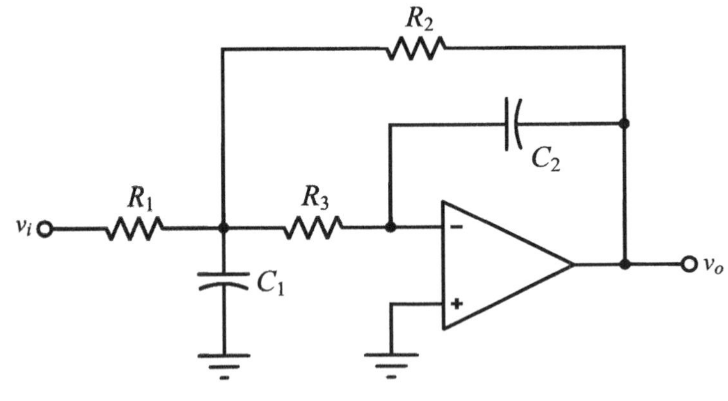

Figure 2.34

Problem 6. *Figure 2.34 shows a low-pass filter. Determine the transfer function and express it in standard form, which is*

$$H(s) = \frac{A_o}{1 + \frac{s}{\omega_o Q} + \frac{s^2}{\omega_o^2}}. \tag{2.49}$$

How are ω_o, Q, and A_o related to the component values? Determine the output noise spectral density of the lowpass filter. The opamp is noisy, as mentioned at the beginning of this problem set.

Problem 7. *Consider a network consisting of resistors, capacitors, and voltage-controlled current sources. The transconductance of the lth VCCS is denoted by $g_{m,l}$. Every resistor is accompanied by its noise source, as usual, and every transconductance is accompanied by a noise current in parallel with the output, with a spectral density that is given by $4kTg_{m,l}$.*

Denote the output noise voltage spectral density of the network by $S_v(f)$. All the resistors in the network are doubled, and transconductors are halved. Determine the output voltage spectral density of the network in terms of $S_v(f)$.

Problem 8. *Consider a network consisting of resistors, capacitors, and voltage-controlled current sources. The transconductance of the ith VCCS is denoted by $g_{m,i}$. Every resistor is accompanied by its noise source, as usual, and every transconductance is accompanied by a noise current in parallel with the output, with a spectral density that is given by $4kTg_{m,i}$.*

Denote the output noise voltage spectral density of the network by $S_{v,out}(f)$. Now, all the capacitors in the network are doubled. Determine the output voltage spectral density of the network now.

Problem 9. *Consider again the bandpass filter of Figure 2.33. The opamps are ideal and noiseless, and so is the amplifier of gain -1. The only noisy elements are the resistors. Assume $Q \gg 1$.*

The opamps have a peak signal swing of ± 1 V at their outputs. The input $|v_i| < 1$ V. It is desired to maximize the dynamic range of the filter. For the purposes of this problem, we define the dynamic range as the maximum possible SNR at the filter output. The peak signal amplitude at the filter output is restricted to 1 V, but can be lower if one of the opamps saturates before the other. The noise power at the filter output is its mean square noise. The dynamic range, therefore, is given by

$$\text{Dynamic range} = \left(\frac{\text{Maximum output signal power}}{\text{Mean-square output noise}} \right). \tag{2.50}$$

The aim of this problem is to illustrate the fundamental difficulty of realizing a high-Q active bandpass filter with a high dynamic range. Choose component values so that all the following constraints are satisfied. Assume that the Q is high, so that the peaks of the low-pass and bandpass outputs occur at the same frequency.

- *Center frequency f_0 is 1 MHz.*
- *$Q = 20$*

- *Gain at f_0 of 1.*
- $C_1 + C = 100\,pF$.

Since there are six component values to be determined but only four constraints, many component choices can realize the desired transfer function.

a. *Choose component values so that the output dynamic range is maximized. What is the maximum dynamic range in dB? Verify your noise calculations using SPICE.*

b. *Repeat part (a) if $C_1 + C = 200\,pF$.*

c. *Repeat part (a) if $Q = 40$, but all other constraints remained as in part (a).*

d. *Repeat part (a) if $f_0 = 2\,MHz$, but all other constraints remained as in part (a).*

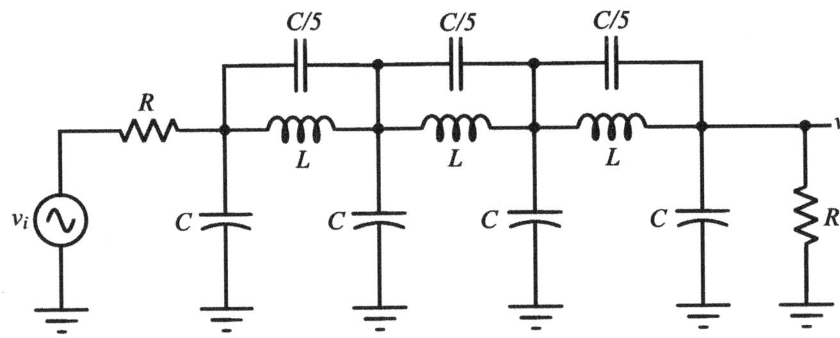

Figure 2.35

Problem 10. *Determine the total integrated output noise of the low-pass filter circuit shown in Figure 2.35. Repeat if the output terminating resistor is removed (i.e., made an open circuit).*

Problem 11. *Consider a one-port consisting of resistors, capacitors, and inductors. We saw that the noise spectral density of the open-circuit port voltage is given by $4kT\,Re\{Z(j2\pi f)\}$, where $Z(j2\pi f)$ denotes the driving point impedance of the one port. Show that the noise spectral density of the short-circuit port current is given by $4kT\,Re\{Y(j2\pi f)\}$, where $Y(j2\pi f)$ denotes the port's driving point admittance.*

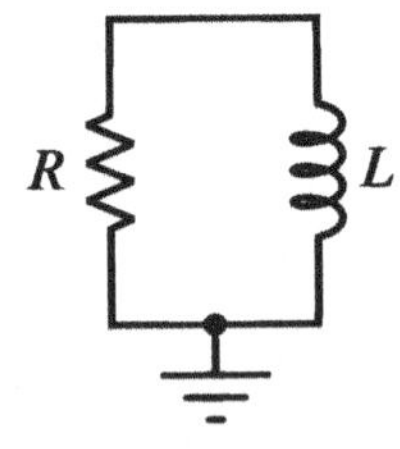

Figure 2.36

Problem 12. *Determine the noise spectral density and mean-square noise of the current through the inductor (Figure 2.36).*

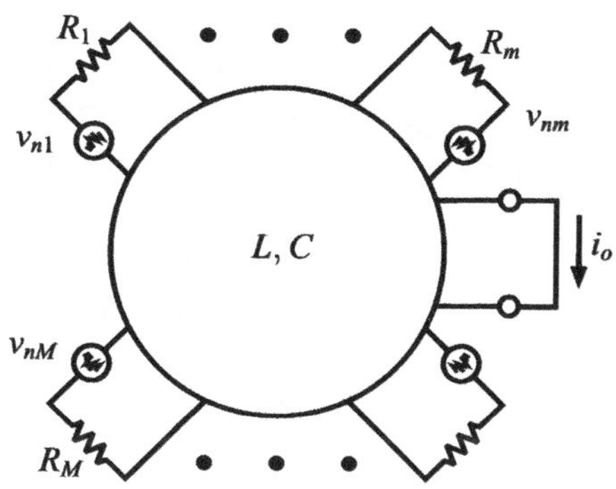

Figure 2.37

Problem 13. *The one-port network of Figure 2.37 consists of inductors, capacitors, and resistors. Determine the mean-square value of the short-circuit current* $\overline{i_o^2}$.

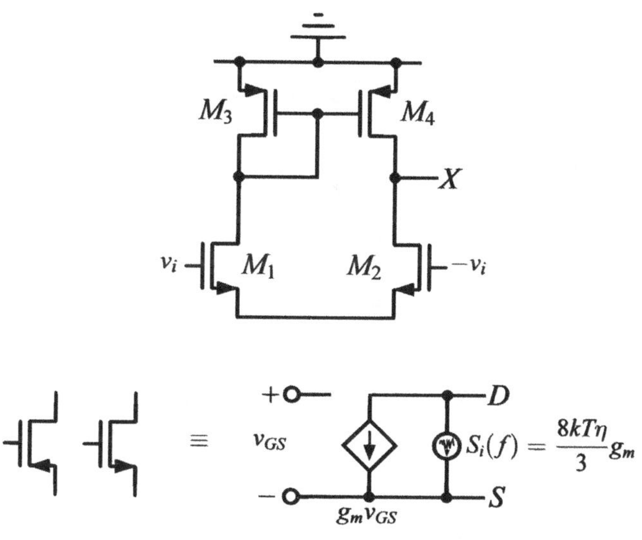

Figure 2.38

Problem 14. *Figure 2.38 shows the incremental diagram of a single-stage CMOS operational transconductance amplifier. The output is node X. Each transistor is represented by the model given. Denote the transconductance of the* l^{th} *transistor by* g_{ml}. *Determine the noise spectral density of the output short-circuit current.*

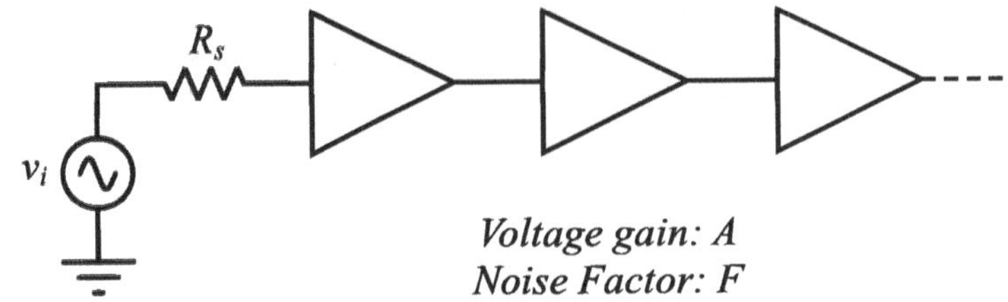

Figure 2.39

Problem 15. *Figure 2.39 shows an infinite cascade of identical amplifiers. The amplifier has a voltage gain denoted by A and a noise factor F when driven by a source with resistance R_s. Furthermore, it has zero output impedance and an infinite input impedance. Determine the noise factor of the cascade. This is called the noise measure of the amplifier.*

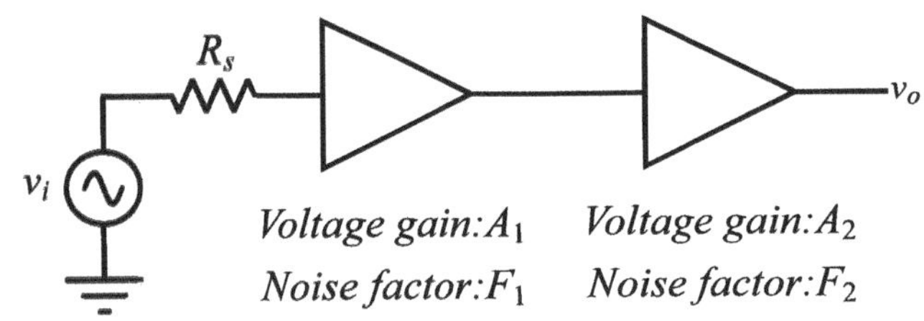

Figure 2.40

Problem 16. *Figure 2.40 shows a cascade of two amplifiers. The voltage gains are denoted by A_1 and A_2. When driven by a source with resistance R_s, their noise factors are F_1 and F_2.*

a. *Determine the noise factor of the cascade.*

b. *It is also possible to achieve the same gain by interchanging the positions of A_1 and A_2. Determine which of these choices ($A_1 - A_2$ or $A_2 - A_1$) will result in a lower noise factor for the cascade?*

c. *Suppose there are three amplifiers with gain/noise factors A_1/F_1, A_2/F_2 and A_3/F_3. Any order will result in a gain $A_1A_2A_3$. How will you order them in a cascade to achieve the lowest noise factor?*

References

1 A. V. der Ziel, *Noise in Solid State Devices and Circuits*. Wiley, 1986.

2 W. B. Davenport and W. L. Root, *An Introduction to the Theory of Random Signals and Noise*. McGraw-Hill, 1958.

3 H. Nyquist, "Thermal agitation of electric charge in conductors," *Physical Review*, vol. 32, no. 1, pp. 110–113, 1928.

4 J. B. Johnson, "Thermal agitation of electricity in conductors," *Physical Review*, vol. 32, no. 1, pp. 97–109, 1928.

5 P. R. Gray, P. J. Hurst, S. H. Lewis, and R. G. Meyer, *Analysis and Design of Analog Integrated Circuits*, 5th. Wiley, 2009.

6 B. Razavi, *Design of Analog CMOS Integrated Circuits*, 2nd. McGraw-Hill, 2017.

7 S. Pavan, "An alternative approach to Bode's noise theorem," *IEEE Transactions on Circuits and Systems II: Express Briefs*, vol. 66, no. 5, pp. 738–742, 2019.

8 T. H. Lee, *The Design of CMOS Radio-Frequency Integrated Circuits*, 2nd. Cambridge University Press, 2004.

9 *Fundamentals of RF and microwave noise figure measurements*, Application Note 57-1 (formerly Agilent/HP), Keysight Technologies, 2010. [Online]. Available: https://www.keysight.com/.

3

Weak Nonlinearities in Circuits

In Chapter 2, we discussed noise in circuits. Noise places a lower limit on the signal strength at the input of a network for the output to be useful. In this chapter, we delve into what happens when the signal strength is too large. One might wonder what the problem is with large signals; the signal-to-noise ratio will only increase, which is desirable. The problem is that in reality, most circuit components exhibit some degree of nonlinearity. As a consequence, networks created using such components will also be nonlinear, albeit often minimized through design efforts. It turns out that nonlinearities cause undesirable artifacts at the output of a system. Furthermore, the strength of these artifacts increases more than linearly with the amplitude of the input. An analogy is the following: suppose one was listening to a friend. If (s)he whispers, it is a problem since the signal-to-noise ratio is too small. On the other hand, it does not help if the friend screams at the top of his/her voice either, as it hurts the ears. The "best" volume for the friend to speak with is one that is not so small, that it is not confused with noise and not so loud as to be annoying.

In electronic circuits, the goal is typically to limit the maximum input signal amplitude to a level where the artifacts due to nonlinearity begin to become annoying. In a digital communication link, for instance, "annoyance" means an increased bit error rate (BER). At such a level, the circuit operates in a weakly nonlinear fashion. The aim of this chapter is to develop techniques to analyze the effect of weak nonlinearities in circuits, primarily focusing on circuits presumed to be linear and time-invariant. To illustrate, consider a simple example: an amplifier. Ideally, we desire this amplifier to exhibit linearity, where the output voltage (v_o) is directly proportional to the input voltage (v_i) by a factor represented as the gain a_1:

$$v_o = a_1 v_i. \tag{3.1}$$

In practice, the output voltage (v_o) will deviate from the ideal linear relationship ($v_o = a_1 v_i$) due to the presence of nonlinear terms. In most cases, the input–output relationship can be modeled as a polynomial over the useful range of inputs according to

$$v_o = a_1 v_i + a_2 v_i^2 + a_3 v_i^3 + \ldots \tag{3.2}$$

where a_2, a_3, etc., are coefficients representing the strength of nonlinear effects. The term $a_1 v_i$ represents the linear contribution, while $a_2 v_i^2$, $a_3 v_i^3$ and subsequent higher-order terms constitute the nonlinear components. When we refer to *weakly nonlinear* behavior, it implies that these nonlinear terms ($a_2 v_i^2$, $a_3 v_i^3$, etc.) are significantly smaller than the linear term ($a_1 v_i$). Mathematically, this can be expressed as:

$$|a_2 v_i^2|,\ |a_3 v_i^3|,\ \ldots \ll |a_1 v_i|. \tag{3.3}$$

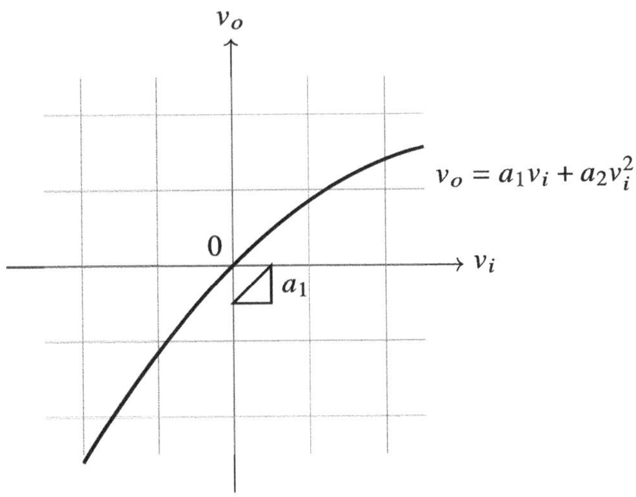

Figure 3.1 Transfer curve of a weakly nonlinear memoryless amplifier with second-order nonlinearity.

In practical circuit design, unless intentional nonlinear behavior is desired, efforts are made to ensure that the nonlinear terms are much weaker compared to the linear term. While deliberately nonlinear circuits exist, where nonlinearity is essential for their intended functionality, the focus of this chapter is on circuits expected to exhibit perfectly linear behavior.

Consider an amplifier whose output voltage (v_o) is the sum of a linear term ($a_1 v_i$) and a quadratic term ($a_2 v_i^2$), with higher-order terms being zero:

$$v_o = a_1 v_i + a_2 v_i^2. \tag{3.4}$$

The transfer curve is shown in Figure 3.1. For very small inputs, the quadratic term is negligible, and the incremental gain $dv_o/dv_i = a_1$. As the input increases, the slope of the transfer curve changes. When $v_i = A\cos(2\pi f t)$ the output is given by

$$v_o = \frac{a_2 A^2}{2} + a_1 A\cos(2\pi f t) + \frac{a_2 A^2}{2}\cos(4\pi f t). \tag{3.5}$$

A DC offset emerges even though the input signal is a sinusoid with a zero average value. This offset arises because, regardless of whether the input is positive or negative, the output is consistently smaller (or larger) than the linear response would suggest. More importantly, a second harmonic distortion is seen, and is quantified by HD_2, which is the ratio of the amplitude of the second harmonic to that of the fundamental [1]:

$$HD_2 = \frac{\left(\frac{|a_2A^2|}{2}\right)}{|a_1A|} = \frac{|a_2A|}{2|a_1|}. \tag{3.6}$$

It is often expressed in decibels (dB) as

$$HD_2(\text{dB}) = 20\log_{10}\left(\frac{|a_2A|}{2|a_1|}\right). \tag{3.7}$$

The strength of the second harmonic $(a_2A^2/2)$ is proportional to the square of the input amplitude. For a 1 dB increase in input amplitude, there is a corresponding 2 dB increase in the second harmonic amplitude, thereby causing HD_2 to increase by 1 dB.

As mentioned earlier, memoryless weakly nonlinear systems can most often be modeled using a polynomial input–output relationship over *the range of useful signals*. In practice, how would we know if the system is operating in a regime where it can be considered weakly nonlinear? The 1 dB/dB increase in HD_2 is often a good test to answer this question.

3.1 Second-Order Intermodulation Distortion

The reader might wonder what all the fuss about distortion is about. Why cannot the effect of the second harmonic simply be filtered? The basic idea is shown in Figure 3.2. Suppose that an ideally linear buffer (gain of 1) is accompanied by second-order distortion. Since we know that an input tone at f_1 will cause an output tone at $2f_1$, we propose to eliminate it by using a low-pass filter as shown in the figure. The filter's bandwidth B is chosen so that the second harmonic lies outside the filter's passband. There are (at least) two problems with this approach (even without considering the complexities involved in realizing the filter). First, what if the frequency was so low that the second harmonic lies within the filter's passband? Next, while a sine wave is used to characterize a system, it is very unlikely that real-world signals are sinusoidal. Consider, for instance, the response of our nonlinear system excited by the sum of two sinusoids:

$$v_i = A\cos(2\pi f_1 t) + A\cos(2\pi f_2 t). \tag{3.8}$$

Using this in (3.4), we obtain

$$\begin{aligned}
v_o = {} & a_1A\{\cos(2\pi f_1 t) + \cos(2\pi f_2 t)\} && \text{(linear output)} \\
& + a_2A^2 && \text{(dc offset)} \\
& + a_2\frac{A^2}{2}\{\cos(4\pi f_1 t) + \cos(4\pi f_2 t)\} && \text{(harmonic distortion)} \\
& + a_2A^2\{\cos(2\pi(f_1 + f_2)t) + \cos(2\pi(f_1 - f_2)t)\} && \text{(intermodulation distortion)}
\end{aligned}$$

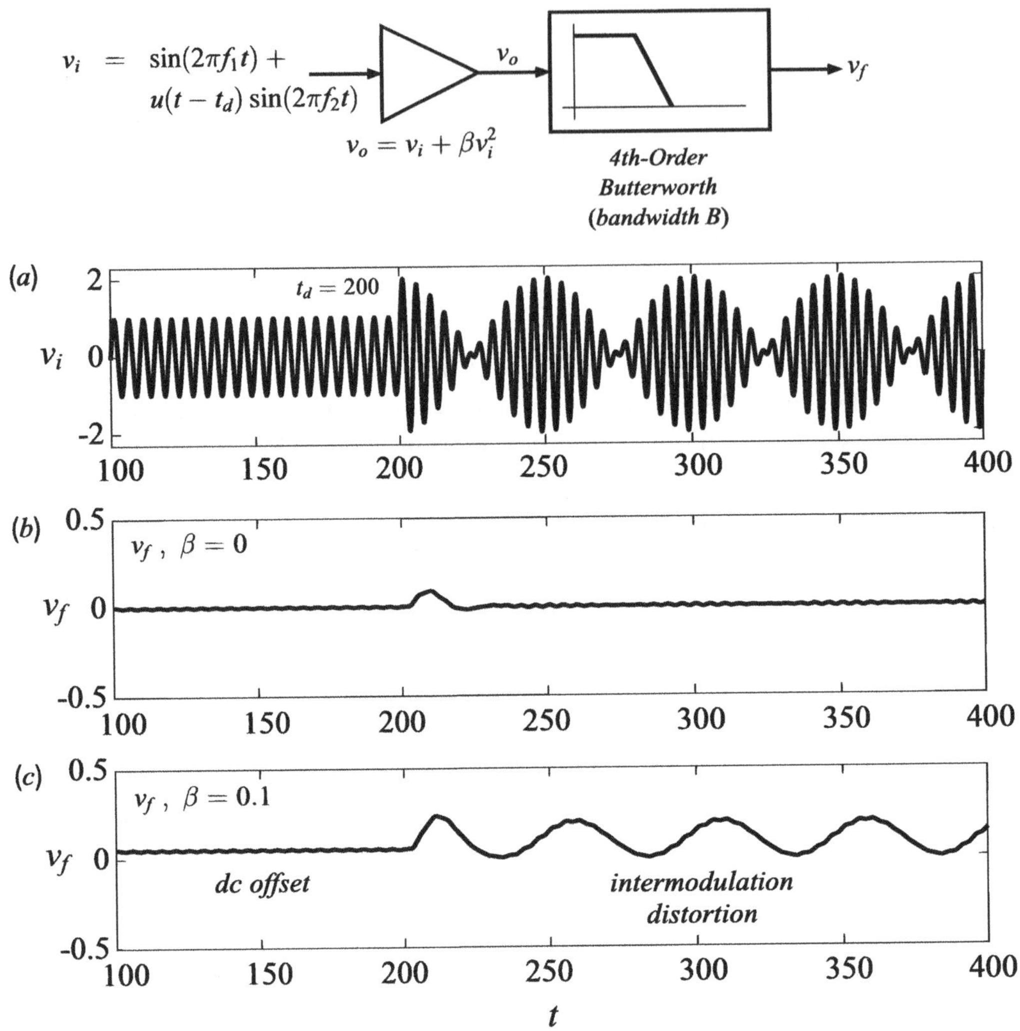

Figure 3.2 Weakly nonlinear buffer followed by a sharp lowpass filter with bandwidth B, excited by a two-tone input. Both f_1 and f_2 are higher than the filter's bandwidth, but $(f_2 - f_1) < B$. (a) The input, where the second tone is turned on after a delay t_d. (b) Filter output for a linear buffer. (c) Filter output when the buffer is weakly nonlinear.

This expression shows that the output signal contains additional frequency components beyond the original input frequencies f_1 and f_2 and their harmonics. Specifically, it includes components at $(f_1 + f_2)$ and $(f_1 - f_2)$. These additional frequency components are known as *intermodulation products* and are a consequence of the nonlinear behavior of the system. When f_1 and f_2 are close together, the difference frequency $(f_1 - f_2)$ can fall within the desired signal bandwidth, making it more problematic than the sum frequency $(f_1 + f_2)$.

The second-order intermodulation distortion (IM_2) quantifies the ratio of the amplitude of the intermodulation tones to that of the fundamental tones in the output signal and is given in decibels by

$$\mathrm{IM_2(dB)} = 20\log\left(\frac{|a_2|A}{|a_1|}\right). \tag{3.9}$$

Like HD_2, increasing the input by 1 dB causes a 1 dB increase in IM_2.

It is instructive to look at time-domain waveforms to "experience" the effects of intermodulation distortion. Consider the signal chain of Figure 3.2. The buffer is excited by two sinusoidal signals of equal amplitude and frequencies f_1 and f_2. The sine wave at f_2 is turned on at a time t_d. The bandwidth of the filter that follows the buffer is B. f_1 and f_2 are larger than B, but $(f_2 - f_1) < B$. Figure 3.2b shows the filter output when the buffer is linear (i.e., $\beta = 0$). v_f is small for $t < t_d$ since $f_1 > B$. When the second sine wave turns on at $t = t_d$, there is a transient after which v_f is the sum of two small-amplitude sine waves, as f_1 and f_2 are both greater than B.

The filter output for $\beta \neq 0$ is shown in Figure 3.2c. For $t < t_d$, where the excitation is a single tone, we see that v_f consists of a DC offset with a small-amplitude signal superposed on it. This corresponds to the tones at f_1 and $2f_1$ attenuated by the filter. For $t > t_d$, we see that v_f is predominantly a large low-frequency sinusoid (along with a dc offset). This is the intermodulation tone at the frequency $(f_2 - f_1)$, which cannot be eliminated by the filter. We thus see that second-order nonlinearity can cause two large out-of-band signals to yield an in-band component. A practical scenario where this could occur is the baseband chain of an RF receiver.

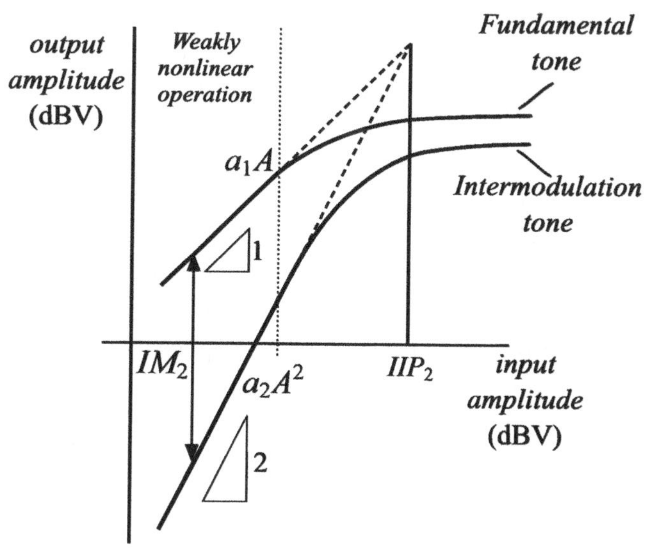

Figure 3.3 Amplitudes of the output fundamental and intermodulation tones as a function of input amplitude.

Suppose we plotted the amplitude of the output tone at f_1 versus that on the input tone (both expressed in dBV[1]). As shown in Figure 3.3, we see a straight line with a slope of 1 dB/dB when the amplifier is operating in its weakly nonlinear region. The amplitude of the intermodulation tones (which is proportional to A^2) on the other hand, is a line of slope 2 dB/dB. *If the amplifier remained* weakly nonlinear for *all input amplitudes,* the two lines when extrapolated would meet at a point. At this input amplitude, given by $|a_1/a_2|$, the strength of the intermodulation tones (at frequencies $(f_2 \pm f_1)$) would be equal to that of the input. The amplitude of the input that results in equal amplitude tones at the input and intermodulation frequencies is denoted by IIP_2, short for input-referred intercept point for second-order intermodulation distortion. It is important to recognize that in practice, IIP_2 is a *fictitious* point, since weakly nonlinear behavior is only valid over a range of input amplitudes; the amplifier will saturate after that. IIP_2 in dBV is given by

$$\text{IIP}_2\,(\text{dBV}) = 20\log\left(\frac{|a_1|}{|a_2|}\right). \tag{3.10}$$

A larger value of IIP_2 indicates a more linear amplifier. Given the input amplitude A and the amplifier's IIP_2, and assuming weakly nonlinear operation, it is straightforward to see that when all quantities are expressed in dB or dBV,

$$\text{IM}_2 = A - \text{IIP}_2. \tag{3.11}$$

Second-order nonlinearity can be characterized either by HD_2 or by the IIP_2. While both describe the same nonlinearity, IIP_2 is often preferred for analysis and comparison. Why? HD_2 requires *two numbers* to specify distortion: the distortion level (e.g., in percent) and the input amplitude at which it is measured. In contrast, IIP_2 is a single figure of merit that characterizes second-order nonlinearity.

IIP_2 models the entire weakly nonlinear operating region where the fundamental term (a_1A) and the second-order term (a_2A^2), when expressed in dBV, follow straight-line relationships versus input amplitude. In this region, extrapolation of these lines to their intersection provides an amplitude-independent description of second-order distortion. It should be noted, however, that the IIP_2 model breaks down at large input amplitudes, where higher-order nonlinearities and saturation effects become significant. In this large-signal regime, HD_2 can still be specified at a particular operating point, whereas IIP_2 no longer accurately represents the system behavior.

---

[1] dBV is the voltage, expressed in dB, normalized to 1 V. If the amplitude is A volts, it is $20\log(A)$ in dBV.

IIP_2 of a MOS Common-Source Amplifier

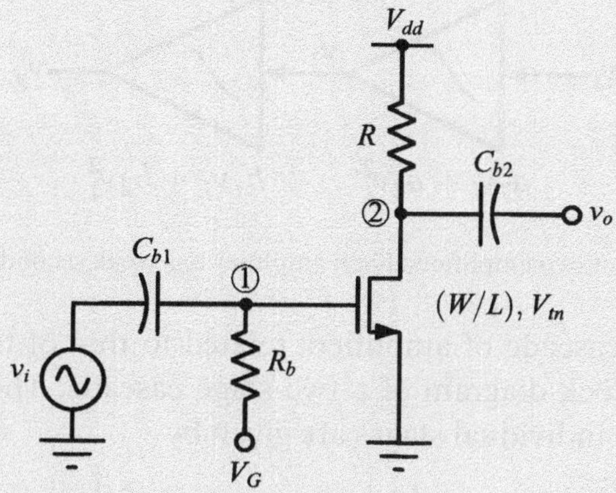

Figure 3.4 A common-source amplifier.

A common-source amplifier is shown in Figure 3.4. C_{b1} and C_{b2} are assumed to be infinite so that they act as level-shifting batteries. The transistor, which obeys the square-law model, is operating in saturation. Channel-length modulation is neglected. What is the IIP_2 of this amplifier?

The voltage at the drain, namely V_2, is given by

$$V_2 = V_{dd} - \frac{1}{2}\mu_n C_{ox}\frac{W}{L}(V_G + v_i - V_{tn})^2 R.$$

The voltage drop across C_{b2} is

$$V_{b2} = V_{dd} - \frac{1}{2}\mu_n C_{ox}\frac{W}{L}(V_G - V_{tn})^2 R.$$

Consequently, $v_o = V_2 - V_{b2}$ is seen to be

$$v_o = -\mu_n C_{ox}\frac{W}{L}(V_G - V_{tn})Rv_i - \frac{1}{2}\mu_n C_{ox}\frac{W}{L}Rv_i^2.$$

In our notation, $|a_1| = \mu_n C_{ox}\frac{W}{L}(V_G - V_{tn})R$, and $|a_2| = \frac{1}{2}\mu_n C_{ox}\frac{W}{L}R$. The IIP_2, therefore, is

$$IIP_2 = 2(V_G - V_{tn}).$$

For instance, a gate-overdrive voltage of 0.5 V results in an IIP_2 of 0 dBV.

3.2 IIP_2 of Cascaded Stages

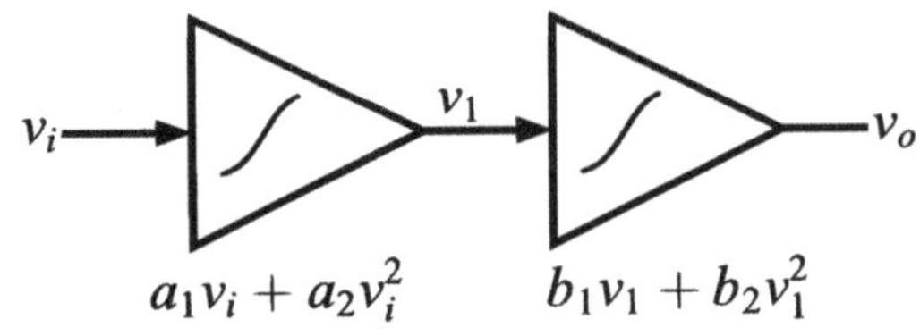

Figure 3.5 Cascade of two amplifiers. Each amplifier has weak second-order nonlinearity.

How is the IIP_2 of a cascade of amplifiers related to that of the individual stages? Figure 3.5 shows a block diagram of a two-stage cascade. The second-order input intercept points of the individual stages are given by

$$\mathrm{IIP}_{2,a} = \frac{|a_1|}{|a_2|}, \qquad \mathrm{IIP}_{2,b} = \frac{|b_1|}{|b_2|}.$$

The output of the cascade, namely v_o, can be expressed in terms of v_i as

$$v_o = b_1(a_1v_i + a_2v_i^2) + b_2(a_1v_i + a_2v_i^2)^2 = a_1b_1v_i + (a_2b_1 + a_1^2b_2)v_i^2 + \cdots. \tag{3.12}$$

Consequently, the IIP_2 of the cascade is seen to be

$$\mathrm{IIP}_{2,ab} = \frac{|a_1b_1|}{|a_2b_1 + a_1^2b_2|}. \tag{3.13}$$

If the signs of a_2b_1 and $a_1^2b_2$ are identical, $|a_2b_1 + a_1^2b_2|$ in the expression above is identical to $|a_2b_1| + |a_1^2b_2|$, allowing us to write $\mathrm{IIP}_{2,ab}$ as

$$\frac{1}{\mathrm{IIP}_{2,ab}} = \frac{1}{\mathrm{IIP}_{2,a}} + \frac{|a_1|}{\mathrm{IIP}_{2,b}}. \tag{3.14}$$

3.3 Differential Operation

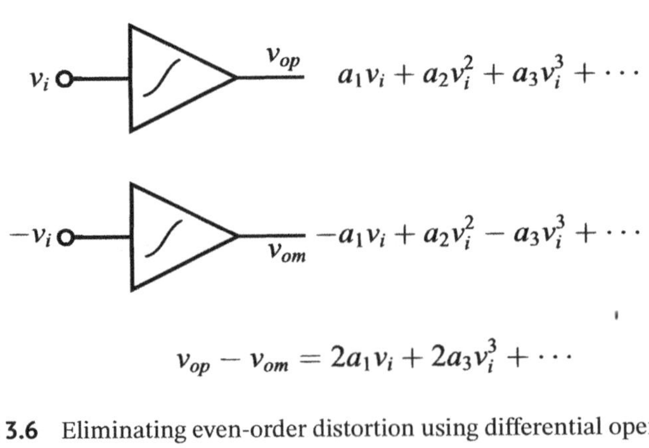

Figure 3.6 Eliminating even-order distortion using differential operation.

We have seen that second-order distortion can be problematic in practical systems. Fortunately, a simple "trick" can eliminate (or greatly reduce) second- and even-order distortion. The idea, called *differential operation,* is illustrated in Figure 3.6. Suppose the input v_i and its inverted version $-v_i$ are available. One amplifier is excited by v_i to result in an output denoted by v_{op}. Another *identical* amplifier is excited by $-v_i$, and its output is denoted by v_{om}. We see that

$$v_{op} = a_1 v_i + a_2 v_i^2 + a_3 v_i^3 + \cdots$$
$$v_{om} = -a_1 v_i + a_2 v_i^2 - a_3 v_i^3 + \cdots$$

The difference between v_{op} and v_{om} is given by

$$v_{op} - v_{om} = 2a_1 v_i + 2a_3 v_i^3 + \cdots . \tag{3.15}$$

It is evident that all even terms are cancelled due to differential operation. Furthermore, any error/disturbance which is common to both amplifier outputs is cancelled when the difference is processed. This is a very important and useful attribute of differential operation. It is thus no wonder that this technique finds widespread use. In practice, however, mismatch between the amplifiers will result in small, residual even-order terms in the transfer characteristic.

To summarize, differential operation either eliminates even-order distortion, or renders it so small as to be negligible. As a consequence, third-order nonlinearity is the dominant source of distortion in practical differential circuits. We study this next.

3.4 Third-Order Nonlinearity

Consider an amplifier whose output voltage (v_o) is the sum of a linear term ($a_1 v_i$) and a cubic term ($a_3 v_i^3$), with higher-order terms being zero.

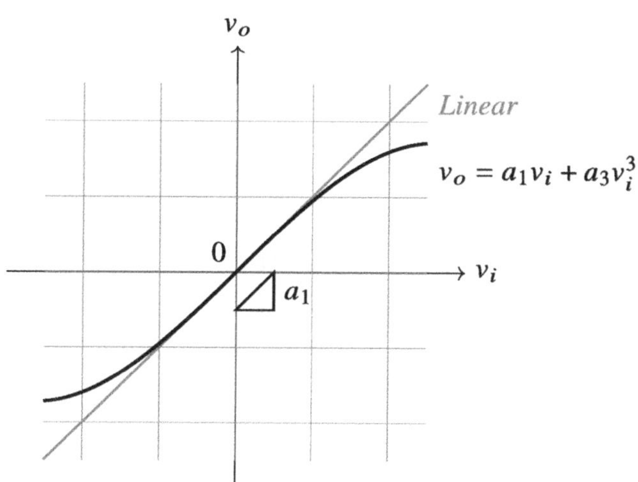

Figure 3.7 Transfer curve of a weakly nonlinear memoryless amplifier with third-order nonlinearity.

$$v_o = a_1 v_i + a_3 v_i^3. \tag{3.16}$$

The transfer curve is shown in Figure 3.7. For very small inputs, the cubic term is negligible, and the incremental gain $dv_o/dv_i = a_1$. When $v_i = A\cos(2\pi ft)$ the output is given by

$$v_o = \left(a_1 A + \frac{3}{4}a_3 A^3\right)\cos(2\pi ft) + \frac{a_3 A^3}{4}\cos(6\pi ft). \tag{3.17}$$

The amplitude of the fundamental at the output is seen to be $(a_1 A + 3a_3 A^3/4)$. In most practical amplifiers, the signs of a_3 and a_1 are opposite. Consequently, the fundamental tone at the output is smaller than what would be expected from a linear amplifier of gain a_1. This is called gain compression, and the amplifier characteristic is termed "compressive." The amplitude at which the output is 1 dB smaller than that of a linear amplifier with gain a_1 is termed the 1 dB compression point. Denoting this by $A_{1\,dB}$, we have

$$10^{-1/20} a_1 A_{1\,\mathrm{dB}} = a_1 A_{1\,\mathrm{dB}} + \frac{3}{4} a_3 A_{1\,\mathrm{dB}}^3. \tag{3.18}$$

which results in

$$A_{1\,\mathrm{dB}} \approx 0.33\sqrt{\frac{4}{3}\frac{|a_1|}{|a_3|}}. \tag{3.19}$$

As expected, third-harmonic distortion also appears at the amplifier's output. It is quantified by HD_3, which is the ratio of the amplitude of the third harmonic to that of the fundamental. Expressed in dB, it is approximately (after neglecting the small $(3/4)a_3A^3$ of the fundamental)

$$HD_3(\mathrm{dB}) \approx 20\log_{10}\left(\frac{|a_3|A^2}{4|a_1|}\right). \tag{3.20}$$

The strength of the third harmonic (a_3A^3) is proportional to the cube of the input amplitude. For a 1 dB increase in the input, the third harmonic increases by 3 dB resulting in a 2 dB increase in HD_3. A 2 dB/dB increase in HD_3 is often a good test to determine if the amplifier is operating in the weakly nonlinear regime.

3.4.1 Third-Order Intermodulation Distortion

Intermodulation distortion occurs when a third-order nonlinearity is subjected to a two-tone signal. Consider, for instance, the response of our nonlinear system excited by the sum of two sinusoids:

$$v_i = A\cos(2\pi f_1 t) + A\cos(2\pi f_2 t). \tag{3.21}$$

Using this in $v_o = a_1 v_i + a_3 v_i^3$, we obtain

$$v_o = \left(a_1 A + \frac{9}{4} a_3 A^3\right)\{\cos(2\pi f_1 t) + \cos(2\pi f_2 t)\} \quad \text{(compressed output)}$$
$$+ a_3 \frac{A^3}{4}\{\cos(6\pi f_1 t) + \cos(6\pi f_2 t)\} \quad \text{(harmonic distortion)}$$
$$+ \frac{3}{4} a_3 A^3 \{\cos(2\pi(2f_1 - f_2)t) + \cos(2\pi(2f_2 - f_1)t\} \quad \text{(intermodulation distortion)}$$
$$+ \frac{3}{4} a_3 A^3 \{\cos(2\pi(2f_1 + f_2)t) + \cos(2\pi(2f_2 + f_1)t)\} \quad \text{(intermodulation distortion)}$$

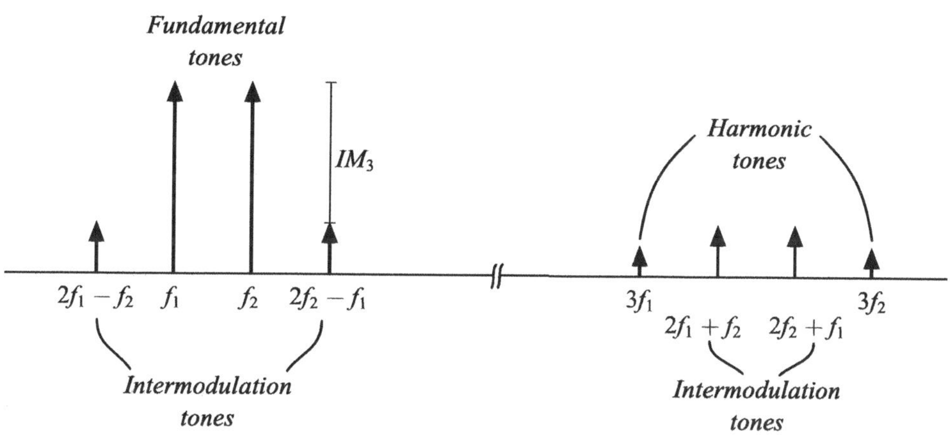

Figure 3.8 Output spectrum of an amplifier with weak third-order nonlinearity for a two-tone input.

A representative spectrum of the amplifier's output is shown in Figure 3.8. As expected, the fundamental tones (i.e., those at f_1 and f_2) undergo gain compression at the amplifier output. Harmonic distortion tones appear at $3f_1$ and $3f_2$. They can presumably be filtered out. In addition, intermodulation tones occur at frequencies $(2f_2 \pm f_1)$ and $(2f_1 \pm f_2)$. Of these, the tones at $(2f_2 - f_1)$ and $(2f_1 - f_2)$ are problematic since they lie close to the input frequencies and cannot be eliminated by filtering. The amplitude of the intermodulation components is given by $(3/4)a_3A^3$. The ratio of the amplitude of the intermodulation tone to that of the fundamental is termed the intermodulation distortion and denoted by IM_3. Expressed in decibels, it is given by

$$\mathrm{IM}_3(\mathrm{dB}) = 20\log_{10}\left(\frac{3|a_3|A^2}{4|a_1|}\right). \tag{3.22}$$

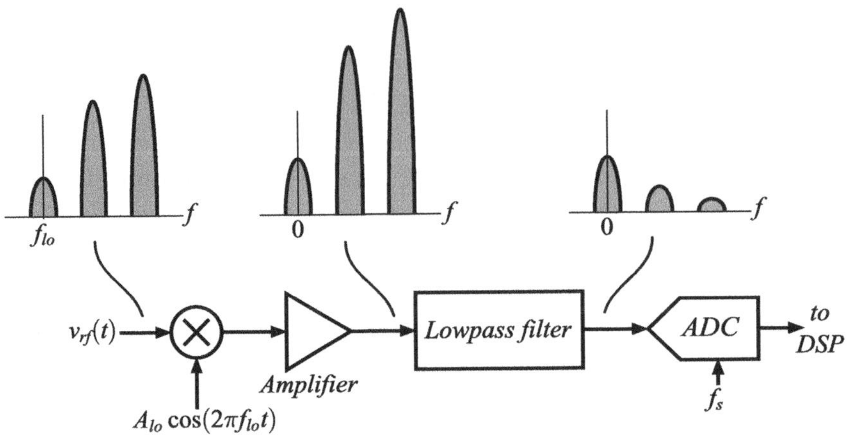

Figure 3.9 (Very) Simplified block diagram of a direct-conversion receiver.

Why is intermodulation distortion problematic in practice? In other words, is there a "use case" where we need to worry about it? It turns out that there are many use cases—we discuss one below. Consider the simplified block diagram of part of an RF receiver shown in Figure 3.9. The input $v_{rf}(t)$ consists of the (small) desired signal centered at f_{lo}. Neighboring channels are occupied by signals that are much larger than the desired signal. The RF input is demodulated to baseband by multiplying with a local oscillator, which we assume to be a sine wave at f_{lo}.[2] The baseband signal undergoes amplification and filtering before being sampled by an analog-to-digital converter (ADC). The input to the amplifier consists of the small desired channel at DC, and two large unwanted signals. The lowpass filter attenuates interferers so that the ADC output predominantly consists of samples of the desired channel.

In practice, however, the mixer, amplifier, and lowpass filter, being implemented with transistors, are weakly nonlinear. We abstract the signal chain as shown in Figure 3.10. The input to the filter is assumed to be demodulated by a linear mixer to result in a baseband signal. The weak nonlinearity of the entire front end (mixer and amplifier) is incorporated into the unity-gain buffer through the βv_i^3 term. The baseband filter has a bandwidth denoted by B. The desired component is assumed to be so small that it is set to zero. The interferers are modeled as tones at f_1 and f_2 that lie outside the filter's passband, while their difference $(2f_1 - f_2) < B$ falls in band.

[2] Only the in-phase path is shown here; the quadrature path where $v_{rf}(t)$ is multiplied by $A_{lo}\sin(2\pi f_{lo}(t)$ and subsequently amplified and filtered is omitted.

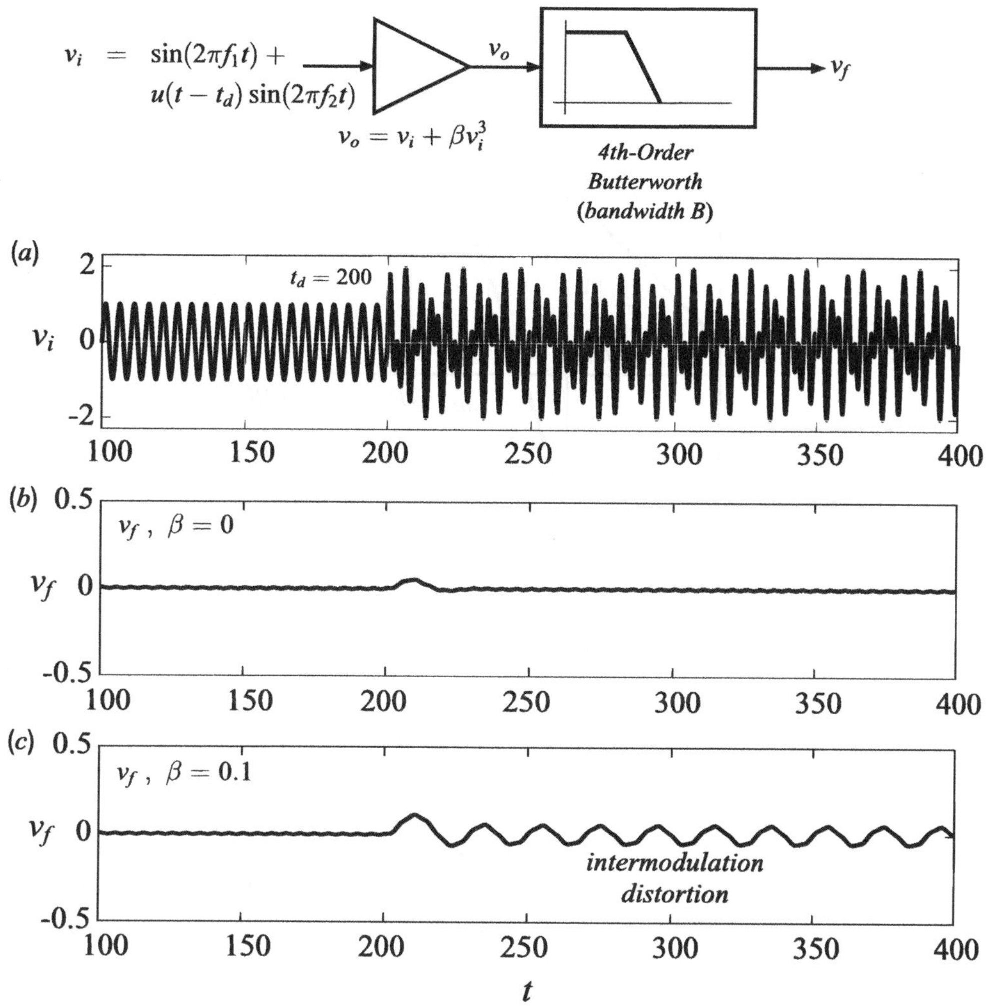

Figure 3.10 Weakly nonlinear buffer followed by a sharp lowpass filter with bandwidth B, excited by a two-tone input. Both f_1 and f_2 are higher than the filter's bandwidth, but $(2f_1 - f_2) < B$ (a) The input, where the second tone is turned on after a delay t_d. (b) Filter output for a linear buffer. (c) Filter output when the buffer is weakly nonlinear.

To "see" the effect of third-order intermodulation, the tone at f_2 is turned on at time t_d, as shown in Figure 3.10a. For $(t < t_d)$, we notice as expected and as shown in Figure 3.10b,c that the output of the filter is very small. This is because $f_1 > B$. At time $t = t_d$, the second interferer turns on. When the buffer is linear (i.e., $\beta = 0$) the filter output remains zero but for a short-lived transient. Since the desired signal is zero, the output of the filter is (almost) zero, as it should be. The situation, however, is different for a nonzero β. For $(t > t_d)$, the intermodulation tone at $(2f_1 - f_2)$ results in a large signal at the filter's output and will be interpreted by the DSP as the desired signal.[3]

[3]In practice, the interferers are random signals. Consequently, the intermodulation component at the output of the baseband filter will appear like noise, and degrade the SNR of the desired signal.

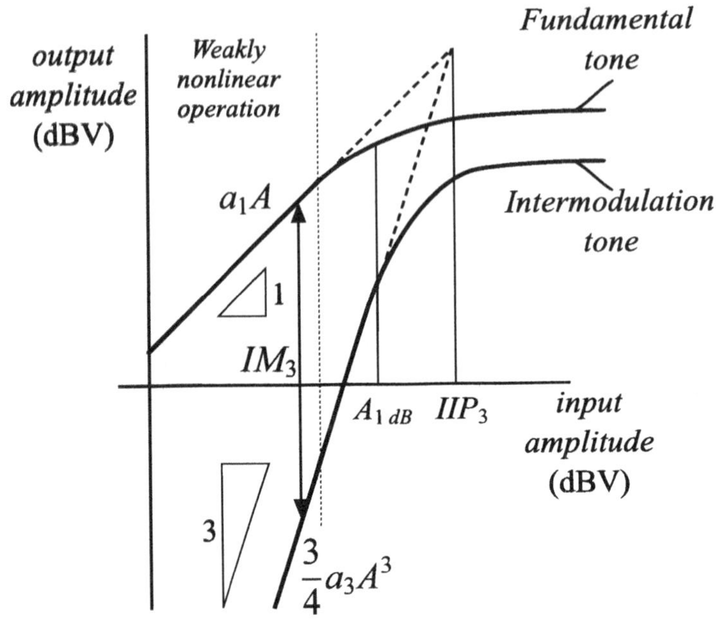

Figure 3.11 Amplitudes of the output fundamental and intermodulation tones as a function of input amplitude.

If we plotted the amplitude of any one of the output tones at f_1 or f_2 (in dBV) as a function of the corresponding input amplitude (also in dBV), we obtain a straight line with a slope of 1 dB/dB when the amplifier operates in a weakly nonlinear fashion. As Figure 3.11 shows, compression and saturation set in for larger amplitudes. The amplitude of the intermodulation tones increases with a slope of 3 dB/dB for small inputs and eventually saturates. If these lines are extrapolated, they intersect at a fictitious point. The abscissa of this point is the input intercept point for third-order intermodulation distortion and is denoted by (IIP_3). If the amplifier is operating in a weakly-nonlinear fashion with a per-tone input amplitude A (in dBV), and the resulting intermodulation distortion is IM_3 (in dB), IIP_3 (in dBV) can be obtained as

$$\mathrm{IIP}_3\ (\mathrm{dBV}) = A\ (\mathrm{dBV}) - \frac{\mathrm{IM}_3}{2}\ (\mathrm{dB}). \tag{3.23}$$

From (3.22), it is straightforward to express IIP_3 in terms of a_1 and a_3 as

$$\mathrm{IIP}_3 = \sqrt{\frac{4|a_1|}{3|a_3|}}. \tag{3.24}$$

From this and (3.19), we see that the 1 dB compression point is about 9.6 dB lower than the IIP_3.

IIP_2 and IIP_3 of a Common-Emitter Amplifier

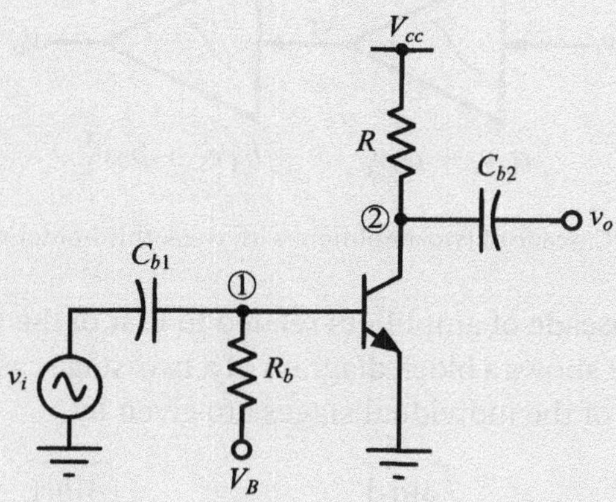

Figure 3.12 A common-emitter amplifier.

A common-emitter amplifier is shown in Figure 3.12. C_{b1} and C_{b2} are assumed to be infinite so that they act as level-shifting batteries. The transistor is operating in the active region, and its base current is assumed to be zero. The Early effect is neglected. The base bias voltage V_B is derived from an appropriate bias network so that the quiescent collector current is I_c. The thermal voltage kT/q is denoted by ϕ_t. What are the IIP_2 and IIP_3 of this amplifier?

The voltage at the collector, namely V_2, is given by

$$V_2 = V_{cc} - I_c R e^{\frac{v_i}{\phi_t}}.$$

The voltage drop across C_{b2} is $V_{cc} - I_c R$. Consequently, $v_o = V_2 - V_{b2}$ is seen to be

$$v_o = I_c R(1 - e^{\frac{v_i}{\phi_t}}) \approx -\frac{I_c R}{\phi_t} v_i - \frac{I_c R}{2\phi_t^2} v_i^2 - \frac{I_c R}{6\phi_t^3} v_i^3 + \cdots.$$

In our notation, $|a_1| = I_c R/\phi_t$, $|a_2| = I_c R/2\phi_t^2$, and $|a_3| = I_c R/6\phi_t^3$. The IIP_2 and IIP_3 are given by

$$IIP_2 = 2\phi_t, \qquad IIP_3 = 2\sqrt{2}\phi_t.$$

3.4.2 IIP$_3$ of Cascaded Stages

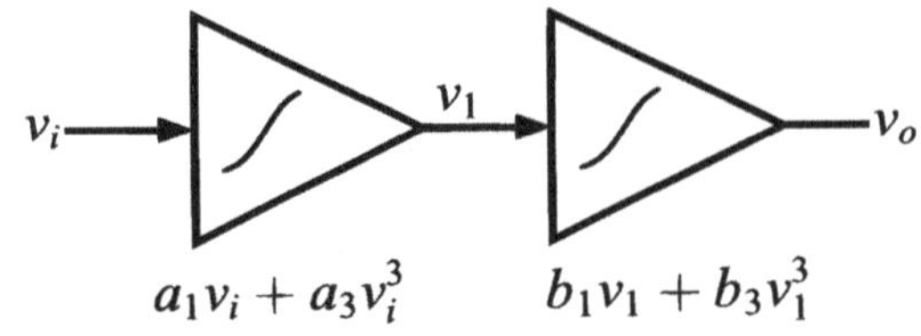

Figure 3.13 Cascade of two amplifiers with weak third-order nonlinearity.

How is the IIP$_3$ of a cascade of amplifiers related to that of the individual stages? Let us find out. Figure 3.13 shows a block diagram of a two-stage cascade. The third-order input intercept points of the individual stages are given by

$$\mathrm{IIP}_{3,a} = \sqrt{\frac{4|a_1|}{3|a_3|}}, \qquad \mathrm{IIP}_{3,b} = \sqrt{\frac{4|b_1|}{3|b_3|}}.$$

The output of the cascade, namely v_o, can be expressed in terms of v_i as

$$v_o = b_1(a_1v_i + a_3v_i^3) + b_3(a_1v_i + a_3v_i^3)^3 = a_1b_1v_i + (a_3b_1 + a_1^3b_3)v_i^3 + \cdots. \tag{3.25}$$

Consequently, the IIP$_3$ of the cascade is seen to be

$$\mathrm{IIP}_{3,ab} = \sqrt{\frac{4|a_1b_1|}{3|a_3b_1 + a_1^3b_3|}}. \tag{3.26}$$

If both stages are expansive or compressive, $|a_3b_1 + a_1^3b_3|$ in the expression above is identical to $|a_3b_1| + |a_1^3b_3|$, allowing us to write IIP$_{3,ab}$ as

$$\frac{1}{\mathrm{IIP}_{3,ab}^2} = \frac{1}{\mathrm{IIP}_{3,a}^2} + \frac{a_1^2}{\mathrm{IIP}_{3,b}^2}. \tag{3.27}$$

If one stage is expansive while the other is compressive, it is possible (for specific values of a_1, a_3, b_1, b_3) that $|a_3b_1 + a_1^3b_3|$ becomes small (or zero), improving the linearity of the cascade. This is an example of *predistortion.*

3.5 DC Transfer-Curve Derivatives: A Quick Nonlinearity Diagnostic

Two-tone tests (IM$_2$/IM$_3$) are the standard way to quantify weak second- and third-order nonlinearity in amplifiers. However, a very useful *complementary* diagnostic is to examine the derivatives of the DC transfer curve, namely $v_o = f(v_i)$. These are much easier to obtain from DC simulation than running multi-tone transient simulations, and it provides insight into which nonlinearity orders are present.

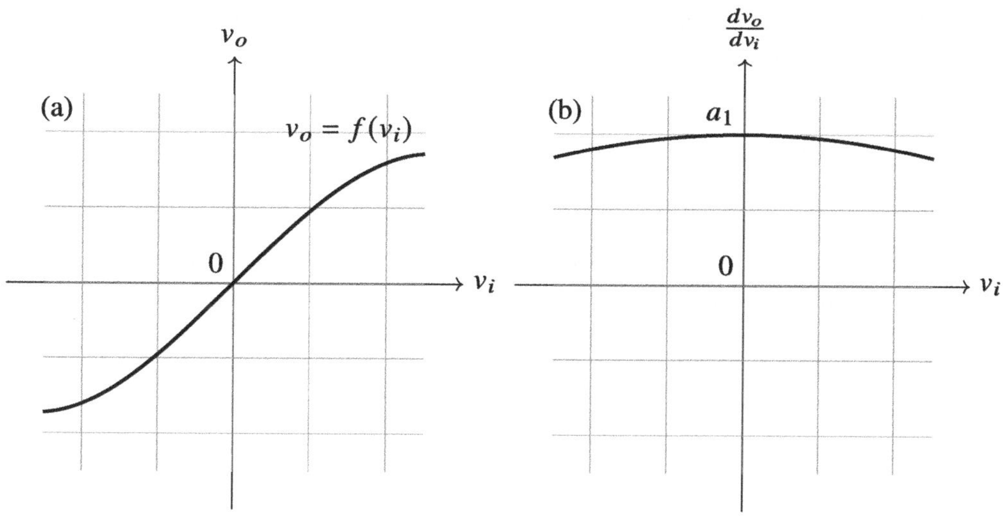

Figure 3.14 (a) Transfer curve of a weakly nonlinear memoryless amplifier with odd-order nonlinearity. (b) The derivative of the transfer curve. A flatter derivative indicates a more linear amplifier.

Suppose the transfer curve can be locally approximated by a polynomial around the operating point:

$$v_o = f(v_i) \approx a_1 v_i + a_2 v_i^2 + a_3 v_i^3 + \cdots$$

Then:

$$\frac{dv_o}{dv_i} \approx a_1 + 2a_2 v_i + 3a_3 v_i^2 + \cdots$$
$$\frac{d^2 v_o}{dv_i^2} \approx 2a_2 + 6a_3 v_i + \cdots$$
$$\frac{d^3 v_o}{dv_i^3} \approx 6a_3 + \cdots$$

For example, if the amplifier is dominated by odd-order nonlinearity (e.g., a cubic term a_3), the dc transfer curve $f(v_i)$ is typically S-shaped and odd-symmetric about the origin, as shown in Figure 3.14a. In that case, the first derivative dv_o/dv_i is an even function of v_i (Figure 3.14b). The difference in the slope of $f(v_i)$ over the useful operating range of inputs is a measure of a_3, and thus third-harmonic distortion. The third derivative $\frac{d^3 v_o}{dv_i^3}$ is approximately constant in the weakly nonlinear region.

We have so far studied the "jargon" of weakly nonlinear amplifiers. The analysis of the output was straightforward; for a sinusoidal input it merely involved using standard trigonometric identities. In practice, one should expect to encounter circuits that are comprised of multiple nonlinear elements. The question that then arises is: How does one determine the output of such a network? The next section describes the method of current injection, which is an incredibly useful tool to have in one's toolkit.

We will derive the method from first principles and apply it—first to simple circuits and finally to an active-RC biquad.

3.6 The Method of Current Injection

In this section, we derive the method of current injection, which is a general and useful technique to determine the behavior of weakly nonlinear networks [2, 3]. As a motivating example, consider the circuit of Figure 3.15. The conductances are weakly nonlinear, and the kth branch current i_k is related to the corresponding branch voltage e_k according to

$$i_k = g_{k1}e_k + g_{k2}e_k^2 + g_{k3}e_k^3 + \cdots \tag{3.28}$$

where

$$|g_{k1}e_k| \gg |g_{k2}e_k^2|,\ |g_{k3}e_k^3|,\ \cdots . \tag{3.29}$$

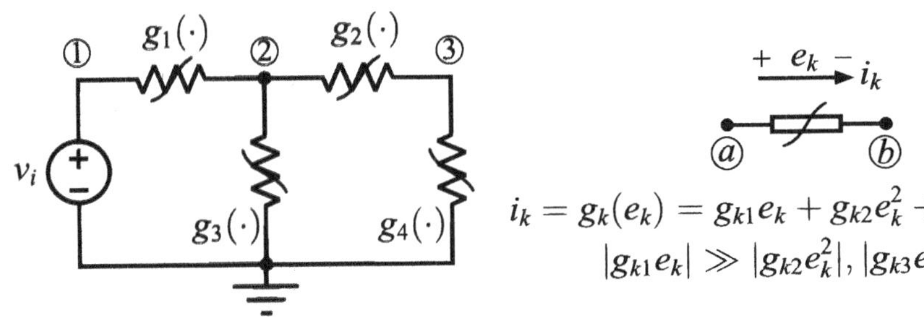

$$i_k = g_k(e_k) = g_{k1}e_k + g_{k2}e_k^2 + g_{k3}e_k^3 + \cdots$$
$$|g_{k1}e_k| \gg |g_{k2}e_k^2|, |g_{k3}e_k^3|, \cdots$$

Figure 3.15 An example weakly nonlinear network.

The question is—how do we "solve" this network? By "solve," we mean: what are the node (and branch) voltages and currents? The general way to proceed is to solve KCL, KVL, and the branch i–v equations. Since i_k is related to e_k in a nonlinear fashion, we will end up with a set of nonlinear equations. Needless to say, solving them would most likely need us to resort to numerical methods. Can we do better if we knew that the components were weakly nonlinear (as in (3.29))? It turns out that we can, as we show below.

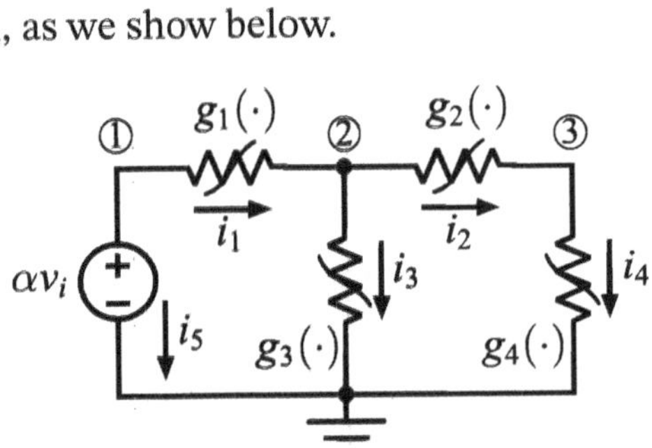

$$e_k = \alpha e_{k(1)} + \alpha^2 e_{k(2)} + \alpha^3 e_{k(3)} + \cdots$$
$$i_k = \alpha i_{k(1)} + \alpha^2 i_{k(2)} + \alpha^3 i_{k(3)} + \cdots$$
$$v_k = \alpha v_{k(1)} + \alpha^2 v_{k(2)} + \alpha^3 v_{k(3)} + \cdots$$

Figure 3.16 The network of Figure 3.15, excited by a source whose magnitude is scaled by a factor α.

Consider the weakly nonlinear network shown in Figure 3.16. The exciting source is now αv_i where α is a scaling factor. How do e_k and i_k vary with α? If the branches were all perfectly linear, i.e., the g_{k2}, g_{k3} etc. were 0, all node/branch voltages and currents would be proportional to α. Since the branches are *weakly* nonlinear, we postulate that a large portion of the node/branch voltages and currents is proportional to α and smaller portions are proportional to α^2, α^3, etc. In other words, the branch voltage e_k can be written as

$$e_k = \alpha e_{k(1)} + \alpha^2 e_{k(2)} + \alpha^3 e_{k(3)} + \cdots \tag{3.30}$$

The quantity $e_{k(1)}$ is called the first-order component, $e_{k(2)}$ the second-order component, etc. The branch voltage e_k is thus seen to be the sum of α times the first-order component, α^2 times the second-order component, α^3 times the third-order component, and so on. The node voltages and branch currents are also expressed as the sum of first-, second-, and higher-order components. i_k is thus written as

$$i_k = \alpha i_{k(1)} + \alpha^2 i_{k(2)} + \alpha^3 i_{k(3)} + \cdots \tag{3.31}$$

The branch voltages and branch currents must, of course, satisfy Kirchhoff's current and voltage laws. At node ②, $i_1 - i_2 - i_3 = 0$. Expressing the currents as the sum of their first-, second-, and third-order components and collecting terms with like powers of α, we have

$$\begin{aligned} &\alpha \left\{i_{1(1)} - i_{2(1)} - i_{3(1)}\right\} \\ &+ \alpha^2 \left\{i_{1(2)} - i_{2(2)} - i_{3(2)}\right\} = 0. \\ &+ \alpha^3 \left\{i_{1(3)} - i_{2(3)} - i_{3(3)}\right\} \end{aligned} \tag{3.32}$$

The key point is to observe that the equation above must be valid for arbitrary values of α. The only way this is possible is if the first-, second-, third-, and higher-order components of the branch currents themselves satisfy KCL. Therefore

$$\begin{aligned} i_{1(1)} - i_{2(1)} - i_{3(1)} &= 0 \\ i_{1(2)} - i_{2(2)} - i_{3(2)} &= 0 \\ i_{1(3)} - i_{2(3)} - i_{3(3)} &= 0. \end{aligned} \tag{3.33}$$

The same arguments hold for KCL at every node. How about the KVL equations? Referring to Figure 3.16, we have for the loop formed by branches 2, 3, and 4

$$e_3 - e_2 - e_4 = 0. \tag{3.34}$$

As we did with the branch currents, we express the branch voltages as the sum of their first-, second- and third-order components and collect terms with like powers of α to obtain

$$\begin{aligned} &\alpha \left\{e_{3(1)} - e_{2(1)} - e_{4(1)}\right\} \\ &+ \alpha^2 \left\{e_{3(2)} - e_{2(2)} - e_{4(2)}\right\} \\ &+ \alpha^3 \left\{e_{3(3)} - e_{2(3)} - e_{4(3)}\right\} = 0. \end{aligned} \tag{3.35}$$

Again, since the equation above must be valid for arbitrary values of α, it follows that the first- (and high-) order branch voltages must individually satisfy KVL. We therefore have

$$\begin{aligned} e_{3(1)} - e_{2(1)} - e_{4(1)} &= 0 \\ e_{3(2)} - e_{2(2)} - e_{4(2)} &= 0 \\ e_{3(3)} - e_{2(3)} - e_{4(3)} &= 0. \end{aligned} \tag{3.36}$$

The same arguments hold for KVL around every loop. To summarize, we conclude the following.

> The first-, second-, and higher-order branch currents i_k and branch voltages e_k satisfy KCL and KVL in their own right.

Recall that when we solved a linear network, we wrote KCL, KVL, and the branch relations and solved the resulting set of simultaneous linear equations. In the weakly nonlinear case that we are considering, we need to determine the $i_{k(1)}, i_{k(2)}, \ldots$ and the $v_{k(1)}, v_{k(2)}, \ldots$. Let us start with determining the first-order branch currents and voltages, namely $i_{k(1)}$ and $e_{k(1)}$. They satisfy KCL and KVL as we have seen earlier. What is the relationship between the $i_{k(1)}$ and $e_{k(1)}$? i_k is related to e_k according to

$$i_k = g_k(e_k) = g_{k1}e_k + g_{k2}e_k^2 + g_{k3}e_k^3 + \cdots$$

Expressing i_k and e_k as the sum of their first-, second-, and higher-order components in the equation above, we have

$$\begin{aligned} \alpha i_{k(1)} + \alpha^2 i_{k(2)} + \alpha^3 i_{k(3)} + \cdots = {} & g_{k1}(\alpha e_{k(1)} + \alpha^2 e_{k(2)} + \alpha^3 e_{k(3)} + \cdots) + \\ & g_{k2}(\alpha e_{k(1)} + \alpha^2 e_{k(2)} + \alpha^3 e_{k(3)} + \cdots)^2 + \\ & g_{k3}(\alpha e_{k(1)} + \alpha^2 e_{k(2)} + \alpha^3 e_{k(3)} + \cdots)^3 + \cdots \end{aligned}$$

Equating terms with like powers of α on both sides we obtain

$$\begin{aligned} i_{k(1)} &= g_{k1}e_{k(1)} \\ i_{k(2)} &= g_{k1}e_{k(2)} + g_{k2}e_{k(1)}^2 \\ i_{k(3)} &= g_{k1}e_{k(3)} + 2g_{k2}e_{k(1)}e_{k(2)} + g_{k3}e_{k(1)}^3 \\ \cdots &= \cdots \end{aligned} \tag{3.37}$$

The first observation is that $i_{k(1)} = g_{k1}e_{k(1)}$, indicating that the first-order branch currents are linearly related to the branch voltages. To determine $i_{k(1)}$ and $e_{k(1)}$, we observe the following:

- The $i_{k(1)}$ satisfy KCL.
- The $e_{k(1)}$ satisfy KVL.
- $i_{k(1)} = g_{k1}e_{k(1)}$.

This means that the first-order branch currents and voltages can be determined as follows. We create a *linear* network by setting the nonlinearities in the conductances to zero, as shown in Figure 3.17. The input voltage source is v_i, which corresponds to the first-order component of the fifth branch (i.e., $e_{5(1)} = v_i$). By solving this linear network, we obtain the first-order branch currents and voltages $i_{k(1)}$ and $e_{k(1)}$.

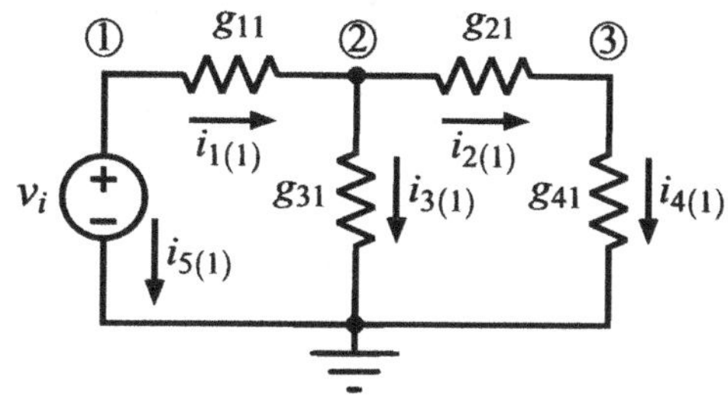

Figure 3.17 The first-order network. Solving this linear network yields all branch voltages $e_{k(1)}$ and currents $i_{k(1)}$.

To obtain the second-order branch voltages and currents, we begin by again noting that

a. The $i_{k(2)}$ satisfy KCL.
b. The $e_{k(2)}$ satisfy KVL.
c. From (3.37), the relationship between the second-order branch currents and voltages is given by $i_{k(2)} = g_{k1}e_{k(2)} + g_{k2}e^2_{k(1)}$. The equivalent circuit of the branch, therefore, is as shown in Figure 3.18.

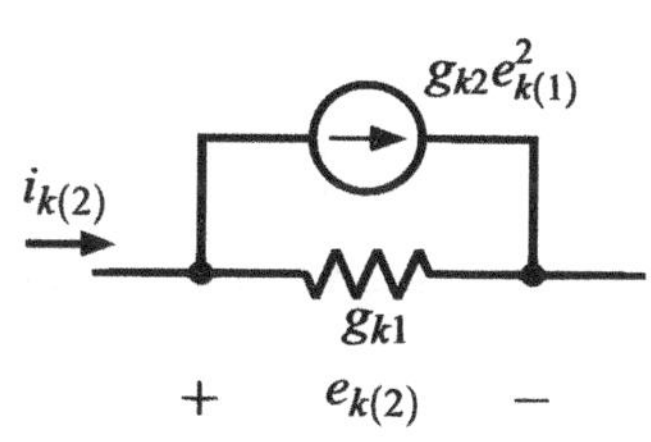

Figure 3.18 Replacement rule for the weakly nonlinear conductance in the second-order network.

The second-order branch voltages and currents are determined using the network of Figure 3.19. Each branch consists of a linear conductance, along with a current source in parallel. This injected current, which depends on the *square* of the first-order branch voltage $e_{k(1)}$, is known since we have already determined $e_{k(1)}$ from the network of Figure 3.17. The branch corresponding to the input voltage v_i is set to zero, since that branch voltage has only a first-order component, namely v_i. Note that this is a linear network (with multiple inputs).

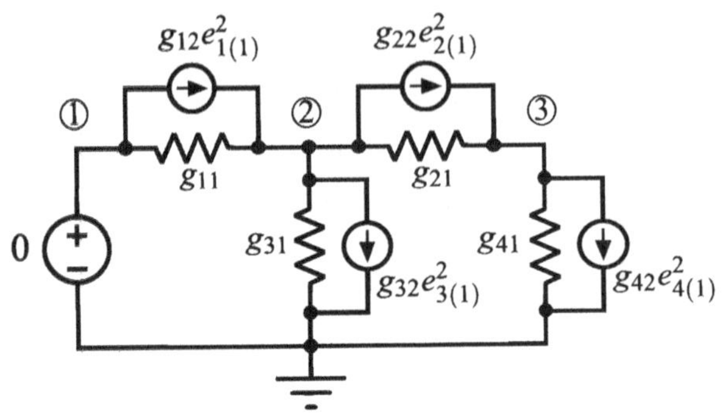

Figure 3.19 The second-order network. The input source is set to 0. The currents injected are proportional to the *squares* of the branch voltages in the first-order network. Solving this linear network yields all branch voltages $e_{k(2)}$ and currents $i_{k(2)}$.

To determine the third-order branch voltages and currents, we proceed in largely the same fashion. The only change is that the branch relation between $i_{k(3)}$ and $e_{k(3)}$ is given by $i_{k(3)} = g_{k1}e_{k(3)} + 2g_{k2}e_{k(1)}e_{k(2)} + g_{k3}e^3_{k(1)}$. When this is drawn as a circuit, Figure 3.20a results. The injected current, given by $2g_{k2}e_{k(1)}e_{k(2)} + g_{k3}e^3_{k(1)}$, is a known quantity since $e_{k(1)}$ and $e_{k(2)}$ have been determined earlier using the circuits of Figures 3.17 and 3.19. Figure 3.20b shows the network that needs to be solved to obtain $e_{k(3)}$ and $i_{k(3)}$. Like the second-order network, this is also a linear circuit driven by multiple current sources.

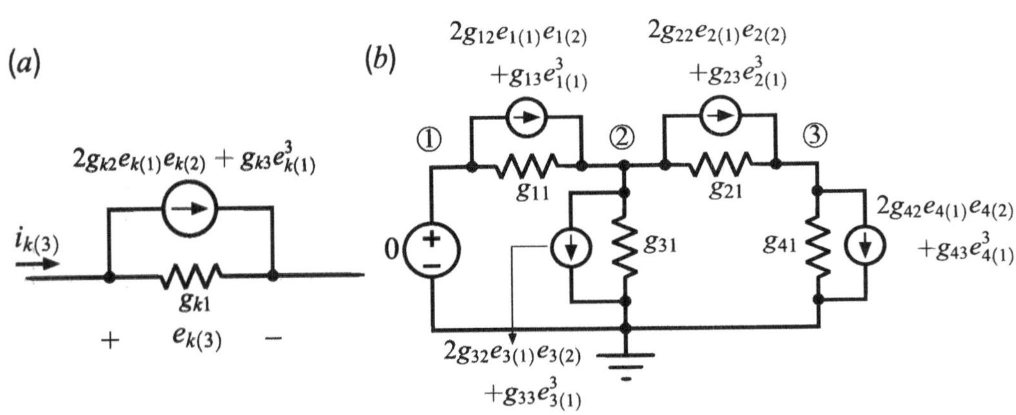

Figure 3.20 (a) Replacement rule for the weakly nonlinear conductance in the third-order network and (b) resulting third-order network. The injected currents depend on the branch voltages in the first- and second-order networks.

The approach described above can be used to derive higher-order nonlinear components of the branch voltages and currents. In practice, however, considering components up to third order is usually sufficient. From the discussion above, we see that the current-injection method has simplified the problem of solving a nonlinear network (in all likelihood needing a numerical solution) to that of solving three linear networks (assuming $e_{k(2)}$ and $e_{k(3)}$ are the dominant distortion terms).

In our analysis so far, our nonlinear elements were conductances. How do we handle other nonlinear components? We begin with a weakly nonlinear capacitor. The branch current i is related to the branch voltage e according to

$$i = (C_1 + C_2 e + C_3 e^2)\frac{de(t)}{dt}. \tag{3.38}$$

i and e are expressed as the sum of their first-, second-, and third-order components. Using $e = \alpha e_{(1)} + \alpha^2 e_{(2)} + \alpha^3 e_{(3)}$ and a similar expression for i in the equation above and collecting terms with like powers of α, we obtain

$$\begin{aligned}\alpha i_{(1)} + \alpha^2 i_{(2)} + \alpha^3 i_{(3)} = \alpha C_1 \frac{de_{(1)}}{dt} + \\ \alpha^2 \left\{ C_1 \frac{de_{(2)}}{dt} + C_2 e_{(1)} \frac{de_{(1)}}{dt} \right\} + \\ \alpha^3 \left\{ C_1 \frac{de_{(3)}}{dt} + C_2 e_{(1)} \frac{de_{(2)}}{dt} + (C_2 e_{(2)} + C_3 e_{(1)}^2) \frac{de_{(1)}}{dt} \right\}.\end{aligned}$$

It thus follows that

$$\begin{aligned} i_{(1)} &= C_1 \frac{de_{(1)}}{dt} \\ i_{(2)} &= \left\{ C_1 \frac{de_{(2)}}{dt} + C_2 e_{(1)} \frac{de_{(1)}}{dt} \right\} \\ i_{(3)} &= \left\{ C_1 \frac{de_{(3)}}{dt} + C_2 e_{(1)} \frac{de_{(2)}}{dt} + (C_2 e_{(2)} + C_3 e_{(1)}^2) \frac{de_{(1)}}{dt} \right\}. \end{aligned}$$

As seen above, in the first-order network used to determine $e_{k(1)}$ and $i_{k(1)}$, the nonlinear capacitor should be replaced by a linear capacitor C_1. In the second-order network, a nonlinear current source $C_2 e_{(1)} de_{(1)}/dt$ appears in parallel with the capacitor as shown in Figure 3.21. The third-order network consists of a current source $C_2 e_{(1)} de_{(2)}/dt + (C_2 e_{(2)} + C_3 e_{(1)}^2) de_{(1)}/dt$ in parallel with a linear capacitor C_1.

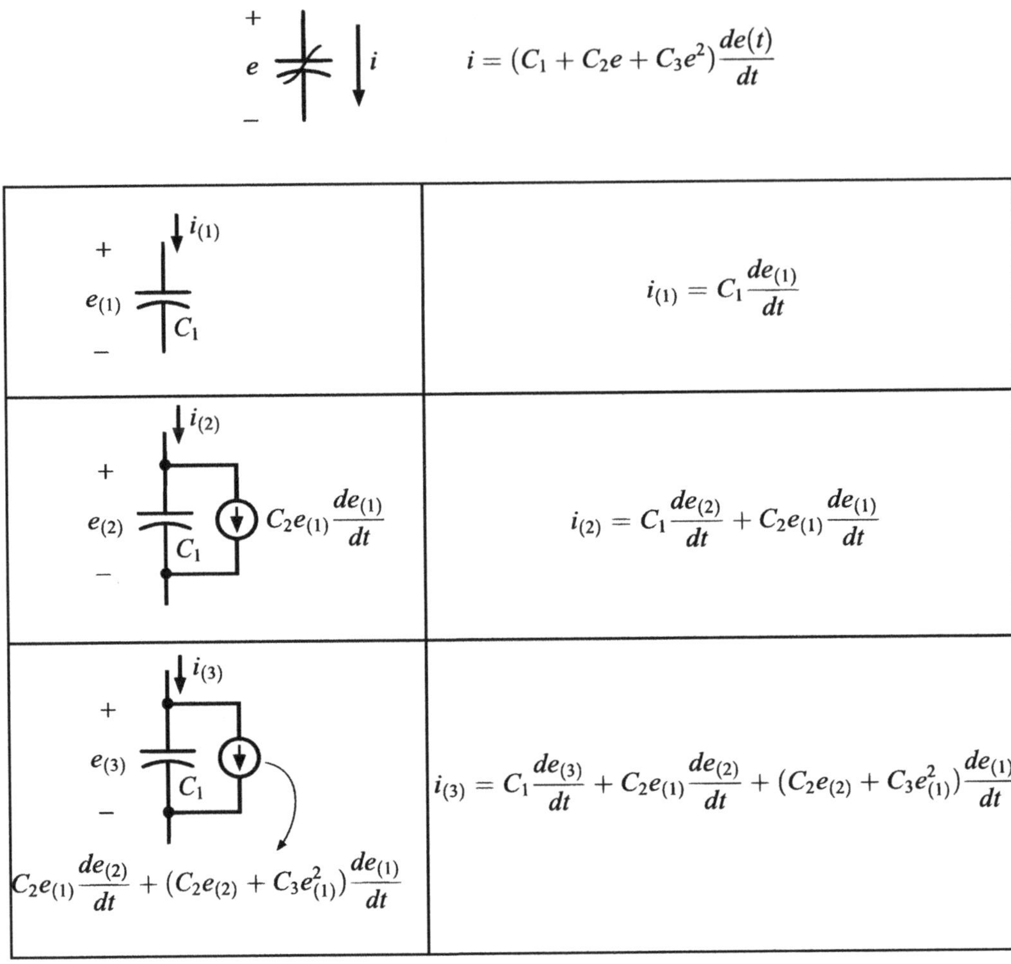

Figure 3.21 Replacement rule for a weakly nonlinear capacitor.

Implementing a Nonlinear Capacitor in a Circuit Simulator

Several applications (especially those that target very low distortion) need capacitors to be extremely linear. Some families of capacitors are inherently nonlinear, and it is important to be able to simulate the effect of capacitor nonlinearity on the performance of circuits. Many simulators do not have nonlinear capacitors as part of their component library. How then does one simulate the effect of nonlinear capacitors in a circuit?

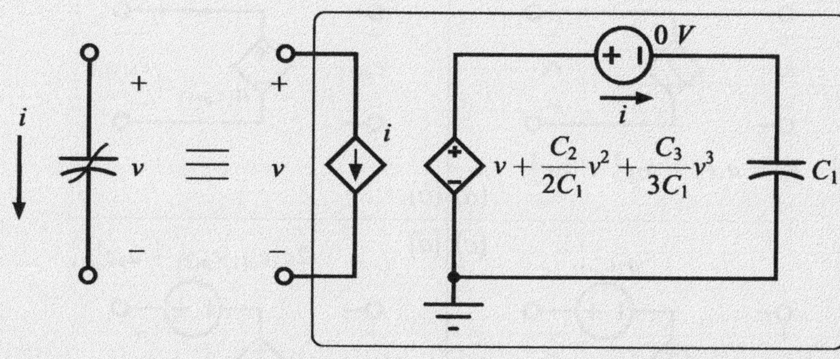

Figure 3.22 Implementing a nonlinear capacitor in a circuit simulator. C_1 is linear.

A way of incorporating nonlinearity into a capacitor is shown in Figure 3.22. Suppose that we wish to model a voltage-dependent capacitance given by $C(v) = C_1 + C_2v + C_3v^2$. A nonlinear VCVS whose output is related to the input as $v_o = v + (C_2/2C_1)v^2 + (C_3/3C_1)v^3$ drives a linear capacitance C_1. The current flowing through the capacitor is given by

$$i = C_1\frac{dv_o}{dt} = (C_1 + C_2v + C_3v^2)\frac{dv}{dt}. \tag{3.39}$$

This current is drawn from the input using a current-controlled current source with gain 1, as shown in Figure 3.22.

Let us now understand the current-injection method in the case of a weakly nonlinear voltage-controlled voltage source. We denote the output voltage by e_b and the input by e_a. e_a and e_b are related according to

$$e_b = a_1e_a + a_2e_a^2 + a_3e_a^3. \tag{3.40}$$

Expressing e_a and e_b as the sum of first-, second-, and third-order components, substituting these expressions into the equation above and equating like powers of α on both sides results in

$$\begin{aligned} e_{b(1)} &= a_1e_{a(1)} \\ e_{b(2)} &= a_1e_{a(2)} + a_2e_{a(1)}^2 \\ e_{b(3)} &= a_1e_{a(3)} + 2a_2e_{a(1)}e_{a(2)} + a_3e_{a(1)}^3. \end{aligned} \tag{3.41}$$

Figure 3.23a shows the nonlinear VCVS and parts (b), (c), and (d) show the linear components to be used in the networks used to determine the first-, second-, and third-order components.

The process described for the VCVS above can be applied to other controlled sources to derive their equivalents in the first-, second-, and higher-order networks. These are left as exercises for the reader.

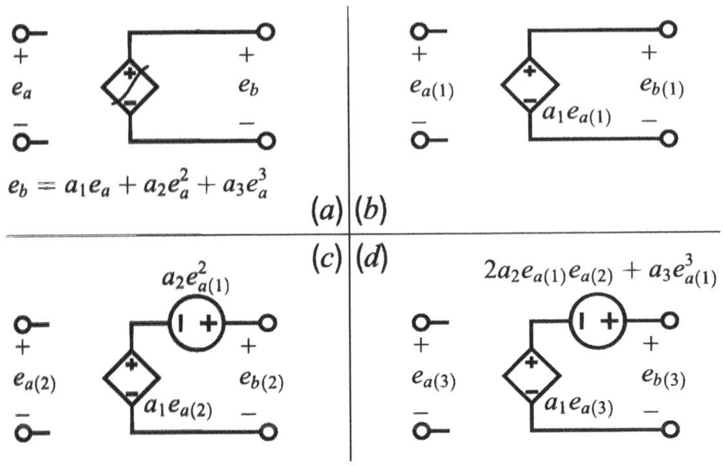

Figure 3.23 (a) Weakly nonlinear VCVS. Replacement rules in the (b) first-order, (c) second-order, and (d) third-order networks.

To summarize, the method of current injection converts the problem of solving a nonlinear network into one of solving multiple linear circuits. Every branch current/-voltage and node voltage is expressed as the sum of a linear component (i.e., one that scales linearly with the input amplitude), a quadratic component that scales as the square of the input amplitude, and a cubic component that scales as the cube of the input amplitude. (Higher-order nonlinear components can also be included if necessary.) To determine these components, we proceed as follows:

a. The linear (or first-order) component is determined by turning all nonlinearities "off." The result is a linear network. By solving this network, all first-order branch voltages/currents are found.

b. To determine the second-order quantities, every nonlinear component is replaced by its second-order equivalent, which consists of a linear component and a nonlinear exciting source. For instance, the second-order equivalent of a nonlinear conductance is a linear conductance in parallel with a nonlinear current. The current is proportional to the *square* of the corresponding branch voltage in the first-order network. The second-order network is thus a linear network with multiple nonlinear currents/voltages "injected." These nonlinear sources depend on the squares of the branch voltages/currents in the first-order network. The input to the circuit is set to zero. Solving this network yields all second-order quantities in the network.

c. The third- and higher-order nonlinear components are determined in a similar fashion. The only difference is the nature of the injected currents/voltages. It turns out that these depend on appropriate powers of quantities in the lower-order circuits (which have already been determined). For instance, the injected quantities in the third-order network depend on branch voltages/currents in the first- and second-order networks.

d. The branch currents and voltages in the weakly nonlinear circuit are simply the sum of the first-, second-, and third-order components found above.

The current-injection method is thus seen to be a technique that enables the evaluation of the output of a weakly nonlinear circuit (which is difficult) by expressing it as the sum of the responses of linear networks. Having learned about the current-injection method, let us now put it to practice.

3.6.1 A Weakly Nonlinear Negative Feedback Loop

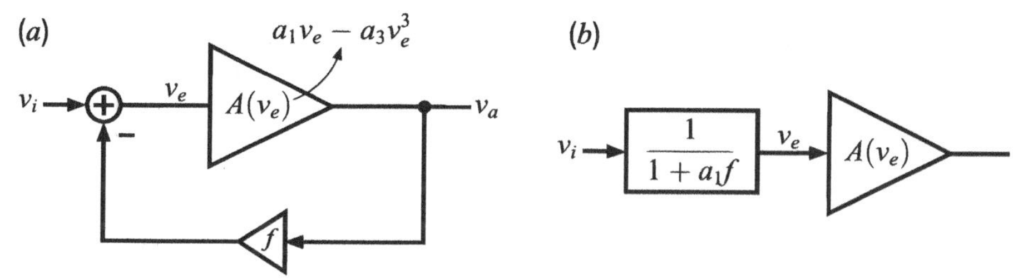

Figure 3.24 Both circuits above have the same small-signal gain. Which is more linear—(a) or (b)?

Consider the circuit "puzzle" shown in Figure 3.24. Part (a) of the figure shows a negative feedback loop. The forward amplifier is weakly nonlinear, afflicted with third-order distortion due to its characteristic $v_a = a_1v_e - a_3v_e^3$. The feedback factor is denoted by f.[4] Furthermore, it turns out that the "linear" loop gain, namely a_1f is $\gg 1$. If the forward amplifier were linear, straightforward analysis shows that

$$v_a = \frac{1}{f}\frac{a_1f}{1+a_1f}v_i$$
$$v_e = \frac{1}{1+a_1f}v_i.$$

Now consider the circuit of Figure 3.24b. The input, after being attenuated by a factor $(1 + a_1f)$, drives the amplifier. If the latter were linear with gain a_1, its output would be

$$v_b = \frac{1}{f}\frac{a_1f}{1+a_1f}v_i = v_a.$$

We thus see that if the amplifier were perfectly linear, the output of both circuits would be identical, since the drive of the amplifier is the same. The question is: which of these circuits is more linear if the amplifier's output is related to v_e by $(a_1v_e - a_3v_e^3)$? At first sight, it would seem that both these circuits would be equally nonlinear, right? Let's find out. v_e in Figure 3.24b is given by $v_i/(1 + a_1f)$. v_b is seen to be

[4] f in this subsection denotes the feedback factor, and is not to be confused with frequency.

$$v_b = \frac{1}{f}\frac{a_1 f}{1+a_1 f}v_i - \frac{a_3}{(1+a_1 f)^3}v_i^3. \tag{3.42}$$

We use the method of current injection to determine v_a in the following step-by-step manner.

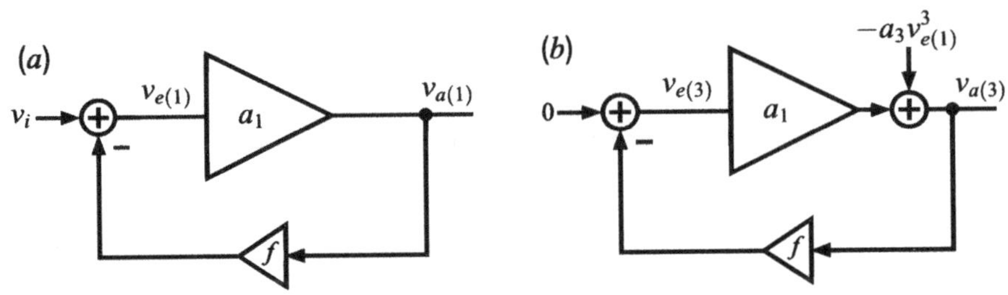

Figure 3.25 Circuits to determine the (a) first- and (b) third-order components of the output and error voltages.

a. Draw the first-order circuit by setting $a_3 = 0$, to result in the linear network shown in Figure 3.25a. Analysis shows that

$$v_{e(1)} = \frac{1}{1+a_1 f}v_i$$
$$v_{a(1)} = \frac{1}{f}\frac{a_1 f}{1+a_1 f}v_i.$$

b. The amplifier does not exhibit second-order nonlinearity. Consequently, all second-order quantities are zero.

c. Next, we draw the network to determine the third-order components of v_e and v_a. v_i is set to zero, and a voltage $-a_3v_{e(1)}^3$ is added at the output of the amplifier as shown in Figure 3.25b. It is straightforward to see that

$$v_{e(3)} = \frac{a_3 f v_{e(1)}^3}{1+a_1 f} = \frac{a_3 f}{(1+a_1 f)^4}v_i^3$$
$$v_{a(3)} = \frac{-a_3 v_{e(1)}^3}{1+a_1 f} = \frac{-a_3}{(1+a_1 f)^4}v_i^3.$$

d. v_e and v_a in the circuit of Figure 3.24a is simply the sum of the first- and third-order components we determined in Figure 3.25a,b, respectively.

$$v_e \approx v_{e(1)} + v_{e(3)} = \frac{1}{1+a_1 f}v_i + \frac{a_3 f}{(1+a_1 f)^4}v_i^3 \tag{3.43}$$
$$v_a \approx v_{a(1)} + v_{a(3)} = \frac{1}{f}\frac{a_1 f}{1+a_1 f}v_i - \frac{a_3}{(1+a_1 f)^4}v_i^3.$$

A remark is in order here. Why do we have the approximate sign in the equations above? The reason is that we have neglected nonlinear components of order higher than three.

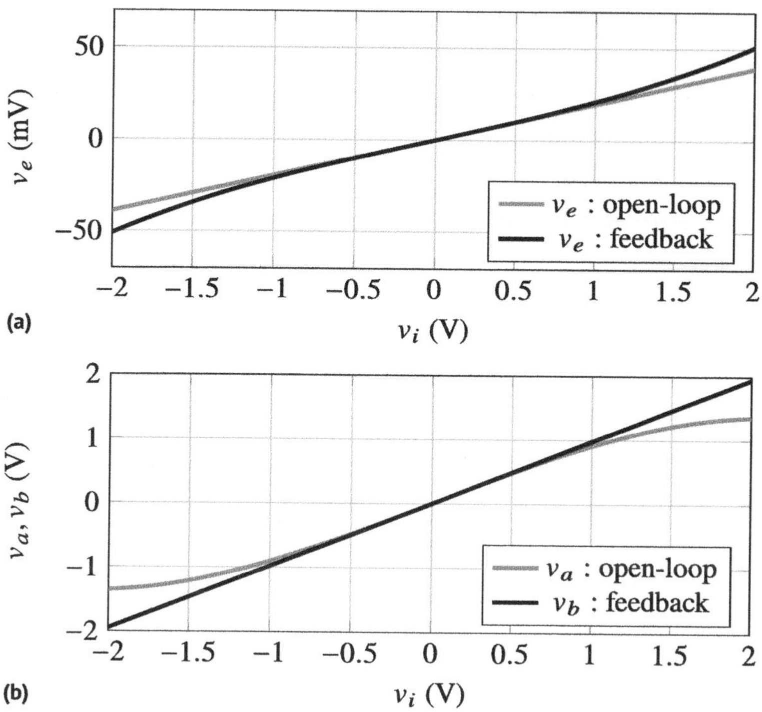

Figure 3.26 Error v_e and output voltages v_a, v_b for the open-loop and feedback structures. $a_1 = 50, f = 1, a_3 = 10^4/V^2$.

Comparing (3.43) and (3.42), we notice (as expected) that the "linear" gain is identical in both cases. However, the circuit with feedback has a third-order nonlinear coefficient, which is *smaller* than that of the open-loop network by a factor of approximately $a_1 f$! Why is this?

To understand this seemingly incredible result, one need look no further than the voltage driving the forward amplifier in both cases. They are

$$v_e = \frac{1}{1 + a_1 f} v_i \qquad \text{(open-loop structure)}$$

$$v_e = \frac{1}{1 + a_1 f} v_i + \frac{a_3 f}{(1 + a_1 f)^4} v_i^3 \qquad \text{(feedback structure)}$$

Again, if $a_3 = 0$ (linear forward amplifier), v_e is identical in both cases. With a nonzero a_3, however, we see that v_e for the feedback structure has a nonlinear component in addition to the linear portion. Interestingly, it is an *expansive* characteristic—unlike the forward amplifier whose transfer curve is compressive. Consequently, a portion of the nonlinearity of the forward amplifier is undone by the expansive nature of v_e. As a result, the feedback structure is far more linear than the open-loop circuit. This property of a negative feedback loop, which causes the error voltage to have a nonlinear characteristic opposite to that of the forward amplifier, is called *predistortion*.

To see the above, we show simulation results for the case where $a_1 = 50$, $f = 1$ and $a_3 = 10^4/V^2$. Figure 3.26a shows v_e for the open-loop and feedback structures as a function of the input voltage v_i. The "expansive" characteristic of v_e is apparent. Figure 3.26b shows the output of the open-loop and feedback circuits. The improved linearity to be had with feedback is evident. The reduction in third-order nonlinearity due to feedback is approximately a factor of the loop gain $a_1 f$.

3.6.2 AC Coupling with a Nonlinear Capacitor

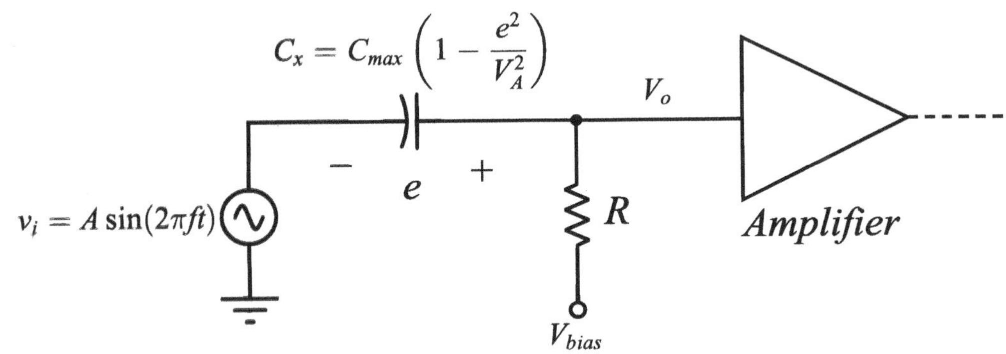

Figure 3.27 v_i is coupled to the amplifier input using an RC network that employs a nonlinear capacitor.

Consider the commonly used coupling circuit shown in Figure 3.27. The input v_i has to be level-shifted by a voltage V_{bias} so that it can be interfaced to the amplifier. A coupling capacitor is used for this purpose. The transfer function from v_i to the amplifier's input is high pass, with a corner frequency determined by R and the capacitor. The assumption, of course, is that the high-pass corner is chosen so that it allows all frequencies of interest in v_i through. Ideally, $V_o = V_{\text{bias}} + v_i$.

Unfortunately, small-size discrete capacitors can be nonlinear. Multi-layer ceramic capacitors (MLCC) are an example; their capacitance reduces with the applied voltage e according to

$$C_x = C_{\max}\left(1 - \frac{e^2}{V_A^2}\right). \tag{3.44}$$

Consequently, the coupling network for Figure 3.27 becomes nonlinear. We now evaluate the distortion at the amplifier input due to capacitor nonlinearity. The input voltage is a sinusoid at a frequency f and amplitude A.

The first step, as with any nonlinear network, is to determine the operating point. We set v_i to zero. The voltage across the capacitor is V_{bias}. The "operating" capacitance is $C_{\max}(1 - (V_{\text{bias}}/V_A)^2))$ as shown in Figure 3.28a.

The next step is to draw the incremental circuit, i.e., the one that relates *changes* in branch voltages to *changes* in branch currents. The resulting network is shown

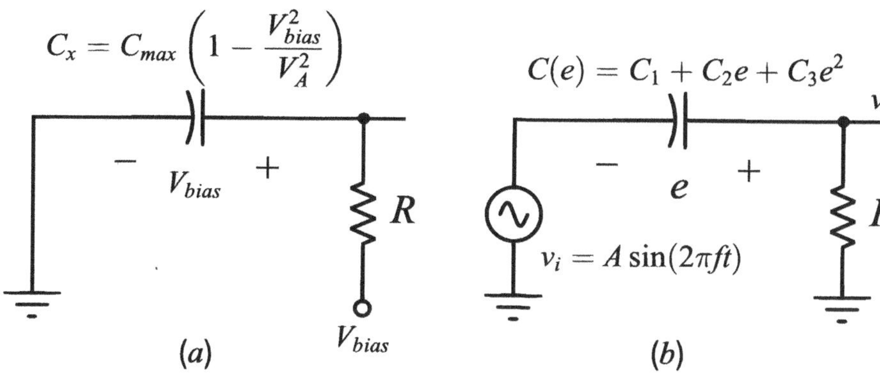

Figure 3.28 (a) Determining the operating point. (b) The incremental circuit, which is weakly nonlinear.

in Figure 3.28b. The input is v_i. Let the incremental voltage across the capacitor be denoted by e. The total voltage is $(V_{\text{bias}} + e)$. C_x is thus seen to be

$$C_x = C_{\max}\left(1 - \frac{(V_{\text{bias}} + e)^2}{V_A^2}\right) = C_{\max}\left(1 - \frac{V_{\text{bias}}^2}{V_A^2}\right) - 2\frac{C_{\max}V_{\text{bias}}}{V_A^2}e - \frac{C_{\max}}{V_A^2}e^2. \quad (3.45)$$

In the notation of Figure 3.22, $C_1 = C_{\max}(1 - (V_{\text{bias}}/V_A)^2)$, $C_2 = -2(C_{\max}V_{\text{bias}}/V_A^2)$, $C_3 = -C_{\max}/V_A^2$.

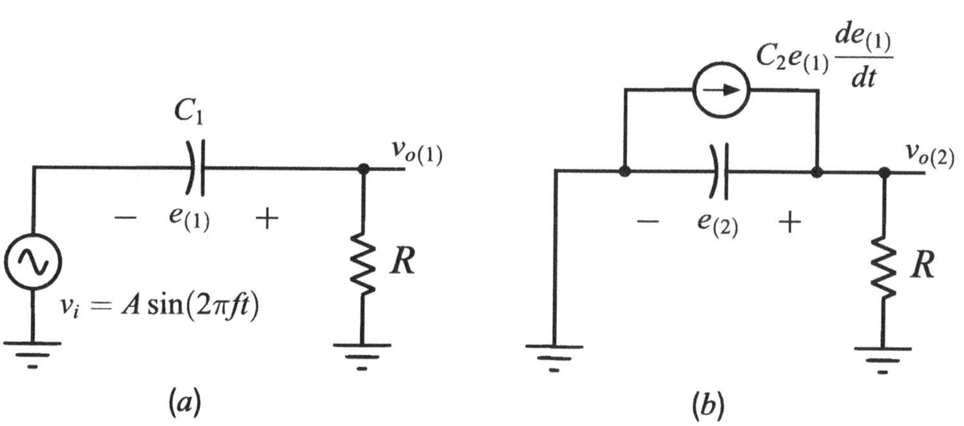

Figure 3.29 Circuits to determine the (a) first- and (b) second-order components of the output voltage.

To determine distortion, we proceed as follows. We form the first-order circuit, i.e., the one where all nonlinearities are turned "off," as shown in Figure 3.29a. Straightforward analysis yields

$$e_{(1)}(t) = -\frac{A}{\sqrt{1 + (2\pi fRC_1)^2}}\sin(2\pi ft + \phi)$$

$$v_{o(1)}(t) = \frac{2\pi fRC_1 A}{\sqrt{1 + (2\pi fRC_1)^2}}\cos(2\pi ft + \phi).$$

where $\phi = -\tan^{-1}(2\pi f RC_1)$. The second-order circuit is shown in Figure 3.29b. The injected current, which is $C_2 e_{(1)} de_{(1)}/dt$, causes $v_{o(2)}$ to be a tone at $2f$ with an amplitude given by

$$A_2 = C_2 \frac{2\pi f}{2(1+(2\pi f RC_1)^2)} \frac{R}{\sqrt{1+(2\pi(2f)RC_1)^2}} A^2. \tag{3.46}$$

The second-harmonic distortion, which is the ratio of the amplitude of the second harmonic to that of the fundamental, is given by

$$HD_2 = \frac{C_2}{2C_1} \frac{1}{\sqrt{1+(2\pi f RC_1)^2)}} \frac{1}{\sqrt{1+(2\pi(2f)RC_1)^2}} A. \tag{3.47}$$

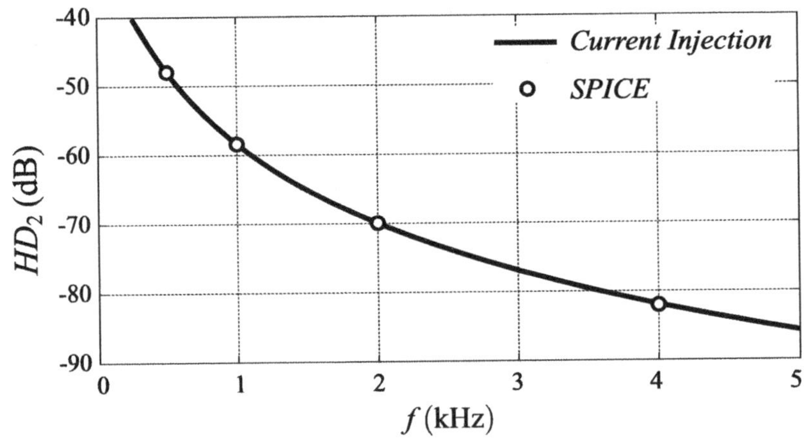

Figure 3.30 Simulated and estimated HD_2 as a function of input frequency. $R = 1\ \text{k}\Omega$, $C_{max} = 0.47\ \mu\text{F}$, $V_A = 10\ \text{V}$, $V_{bias} = 2\ \text{V}$, $A = 1\ \text{V}$.

Let us see how the method of current injection compares with a full-blown transient simulation in SPICE (which solves nonlinear differential equations governing the circuit). We consider a case where $C_{max} = 0.47\ \mu\text{F}$, $V_A = 10\ \text{V}$, $R = 1\ \text{k}\Omega$ and $V_{bias} = 2\ \text{V}$. The resulting high-pass corner is about 330 Hz. $v_i = \sin(2\pi f t)$. Figure 3.30 compares HD_2 obtained from SPICE with (3.47). Excellent agreement is seen. We notice that HD_2 decreases by 12 dB (a factor of 4) on doubling the frequency. Why does this make sense?

For $f \gg 1/(2\pi RC_1)$, when f is doubled, the voltage across the capacitor is reduced by a factor of 2. The magnitude of the injected current, which is proportional to $e_{(1)} de_{(1)}/dt$ reduces by 2×. This current is injected into an RC-parallel network, whose impedance is 2× smaller. Consequently, the amplitude of $v_{o(2)}$ is lower by a factor of 4.

3.6.3 Distortion in a Common-Collector Amplifier

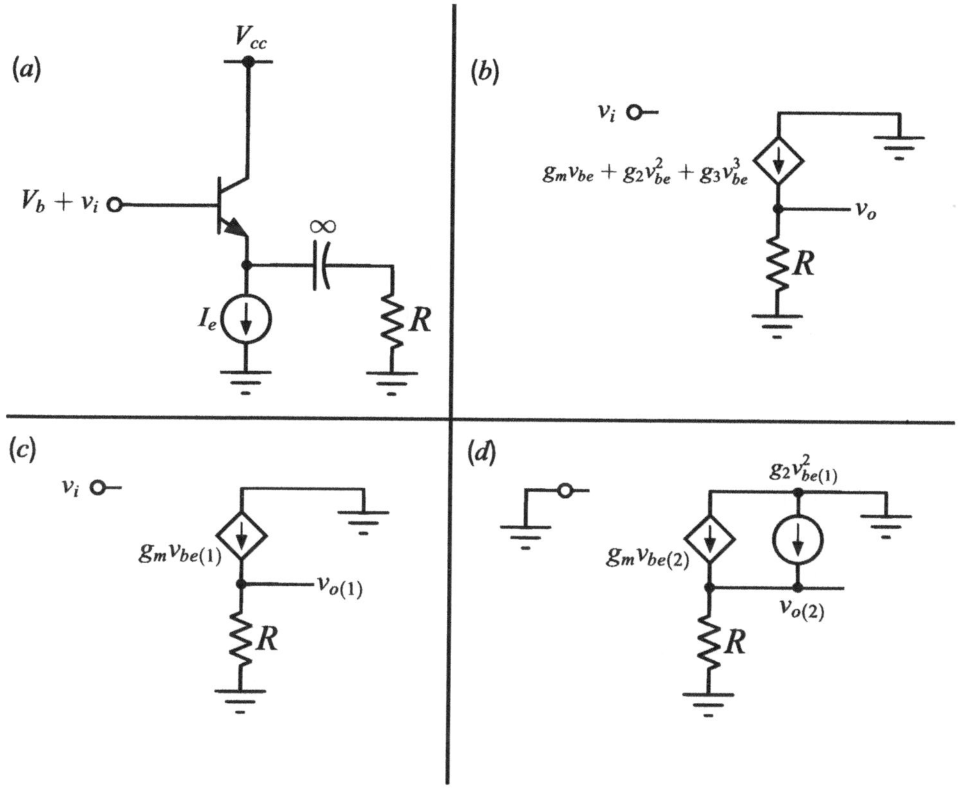

Figure 3.31 (a) A common-collector amplifier. (b) Incremental equivalent, which is weakly nonlinear. Circuit to determine (c) first- and (d) second-order components.

Consider the common-collector amplifier shown in Figure 3.31a. The transistor, biased with an emitter current I_e, is assumed to have an infinite β. The capacitor, which couples the emitter to the load-resistor R, is linear and infinite. The input signal v_i is at the base, which is biased at V_b. The question is how the voltage across R varies in response to v_i.

Since the transistor's β is infinite, the incremental model of the amplifier is the one shown in Figure 3.31b. The (incremental) collector current is expressed as a polynomial function of v_{be}, and is given by [4]

$$i_c = g_m v_{be} + g_2 v_{be}^2 + g_3 v_{be}^3 + \cdots \tag{3.48}$$

where

$$g_m = \frac{I_e}{\phi_t}, \quad g_2 = \frac{I_e}{2\phi_t^2}, \quad g_3 = \frac{I_e}{6\phi_t^3}. \quad (\phi_t \text{ denotes the thermal voltage}) \tag{3.49}$$

The first-order circuit is given in Figure 3.31c. Straightforward analysis yields

$$v_{o(1)} = \frac{g_m R}{1 + g_m R} v_i\,, \quad v_{be(1)} = \frac{1}{1 + g_m R} v_i. \tag{3.50}$$

Figure 3.31d shows the second-order network. v_i is set to zero, and a current $g_2 v_{be(1)}^2$ is injected in parallel with the controlled current source. Analysis shows that

$$v_{o(2)} = g_2 \frac{R}{1 + g_m R} v_{be(1)}^2 = \frac{g_2 R}{(1 + g_m R)^3} v_i^2. \tag{3.51}$$

The third-order terms can be similarly found and are left as an exercise. Using the expressions for g_m and g_2 above, v_o can be written as

$$v_o \approx v_{o(1)} + v_{o(2)} + \cdots = \frac{g_m R}{1 + g_m R} v_i + \frac{1}{2\phi_t} \frac{g_m R}{(1 + g_m R)^3} v_i^2 + \cdots \tag{3.52}$$

We have so far analyzed the effect of distortion in simple (toy?) circuits. While the results were obtained in a straightforward manner, it might seem that the same results might have been obtained even by brute-force analysis. It is thus natural to wonder if the current-injection technique works for more "complicated" networks. In Section 3.6.4, we see the application of current injection to an active-RC biquad where the OTAs are two-stage feedforward structures. In the analysis, we will use what we have learned about OTA-RC filters and inter-reciprocity in addition to the current-injection technique.

3.6.4 Case Study: Distortion in an Active-RC Biquad

In this section, we analyze the effect of weak OTA nonlinearities on the performance of an active-RC filter.[5] Our objective is to evaluate the linearity of the filter while incorporating third-order[6] nonlinearity. There are many ways of quantifying nonlinearity; in this section, we use intermodulation distortion as a metric. We proceed by applying a stimulus of two equal-amplitude tones at frequencies $f \pm \Delta f$ (with $\Delta f \ll f$) and measuring the strength of the resulting intermodulation tones at the output of the signal chain. These tones will be at frequencies $f \pm 3\Delta f$. We use the method of current injection discussed in this chapter. We first give a quick overview of the technique for the benefit of the reader [5].

Consider the first-order lowpass filter example shown in Figure 3.32a. The OTA is weakly nonlinear according to $i = g_m v_x - g_3 v_x^3$. Assuming that the system is weakly nonlinear, v_o can be approximately determined as follows.

- Create a circuit where all the nonlinearities are "turned off." In our example, this means that the weakly nonlinear transconductor is replaced by a linear one with transconductance g_m, as shown in Figure 3.32b. The resulting linear circuit is

[5] See Chapter 7 for a background on filters.
[6] Since practical designs are fully differential, third-order nonlinearity is dominant.

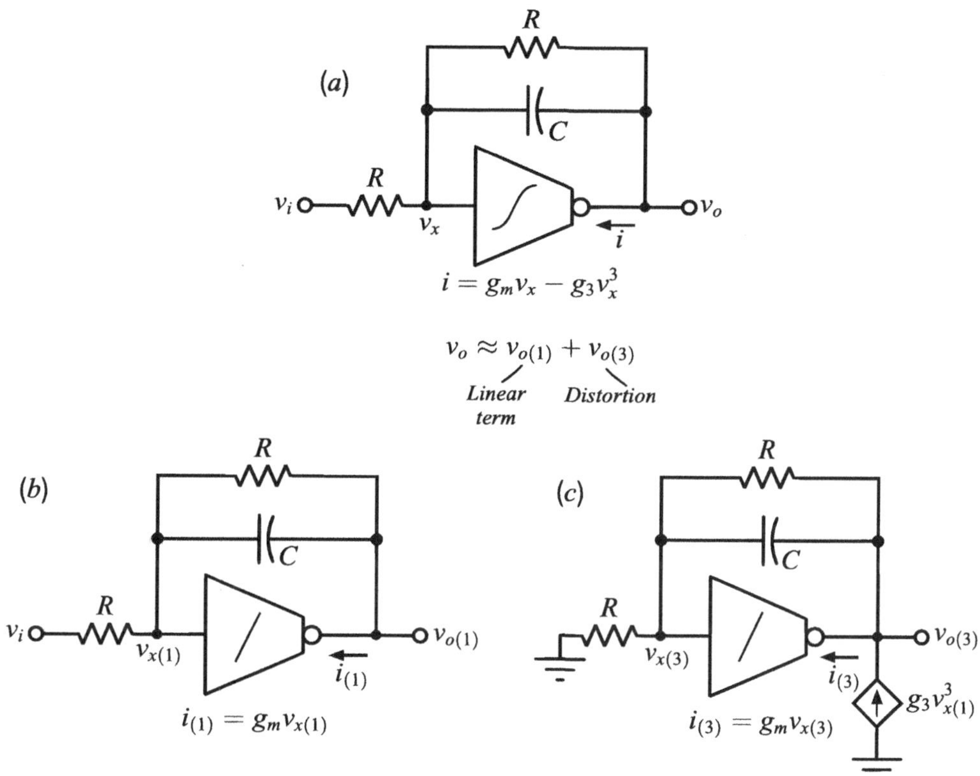

Figure 3.32 (a) A weakly nonlinear first-order OTA-RC filter. The OTA is nonlinear, with $|g_m v_x| \gg |g_3 v_x^3|$. (b), (c) Current-injection method to find distortion due to a weakly nonlinear element in a network. Both networks are linear.

excited with the input $v_i(t)$, and the "linear" output and virtual-ground voltages are denoted by $v_{o(1)}(t)$ and $v_{x(1)}(t)$ respectively.

- For the linear circuit created in step (a) above, set the input excitation to zero, and *inject* the nonlinear current $g_3 v_{x(1)}^3(t)$ at the output of the transconductor as shown in Figure 3.32c. The resulting output is denoted by $v_{o(3)}(t)$.
- The output of the weakly nonlinear circuit in Figure 3.32a is approximately given by

$$v_o(t) \approx v_{o(1)}(t) + v_{o(3)}(t). \tag{3.53}$$

From the discussion above, it is apparent that $v_{o(3)}(t)$ consists of all the nonlinear components of $v_o(t)$.

3.6.5 Intermodulation Distortion in a Tow-Thomas Biquad due to OTA Weak Nonlinearities

Consider a second-order low-pass filter realized as a Tow-Thomas biquadratic section. The single-ended equivalent of the filter is given in Figure 3.33a. The inversion in the feedback path, shown as a gain of −1, is actually achieved by interchanging wires in the differential implementation. The ADC following the filter is assumed to be ideal. With ideal OTAs, the filter transfer function can be shown to be

$$H_l(s) = \frac{1}{1 + \frac{s}{2\pi f_o Q} + \frac{s^2}{(2\pi f_o)^2}}, \tag{3.54}$$

where $2\pi f_o = 1/RC$. $H(s)$ is independent of α and β. The transfer function at node ④ is given by

$$H_b(s) = -\frac{\beta}{Q}\left(1 + \frac{sQ}{2\pi f_o}\right) H(s). \tag{3.55}$$

The peak gain of $|H_b(j2\pi f)|$ is controlled by β. For maximum dynamic range, β is chosen so that the peak magnitude of $H_b(j2\pi f)$ is unity. Increasing the "free variable" α impedance scales the damped integrator, which increases filter noise but allows the power dissipation of OTA-b to reduce. α is typically chosen so that it minimizes the biquad's power-noise product.

In our analysis, we assume that the OTAs are two-stage feedforward compensated structures as depicted in Figure 3.33a. The transconductors used in the OTAs are weakly nonlinear, as shown in the figure. We assume that the current of a transconductor is to be modeled as per $i = g(v - \eta v^3)$. The transconductance g will vary across transconductors, but the η in our analysis is assumed to be the same. This simplifies notation but does not lead to a loss of generality. The parasitic capacitances in OTA-a are denoted by c_{p1}, c_{p2}, and c_{p3}, corresponding to those at nodes ②, ③, and ④ respectively. The parasitics in OTA-b are labeled as c_{q1}, c_{q2}, and c_{q3}.

To determine the effect of weak transconductor nonlinearities, we apply the method of current injection as follows. The discussion below and Figure 3.33 refer to nonlinear currents pertaining to OTA-a. The same applies to those originating in OTA-b.

a. We excite the biquad with two closely spaced tones, each with amplitude A and frequencies $f \pm \Delta f$, where $\Delta f \ll f$. To determine the linear component of the output, all nonlinearities are turned off, resulting in the linear network shown in Figure 3.33a. The transfer functions from the input v_i to nodes ② and ③ are denoted by $H_2(s)$ and $H_3(s)$, respectively. Since $\Delta f \ll f$, the amplitudes of these tones at nodes ② and ③ are approximately given by $|H_2(j2\pi f)|A$ and $|H_3(j2\pi f)|A$, respectively.

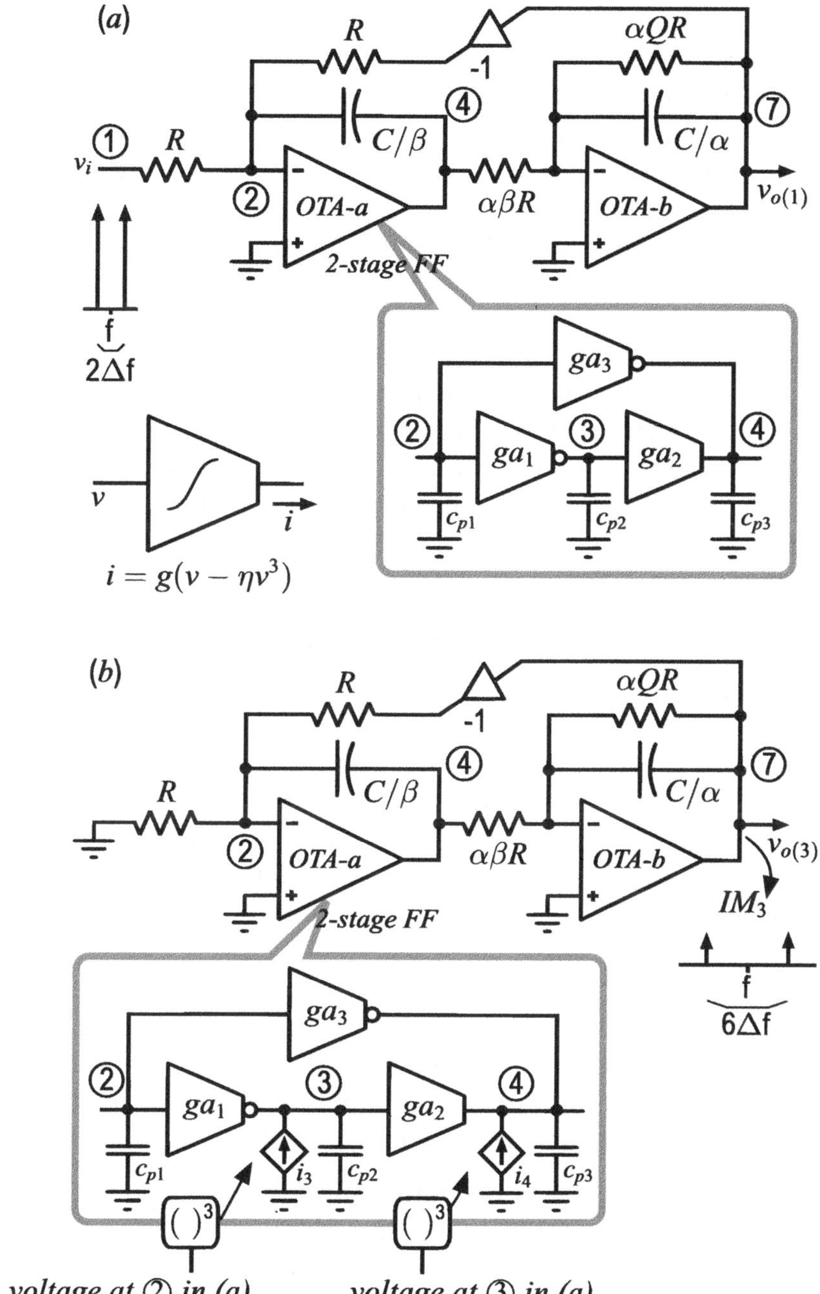

Figure 3.33 (a) The linear model of the Tow-Thomas filter, which employs two-stage, feed-forward compensated OTAs, named OTA-a and OTA-b. The circuit is excited by two closely spaced tones. The output is denoted by $v_{o(1)}$. (b) Nonlinear currents are injected into a linear filter with the main input set to zero. The resulting output is $v_{o(3)}$. The output of the weakly nonlinear filter is $\approx v_{o(1)} + v_{o(3)}$.

b. To determine the distortion terms, the waveforms at nodes ② and ③ in Figure 3.33a are cubed, multiplied by $\eta\, ga_1$ and $\eta\, ga_2$, respectively, and injected into the linear network with v_i set to zero as shown in Figure 3.33b.[7] The cubing of the dual tone signals in Figure 3.33a leads to the injection of intermodulation tones at $f \pm 3\Delta f$ into the network of Figure 3.33b, which appear in the output $v_{o(3)}$.

c. The strength of the intermodulation tones in $v_{o(3)}$ is determined as follows. We derive below the expression for distortion due to ga_2; the process is similar for the other transconductors. In the analysis that ensues, we denote the transfer functions from i_3 and i_4 to $v_{o(3)}$ (Figure 3.33b) by $Z_3(s)$ and $Z_4(s)$, respectively. As described earlier, the amplitude of the fundamental tones at ③ in Figure 3.33a is given by $|H_3(j2\pi f)|A$. Consequently, the amplitude of the distortion currents injected into the network in Figure 3.33b is $(3/4)\eta\, ga_2|H_3(j2\pi f)|^3A^3$. The resulting amplitude of the distortion tones in $v_{o(3)}$ is thus seen to be $(3/4)\eta\, ga_2|H_3^3(j2\pi f)Z_4(j2\pi f)|A^3$. The input-referred intermodulation distortion due to ga_2's weak nonlinearity is thus given by

$$IM_3 = \frac{3}{4}\eta\, ga_2|H_3^3(j2\pi f)Z_4(j2\pi f)|A^2. \tag{3.56}$$

The same process is followed for all the transconductors in the filter.

From the discussion above, it is apparent that we need to determine the voltage transfer functions from the input to the internal nodes of the filter. This will enable us to determine the strengths of the nonlinear currents that need to be injected into the network of Figure 3.33b. Furthermore, we also need to find the transfer functions from the injected currents to the filter's output. Both sets of transfer functions are estimated next.

Figure 3.34 shows the single-ended schematic used to determine the transfer functions from the input to the internal nodes. For the time being, neglect the current i_4. The system is of the sixth order, and making sense of the exact expressions for the transfer functions is all but impossible. To simplify analysis and build intuition, we use approximations that are valid in a well-designed filter. We proceed as follows:

a. The filter transfer function to the output (namely, node ⑦) is low-pass and is denoted by $H_l(s)$. If the OTAs are ideal, $H_l(s)$ is given by (3.54). In practice, the OTAs are not ideal and cause the transfer function to deviate from $H_l(s)$. The passive components are slightly adjusted to restore the transfer function. For the purposes of our analysis, we assume that the changes in component values are very small and can be neglected. By the same argument, the transfer function to ④, namely $H_4(s)$ is $\approx H_b(s)$ given in (3.55).

---

[7] A current $\eta ga_3 v_2^3$, corresponding to the nonlinear current generated by the feedforward transconductor g_{m3} is also injected into ④, but is not shown in the figure to reduce clutter.

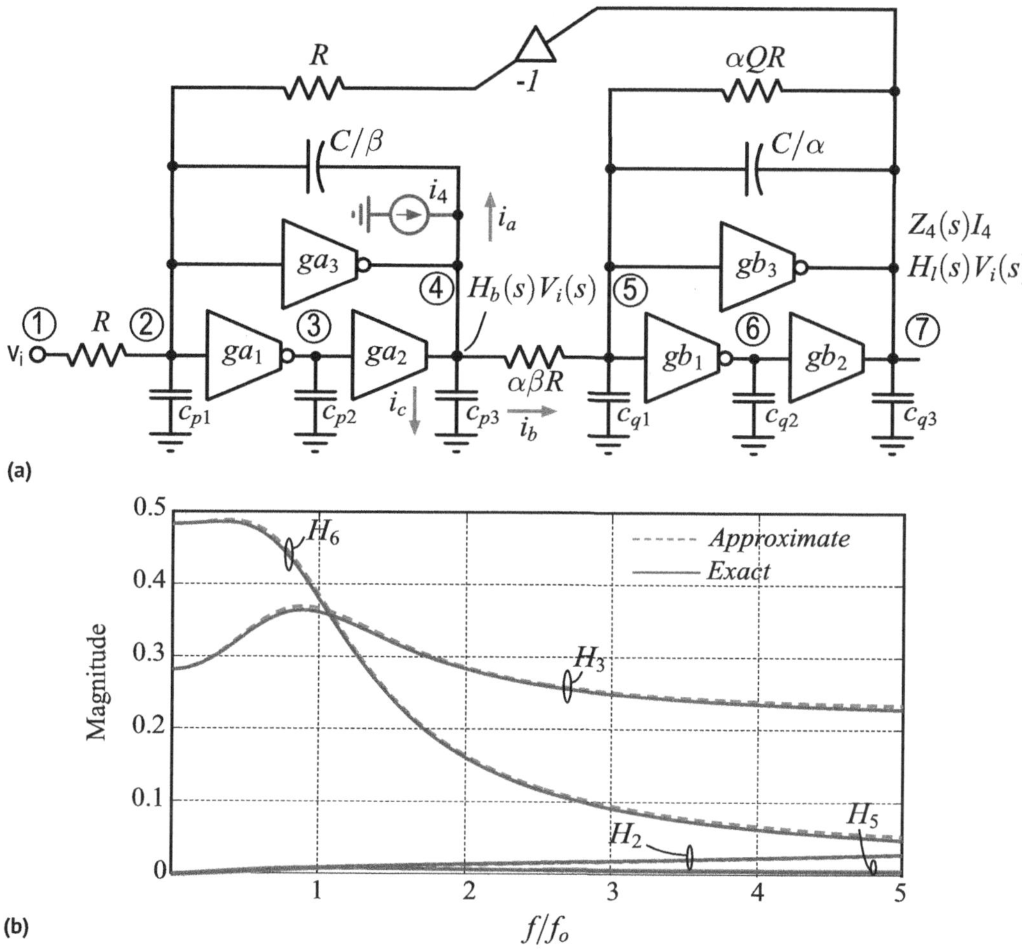

Figure 3.34 The Tow-Thomas biquad employing two-stage feedforward-compensated OTAs and the comparison of the exact and approximate transfer functions from v_i to the internal nodes ②, ③, ⑤, and ⑥.

b. Transconductors ga_1, ga_2, ga_3 represent the two-stage feedforward-compensated OTA-a, while gb_1, gb_2, gb_3 form OTA-b. In a well-designed OTA, the gain of the high-gain path, namely ga_1–ga_2 is much higher than that of the high-speed path through ga_3 up to the unity-gain frequency given by ga_3/c_{p3}. Furthermore, in a well-designed active filter, the OTA's unity-gain frequency is chosen to be much higher than the bandwidth-Q product of the filter. Given this, it is clear that for inputs up to almost the unity-gain frequency of the OTA, the two-stage path supplies most of the OTA's current. Using this observation, and assuming that the potential of ② (the virtual ground node) is very small, we see that the current supplied by the transconductor ga_2 in Figure 3.34a, namely $H_3(s)ga_2$ is approximately the sum of i_a, i_b, i_c, and can be written as

$$\underbrace{H_b(s)\left(\frac{1}{\alpha\beta R}+s(C/\beta+c_{p3})\right)}_{i_a+i_b+i_c}\approx H_3(s)\,ga_2. \tag{3.57}$$

From this, we have

$$H_3(s)\approx H_b(s)\frac{1}{ga_2}\left(\frac{1}{\alpha\beta R}+s(C/\beta+c_{p3})\right). \tag{3.58}$$

The dc gain of $H_b(s)$ is $-\beta/Q$, which means that $H_3(0)=-1/(\alpha Q\,ga_2R)$. Inspecting the internals of the two-stage OTA, it is apparent that v_3 is the integral of v_2. We thus have

$$H_2(s)=-\frac{sc_{p2}}{g_{a1}}H_3(s). \tag{3.59}$$

Since ga_1/c_{p2} is much larger than the filter's center frequency f_o, it follows that $|H_2(j2\pi f)|\ll|H_3(j2\pi f)|$ in the signal band. Using similar arguments for OTA-b, it is straightforward to see that

$$\begin{aligned} H_6(s)\ &\approx H_l(s)\frac{1}{gb_2}\left(\frac{1}{\alpha QR}+\frac{1}{R}+s(C/\alpha+c_{q3})\right) \\ H_5(s)\ &=-\frac{sc_{q2}}{g_{b1}}H_6(s). \end{aligned} \tag{3.60}$$

The dc gain of $H_6(s)$ is seen to be $1/(\alpha\,Q\,gb_2R)+1/(gb_2R)$, and $|H_5(j2\pi f)|\ll|H_6(j2\pi f)|$ in the signal band.

Simulations confirm the analysis above. Figure 3.34 compares the magnitudes of our approximations to H_2, H_3, H_5, and H_6 with the corresponding exact transfer functions. $H_l(s)$ is Butterworth, with $\alpha=1$ and $\beta=0.69$. $C=1/(2\pi f_oR)$, where f_o denotes the 3 dB bandwidth of the filter. ga_1, ga_2, ga_3 were all chosen to be $5/R$, while c_{p1}, c_{p2}, c_{p3} were all chosen as $0.125/(2\pi f_oR)$. OTA-a and OTA-b were assumed to be identical. The unity-gain frequency of the OTAs is thus approximately given by $ga_3/(c_{p1}+c_{p3})=20\pi f_o$. The approximate transfer functions derived above are seen to be in good agreement with the exact ones. They are collected in Table 3.1 for ready reference.

c. Our next task is to determine the transfer functions from currents injected at nodes ③, ④, ⑥, and ⑦ in Figure 3.34a to the output. These transfer functions, which have the dimensions of impedance, are denoted by $Z_3(s), \dots, Z_7(s)$. Since there are many inputs and one output, it is convenient to use the interreciprocal (or adjoint) network. Recall from Chapter 1 that the adjoint network is obtained by interchanging the controlling and controlled ports of the transconductors in the original network, while the resistors and capacitors remain unchanged. The resulting circuit is shown in Figure 3.35. By the properties of the adjoint network

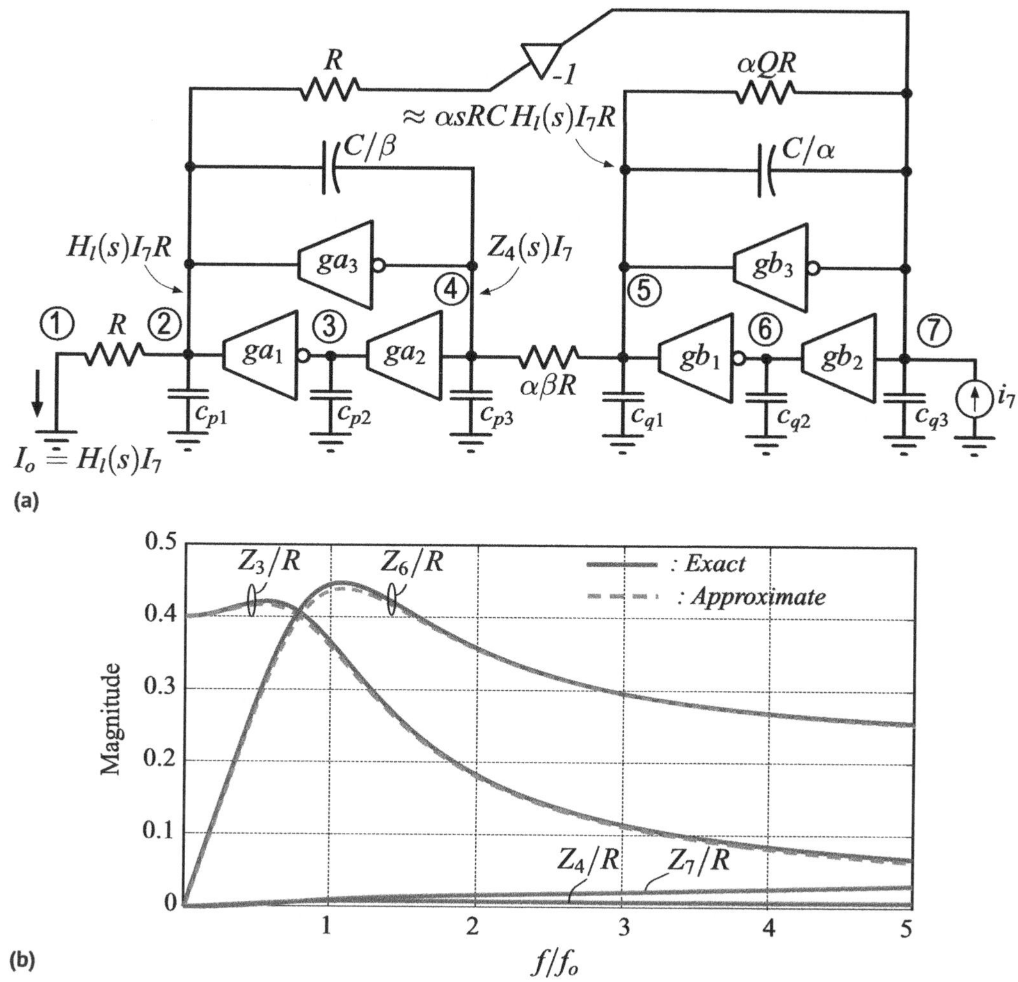

Figure 3.35 The inter-reciprocal(adjoint) network to determine the transfer impedances from the currents injected in the filter of Figure 3.35 to its output and comparison of the exact and approximate transfer impedances.

[a.] The voltage transfer function $V_7(s)/V_i(s) = H_l(s)$ in the original network is identical to the current transfer function $I_o(s)/I_7(s)$ in the adjoint network.

[b.] The current-to-output transfer functions $V_7(s)/I_3(s)$, $V_7(s)/I_4(s)$ etc. in the original network of Figure 3.35 are identical to the transfer functions from $I_7(s)$ to the voltages at ③, ④ etc. in the adjoint network of Figure 3.35. We again use the observation that the high-gain path in the two-stage OTA supplies most of the current for frequencies below the OTA unity-gain frequency. With this, the transimpedances can be easily approximated. We omit the algebra here; the final results are given in Table 3.2.

Figure 3.35b compares the approximate magnitudes of Z_3/R, Z_4/R, Z_6/R and Z_7/R with the exact responses. Again, good agreement is seen.

Table 3.1 Approximate expressions for the transfer impedances in Figure 3.34.

	Transfer function
$H_7(s) = \dfrac{V_7(s)}{V_i(s)}$	$H_l(s)$
$H_6(s) = \dfrac{V_6(s)}{V_i(s)}$	$H_l(s)\dfrac{1}{gb_2}\left(\dfrac{1}{\alpha QR} + \dfrac{1}{R} + s(C/\alpha + c_{q3})\right)$
$H_5(s) = \dfrac{V_5(s)}{V_i(s)}$	$-\dfrac{sc_{q2}}{g_{b1}}H_6(s)$
$H_4(s) = \dfrac{V_4(s)}{V_i(s)}$	$H_b(s)$
$H_3(s) = \dfrac{V_3(s)}{V_i(s)}$	$H_b(s)\dfrac{1}{ga_2}\left(\dfrac{1}{\alpha\beta R} + s(C/\beta + c_{p3})\right)$
$H_2(s) = \dfrac{V_2(s)}{V_i(s)}$	$-\dfrac{sc_{p2}}{g_{a1}}H_3(s)$

Table 3.2 Approximate expressions for the transfer impedances in Figure 3.34, obtained by solving the adjoint network of Figure 3.35.

Original network	Adjoint network	Transfer impedance
$Z_2(s)$	$\dfrac{V_2(s)}{I_7(s)}$	$H_l(s)R$
$Z_3(s)$	$\dfrac{V_3(s)}{I_7(s)}$	$-(H_l(s)/ga_1)\,(2 + sR(C/\beta + c_{p1}))$
$Z_4(s)$	$\dfrac{V_4(s)}{I_7(s)}$	$Z_3(s)(sc_{p2}/ga_2)$
$Z_5(s)$	$\dfrac{V_5(s)}{I_7(s)}$	$-\alpha sRCH_l(s)R$
$Z_6(s)$	$\dfrac{V_6(s)}{I_7(s)}$	$-\dfrac{sCRH_l(s)}{gb_1}\left(\dfrac{1}{Q} + \dfrac{1}{\beta} + sR(C + \alpha c_{q1})\right)$
$Z_7(s)$	$\dfrac{V_7(s)}{I_7(s)}$	$Z_6(s)(sc_{p2}/gb_2)$

As discussed earlier in this section, the intermodulation distortion due to a particular transconductor is proportional to the product of the appropriate $|H^3|$ and $|Z|$. For example, distortion introduced by ga_2 and gb_2 are proportional to $|H_3^3 Z_4|$ and $|H_6^3 Z_7|$, respectively. Since the nonlinear currents injected by ga_1 and ga_3 depend on H_2^3, which is much, much smaller in magnitude when compared to H_3^3, it is apparent that the dominant contributor to distortion in OTA-a is the transconductor ga_2. Similarly, gb_2 is the dominant contributor to distortion in OTA-b.

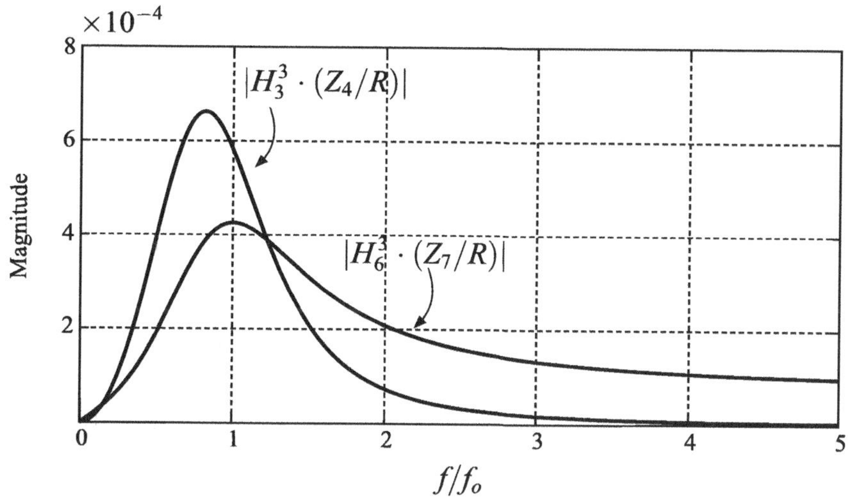

Figure 3.36 Magnitudes of $H_3^3(Z_4/R)$ and $H_6^3(Z_7/R)$ for the filter. All other terms are very small in comparison.

Figure 3.36 shows the relative magnitudes of $H_3^3 Z_4/R$ and $H_6^3 Z_7/R$. As described earlier, these are the dominant terms that one needs to be concerned about. IM_3, which is the ratio of the power of the intermodulation tones to that of the fundamental is given by

$$\mathrm{IM}_{3,a2} = \frac{3ga_2\eta A^2}{4|H_l(j2\pi f)|}|H_3^3(j2\pi f)Z_4(j2\pi f)| \quad \text{due to } g_{a2} \tag{3.61}$$

$$\mathrm{IM}_{3,b2} = \frac{3gb_2\eta A^2}{4|H_l(j2\pi f)|}|H_6^3(j2\pi f)Z_7(j2\pi f)| \quad \text{due to } g_{b2}$$

Figure 3.37 shows SPICE simulation results for IM_3 and compares it with our (approximate) analytical results. The two closely-spaced input tones have an average frequency f, which is swept from low frequencies up to twice the 3 dB bandwidth. The amplitude of each tone is 0.25 V. The transconductors are weakly nonlinear, with η chosen as 4/3. The curves show our analytical estimates assuming only OTA-a and OTA-b are nonlinear (neglecting the contributions of ga_1 and ga_3). The markers show the IM_3 values obtained from simulation. Good agreement is seen, confirming our approximations and intuition.

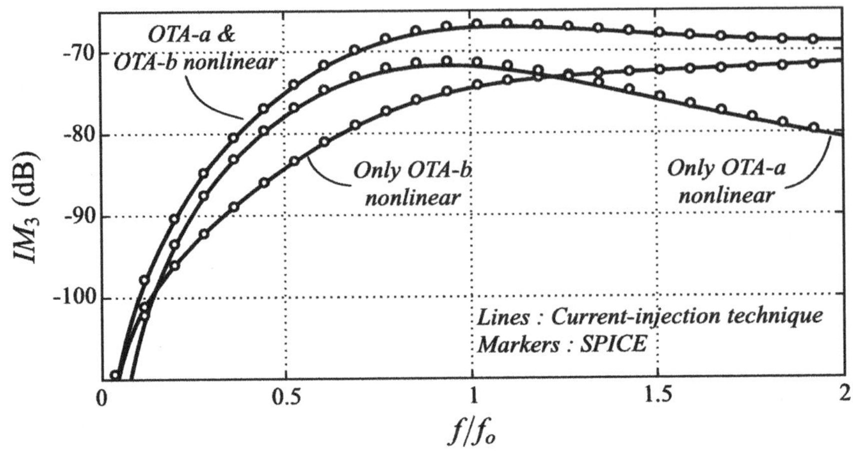

Figure 3.37 Comparison of our (approximate) analytical expressions for IM_3 in a filter with circuit simulations. $\eta = (4/3)/V^2$, and each input tone has an amplitude of 0.25 V.

What happens when both OTA-a and OTA-b are both weakly nonlinear? The intermodulation tones generated by ga_2 and gb_2 have to be added, keeping in mind the appropriate phase shifts experienced by them. Denoting the frequencies of the two input tones by $(f - \Delta f)$ and $(f + \Delta f)$ and their amplitudes by A, we see that the tones at node ③ can be approximated as

$$v_3(t) = A|H_3(j2\pi f)|\cos(2\pi(f - \Delta f)t + \phi) + A|H_3(j2\pi f)|\cos(2\pi(f + \Delta f)t + \phi)).$$

where $\phi = \angle H_3(j2\pi f)$. The injected intermodulation components, which arise from $v_3^3(t)$, and are at frequencies $(f \pm 3\Delta f)$ have an amplitude proportional to A^3, and *a phase equal to* ϕ. It is tempting, albeit incorrect, to think that the phase would be 3ϕ. Straightforward expansion of $v_3^3(t)$ will show this. In phasor notation, this corresponds to the term $|H_3^2(j2\pi f)|H_3(j2\pi f)$. The strength of the intermodulation tones that result when both transconductors are weakly nonlinear is thus given by

$$\boxed{\mathrm{IM}_3 = \frac{3\eta A^2}{4|H_i|} ga_2 H_3 |H_3|^2 Z_4 + gb_2 H_6 |H_6|^2 Z_7.} \tag{3.62}$$

The curve in Figure 3.37 is the expression above, while the markers are those obtained from SPICE. Again, good agreement is seen.

Chapter Summary

- **Weak nonlinearities.** Circuits are said to operate with weak nonlinearity when the distortion components are small in magnitude in relation to the linear terms.
- **Harmonic and intermodulation distortion.** We learned about even- and odd-order harmonic and intermodulation distortion. Furthermore, we recognized that HD_2 and IM_2 are proportional to input amplitude, while HD_3 and IM_3 are proportional to the square of the input amplitude.
- **Method of current injection.** A simple, powerful, and elegant technique to (approximately) solve weakly nonlinear networks. The method converts the problem of solving a nonlinear network into that of analyzing multiple linear networks—one for each nonlinear component. We applied this to several simple (and not so simple) circuits.

Problems

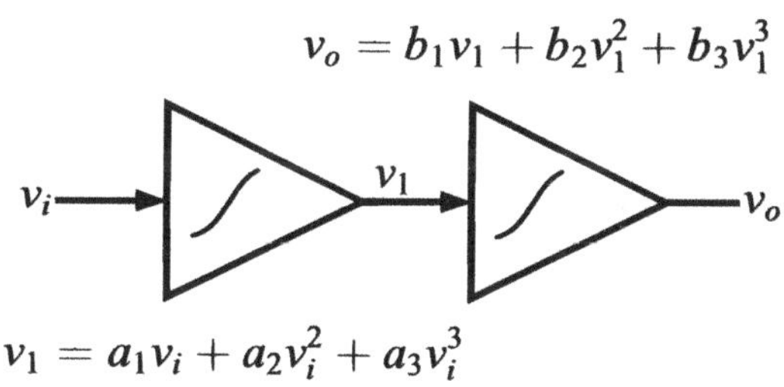

Figure 3.38

Problem 1. *Figure 3.38 shows a cascade of two amplifiers. Their second- and third-order input intercept points are denoted by IIP_{2a}, IIP_{2b}, IIP_{3a}, and IIP_{3b}. Determine the second- and third-order input intercept points for the cascade.*

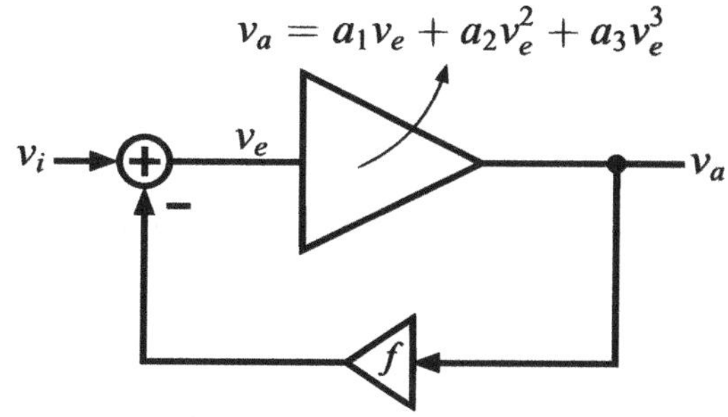

Figure 3.39

Problem 2. *Figure 3.39 shows a weakly nonlinear amplifier enclosed in a feedback loop. Determine v_a in terms of v_i.*

Problem 3. *In the feedback amplifier of the previous problem, assume that v_i was sinusoidal with an amplitude chosen so that v_a has a fundamental component with an amplitude A.*

Suppose that the forward amplifier was operated in an open-loop fashion, and the input was so adjusted that the output had the same fundamental amplitude A. Compare HD_2 and HD_3 obtained in the feedback and open-loop cases.

Problem 4. *A weakly nonlinear amplifier with a transfer curve $v_o = a_1v_i + a_2v_i^2 + a_3v_i^3$ is excited by two closely spaced tones at frequencies f_1 and f_2 with amplitudes A_1 and A_2, respectively. Determine the amplitudes of the third-order intermodulation tones at frequencies $(2f_1 - f_2)$ and $(2f_2 - f_1)$ at the output of the amplifier.*

Problem 5. *An amplifier has a transfer characteristic given by*

$$v_o = V_{max} \tanh\left(\frac{v_i}{V_x}\right). \tag{3.63}$$

Another amplifier has a weakly nonlinear characteristic, which in the range of applied inputs can be approximated by

$$v_o = a_1v_i - a_3v_i^3. \tag{3.64}$$

How would you choose V_{max} and V_x so that the two amplifiers yield the same distortion for small inputs?

(While running simulations, one should use the (3.63) to avoid convergence problems that will arise with (3.64) on account of its non-monotonicity.)

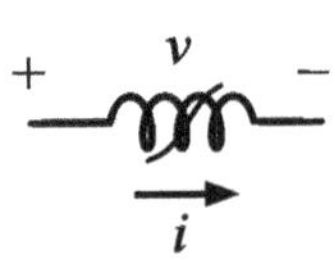

$$\phi = f(i) = L_o(i + k_2i^2 + k_3i^3)$$

Figure 3.40

Problem 6. *A weakly nonlinear inductor is one where the flux is nonlinearly related to current as shown in Figure 3.40.*

a. *How will you implement this in a circuit simulator that does not have a nonlinear inductor as a part of its library of built-in elements?*

b. *How will such an inductor appear in the first-, second-, and third-order circuits?*

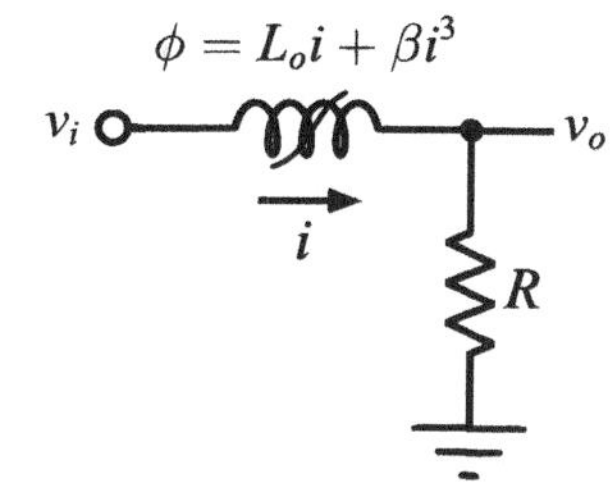

Figure 3.41

Problem 7. *Figure 3.41 shows a first-order lowpass filter. Unfortunately, the inductor is weakly nonlinear, with a ϕ–i relationship as shown. $\beta = 16\,H/A^2$, $R = 1\ k\Omega$, and the small-signal bandwidth is 100 kHz. $v_i(t) = A\sin(2\pi f t)$, with $A = 1$ V. Evaluate the third-harmonic distortion as a function of input frequency from 10 kHz to 150 kHz and compare your results with those obtained from a circuit simulator.*

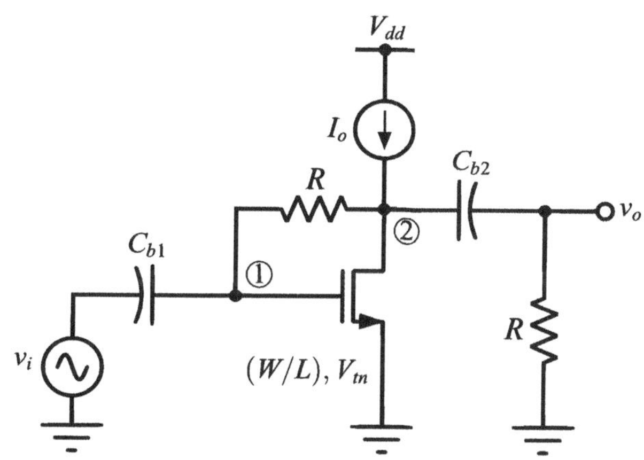

Figure 3.42

Problem 8. *Figure 3.42 shows a transistor amplifier. Neglecting channel-length modulation and assuming square-law behavior in saturation, determine the amplifier's IIP_2 and IIP_3. C_{b1} and C_{b2} can be assumed to be infinite so that they act like batteries.*

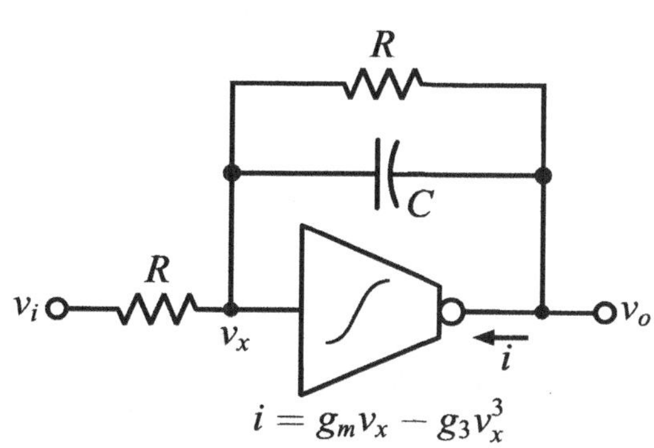

Figure 3.43

Problem 9. *Figure 3.43 shows a first-order OTA-RC filter. The OTA is weakly nonlinear as indicated.* $g_m R \gg 1$, *and* v_i *is a two-tone signal with the spacing between the tones being very small. Determine the* IIP_3 *as a function of the average frequency of the input tones.*

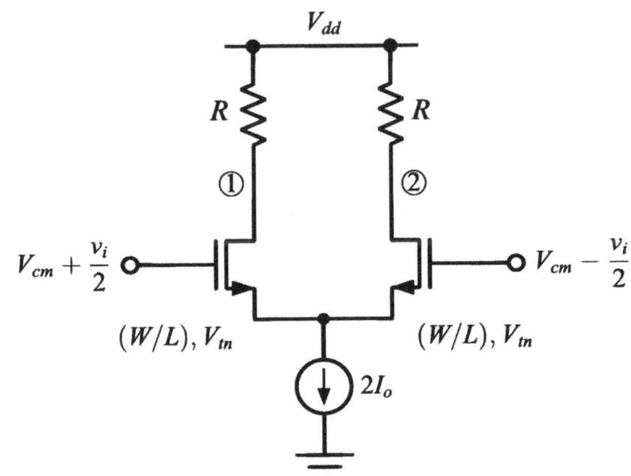

Figure 3.44

Problem 10. *Figure 3.44 shows a differential pair. The transistors operate in saturation and obey the square law. Neglect channel-length modulation. The output is* $v_o = (v_1 - v_2)$. *The transfer characteristic, for small* v_i, *is modeled by a polynomial as*

$$v_o = a_1 v_i + a_3 v_i^3.$$

Determine a_1 *and* a_3.

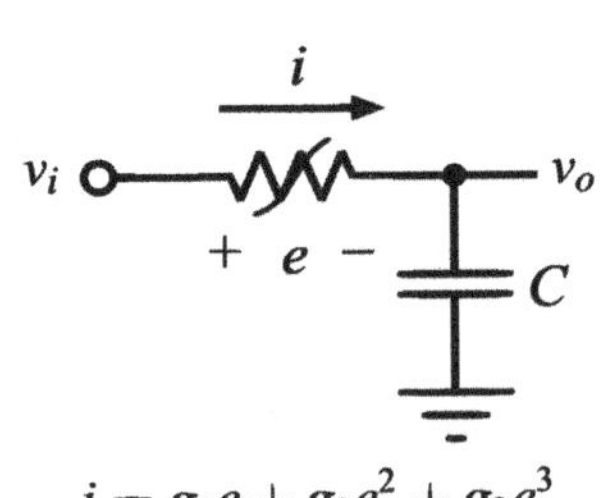

Figure 3.45

Problem 11. *This problem verifies the theory of current injection using simulations and is intended to give you a sense of the numbers involved and how well the method performs. Use the weakly nonlinear RC filter shown in Figure 3.45 to demonstrate the method. Use LTSPICE to set up the network. You should use the arbitrary behavioral* *controlled current source element in LTSPICE to implement the nonlinear conductance, as well as the various nonlinear currents needed in the current injection method.*

Use the following: $v_i = \sin(2\pi f_i t)$ *where* $f_i = 1$ *kHz,* $C = 159.15$ *nF,* $g_1 = 10^{-3}$ *S.*

a. *For this part, set* $g_2 = 10^{-5}$ *S/V and* $g_3 = 0$*. Plot the simulated* $v_o(t)$ *for* t *between 0 and 10 mS. On the same plot, show* $v_{o(1)}(t)$*. On a separate plot, show* $v_o(t) - v_{o(1)}(t)$*,* $v_o(t) - v_{o(1)}(t) - v_{o(2)}(t)$ *and* $v_o(t) - v_{o(1)}(t) - v_{o(2)}(t) - v_{o(3)}(t)$*. What do you observe?*

b. *For this part, set* $g_2 = 10^{-5}$ *S/V and* $g_3 = 10^{-5}$ *S/V*^{2}*. Plot the simulated* $v_o(t)$ *for* t *between 0 and 10 ms. On the same plot, show* $v_{o(1)}(t)$*. On a separate plot, show* $v_o(t) - v_{o(1)}(t)$*,* $v_o(t) - v_{o(1)}(t) - v_{o(2)}(t)$ *and* $v_o(t) - v_{o(1)}(t) - v_{o(2)}(t) - v_{o(3)}(t)$*. What do you observe?*

c. *For this part, set* $g_2 = 0$ *and* $g_3 = 10^{-4}$ *S/V*^{2}*. Plot the simulated* $v_o(t)$ *for* t *between 0 and 10 ms. On the same plot, show* $v_{o(1)}(t)$*. On a separate plot, show* $v_o(t) - v_{o(1)}(t)$*,* $v_o(t) - v_{o(1)}(t) - v_{o(2)}(t)$ *and* $v_o(t) - v_{o(1)}(t) - v_{o(2)}(t) - v_{o(3)}(t)$*. What do you observe?*

References

1 B. Razavi, *RF Microelectronics*, 2nd. Prentice Hall, 2011.

2 J. J. Bussgang, L. Ehrman, and J. W. Graham, "Analysis of nonlinear systems with multiple inputs," *Proceedings of the IEEE*, vol. 62, no. 8, pp. 1088–1119, 1974.

3 P. Wambacq and W. Sansen, *Distortion Analysis of Analog Integrated Circuits*. Kluwer Academic Publishers, 1998.

4 D. O. Pederson and K. Mayaram, *Analog Integrated Circuits for Communication: Principles, Simulation and Design*, 2nd. Springer Science & Business Media, 2008.

5 C. Kumar and S. Pavan, "Weak nonlinearities in active filters and continuous-time pipeline ADCs," *IEEE Transactions on Circuits and Systems I: Regular Papers*, vol. 72, no. 12, pp. 7565–7577, 2025.

4

Introduction to Distributed Circuits

If the dimensions of the circuit and its elements are much smaller than the wavelength of the fastest signal being processed, then we can assume that the current flowing at all points of a branch is the same and that the voltage along any wire with zero resistance is constant. The circuit can then be analyzed and designed using only Kirchhoff's current and voltage laws, and it is not necessary to use Maxwell's more generally valid equations. The wavelength of an electromagnetic signal of frequency f is given by $\lambda = v/f$, where v is the propagation speed of the signal. For a frequency of 3 GHz, this yields $\lambda = 3 \times 10^8$ m/s/3 $\times 10^9$ Hz $= 10^{-1}$ m $=$ 10 cm. On an integrated circuit, however, components have dimensions much smaller than this. It is therefore logical to ask the question: why does an IC designer need to worry about distributed circuits? The reasons are the following. While building blocks on chip can be treated as lumped circuits at most commonly used frequencies, chips often need to interface to external components. For instance, an RF transceiver chip has to connect to external filters and/or the antenna. The lengths of these connections are typically a significant fraction of the wavelength, necessitating treatment as distributed elements. A serial link transceiver chip might have to transmit/receive data at several gigabits per second from another chip located many inches away. The interconnect has to be modeled as a distributed element. Another practical aspect is that of measurements. While the on-chip building blocks are small by themselves, they need to be characterized, requiring connections to measurement equipment. These cables can be several feet long and need to be treated as distributed elements.

Low-frequency characterization of networks involves the determination of the admittance (Y) or impedance (Z) matrices. Measuring these parameters requires the network ports to be terminated in open or short circuits. When components of the test setup have to be treated as distributed elements, it turns out that making measurements with open- or short-circuit ports is difficult, if not impossible. Furthermore, voltages and currents are hard to measure at these frequencies, as they vary with position along the conductor. These practical problems motivate the introduction of another way of characterizing networks, namely the scattering parameters.

As the title of this chapter indicates, the intent of this chapter is to introduce the basics of ideal transmission-line networks and methods of analyzing them. Furthermore, we describe the fundamentals of the scattering matrix and its relationship with the more familiar admittance/impedance matrices. We conclude by introducing the principle behind a vector network analyzer, which is an instrument used to measure the scattering parameters.

4.1 The Transmission Line

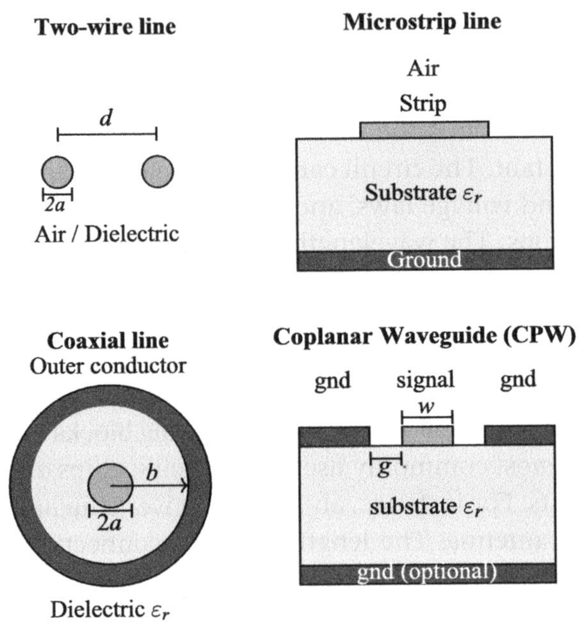

Figure 4.1 Cross sections of some commonly used transmission lines. The lines extend into and out of the plane of the page, which are also the directions of propagation of the traveling waves.

Simply speaking, a transmission line is a structure intended to transport electromagnetic waves from one point to another. In its basic form, it has a distributed two-port structure containing two conductors separated by an insulating layer, which may be air. Figure 4.1 shows the cross-section of some possible implementations [1, 2]. The structures extend into and out of the plane of the page, which are also the directions of propagation of the traveling waves. The microstrip line and coplanar waveguide can be fabricated on a printed circuit board or on an integrated circuit. A coaxial cable has an inner conductor and an outer conducting shield. Thanks to this structure, it is immune to external electromagnetic interference, and is used extensively. The insulator, needed for mechanical stability, is often chosen to be Teflon. In this chapter, we assume that all transmission lines are lossless, i.e., the metals are perfect conductors and the dielectric is a perfect insulator. Furthermore, the cross-section and other physical properties of the line are assumed to remain the same along the length of the line. Such a line is called the uniformly distributed lossless transmission line. Next, we analyze signal propagation along such a line.[1]

4.1.1 Analysis of the Uniformly Distributed Lossless Transmission Line

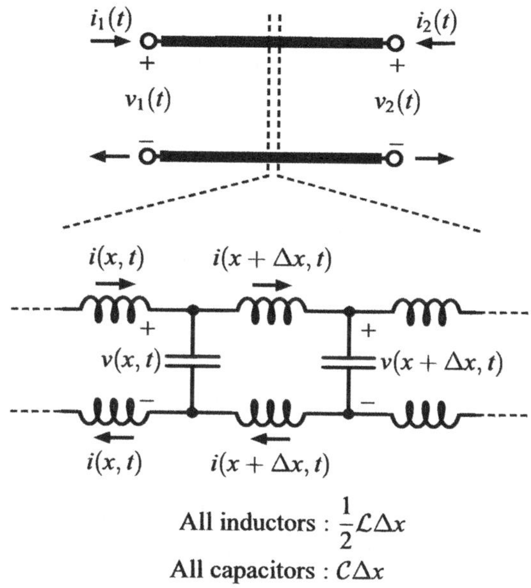

Figure 4.2 Circuit model for a lossless uniformly distributed transmission line. $\mathcal{L}$ and $\mathcal{C}$ denote the inductance and capacitance per unit length, respectively.

Figure 4.2 shows a uniformly distributed lossless transmission line and the circuit model of a section of length Δx at a position x. The voltage and current are functions of distance and time. The inductance and capacitance per unit length are denoted by $\mathcal{L}$ and $\mathcal{C}$, respectively. Writing KCL and KVL for the section, we see that

$$v(x+\Delta x, t) = v(x,t) - \mathcal{L}\Delta x \frac{\partial i(x+\Delta x, t)}{\partial t}$$
$$i(x+\Delta x, t) = i(x,t) - \mathcal{C}\Delta x \frac{\partial v(x,t)}{\partial t}.$$

As $\Delta x \to 0$, these reduce to

$$\begin{aligned} \frac{\partial v(x,t)}{\partial x} &= -\mathcal{L}\frac{\partial i(x,t)}{\partial t} \\ \frac{\partial i(x,t)}{\partial x} &= -\mathcal{C}\frac{\partial v(x,t)}{\partial t}. \end{aligned} \tag{4.1}$$

[1]Wave propagation we consider is the transverse electromagnetic (TEM) mode. The electric and magnetic fields are in the plane of the paper, and the direction of propagation is perpendicular to the plane of the paper.

These are two coupled first-order partial differential equations in $v(x,t)$ and $i(x,t)$. To solve for $v(x,t)$, we eliminate $i(x,t)$ by differentiating the first equation above with respect to x, the second with respect to t, and using $\partial^2 i(x,t)/\partial x \partial t = \partial^2 i(x,t)/\partial t \partial x$ to obtain the wave equation

$$\boxed{\frac{\partial^2 v(x,t)}{\partial x^2} = \mathcal{L}\mathcal{C}\frac{\partial^2 v(x,t)}{\partial t^2}.} \tag{4.2}$$

To solve for $v(x,t)$, we observe the following. Consider any function of the form $f(x \pm \mathcal{c}t)$. Notice that

$$\begin{aligned}\frac{\partial^2 f(x \pm \mathcal{c}t)}{\partial x^2} &= f''(x \pm \mathcal{c}t)\\ \frac{\partial^2 f(x \pm \mathcal{c}t)}{\partial t^2} &= \mathcal{c}^2 f''(x \pm \mathcal{c}t),\end{aligned} \tag{4.3}$$

where the double prime refers to the second derivative of $f(\cdot)$ with respect to its argument $(x + \mathcal{c}t)$. From (4.2) and (4.3), we see that $f(x \pm \mathcal{c}t)$ satisfies the wave equation if $\mathcal{c} = 1/\sqrt{\mathcal{L}\mathcal{C}}$. It is evident that $\mathcal{c}$ has the units of velocity. Since $f()$ is arbitrary, we can write $v(x,t)$ as a linear combination of a function $f_1(x - \mathcal{c}t)$ and another function $f_2(x + \mathcal{c}t)$. What does $f_1(x - \mathcal{c}t)$ look like? To see this, suppose we took snapshots of $f_1(x - \mathcal{c}t)$ at times $t = 0$ and $t = t_1$. This will be $f_1(x)$ at $t = 0$ and $f_1(x - \mathcal{c}t_1)$ at $t = t_1$; i.e., it appears as if $f_1()$ has traveled to the right by a distance $\mathcal{c}t_1$. $f_1()$, therefore, is referred to as the forward wave. To remind ourselves that it is a voltage that moves forward, we choose to call it v^+. By the same logic, $f_2()$ is a voltage wave that moves backward (in the negative x direction), which we call v^-. We can therefore express $v(x,t)$ as[2]

$$v(x,t) = v^+(x - \mathcal{c}t) + v^-(x + \mathcal{c}t). \tag{4.4}$$

What is the current $i(x,t)$? It can be obtained by using (4.1) as the sum of a forward current i_+ (due to v^+) and a backward current i_- (due to v^-).

$$i(x,t) = \underbrace{\frac{1}{Z_o}v^+(x - \mathcal{c}t)}_{i^+(x,t)} - \underbrace{\frac{1}{Z_o}v^-(x + \mathcal{c}t)}_{i^-(x,t)}, \tag{4.5}$$

where

$$\boxed{Z_o = \sqrt{\frac{\mathcal{L}}{\mathcal{C}}},} \tag{4.6}$$

is termed the *characteristic impedance* of the transmission line. To summarize,

[2] This is the d'Alembert solution to the wave equation. Alternative (and equivalent) forms that satisfy the wave equation are $f(\pm x/\mathcal{c} \pm t)$.

a. The voltage at any point on a uniformly distributed transmission line can be expressed as the sum of a forward-going wave v^+ and a backward-going wave v^-, which travel in the positive and negative x directions, respectively, with a velocity $c = 1/\sqrt{\mathcal{L}\mathcal{C}}$.

b. At any point along the line, the forward-going wave is associated with a current $i^+ = v^+/Z_o$, while the backward-going wave is associated with a current $i^- = -v^-/Z_o$. The total current at a point is the sum of the forward and backward currents.

c. The transmission line is specified by its characteristic impedance Z_o, and its length l.

d. The time taken for a wave to travel the length of the line is given by $\tau = l\sqrt{\mathcal{L}\mathcal{C}}$. Recall that $\mathcal{L}$ and $\mathcal{C}$ are the inductance and capacitance of the line *per unit length*. Rather than specify $\mathcal{L}$, $\mathcal{C}$, and the length of the line l, one may alternatively specify the line by its characteristic impedance and its delay, namely Z_o and τ, respectively.

It is instructive to determine the total inductance and capacitance of a transmission line of length l, which we denote by L_{tot} and C_{tot} respectively. $L_{tot} = \mathcal{L}l$ and $C_{tot} = \mathcal{C}l$. Using $\tau = l\sqrt{\mathcal{L}\mathcal{C}} = \sqrt{L_{tot}C_{tot}}$ and $Z_o = \sqrt{\mathcal{L}/\mathcal{C}} = \sqrt{L_{tot}/C_{tot}}$, we obtain

$$\boxed{L_{tot} = Z_o\tau, \qquad C_{tot} = \frac{\tau}{Z_o}.} \tag{4.7}$$

A lumped inductor of value L can, therefore, be thought of as an infinitesimal transmission line with infinite characteristic impedance where $Z_o\tau = L$. Along the same lines, a lumped capacitor of value C can be thought of as an infinitesimal transmission line with a characteristic impedance tending to zero, with $\tau/Z_o = C$.

What is the capacitance of a cable?

A student needs to view the digital waveform at a node inside her chip. A "strong" on-chip CMOS inverter, designed to drive a maximum load of 100 pF, senses this waveform and connects to one of the chip's output pads. The student intends to connect the pad to an oscilloscope so as to capture the waveform. She uses an 80 cm-long coaxial cable to connect from the test PCB to the scope. The oscilloscope's input can be considered high impedance. Will the on-chip inverter be able to drive the cable's capacitance?

Recall that the total capacitance of the cable, which is a transmission line, is given by τ/Z_o, where τ is the time taken for a wave to travel the length of the cable, and Z_o is its characteristic impedance. To find capacitance, we note that the "standard" Z_o used for cables is 50 Ω. To estimate τ, we proceed as follows. Assuming that

(Continued)

(Continued)

the space between the conductors of the cable is filled with Teflon, whose dielectric constant is 2.25, the velocity of light in the medium is $(3/\sqrt{2.25}) \times 10^8$ m/s = 20 cm/ns. τ is thus seen to be 4 ns. $C = \tau/Z_o = 80$ pF. The CMOS driver should therefore be able to drive the cable.

4.2 Transmission-Line Circuit Analysis

Having understood the basic aspects of wave propagation on a lossless transmission line, let us turn our attention to what happens at the boundaries of the line.

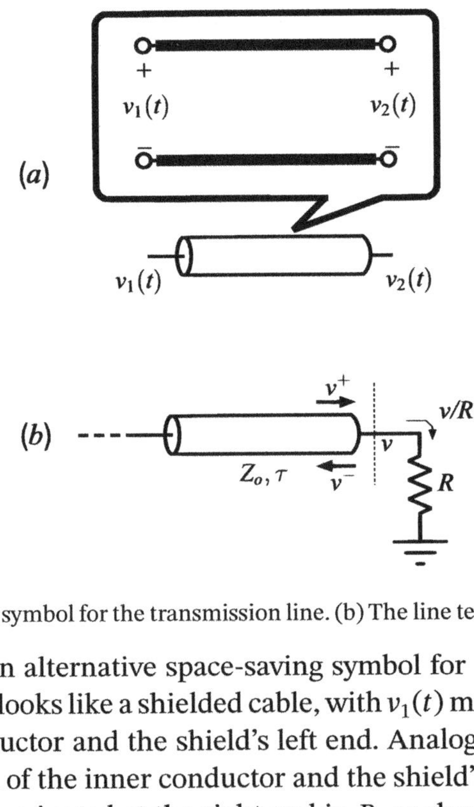

Figure 4.3 (a) Alternative symbol for the transmission line. (b) The line terminated with a resistance R.

Figure 4.3a shows an alternative space-saving symbol for a transmission line that we use in this book. It looks like a shielded cable, with $v_1(t)$ measured between the left end of the inner conductor and the shield's left end. Analogously, $v_2(t)$ is measured between the right end of the inner conductor and the shield's right end. Suppose the transmission line is terminated at the right end in R, as shown in Figure 4.3b. What constraints does the termination impose? The voltage and current must, of course, remain identical to the left and right of the line and resistor interface. Denoting the voltage across R by v, we have

$$\begin{aligned} v^+ + v^- &= v \\ \frac{1}{Z_o}(v^+ - v^-) &= \frac{v}{R}, \end{aligned} \tag{4.8}$$

from which we see that

$$\frac{v^-}{v^+} = \frac{R - Z_o}{R + Z_o} \ (\equiv \Gamma). \tag{4.9}$$

Termination of the line with R, therefore, relates v^- and v^+ at the end of the line. v^-/v^+ is called the *reflection coefficient* and is denoted by the symbol Γ. If R is replaced by an impedance $Z(s)$, it is more convenient to work with Laplace-transformed quantities v^+ and v^-, and Γ is given by

$$\boxed{\Gamma(s) = \frac{V^-(s)}{V^+(s)} = \frac{Z(s) - Z_o}{Z(s) + Z_o}.} \tag{4.10}$$

Note that the V^+ and V^- in the equation above refer to the transforms of the forward- and backward waves at the ends of the line. We note the following special cases:

a. If $Z(s) = Z_o$, $\Gamma = 0$. The termination is referred to as being *matched* to the transmission line.
b. If $Z(s)$ is an open circuit, $\Gamma = 1$.
c. If $Z(s)$ is a short circuit, $\Gamma = -1$.

Suppose that the circuit of Figure 4.3 is operating in the sinusoidal steady state at a frequency f. The power dissipated in the terminating impedance is given by

$$P = \frac{1}{2}\mathrm{Re}(VI^*) = \frac{1}{2Z_o}\mathrm{Re}\{(V^+ + V^-)((V^+)^* - (V^-)^*)\} = \frac{1}{2Z_o}(|V^+|^2 - |V^-|^2). \tag{4.11}$$

Since $V^- = \Gamma V^+$, it follows that

$$P = \frac{|V^+|^2}{2Z_o}(1 - |\Gamma|^2). \tag{4.12}$$

This result can be interpreted as follows. The forward wave (which is also called the incident wave) is associated with a power $|V^+|^2/2Z_o$. A fraction $|\Gamma|^2$ of this power is reflected back, and the rest is dissipated in the load. For a matched termination, no power is reflected—all of the incident power is dissipated in the load. On the other hand, if the terminating impedance is an open or a short, $|\Gamma| = 1$, and all the incident power is reflected back into the line.

The discussion above also indicates the following. If the terminating impedance is passive, the power dissipated $P \geq 0$. Consequently, $|\Gamma(j2\pi f)| \leq 1$. If the impedance is lossless, $Z(j2\pi f)$ is purely imaginary and $|\Gamma| = 1$.

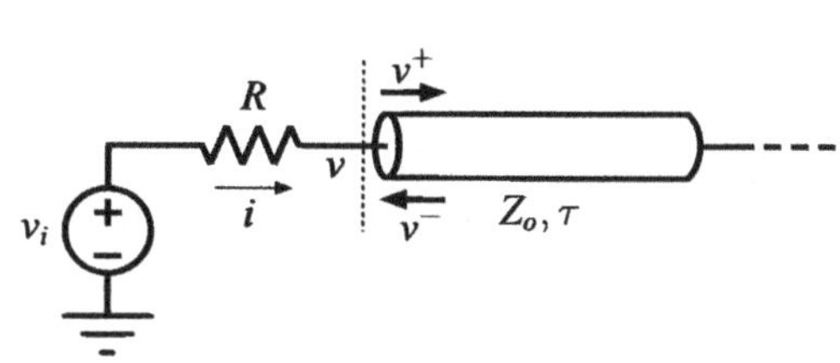

Figure 4.4 The line driven by a source with resistance R.

Let us now examine a transmission line excited by a source with resistance R, as shown in Figure 4.4. Equating voltages and currents across the interface (indicated by the dotted line), we obtain

$$v_i - iR = v^+ + v^-$$
$$i = \frac{1}{Z_o}(v^+ - v^-). \tag{4.13}$$

By eliminating i from the equations above, we see that v^+ can be related to v^- according to

$$v^+ = v_i \frac{Z_o}{Z_o + R} + v^- \frac{R - Z_o}{R + Z_o}. \tag{4.14}$$

Why does this expression make intuitive sense? The first term is justified due to the following. Suppose the line was terminated at the load end with a resistor of value Z_o. There would then be no reflected wave (i.e., $v^- = 0$). The voltage and current at the source end of the transmission line would simply be v^+ and v^+/Z_o, respectively, which indicates that the impedance looking into the line is simply Z_o. v^+ is thus the output of the potential divider formed by R and Z_o. The second term in (4.14) makes sense due to the following. Suppose $v_i = 0$. v^- is incident on the junction between the line and the source resistance R, so v^+ must be given by Γv^-, where Γ is the reflection coefficient at the source and is given by $(R - Z_o)/(R + Z_o)$. If the output impedance of the source is denoted by $Z(s)$, we work with Laplace-transformed quantities to write

$$V^+ = V_i \frac{Z_o}{Z_o + Z(s)} + V^- \frac{Z(s) - Z_o}{Z(s) + Z_o}. \tag{4.15}$$

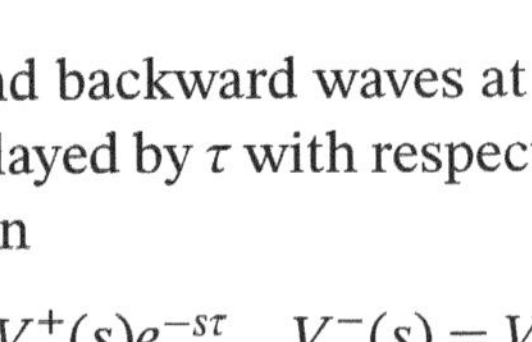

Figure 4.5 Deriving the relationships between the forward and backward waves at the two ends of the line.

We now relate the forward and backward waves at the two ends of the line shown in Figure 4.5. v_{2+} and v_{1-} are delayed by τ with respect to v_{1+} and v_{2-}, respectively. We thus have in the Laplace domain

$$V_2^+(s) = V_1^+(s)e^{-s\tau}, \quad V_1^-(s) = V_2^-(s)e^{-s\tau}. \tag{4.16}$$

We are now ready to analyze the network of Figure 4.6, where the aim is to find the voltage at the load end of the line, namely v_2. We work with Laplace transformed quantities. To reduce clutter in the expressions, we refer to $Z_l(s)$ as Z_l etc. The source and load reflection coefficients, denoted by Γ_s and Γ_l, respectively are given by

$$\Gamma_s = \frac{Z_s - Z_o}{Z_s + Z_o}, \quad \Gamma_l = \frac{Z_l - Z_o}{Z_l + Z_o}. \tag{4.17}$$

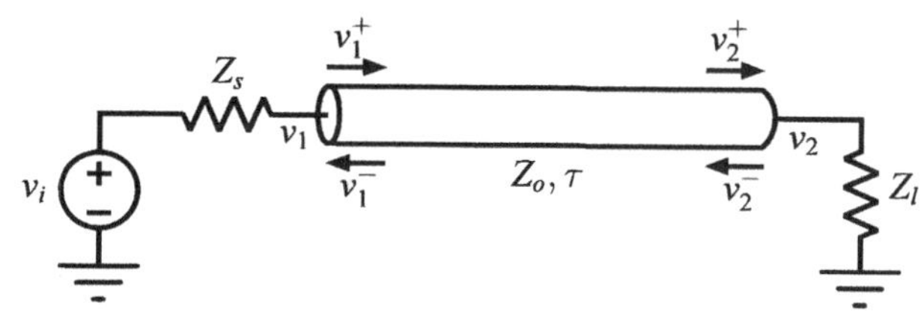

Figure 4.6 A source with impedance $Z_s(s)$ driving a line terminated by $Z_l(s)$.

We have the following system of equations in as many unknowns.

$$V_1^+ = V_i \frac{Z_o}{Z_o + Z_s} + \Gamma_s V_1^-$$
$$V_2^- = \Gamma_l V_2^+$$
$$V_2^+ = V_1^+ e^{-s\tau}$$
$$V_1^- = V_2^- e^{-s\tau}.$$

From the equations above, we obtain

$$V_2 = V_2^+(1 + \Gamma_l) = V_i \frac{Z_o e^{-s\tau}(1 + \Gamma_l)}{(Z_o + Z_s)(1 - \Gamma_l \Gamma_s e^{-2s\tau})}$$
$$V_1 = V_i \frac{Z_o(1 + \Gamma_l e^{-2s\tau})}{(Z_o + Z_s)(1 - \Gamma_l \Gamma_s e^{-2s\tau})}. \tag{4.18}$$

It is instructive to consider some special cases.

Source and Load Impedances Matched to the Line ($Z_s = Z_l = Z_o$)

The source and load are matched to the line. Consequently, $\Gamma_l = \Gamma_s = 0$. From (4.18), we have

$$V_2 = V_2^+ = \frac{1}{2} V_i e^{-s\tau}. \tag{4.19}$$

This makes sense due to the following. Since Z_l is matched to the line, $V_2^- = V_1^- = 0$, and the looking-in impedance at the source end of the line is Z_o. Since $Z_s = Z_o$, $V_1 = V_1^+ = V_i/2$. V_2^+ is a τ-delayed version of V_1^+. Consequently, $V_2 = (1/2)V_i e^{-s\tau}$.

Load Impedance Matched to the Line: Source Impedance is Zero ($Z_l = Z_o, Z_s = 0$)

Since the load is matched, $\Gamma_l = 0$, and $\Gamma_s = -1$ since $Z_s = 0$. From (4.18), we have

$$V_2 = V_2^+ = V_i e^{-s\tau}. \tag{4.20}$$

The load voltage is the same as V_i, except for a delay τ, which is the time needed for the forward wave to propagate to the load end. Due to the singly-terminated line, the load voltage is twice that in the doubly-terminated case.

Source Impedance Matched to the Line, Load is Open ($Z_s = Z_o, Z_l = \infty$)

The source is matched to the line. Consequently, $\Gamma_s = 0$, and $\Gamma_l = 1$ since the load is an open circuit. From (4.18), we have

$$V_2 = (1 + \Gamma_l)V_2^+ = V_i e^{-s\tau}. \tag{4.21}$$

The load voltage is the same as V_i, except for a delay τ, which is the time needed for the forward wave to propagate to the load end.

Source Impedance Matched to the Line, Arbitrary Load ($Z_s = Z_o$)

The source is matched to the line. Consequently, $\Gamma_s = 0$. From (4.18), we have

$$V_2 = (1 + \Gamma_l)V_2^+ = \frac{1}{2}V_i e^{-s\tau}(1 + \Gamma_l) = V_i e^{-s\tau}\frac{Z_l}{Z_l + Z_o}. \tag{4.22}$$

This makes sense due to the following. The Thevenin voltage seen by the load, which is the load voltage when it is an open circuit, is $V_2 = V_i e^{-s\tau}$ as we have seen previously. The Thevenin impedance seen by the load is Z_o since $Z_s = Z_o$. V_2 is thus given by $V_i e^{-s\tau} Z_l/(Z_l + Z_o)$.

Arbitrary Source and Load Impedance

Suppose Z_l and Z_s are not matched to the line. From (4.18), we have

$$V_2 = V_i(1 + \Gamma_l)\frac{Z_o e^{-s\tau}}{(Z_o + Z_s)(1 - \Gamma_l\Gamma_s e^{-2s\tau})}. \tag{4.23}$$

The equation above can be equivalently written as follows, where the $1/(1 - \Gamma_l\Gamma_s e^{-2s\tau})$ term in the denominator of the equation above is expanded as a geometric series.

$$\begin{aligned}
V_2 &= V_i(1 + \Gamma_l)\frac{Z_o e^{-s\tau}}{(Z_o + Z_s)}\left\{1 + \Gamma_l\Gamma_s e^{-2s\tau} + (\Gamma_l\Gamma_s)^2 e^{-4s\tau} + \cdots\right\} \\
&= V_i\frac{Z_o e^{-s\tau}}{(Z_o + Z_s)}(1 + \Gamma_l) + \\
&\quad V_i\frac{Z_o e^{-s\tau}}{(Z_o + Z_s)}\Gamma_l\Gamma_s e^{-2s\tau}(1 + \Gamma_l) + \\
&\quad V_i\frac{Z_o e^{-s\tau}}{(Z_o + Z_s)}(\Gamma_l\Gamma_s)^2 e^{-4s\tau}(1 + \Gamma_l) + \cdots
\end{aligned} \tag{4.24}$$

V_2 can be interpreted as follows.

a. $V_i Z_o/(Z_o + Z_s)$ $(= V_1^+)$ is the forward wave launched by the source onto the line. It reaches the end of the line after a time τ and gets reflected by the load with a coefficient Γ_l. The resulting output due to this is given by $(V_i Z_o/(Z_o + Z_s))e^{-s\tau}(1 + \Gamma_l)$, and is the first term in (4.24).

b. The component reflected at the load, which is given by $(V_i Z_o/(Z_o + Z_s))e^{-s\tau}\Gamma_l$, travels to the source end, gets reflected by a coefficient Γ_s, and arrives at the load end a round-trip time 2τ later, where it is reflected by a coefficient Γ_l again. The resulting contribution to V_2 is thus seen to be $(V_i Z_o/(Z_o + Z_s))e^{-3s\tau}\Gamma_l\Gamma_s(1 + \Gamma_l)$. This is the second term in (4.24). This process continues ad infinitum. The intuition above can be illustrated using what is called the *bounce diagram*, shown in Figure 4.7.

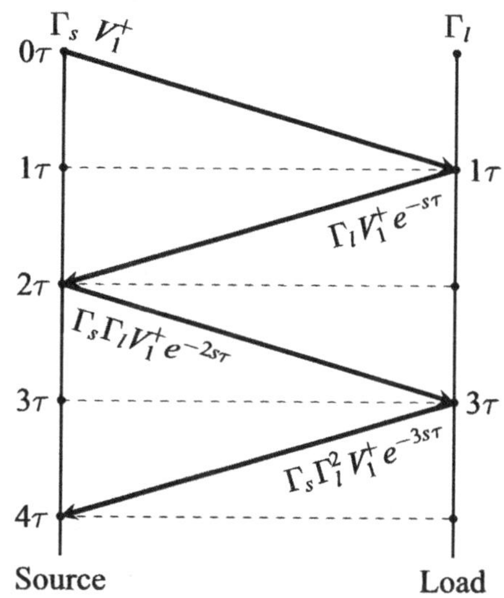

Figure 4.7 The bounce diagram. τ denotes the delay of the line, and Γ_l and Γ_s denote the load and source reflection coefficients, respectively.

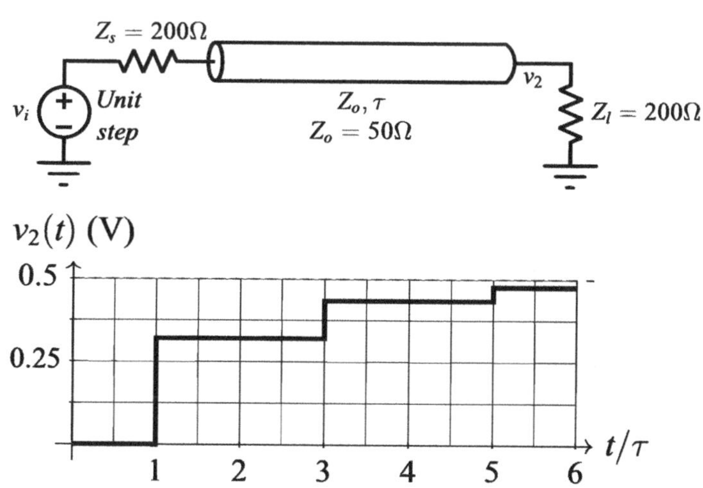

Figure 4.8 Step response for $Z_l = Z_s = 200\ \Omega$.

Figure 4.8 shows the $v_2(t)$ waveform for an input step when $Z_s = Z_l = 200\ \Omega$ and $Z_o = 50\ \Omega$. $v_2(t)$ first jumps at $t = \tau$, with subsequent jumps occurring at $3\tau,\ 5\tau, \ldots$. As $t \to \infty$, $v_2(t) \to 0.5$.

Simulation Corner

How does one "see" the forward- and backward-going waves in a simulator? In Figure 4.9a for instance, the load is shorted, so $v_2 = 0$. Suppose we were interested in what $v_2^+(t)$ looks like. We only have access to v_2 and $i_2(t)$. The voltage and current must be continuous across the transmission-line boundary; consequently, we have

$$\begin{aligned} v_2 &= v_2^+ + v_2^- \\ i_2 &= \frac{1}{Z_o}(v_2^+ - v_2^-). \end{aligned} \tag{4.25}$$

This means that v_2^+ and v_2^- can be obtained as $v_2^+ = (1/2)(v_2 + i_2 Z_o)$ and $v_2^- = (1/2)(v_2 - i_2 Z_o)$, using controlled sources as shown in Figure 4.9a.

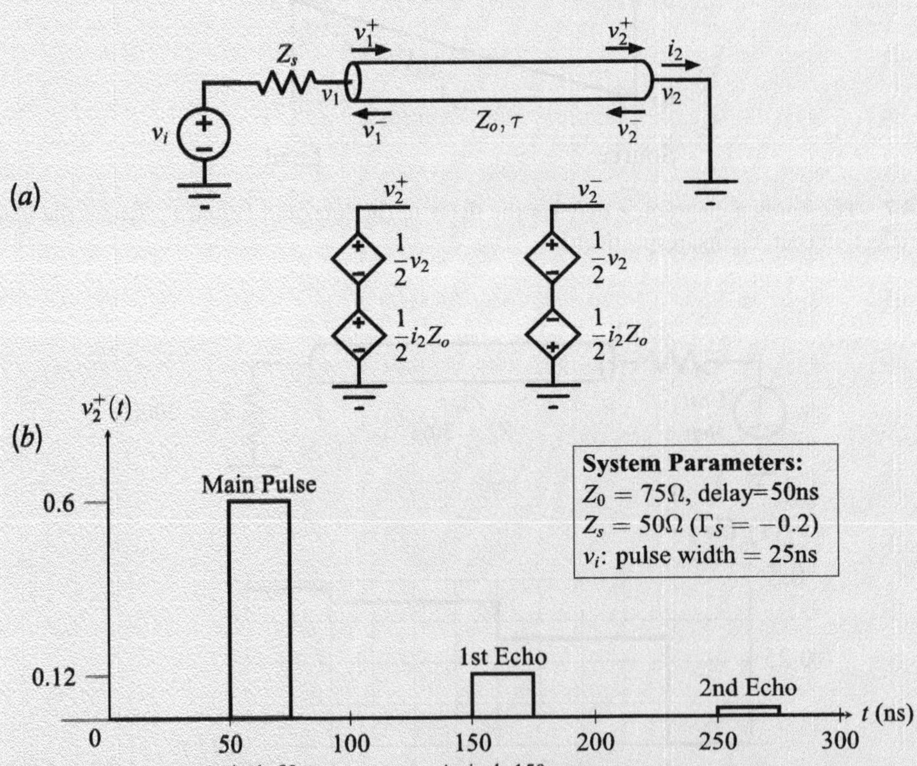

Figure 4.9 (a) Simulation setup to view the forward- and backward-going waves. Since only v_2 and i_2 are accessible, controlled sources are used to generate $v_2^+(t)$ and $v_2^-(t)$. (b) Example $v_2^+(t)$ for a 50 ns-long 75 Ω line driven by a unit 25 ns-wide pulse, with $Z_s = 50\ \Omega$.

Figure 4.9b shows example simulation results, with $Z_o = 75\ \Omega$, $Z_s = 50\ \Omega$, and $\tau =$ 50 ns. v_i is a unit-height 25 ns-wide pulse. $v_2(t)$ is zero as the load end of the line is shorted to ground. $v_2^+(t)$ however, is the series of pulses shown in Figure 4.9b. The main pulse, which has a height of $Z_o/(Z_o + Z_s) = 0.6$, reaches the load end of the line at time $t = \tau$. It gets reflected with a coefficient -1, reaches the source-side interface where it is reflected back with a coefficient $\Gamma_s = -0.2$ and hits the load end at a time $t = \tau + 2\tau = 150$ ns. The height of this "first echo" is $0.6 \times 0.2 = 0.12$. The second echo occurs at $t = 5\tau = 250$ ns.

Voltage and Current on a Transmission Line, and the Voltage Standing Wave Ratio (VSWR)

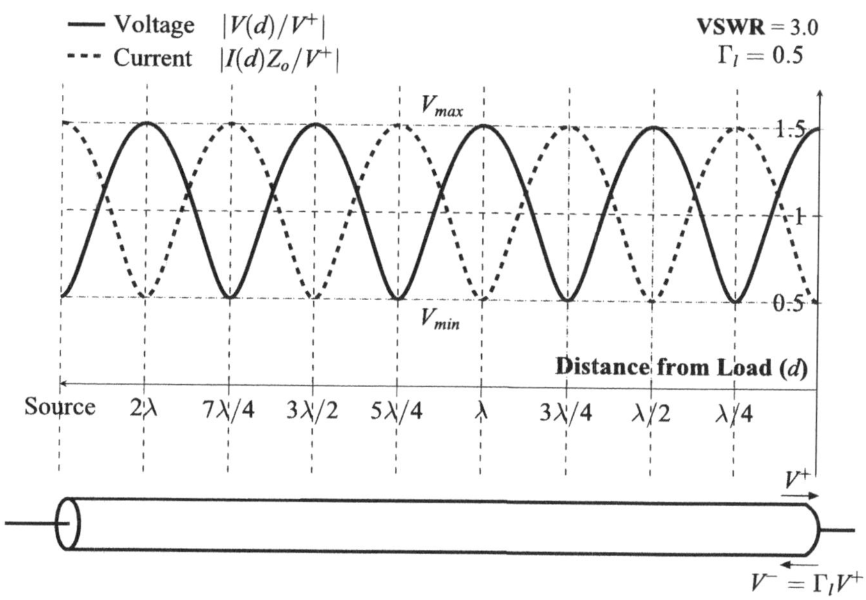

Figure 4.10 Normalized amplitude of voltage and current along a transmission line. The load reflection coefficient is 0.5.

As we have already seen, the voltage at any point on a transmission line can be expressed as the sum of the forward- and backward-going waves. This means that the voltage waveform on the line varies with the position at which it is measured. A similar argument holds for the current. We illustrate this when the line is operating in the sinusoidal steady state. As an example, consider a line terminated with a resistive load, so that the load reflection coefficient is 0.5 as shown in Figure 4.10. We denote the amplitude of the forward-going wave at the load end by V^+. Due to the reflection at the load end, the amplitude of the voltage there is $V^+ + V^- = V^+ + \Gamma_l V^+ = 1.5V^+$. As we travel towards the source, the phase of the forward-going wave leads with respect to V^+, while that of V^- lags with respect to V^-. At a distance $\lambda/4$ away from the load, where λ denotes the wavelength, the phase lead and lag of the forward- and backward-going waves are 90°. Consequently, the relative phase between them is 180°. The forward and reverse going waves interfere destructively, resulting in a voltage with amplitude $(V^+ - V^-) = 0.5V^+$.

A further $\lambda/4$ away, the relative phase between the forward- and backward-going waves is again zero, and there is a voltage maximum due to the constructive interference between the forward and backward-going waves. The ratio of the maximum and minimum voltage amplitudes on the line is called the voltage standing-wave ratio (VSWR) and is given by

$$\text{VSWR} = \frac{1 + |\Gamma_l|}{1 - |\Gamma_l|}. \tag{4.26}$$

In our example, where $\Gamma_l = 0.5$, the VSWR=3. VSWR = 1 for a matched load and infinite when $|\Gamma_l| = 1$. A similar conclusion can be drawn for the amplitude of the current along the line [3]. It is evident that a voltage maximum must coincide with a current minimum and vice versa, as shown in Figure 4.10.

4.2.1 Impedance Transformation Due to a Lossless Line

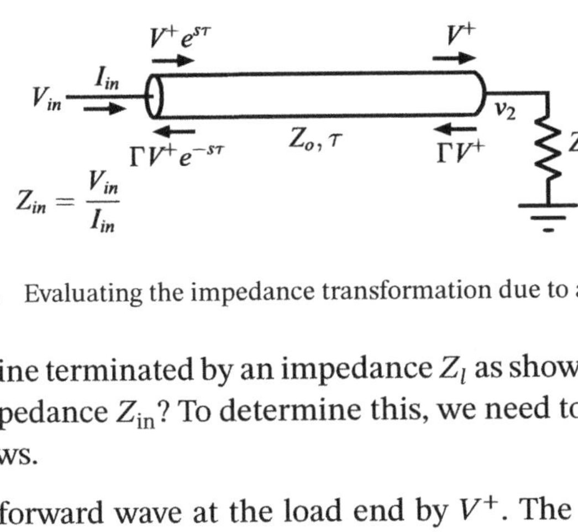

Figure 4.11 Evaluating the impedance transformation due to a lossless line.

Consider a lossless line terminated by an impedance Z_l as shown in Figure 4.11. What is the looking-in impedance Z_{in}? To determine this, we need to find the ratio $V_{\text{in}}/I_{\text{in}}$. We proceed as follows.

a. We denote the forward wave at the load end by V^+. The reflected wave is thus given by ΓV^+ where $\Gamma = (Z_l - Z_o)/(Z_l + Z_o)$.

b. The forward and backward waves at the input of the line are given by $V^+e^{s\tau}$ and $\Gamma V^+e^{-s\tau}$, respectively.

c. The voltage and current at the input of the line, namely V_{in} and I_{in}, are thus seen to be

$$V_{\text{in}} = V^+e^{s\tau} + \Gamma V^+e^{-s\tau}$$

$$I_{\text{in}} = \frac{1}{Z_o}(V^+e^{s\tau} - \Gamma V^+e^{-s\tau}).$$

This results in

$$Z_{\text{in}} = Z_o \frac{1 + \Gamma e^{-2s\tau}}{1 - \Gamma e^{-2s\tau}}. \tag{4.27}$$

Using $\Gamma = (Z_l - Z_o)/(Z_l + Z_o)$, we can rewrite this as

$$Z_{\text{in}} = Z_o \frac{Z_l + Z_o \tanh(s\tau)}{Z_o + Z_l \tanh(s\tau)}. \tag{4.28}$$

For a sinusoidal excitation, $s = j2\pi f$, and $\tanh(s\tau) = j\tan(2\pi f\tau)$. Using this in the equation above, we have

$$\boxed{Z_{\text{in}}(j2\pi f) = Z_o \frac{Z_l(j2\pi f) + jZ_o \tan(2\pi f\tau)}{Z_o + jZ_l(j2\pi f)\tan(2\pi f\tau)}}. \tag{4.29}$$

It is instructive to consider a few special cases of Z_l.

a. If $Z_l = Z_o$, the line is matched to Z_l, and $Z_{\text{in}} = Z_o$ as expected.
b. Suppose $Z_l = 0$. From (4.29), we see that $Z_{\text{in}}(j2\pi f) = jZ_o \tan(2\pi f\tau)$. Z_{in} is imaginary, as the circuit has no dissipative elements in it. Figure 4.12 shows the reactance of the short-circuited line as a function of frequency.

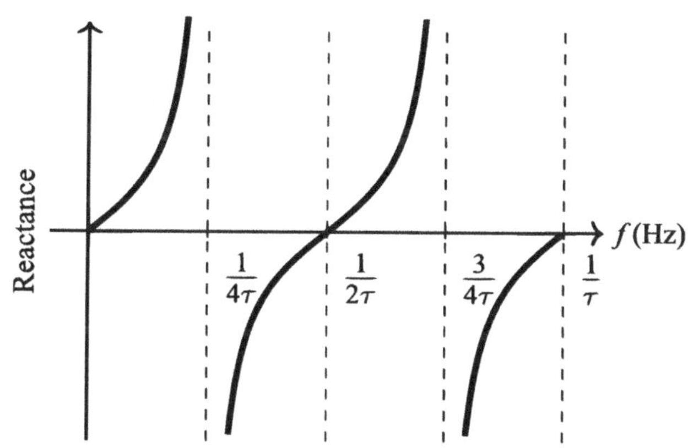

Figure 4.12 Reactance of a short-circuited line as a function of frequency.

For frequencies $f \ll 1/\tau$, $\tan(2\pi f\tau) \approx 2\pi f\tau$, so $Z_{\text{in}}(j2\pi f) \approx j2\pi f Z_o \tau$. Why does this make intuitive sense? To see this, consider the shorted transmission line of Figure 4.13a. The equivalent distributed LC network is shown in part (b) of the figure. At low frequencies, no current flows through the capacitors. The voltage across the current source is that across the inductors, which are all in series. The total inductance, as we have seen earlier, is $Z_o\tau$. The impedance of the shorted line at low frequencies, therefore, is seen to be $j2\pi f(Z_o\tau)$. Furthermore, the line behaves like a short circuit at all frequencies that are multiples of $1/(2\tau)$. However, at odd multiples of the frequency $1/4\tau$, the short-circuited line appears open!

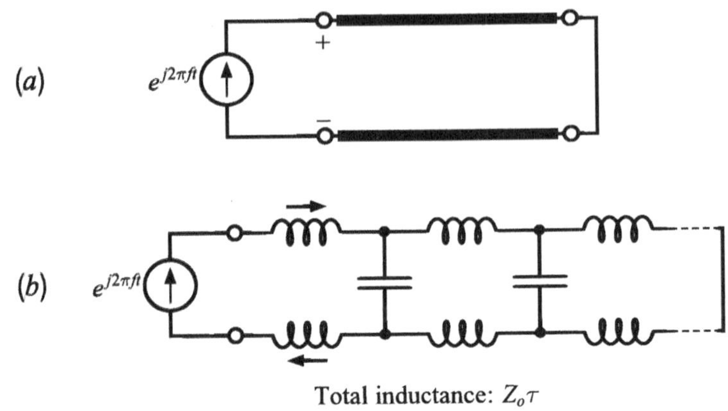

Figure 4.13 (a) A shorted transmission line. (b) At low frequencies, no current flows through the capacitors. The voltage across the current source is that across the inductors.

If, on the other hand, the line were terminated in an open circuit, the input impedance (as seen by using $Z_l \to \infty$ in (4.29)) is given by $Z_{\text{in}}(j2\pi f) = -jZ_o \cot(2\pi f\tau)$. At low frequencies, Z_{in} tends to $Z_o/(j2\pi f\tau)$, indicating that it behaves like a capacitor of value τ/Z_o, which is the total capacitance of the line. Figure 4.14 shows the reactance of the open-circuited line as a function of frequency. The line behaves like an open circuit at all integer multiples of the frequency $1/2\tau$. However, at odd multiples of $1/4\tau$, the line behaves like a short circuit!

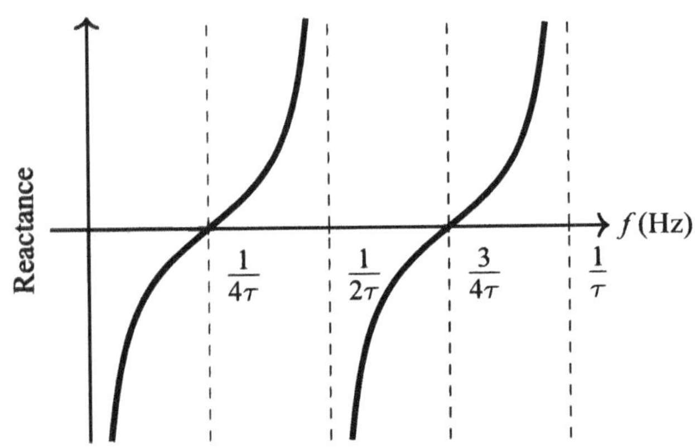

Figure 4.14 Reactance of an open-circuited line as a function of frequency.

A lumped network that emulates the response of open (or shorted) transmission lines would have an infinite number of poles and zeros, and therefore, an infinite number of inductors/capacitors. This is consistent with the observation that a transmission line network has an infinite number of poles, since the line can be thought of as being comprised of an infinite number of infinitesimally small inductors and capacitors.

c. If we operate at a frequency such that $2\pi f\tau = \pi/2$ i.e., $f = 1/4\tau$, we have

$$Z_{\text{in}}(j2\pi f) = \frac{Z_o^2}{Z_l(j2\pi f)}, \tag{4.30}$$

indicating that the impedance Z_l in inverted. The delay of the line is τ. If $f = 1/4\tau$, it follows that the length of the line is a quarter wavelength.[3] Such a line is often referred to as a quarter-wave transformer.

4.3 The Uniformly Distributed RC Line

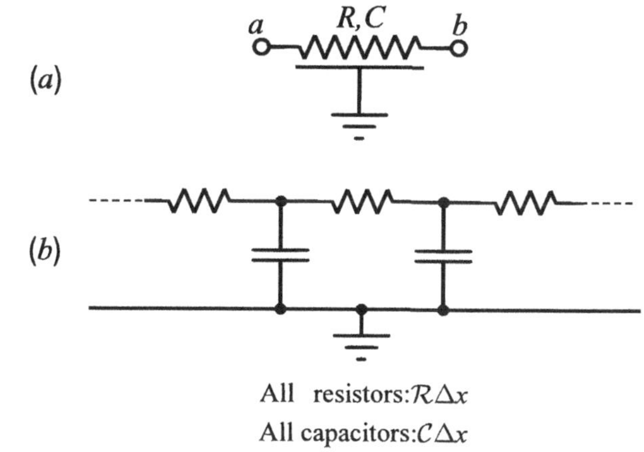

Figure 4.15 (a) Symbol of a uniformly distributed RC line, often referred to as a $\overline{\text{URC}}$ element. The total resistance is R, and the total capacitance is C. (b) Model for a small section of the line.

Related to the transmission line is the distributed-RC line. Why is this worth studying? It turns out that a resistor realized on an integrated circuit is unfortunately accompanied by parasitic capacitance to the substrate, which is distributed uniformly over the length of the resistor. Such a structure is called a uniformly distributed RC line, and referred to as a $\overline{\text{URC}}$ element [4]. The symbol is shown in Figure 4.15a. The total resistance and capacitance are denoted by R and C, respectively.

Like we did with a transmission line, a $\overline{\text{URC}}$ element can be modeled by an infinitely large number of sections, as shown in Figure 4.15b. The resistance and capacitance of each section are $\mathcal{R}\Delta x$ and $\mathcal{C}\Delta x$, respectively, where $\mathcal{R}$ and $\mathcal{C}$ denote the resistance and capacitance per unit length.

[3] $c = f\lambda$, and $\tau = l/c$, where l is the length of the line and c is the velocity of the wave along the line. l is thus given by $l = (f\tau)\lambda$.

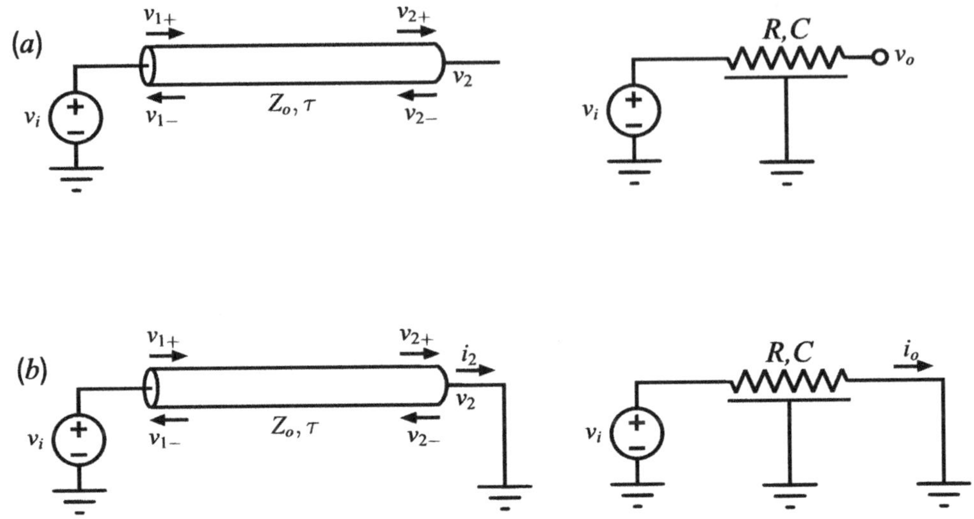

Figure 4.16 Analysis of (a) an open-circuit URC line and (b) a URC line shorted at one end.

It turns out that we can leverage our analysis of transmission lines to understand the behavior of the $\overline{\text{URC}}$ element. The idea is to replace $\mathcal{L}s$ with $\mathcal{R}$ (or equivalently, Ls with R) in the relevant expression derived with the transmission line. This is best illustrated with examples. Consider the $\overline{\text{URC}}$ line driven by a voltage source v_i, as in Figure 4.16a. What is the transfer function $V_o(s)/V_i(s)$? To determine this, consider the transmission-line analog shown in the left. We see that the following hold:

$$V_2^- = V_2^+,\quad V_1^+ = V_2^+ e^{s\tau},\quad V_1^- = V_2^- e^{-s\tau},\quad V_1^+ = -V_1^- + V_i. \tag{4.31}$$

Solving for V_2^+, recognizing that $V_2 = 2V_2^+$, and using $\tau = \sqrt{LC}$, we obtain

$$\frac{V_2(s)}{V_i(s)} = \frac{1}{\cosh(s\sqrt{LC})}. \tag{4.32}$$

For the $\overline{\text{URC}}$ line, we simply replace L in the transfer function by R/s to obtain

$$\frac{V_2(s)}{V_i(s)} = \frac{1}{\cosh(\sqrt{sRC})}. \tag{4.33}$$

At low frequencies, $\cosh(\sqrt{sRC}) \approx 1 + (sRC/2)$. Consequently, the transfer function can be approximated by $1/(1 + (sCR/2))$.

Consider now the short-circuited $\overline{\text{URC}}$ line of Figure 4.16b. To determine the transfer function from v_i to i_o, we consider the corresponding transmission-line circuit shown towards the left. We have

$$V_2^- = -V_2^+,\quad V_1^+ = V_2^+ e^{s\tau},\quad V_1^- = V_2^- e^{-s\tau},\quad V_1^+ = -V_1^- + V_i. \tag{4.34}$$

Solving for V_2^+, recognizing that $I_2 = 2V_2^+/Z_o$, and using $\tau = \sqrt{LC}$, $Z_o = \sqrt{L/C}$, we obtain

$$\frac{I_2(s)}{V_i(s)} = \sqrt{\frac{C}{L}}\frac{1}{\sinh(s\sqrt{LC})}. \tag{4.35}$$

For the $\overline{\text{URC}}$ line, we simply replace L in the transfer function by R/s to obtain

$$\frac{I_2(s)}{V_i(s)} = \frac{1}{R}\frac{\sqrt{sCR}}{\sinh(\sqrt{sCR})}. \tag{4.36}$$

At low frequencies, $\sinh(\sqrt{sCR}) \approx \sqrt{sCR}(1 + (sCR)/6)$. The transfer function can thus be approximated by $(1/R)/(1 + (sCR/6))$.

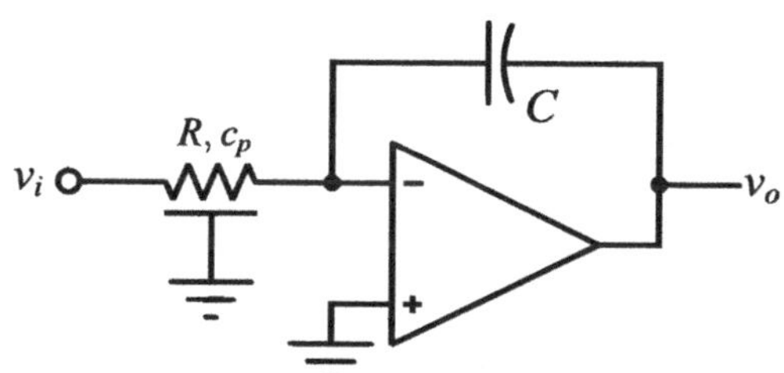

Figure 4.17 An active-RC integrator: the resistor is associated with an undesired distributed parasitic capacitance c_p.

The distributed parasitic capacitance can be problematic in many circuits. Figure 4.17 shows an active-RC integrator. The resistor is accompanied by an undesired parasitic capacitance c_p. Assuming an ideal opamp, and using our previous analysis, the integrator's transfer function can be seen to be

$$\frac{V_o(s)}{V_i(s)} \approx -\frac{1}{sCR}\frac{1}{\left(1 + s\frac{Rc_p}{6}\right)}. \tag{4.37}$$

It is evident that the effect of c_p is to introduce an extra delay of $Rc_p/6$ in the integrator's response. As we will see in Chapter 7, this modifies the frequency response of an active-RC filter.

4.3.1 Modeling Pitfalls

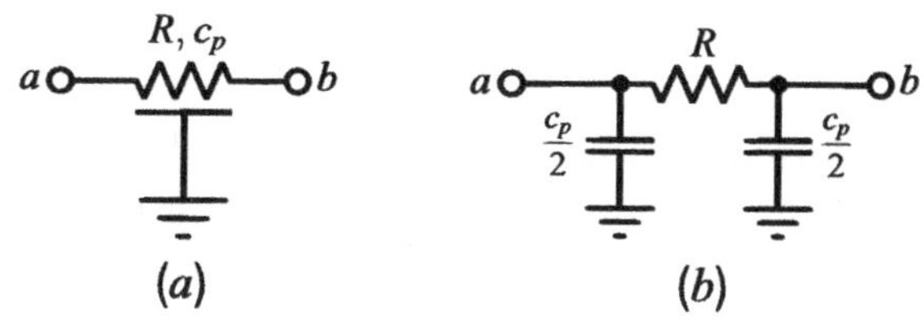

Figure 4.18 (a) A $\overline{\text{URC}}$ line. (b) A three-element lumped model.

In a simulator, implementing components whose transfer functions involve irrational functions of s is challenging. Consequently, lumped approximations for the $\overline{\text{URC}}$ line are used. It is not uncommon to use a three-element model for the line, as shown in Figure 4.18. The total capacitance c_p is split into two halves and placed at each terminal of the resistor. On the face of it, this seems reasonable; the DC current between ⓐ and ⓑ is correctly modeled. If ⓑ is open, the capacitance from ⓐ to ground is c_p in the model, as it should be. However, this model can be erroneous (and even downright dangerous), as we will illustrate with Figure 4.19, which shows a second-order active filter (the negative R indicates inversion in the fully differential implementation). The opamps are assumed to be ideal.

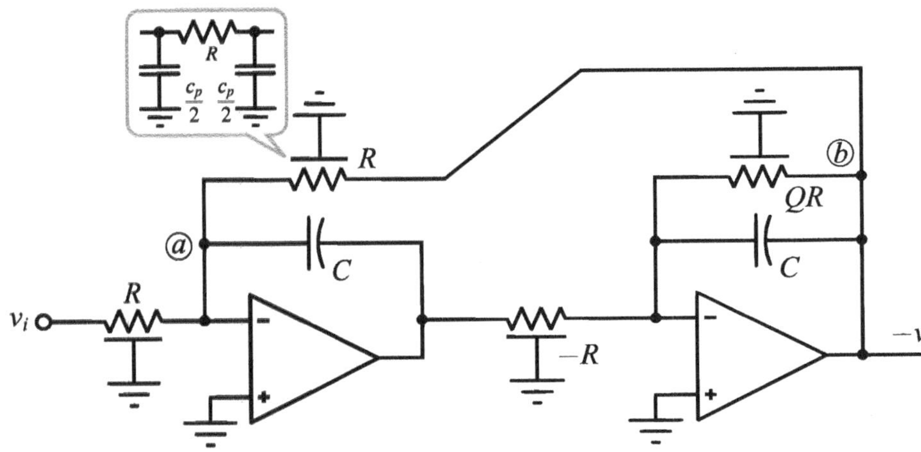

Figure 4.19 A circuit example where the three-element lumped model of Figure 4.18b can be terribly misleading.

Every resistor is accompanied by a distributed capacitance c_p. From (4.37), we know that the effect of distributed capacitance is to add an extra delay (phase shift) to each integrator, thereby modifying the filter's transfer function. Let us examine the consequences of using the lumped model. As shown by the callout in Figure 4.19, modeling the $\overline{\text{URC}}$ line by the 3-element lumped equivalent places the parasitic capacitances $c_p/2$ at ⓐ and ⓑ. Since ⓐ is a virtual ground node, no current flows through the parasitic capacitance connected there. ⓑ is the output of an opamp, which is a voltage source. The capacitance $c_p/2$ at ⓑ, therefore, is of no consequence. The same reasoning applies to the distributed parasitics of all resistors. It thus appears, if one employs the model, that c_p does not affect the performance of the circuit. In reality, however, the extra delays due to c_p will enhance the quality factor of the biquad and can even render it unstable! To prevent unpleasant surprises at a later date, one must check the lumped circuit used to model the $\overline{\text{URC}}$ line. If the simple-minded model of Figure 4.18b is supplied by the fabrication foundry, the resistor should be implemented as a series combination of smaller segments, which in effect distributes the total capacitance over multiple sections.

4.4 Scattering Parameters and the Scattering Matrix

A linear two-port network can be described, as far as port behavior is concerned, by its two-port parameters. Many such sets can be conceived: the reader is no doubt aware of the y, z, h, and g parameter sets. Furthermore, one set of parameters can be derived from another, so it is natural to wonder why one needs *yet another* set of parameters to describe the same two-port. It turns out that the motivation for the parameter set that we describe in this section, namely the *scattering parameters*, comes from practical problems encountered while making measurements.

Recall that the y-matrix description of a two-port is given by

$$\begin{bmatrix} I_1 \\ I_2 \end{bmatrix} = \begin{bmatrix} y_{11} & y_{12} \\ y_{21} & y_{22} \end{bmatrix} \begin{bmatrix} V_1 \\ V_2 \end{bmatrix}. \tag{4.38}$$

The voltages applied across the ports, namely V_1 and V_2, are the independent quantities; and the resulting currents I_1 and I_2 are dependent on the applied voltages. How does one measure the y parameters? From their definition, we have

$$y_{11} = \left.\frac{I_1}{V_1}\right|_{V_2=0}, \quad y_{21} = \left.\frac{I_2}{V_1}\right|_{V_2=0}, \quad y_{12} = \left.\frac{I_1}{V_2}\right|_{V_1=0}, \quad y_{22} = \left.\frac{I_2}{V_2}\right|_{V_1=0}, \tag{4.39}$$

indicating that these are short-circuit measurements.

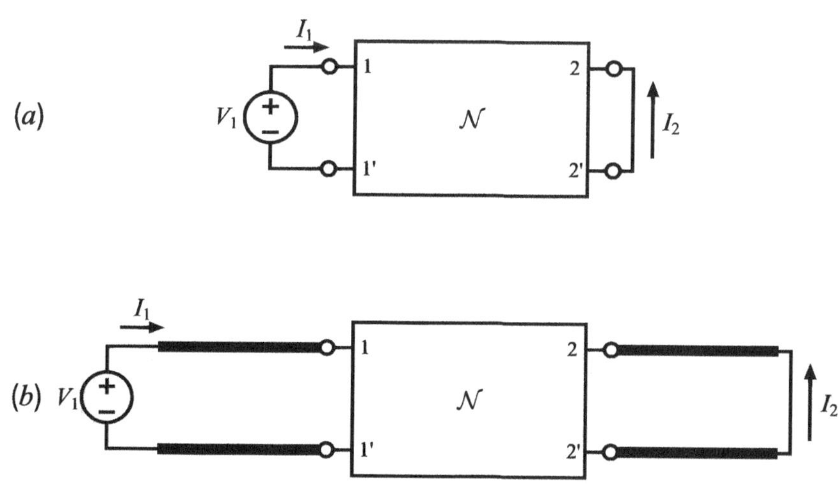

Figure 4.20 (a) Ideal setup to measure y_{11} and y_{21}. (b) A more realistic setup. The transmission lines represent cables needed to connect the ports to the voltage source and the short circuit.

As shown in Figure 4.20, y_{11} and y_{21} are measured by exciting port 1 with a voltage source v_1, terminating port 2 in a short circuit, and calculating y_{11} and y_{21} from the equations above. While this might seem straightforward from the schematic representation shown in the figure, it is very problematic in practice due to the following. It is physically impossible to place the voltage source exactly between terminals 1 and 1′.

Likewise, it is difficult, if not impossible, to terminate port 2 with a short circuit *right at its terminals.* This discounts other difficulties like creating an ideal voltage source and making current measurements right at the ports. A more realistic model for the test setup is shown in Figure 4.20b. The ports are connected to v_1 and the short-circuit termination via cables. For simplicity, we assume that these are ideal transmission lines. How are the measurements affected by the extra elements added by the test setup?

From the discussion in the previous section, we saw that a shorted transmission line behaves like a short circuit only at specific frequencies. To measure the y-parameters, the impedance at the ports should be zero—but the shorted transmission lines can behave like a short circuit, open circuit, or anything in between, depending on the length of the line and frequency of operation (recall that the impedance presented by the line is of the form $jZ_o\tan(2\pi f\tau)$). This will of course cause gross measurement errors.

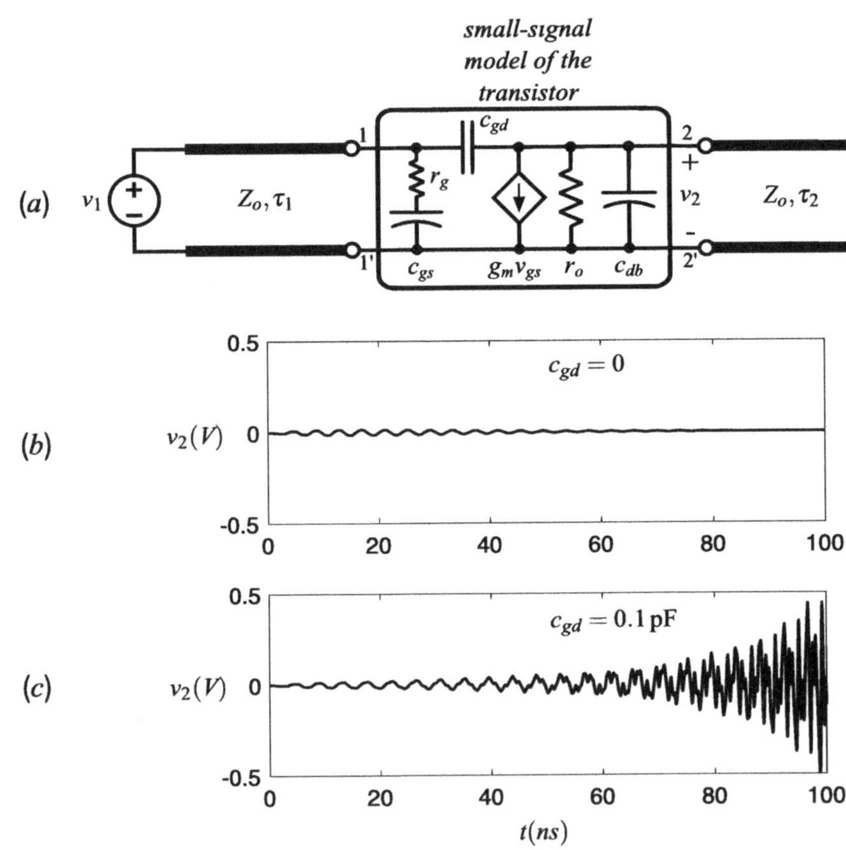

Figure 4.21 (a) Realistic setup to measure y_{11} and y_{21} for a transistor: $g_m = 40$ mS, $c_{gs} = c_{db} = 1$ pF, $r_o = 5000\ \Omega$, $r_g = 1000\ \Omega$, $Z_o = 50\ \Omega$, $\tau_1 = \tau_2 = 1$ ns. (b) $v_2(t)$ with $c_{gd} = 0$, $v_1(t)$ is a 1 mV step. (c) $c_{gd} = 0.1$ pF, $v_1(t)$ is a 1 mV step.

A more serious problem can occur with active two ports, like a transistor, as shown in Figure 4.21a. The shorted transmission lines behave like high-Q resonators at the input and output ports. Due to the feedback inherent in the two port (caused by c_{gd} in Figure 4.21), and the high frequency-selective gain due to g_m and the impedance of the resonators, the test setup can become unstable and break into oscillations. Figure 4.21b shows the simulated $v_2(t)$ when $v_1(t)$ is a 1 mV step, for $c_{gd} = 0$. It eventually settles to zero, as it should. With $c_{gd} = 0.1$ pF, however, feedback renders the entire test setup unstable and $v_2(t)$ grows exponentially as shown in Figure 4.21c.

Related to the transmission-line connections at the ports is another practical problem, this time having to do with measuring voltage and current. Recall from Section 4.2 that the voltage and current waveforms on a transmission line vary with position. This is because the voltage/current at any location on the line is the sum of the forward- and reverse-going voltage/current waves. Moving the "measurement point" even by a small amount, therefore, will result in errors in the measured v and i, and consequently, the y parameters.

To summarize, the root cause of the problem is the following.

a. To make y-parameter measurements, the network ports have to be terminated with short circuits.

b. Due to practical physical considerations, the ports of the network $\mathcal{N}$ have to be connected to the exciting sources and short-circuit terminations through transmission lines.

c. The impedance presented by the shorted transmission lines, however, is a function of cable length and frequency. As a result, the ports are not correctly terminated.

d. Measuring voltages/currents on a transmission line is problematic, as they depend on the location of measurement.

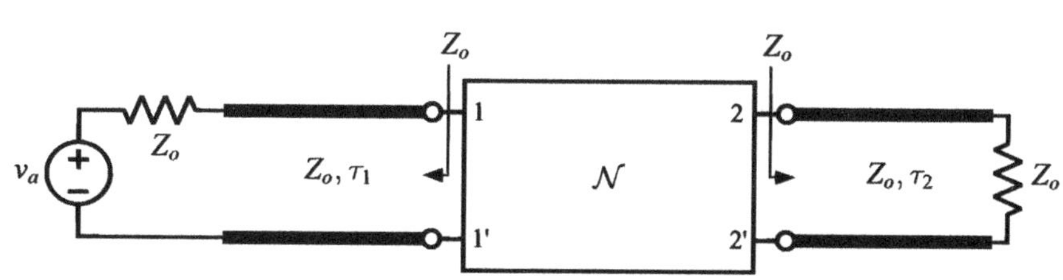

Figure 4.22 Model for the test setup where the transmission lines are terminated in their characteristic impedances.

Ideally, we would like that the impedance presented by the cables (i.e., transmission lines) at the ports is *independent* of frequency and the length of the line. How can we achieve this? From Section 4.1, this is possible if we terminated the lines with their characteristic impedance Z_o, and not short circuits, as shown in Figure 4.22. Regardless of τ_1 and τ_2, the impedance seen at the ports 2–2′ and 1–1′ is Z_o, thanks to the termination resistors. Consequently, potential instability that might occur with active networks is also avoided. Terminating the transmission lines solves the problem of cable length and frequency-dependence of the impedance at the ports.

How about the problem of location-dependent voltage and current that we encountered while measuring y parameters? This can be addressed in the following manner. The basic idea is to recognize (assuming sinusoidal operation) that though the voltage and current on a transmission line vary with position, the forward- and backward-going waves remain the same (except for a phase shift along the line). If we thought of the waves incident on the two ports as the independent quantities and those reflected by the ports as dependent on the incident waves, we wouldn't have to worry about the voltages/currents varying as a function of position along the line. One might wonder how one can separate out the forward- and backward-going waves along a transmission line. Fortunately, it turns out that it is possible using a passive microwave device called the directional coupler, which essentially performs the operations of the controlled sources in Figure 4.9.

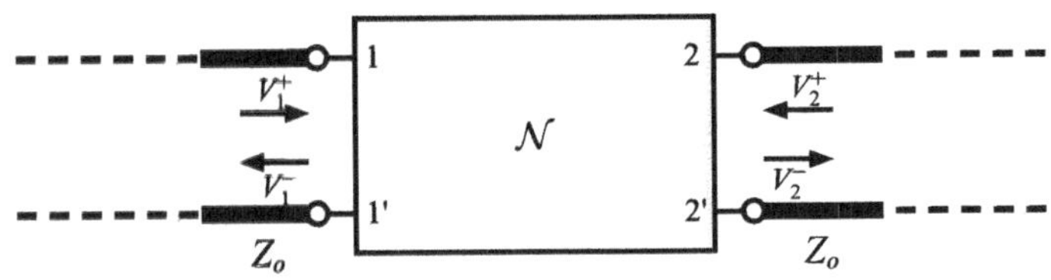

Figure 4.23 The scattering matrix relates the reflected waves to the waves incident on the ports.

We can thus write

$$\begin{bmatrix} V_1^- \\ V_2^- \end{bmatrix} = \begin{bmatrix} S_{11} & S_{12} \\ S_{21} & S_{22} \end{bmatrix} \begin{bmatrix} V_1^+ \\ V_2^+ \end{bmatrix}. \tag{4.40}$$

$$S_{11} = \left.\frac{V_1^-}{V_1^+}\right|_{V_2^+=0}, \quad S_{21} = \left.\frac{V_2^-}{V_1^+}\right|_{V_2^+=0}, \quad S_{12} = \left.\frac{V_1^-}{V_2^+}\right|_{V_1^+=0}, \quad S_{22} = \left.\frac{V_2^-}{V_2^+}\right|_{V_1^+=0}. \tag{4.41}$$

S_{11} is the ratio of the reflected wave at port 1 (V_1^-) to the incident wave at port 1 (V_1^+), when $V_2^+ = 0$. What does it mean to say $V_2^+ = 0$? Referring to Figure 4.23, it is evident that V_2^+ will be zero if the transmission line at port 2 is terminated with Z_o. This way, none of V_2^-, which is incident on the termination, will be reflected back, resulting in $V_2^+ = 0$. S_{11} is therefore, the reflection coefficient at port 1 under the condition that port 2 is matched. Along the same lines, S_{21} is the ratio of V_2^- to V_1^+ with a matched port 2. Figure 4.24 shows the setup to measure S_{11} and S_{21}. S_{22} and S_{12} are determined by setting $V_a = 0$ and activating V_b.

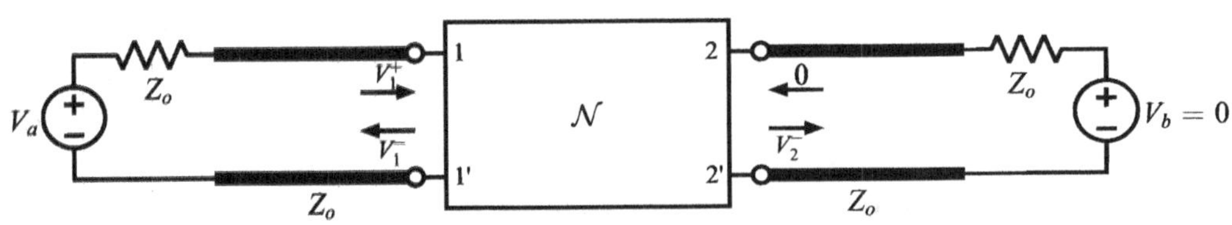

Figure 4.24 Setup to measure S_{11} and S_{21}. S_{12} and S_{22} are measured by setting $V_a = 0$ and exciting port 2.

The scattering parameters can be similarly defined for a network with N ports. The resulting matrix is a $N \times N$ matrix.

$$S_{ab} = \left.\frac{V_a^-}{V_b^+}\right|_{\text{all ports matched, only port b excited}} \tag{4.42}$$

Example

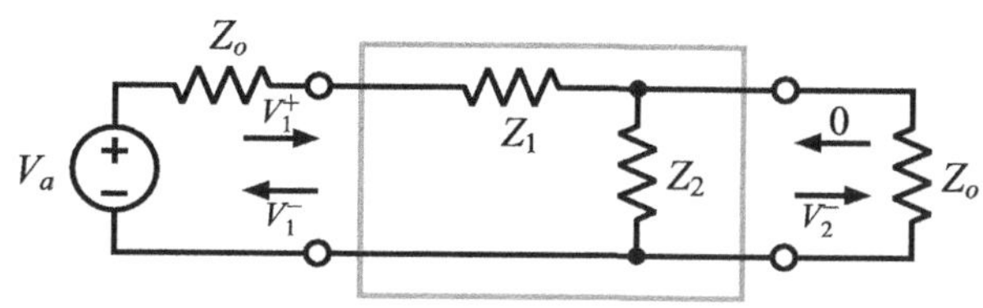

Figure 4.25 Finding S_{11} and S_{21} of the divider network.

Let us find the scattering matrix for the divider network shown in Figure 4.25. To find S_{11}, we terminate port 2 with Z_o and find the reflection coefficient at port 1. The impedance seen looking into port 1 is seen to be

$$Z_{\text{in1}} = Z_1 + Z_2 \parallel Z_o. \tag{4.43}$$

S_{11} is thus seen to be

$$S_{11} = \frac{Z_{\text{in1}} - Z_o}{Z_{\text{in1}} + Z_o}. \tag{4.44}$$

S_{21} is given by V_2^-/V_1^+. Since $V_2^+ = 0$, $V_2 = (V_2^+ + V_2^-) = V_2^-$. Furthermore, since $V_1^- = S_{11}V_1^+$, it follows that $V_1 = (V_1^+ + V_1^-) = V_1^+(1 + S_{11})$. V_2/V_1 is seen to be $(Z_2 \parallel Z_o)/(Z_1 + (Z_2 \parallel Z_o))$. Consequently, we have

$$S_{21} = \frac{V_2^-}{V_1^+} = \frac{V_2(1 + S_{11})}{V_1} = \frac{Z_2 \parallel Z_o}{Z_1 + (Z_2 \parallel Z_o)}(1 + S_{11}) = \frac{Z_2 \parallel Z_o}{Z_1 + (Z_2 \parallel Z_o)} \frac{2Z_{in1}}{Z_{in1} + Z_o} \tag{4.45}$$

where Z_{in1} is given by (4.43). Proceeding along similar lines, S_{22} and S_{12} can be determined. They turn out to be

$$S_{22} = \frac{V_2^-}{V_2^+} = \frac{Z_{\text{in2}} - Z_o}{Z_{\text{in2}} + Z_o} \tag{4.46}$$

where

$$Z_{\text{in2}} = (Z_o + Z_1) \parallel Z_2. \tag{4.47}$$

$$S_{12} = \frac{V_1^-}{V_2^+} = \frac{Z_o}{Z_1 + Z_o} \frac{2Z_{\text{in2}}}{Z_{\text{in2}} + Z_o}. \tag{4.48}$$

The scattering matrix is thus given by

$$S = \begin{bmatrix} \dfrac{Z_{\text{in1}} - Z_o}{Z_{\text{in1}} + Z_o} & \dfrac{Z_o}{Z_1 + Z_o} \dfrac{2Z_{\text{in2}}}{Z_{\text{in2}} + Z_o} \\ \dfrac{Z_2 \parallel Z_o}{Z_1 + (Z_2 \parallel Z_o)} \dfrac{2Z_{\text{in1}}}{Z_{\text{in1}} + Z_o} & \dfrac{Z_{\text{in2}} - Z_o}{Z_{\text{in2}} + Z_o} \end{bmatrix}. \tag{4.49}$$

The reader should verify that $S_{12} = S_{21}$, though their expressions look different. Notice that the scattering parameters depend on the characteristic impedance of the transmission lines at the ports, namely Z_o. The S-matrix will change with Z_o, even when the network itself remains the same.

Example

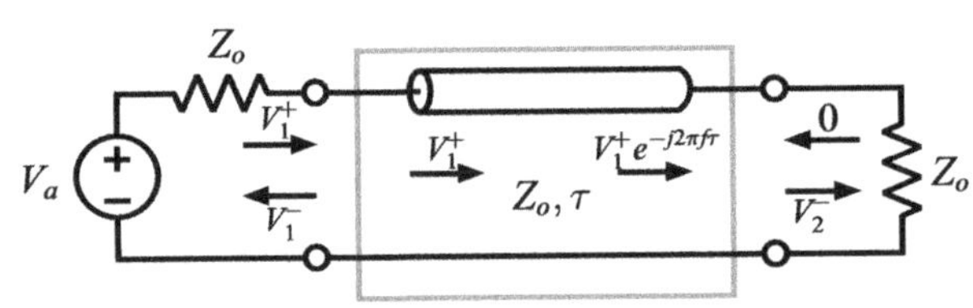

Figure 4.26 Finding S_{11} and S_{21} of a transmission line of characteristic impedance Z_o and delay τ.

Let us find the scattering matrix for the transmission line shown in Figure 4.26. To find S_{11}, we terminate port 2 with Z_o and find the reflection coefficient at port 1. Clearly, the input impedance looking into port 1 is Z_o, resulting in $S_{11} = 0$. The forward wave on the line is V_1^+ at port 1. At port 2, it is delayed by τ to result in $V_1^+ e^{-j2\pi f\tau}$. Since the line is terminated with Z_o, it follows that $V_2^- = V_1^+ e^{-j2\pi f\tau}$. Furthermore, the network is symmetric. Consequently,

$$S = \begin{bmatrix} 0 & e^{-j2\pi f\tau} \\ e^{-j2\pi f\tau} & 0 \end{bmatrix}. \tag{4.50}$$

Example

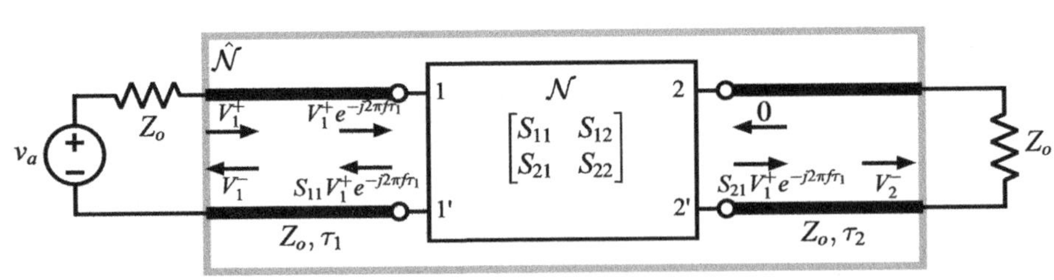

Figure 4.27 The S-matrix of $\mathcal{N}$ is known. What is the S-matrix of $\hat{\mathcal{N}}$?

Suppose the scattering matrix of a network $\mathcal{N}$ is known. $\hat{\mathcal{N}}$ is formed by cascading ports 1 and 2 with transmission lines with characteristic impedance Z_o and delays

τ_1 and τ_2, respectively, as shown in Figure 4.27. What are the scattering parameters of $\hat{\mathcal{N}}$?

The forward wave at port 1 of $\hat{\mathcal{N}}$ is denoted by V_1^+. The wave incident at port 1 of $\mathcal{N}$ is given by $V_1^+ e^{-j2\pi f\tau_1}$. Since $V_2^+ = 0$, it follows that the reflected wave at port 1 of $\mathcal{N}$ is given by $S_{11}V_1^+ e^{-j2\pi f\tau_1}$. Since the delay of the line is τ_1, V_1^- is a delayed version of $S_{11}V_1^+ e^{-j2\pi f\tau_1}$. We thus have

$$V_1^- = S_{11}e^{-j4\pi f\tau_1}V_1^+. \tag{4.51}$$

The "reflected wave" from port 2 of $\mathcal{N}$ is given by $S_{21}V_1^+ e^{-j2\pi f\tau_1}$. It is delayed by τ_2 when it gets to port 2 of $\hat{\mathcal{N}}$. As a result,

$$V_2^- = S_{21}e^{-j2\pi f(\tau_1+\tau_2)}V_1^+. \tag{4.52}$$

Proceeding along similar lines, it is straightforward to see that

$$V_2^- = S_{22}e^{-j4\pi f\tau_2}V_2^+ \tag{4.53}$$

and

$$V_1^- = S_{12}e^{-j2\pi f(\tau_1+\tau_2)}V_2^+. \tag{4.54}$$

The S-matrix of $\hat{\mathcal{N}}$, denoted by $\hat{S}$, is thus given by

$$\hat{S} = \begin{bmatrix} S_{11}e^{-j4\pi f\tau_1} & S_{12}e^{-j2\pi f(\tau_1+\tau_2)} \\ S_{21}e^{-j2\pi f(\tau_1+\tau_2)} & S_{22}e^{-j4\pi f\tau_2} \end{bmatrix}. \tag{4.55}$$

As seen above, the magnitudes of the S parameters have not changed due to the transmission lines added in tandem with the ports.

4.4.1 Deriving the Scattering Matrix from the Admittance Matrix

Suppose that we knew the y-matrix of an N-port network. How would we obtain the scattering matrix from the y-matrix? We proceed as follows.

$$\mathbf{I} = \mathbf{YV}, \tag{4.56}$$

where $\mathbf{I}$ and $\mathbf{V}$ denote the column vector of port current- and port voltage-phasors, respectively. As discussed earlier, the S-parameters depend on the characteristic impedance of the transmission lines connected to the ports. We assume that all the lines have the same characteristic impedance Z_0. Denoting the vector of incident and reflected waves by $\mathbf{V}^+$ and $\mathbf{V}^-$ respectively, we have

$$\mathbf{V} = \mathbf{V}^+ + \mathbf{V}^-, \qquad \mathbf{I} = \frac{1}{Z_0}\left(\mathbf{V}^+ - \mathbf{V}^-\right). \tag{4.57}$$

Using $\mathbf{I} = \mathbf{YV}$ and substituting the wave expressions:

$$\frac{1}{Z_0}(\mathbf{V}^+ - \mathbf{V}^-) = \mathbf{Y}(\mathbf{V}^+ + \mathbf{V}^-). \tag{4.58}$$

Rearranging, we have

$$\mathbf{V}^- = (\mathbf{1} + Z_0\mathbf{Y})^{-1}(\mathbf{1} - Z_0\mathbf{Y})\mathbf{V}^+. \tag{4.59}$$

where $\mathbf{1}$ denotes the $N \times N$ identity matrix. By definition, $\mathbf{V}^- = \mathbf{S}\mathbf{V}^+$. Therefore,

$$\boxed{\mathbf{S} = (\mathbf{1} + Z_0\mathbf{Y})^{-1}(\mathbf{1} - Z_0\mathbf{Y}).} \tag{4.60}$$

A reinterpretation of (4.60) allows the simplest derivation of the scattering matrix from the circuit diagram. (4.60) can be written as

$$\mathbf{S} = \underbrace{(Y_o\mathbf{1} + \mathbf{Y})}_{\mathbf{Y}_a}^{-1}(Y_o\mathbf{1} - \mathbf{Y}) \tag{4.61}$$

where $Y_o = 1/Z_o$. Denoting $(Y_o\mathbf{1} + \mathbf{Y})$ by $\mathbf{Y}_a$, (4.61) can be recast as

$$\mathbf{S} = 2Y_o\mathbf{Y}_a^{-1} - \mathbf{1}. \tag{4.62}$$

From the definition, it is evident that $\mathbf{Y}_a$ is the admittance matrix of the augmented network where resistors of value $Z_o = 1/Y_o$ are placed in parallel with every port. To summarize, to find the S-parameters of a network, we could proceed as follows:

a. Add resistors of value Z_o in parallel with every port of the network. The result is the *augmented* network.
b. Determine the admittance matrix $\mathbf{Y}_a$ of the augmented network.
c. Use $\mathbf{Y}_a$ obtained from the step above in (4.62) to determine the scattering matrix of the original network.

$\mathbf{S}$ can be similarly derived from the impedance matrix $\mathbf{Z}$.

Scattering Matrix of the Π-Pad Attenuator

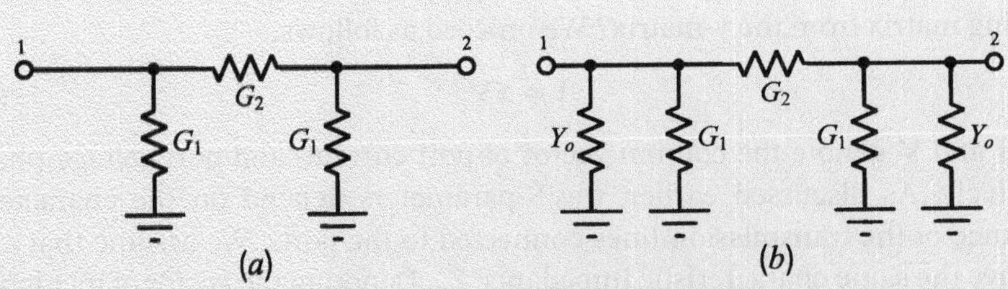

Figure 4.28 (a) A Π-Pad attenuator. (b) The augmented network, whose admittance parameters are used to find the scattering matrix of the attenuator $Y_o = 1/Z_o$.

A Π-pad attenuator is the symmetric two-port network shown in Figure 4.28a. Its scattering parameters need to be $S_{11} = S_{22} = 0$ and $S_{12} = S_{21} = k < 1$.

$$\mathbf{Y} = \begin{bmatrix} (G_1 + G_2) & -G_2 \\ -G_2 & (G_1 + G_2) \end{bmatrix}, \qquad \mathbf{Y}_a = \begin{bmatrix} (G_1 + G_2 + Y_o) & -G_2 \\ -G_2 & (G_1 + G_2 + Y_o) \end{bmatrix}. \tag{4.63}$$

Using $\mathbf{Y}_a$ in (4.62), we obtain

$$\mathbf{S} = \frac{1}{\Delta}\begin{bmatrix} 2Y_o(G_1 + G_2 + Y_o) - \Delta & 2Y_oG_2 \\ 2Y_oG_2 & 2Y_o(G_1 + G_2 + Y_o) - \Delta \end{bmatrix} \tag{4.64}$$

where Δ is given by $(G_1 + G_2 + Y_o)^2 - G_2^2$.

Equating $S_{11} = S_{22} = 0$ and $S_{12} = S_{21} = k$ gives

$$G_1 = Y_o\,\frac{1-k}{1+k}, \qquad G_2 = \frac{2kY_o}{1-k^2}. \tag{4.65}$$

For $k = 1/\sqrt{2}$, the circuit provides an attenuation of $10\log_{10}(1/k^2) = 3$ dB. If $Z_o = 1/Y_o = 50\ \Omega$, the resistors have the values $R_1 = 1/G_1 = 291.3\ \Omega$ and $R_2 = 1/G_2 = 17.7\ \Omega$. For $k = 1/2$, the attenuation is 6 dB, and the resistor values are $R_1 = 150\ \Omega$ and $R_2 = 37.5\ \Omega$.

4.5 Some Properties of Scattering Matrices

4.5.1 The Scattering Matrix of a Lossless Network

Recall that the power incident on and reflected from the lth port are given by $(1/2)|V_l^+|^2/Z_o$ and $(1/2)|V_l^-|^2/Z_o$ respectively. The factor (1/2) is because we assume that the magnitude of the phasors represents the peak values of the sinusoid. Denoting the column vectors of incident and reflected waves by $\mathbf{V}^+$ and $\mathbf{V}^-$, respectively, the total incident and reflected powers are as follows:

$$\text{Total incident power}: \frac{1}{2Z_o}(\mathbf{V}^+)^\dagger\mathbf{V}^+$$

$$\text{Total reflected power}: \frac{1}{2Z_o}(\mathbf{V}^-)^\dagger\mathbf{V}^- = \frac{1}{2Z_o}(\mathbf{S}\mathbf{V}^+)^\dagger\mathbf{S}\mathbf{V}^+ = \frac{1}{2Z_o}(\mathbf{V}^+)^\dagger\mathbf{S}^\dagger\mathbf{S}\mathbf{V}^+,$$

where the $\dagger$ denotes the conjugate transpose (Hermitian). For a lossless network, the total incident and reflected powers must be equal for an arbitrary $\mathbf{V}^+$. It thus follows that

$$\mathbf{S}^\dagger\mathbf{S} = \mathbf{1} \tag{4.66}$$

where $\mathbf{1}$ denotes the identity matrix. Thus, the scattering matrix of a lossless network is unitary.

4.5.2 The Scattering Matrix of a Reciprocal Network

We know that the admittance matrix $\mathbf{Y}$ of a reciprocal network is symmetric, i.e., $\mathbf{Y} = \mathbf{Y}^T$. From (4.62), we know that $\mathbf{S} = 2Y_o\mathbf{Y}_a^{-1} - \mathbf{1}$ where, $\mathbf{Y}_a = Y_o\mathbf{1} + \mathbf{Y}$. Furthermore, $\mathbf{Y}_a^T = \mathbf{Y}_a$. Consequently, we see that

$$\mathbf{S}^T = 2Y_o(\mathbf{Y}_a^T)^{-1} - \mathbf{1} = 2Y_o(\mathbf{Y}_a)^{-1} - \mathbf{1} = \mathbf{S}. \tag{4.67}$$

Thus, the scattering matrix of a reciprocal network is symmetric.

Scattering Matrix of the Gyrator

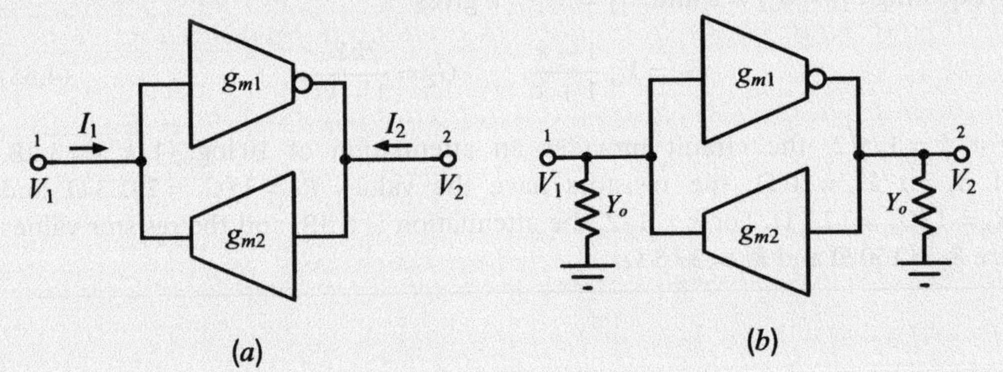

Figure 4.29 (a) A gyrator. (b) The augmented network, whose admittance parameters are used to find the scattering matrix of the gyrator, $Y_o = 1/Z_o$.

A two-port network called the *gyrator* is shown in Figure 4.29a. It satisfies the equations

$$I_2 = g_{m1}V_1, \qquad I_1 = -g_{m2}V_2. \tag{4.68}$$

The augmented network, formed by adding conductances of value Y_o in parallel with the ports, is shown in Figure 4.29b. To determine its S-parameters, we write $\mathbf{Y}$ and $\mathbf{Y}_a$, which are the admittance matrices of the original and augmented networks, respectively.

$$\mathbf{Y} = \begin{bmatrix} 0 & -g_{m2} \\ g_{m1} & 0 \end{bmatrix}, \qquad \mathbf{Y}_a = \begin{bmatrix} Y_o & -g_{m2} \\ g_{m1} & Y_o \end{bmatrix}. \tag{4.69}$$

Using $\mathbf{Y}_a$ in (4.62), we obtain

$$\mathbf{S} = \frac{1}{\Delta}\begin{bmatrix} 2Y_o^2 - \Delta & 2Y_o g_{m2} \\ -2Y_o g_{m1} & 2Y_o^2 - \Delta \end{bmatrix} \tag{4.70}$$

where Δ is given by $(Y_o^2 + g_{m1}g_{m2})$. If $g_{m1} = g_{m2} = Y_o$, the scattering matrix reduces to

$$\mathbf{S} = \begin{bmatrix} 0 & 1 \\ -1 & 0 \end{bmatrix}$$

which indicates that the ports are matched and the network is lossless and nonreciprocal: it introduces a phase inversion in the forward signal path.

4.6 Measuring S-Parameters: The Vector Network Analyzer

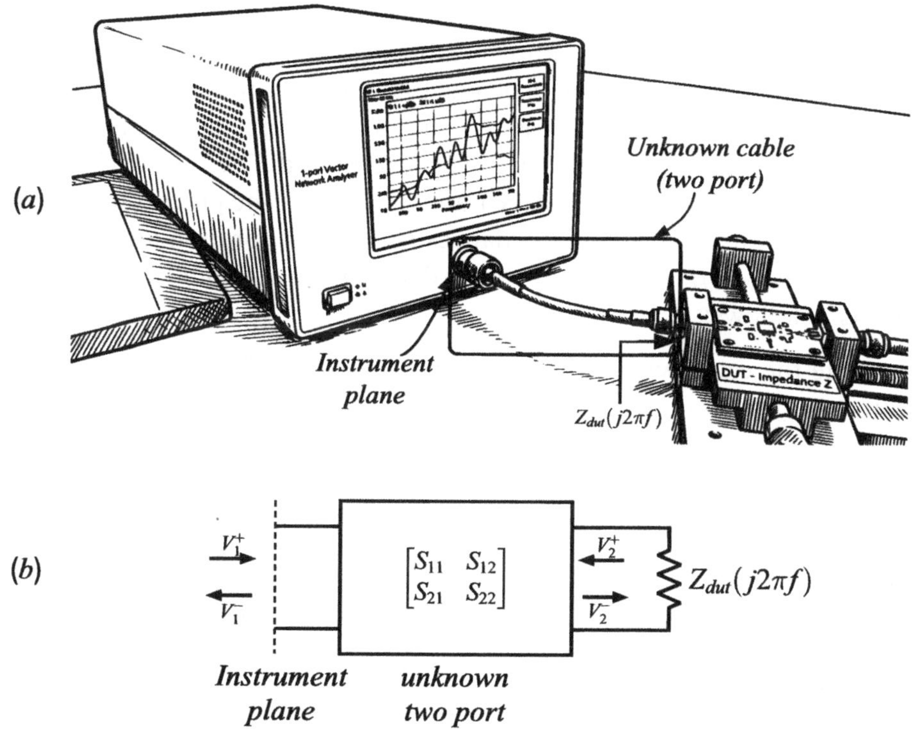

Figure 4.30 (a) Cartoon of a one-port vector network analyzer measuring an unknown impedance. (b) Model for the test setup. $Z_{\text{dut}}(j2\pi f)$ needs to be measured, but the analyzer can only measure the reflection coefficient at the instrument plane.

A vector network analyzer (VNA) is an instrument used to measure scattering parameters of an N-port. The simplest case, where $N = 1$, is referred to as a one-port network analyzer and can only measure S_{11} (or equivalently, the driving point impedance at a port). A cartoon depicting a typical test setup is shown in Figure 4.30a. The instrument is connected to the unknown impedance that needs to be measured, namely $Z_{dut}(j2\pi f)$, by a cable. The properties of the cable are not known a priori. The VNA can only measure the reflection coefficient (as a function of frequency) at the instrument plane, but what we want is (equivalently) the reflection coefficient of the device under test (DUT)!

The model of the measurement setup is shown in Figure 4.30b. The VNA measures $V_1^-(j2\pi f)/V_1^+(j2\pi f)$ as a function of frequency. $Z_{dut}(j2\pi f)$, which needs to be determined, appears at port 2 of an unknown two-port. Since $V_2^+ = \Gamma_{dut} V_2^-$, straightforward analysis shows that the reflection coefficient Γ measured at the instrument plane is given by

$$\Gamma = \frac{V_1^-}{V_1^+} = S_{11} + \frac{S_{12}S_{21}\Gamma_{dut}}{1 - \Gamma_{dut}S_{22}}. \tag{4.71}$$

If the S-parameters of the two-port were known a priori, it is easy to see that $\Gamma_{dut}(j2\pi f)$ (and hence $Z(j2\pi f)$) can be determined using (4.71). But how does one figure out the properties of the unknown cable between the instrument and DUT? The basic idea is the following. We recognize that there are three unknowns, namely S_{11}, $S_{12}S_{21}$, and S_{22}. By measuring Γ using three *known* impedances in place of Z_{dut}, the properties of the unknown network can be determined. The "known" impedances are: a short circuit (0), an open circuit (∞), and $Z_o = 50\ \Omega$. The process of determining S_{11}, $S_{12}S_{21}$, and S_{22} using known impedances is called calibration and is performed before measuring $Z_{dut}(j2\pi f)$. The process to determine $Z_{dut}(j2\pi f)$ at each frequency of interest is as follows:

a. Terminate the cable with a short circuit, i.e., use $Z = 0$ in place of the DUT in Figure 4.30a, and record the reflection coefficient, denoted by Γ_0. We see that

$$\Gamma_0 = S_{11} - \frac{S_{12}S_{21}}{1 + S_{22}}, \tag{4.72}$$

obtained by setting $\Gamma_{dut} = -1$ in (4.71).

b. Terminate the cable with an open circuit, i.e., use $Z = \infty$ in place of the DUT in Figure 4.30a, and record the reflection coefficient, denoted by Γ_∞. We see that

$$\Gamma_\infty = S_{11} + \frac{S_{12}S_{21}}{1 - S_{22}}, \tag{4.73}$$

obtained by setting $\Gamma_{dut} = 1$ in (4.71).

c. Terminate the cable with Z_o, i.e., use $Z = 50\ \Omega$ in place of the DUT in Figure 4.30a, and record the reflection coefficient, denoted by Γ_r. We see that

$$\Gamma_r = S_{11}, \tag{4.74}$$

obtained by setting $\Gamma_{dut} = 0$ in (4.71).

d. Calculate S_{11}, $S_{12}S_{21}$ and S_{22} using the relations Γ_0, Γ_∞ and Γ_r. This is possible, as there are three equations in as many unknowns.

e. Terminate the cable with the unknown impedance Z_{dut}, and record the reflection coefficient, denoted by Γ_u.

$$\Gamma_u = S_{11} + \frac{S_{12}S_{21}\Gamma_{\text{dut}}}{1 - \Gamma_{\text{dut}}S_{22}}. \tag{4.75}$$

Γ_{dut} can be determined from the equation above since S_{11}, $S_{12}S_{21}$, and S_{22} have been determined in the previous step.

The discussion above motivated the need for calibration and the procedure for a one-port VNA. We chose an instrument with a single port for simplicity. Two-port VNAs are perhaps the most commonly used analyzers. Like in the single-port analyzer, calibration is needed to de-embed the test setup, meaning that the effect of nonidealities are "removed" from the measurement. The calibration procedure is a lot more involved; it turns out that 12 parameters are needed to account for nonidealities of the setup and measurement. The reader is referred to advanced texts on microwave measurements [5] for the details.

Chapter Summary

- **Lumped vs. Distributed Regime:** A circuit must be treated as a distributed system when its physical dimensions are a significant fraction of the signal wavelength ($\lambda = c/f$).
- **Transmission Line Parameters:** A uniform lossless transmission line is characterized by its inductance and capacitance per unit length ($\mathcal{L}$ and $\mathcal{C}$). Key properties include:
  - **Characteristic Impedance:** $Z_o = \sqrt{\mathcal{L}/\mathcal{C}}$, which defines the ratio of voltage to current for a single traveling wave.
  - **Propagation Velocity:** $c = 1/\sqrt{\mathcal{L}\mathcal{C}}$, the speed at which the electromagnetic wave travels.
- **Wave Propagation:** Signal behavior is modeled as the superposition of a forward-traveling wave (v^+) and a backward-traveling wave (v^-). The interaction of these waves at boundaries is governed by the reflection coefficient.
- **Reflections and Matching:** At a load impedance Z_L, the reflection coefficient is given by $\Gamma = (Z_L - Z_o)/(Z_L + Z_o)$.
  - **Matched Load:** $Z_L = Z_o \Rightarrow \Gamma = 0$ (no reflections, maximum power transfer).
  - **Open/Short Circuit:** Results in total reflection ($|\Gamma| = 1$), leading to standing wave patterns.
- **Transient and Steady-State Analysis:**
  - **Bounce Diagrams:** Used to track the "history" of reflections in the time domain for step or pulse inputs.
  - **Voltage Standing Wave Ratio (VSWR):** Measures the ratio of maximum to minimum voltage magnitudes along the line in the sinusoidal steady state.

(Continued)

(Continued)

- **Scattering (S) Parameters:** Unlike Z or Y parameters, S-parameters characterize a network based on incident and reflected waves relative to a reference impedance (usually 50 Ω). This is the standard for high-frequency measurements, as it avoids the need for ideal open or short circuits.
- **Measurement and Calibration:** Vector Network Analyzers (VNAs) measure S-parameters. A calibration procedure is performed to remove the effect of the cables and connectors.

Problems

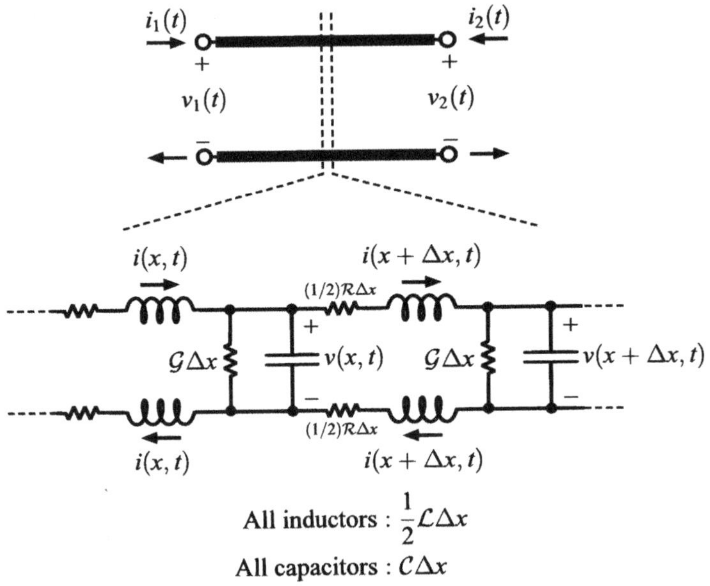

Figure 4.31

Problem 1. *Figure 4.31 shows a lossy uniformly distributed line. The inductance is accompanied by a loss, modeled as a resistance $\mathcal{R}$ per unit length, and the capacitance is lossy with a conductance $\mathcal{G}$ per unit length. Determine the characteristic impedance. What condition should be satisfied for it to be real?*

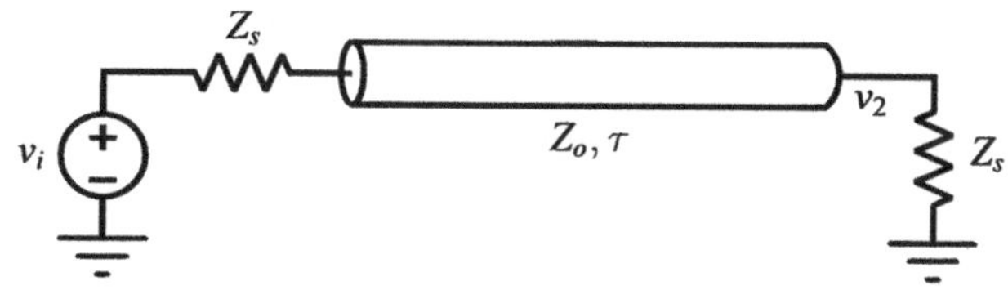

Figure 4.32

Problem 2. *In Figure 4.32, v_i is a unit step. Sketch $v_2(t)$. Plot $v_2(t)$ in the following limiting cases.*

a. $Z_o \to \infty$ *with* $Z_o\tau \to L$.

b. $Z_o \to 0$ *with* $\tau/Z_o \to C$.

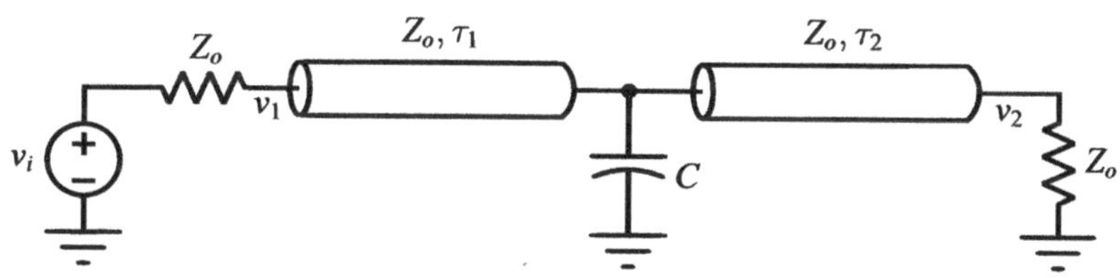

Figure 4.33

Problem 3. *In the circuit of Figure 4.33, v_i is a unit step. Plot $v_1(t)$ and $v_2(t)$.*

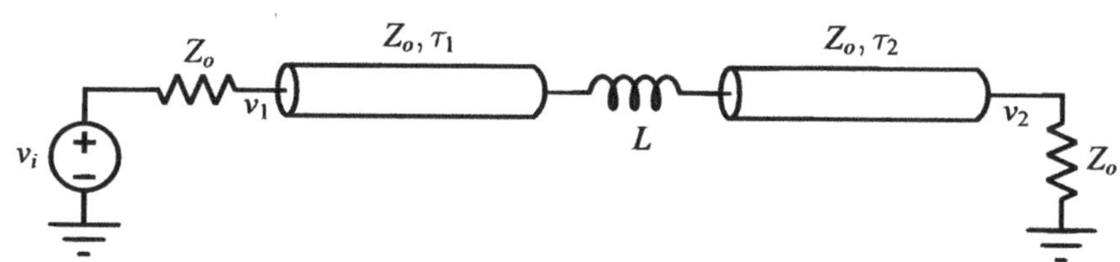

Figure 4.34

Problem 4. *In the circuit of Figure 4.34, v_i is a unit step. Plot $v_1(t)$ and $v_2(t)$.*

Problem 5. *A transmission line with a characteristic impedance Z_o is terminated by an impedance $(R + jX)$, with $R \geq 0$. The normalized impedance is given by $r + jx = (R/Z_o) + j(X/Z_o)$. The reflection coefficient in terms of the normalized impedance is given by*

$$\Gamma = \frac{r + jx - 1}{r + jx + 1}.$$

a. *Plot the locus of Γ as a function of r for a fixed x. Draw these loci for different values of x.*

b. *Plot the locus of Γ as a function of x for a fixed r. Draw these loci for different values of r.*

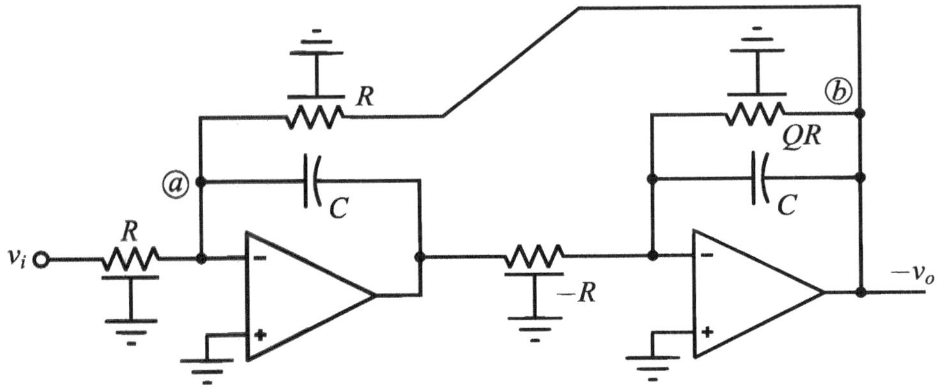

Figure 4.35

Problem 6. *Figure 4.35 shows a biquad. All resistors are associated with a parasitic distributed capacitance, with the product being denoted by τ_p. Assume that τ_p is identical for all resistors. Determine ω_o and quality factor for $\tau_p \ll RC$. The opamps are ideal.*

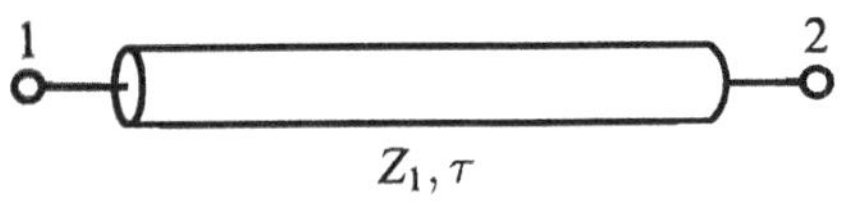

Figure 4.36

Problem 7. *Determine the scattering matrix* **S** *of the transmission line of Figure 4.36, with respect to a reference impedance of Z_o.*

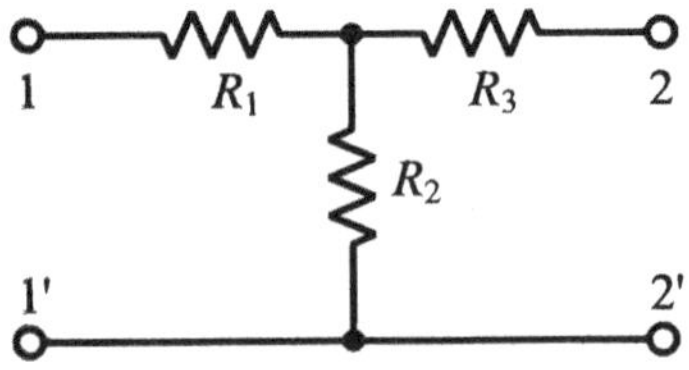

Figure 4.37

Problem 8. *Determine the scattering matrix* **S** *of the T-attenuator of Figure 4.37. Choose the resistors so that* $S_{11} = S_{22} = 0$*, and* $S_{12} = S_{21} = k$.

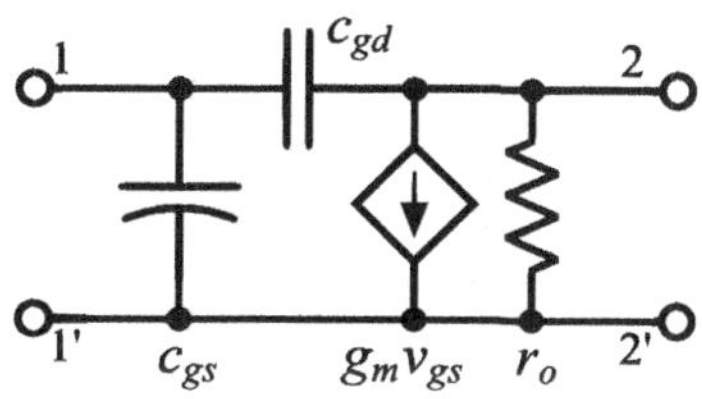

Figure 4.38

Problem 9. *Determine the scattering matrix* **S** *of the small-signal model of the transistor shown in Figure 4.38.*

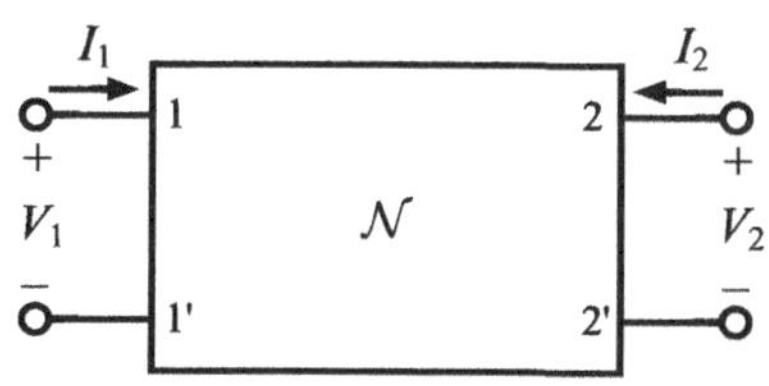

Figure 4.39

Problem 10. *A two-port* $\mathcal{N}$ *(see Figure 4.39) can also be characterized by the so-called chain matrix, denoted by* **ABCD***, defined as follows:*

$$\begin{bmatrix} V_1 \\ I_1 \end{bmatrix} = \begin{bmatrix} A & B \\ C & D \end{bmatrix} \begin{bmatrix} V_2 \\ -I_2 \end{bmatrix}.$$

Suppose the scattering parameters of $\mathcal{N}$ *are known. Determine the chain matrix. If* $\mathcal{N}$ *is reciprocal, what constraints does that impose on the chain parameters?*

Problem 11. *Two 2-port networks have the scattering matrices given by* $\mathbf{S}_a$ *and* $\mathbf{S}_b$*. Determine the scattering matrix of the cascade. Suppose their chain matrices are denoted by* $\mathbf{ABCD}_a$ *and* $\mathbf{ABCD}_b$*. What is the chain matrix of the cascade?*

Problem 12. *A two-port network has the impedance matrix* **Z***. Determine the scattering matrix in terms of* **Z***.*

Problem 13. *Derive equation (4.71).*

Problem 14. *Derive a closed-form expression that gives* Γ_{dut} *in terms of* Γ_o*,* Γ_∞ *and* Γ_r *in Section 4.6.*

Problem 15. *Write down the form of the scattering matrix of a reciprocal three-port with all its ports matched (meaning the reflection coefficient at any port is 0 if the other ports are terminated with Z_o). Show that the three-port cannot be lossless.*

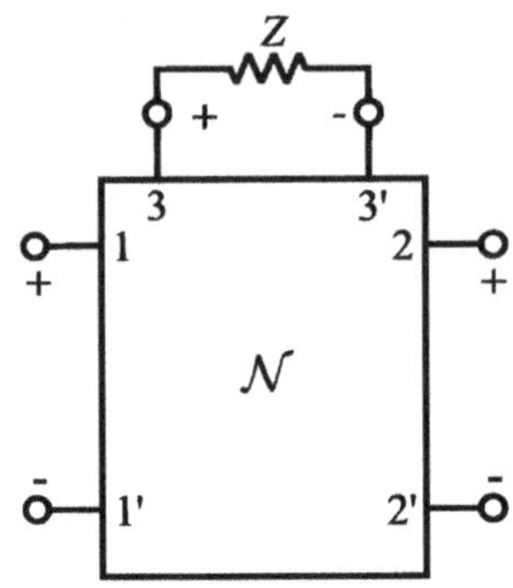

Figure 4.40

Problem 16. *The three-port $\mathcal{N}$ in Figure 4.40 with a known scattering matrix is lossless and reciprocal. Port 3 is terminated with an impedance Z, so that $S_{12} = S_{21} = 0$. How will you determine Z?*

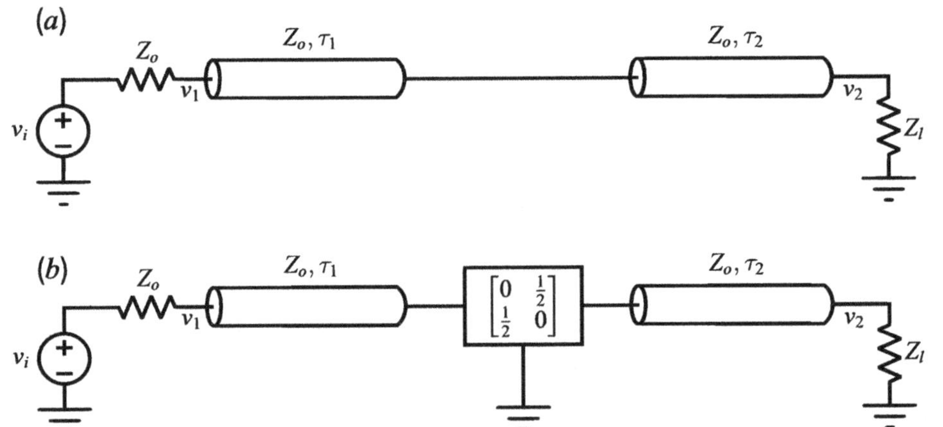

Figure 4.41

Problem 17. *In Figure 4.41, $v_i(t)$ is a unit step, and $Z_l = 2Z_o$. Plot $v_1(t)$ and $v_2(t)$ in the network of Figure 4.41a.*

A network with S-parameters shown in the box is inserted between the lines to result in the network of Figure 4.41b. Plot $v_1(t)$ and $v_2(t)$ now. What do you notice and why?

References

1 D. M. Pozar, *Microwave Engineering*, 4th. John Wiley & Sons, 2011.
2 R. E. Collin, *Foundations for Microwave Engineering*, 2nd. IEEE Press / Wiley-Interscience, 2001.
3 P. A. Rizzi, *Microwave Engineering: Passive Circuits*. Prentice Hall, 1988.
4 G. C. Temes and J. W. LaPatra, *Introduction to Circuit Synthesis and Design*. McGraw-Hill, 1977.
5 J. P. Dunsmore, *Handbook of Microwave Component Measurements: with Advanced VNA Techniques*, 2nd. John Wiley & Sons, 2020.

5

Basics of Linear Time-Varying Circuits and Systems

5.1 Motivation

The first question that arises in a practical mind like that of the reader is, "Why should I learn, or even care, about time-varying circuits?" To answer this, we need not look any further than our pockets and reach for the phone. But a bit of history first. Early radios were of the *tuned RF* type, whose block diagram is shown in Figure 5.1. The small RF signal received by the antenna is processed by a combination of filters and amplifiers. The filters are needed to attenuate interferers and adjacent channels. Doing this in one shot, with a high-order passive bandpass filter, is not a practical option due to the following. A high-order bandpass filter will also have a large passband loss associated with it. The desired signal, which is already small, will get further attenuated at the filter's output, making it weaker than the input-referred noise of the subsequent amplifier.

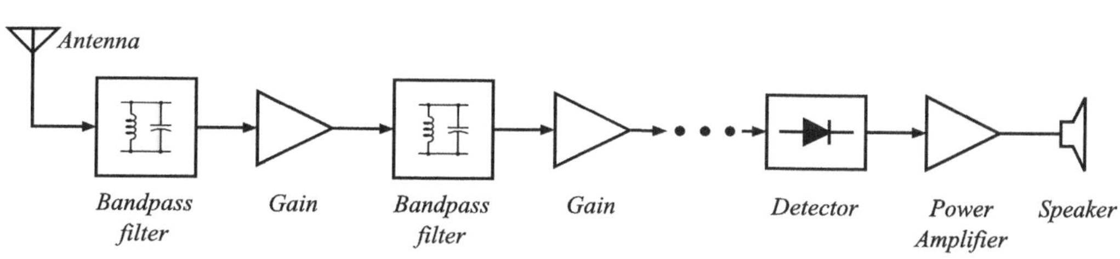

Figure 5.1 Block diagram of a tuned-RF radio receiver.

To address this, low-order bandpass filters are cascaded with gain stages so as to eventually obtain what is equivalently a very narrowband filter and an adequately high gain. The output of the RF stage excites a nonlinear detector, whose function is to detect the low-frequency envelope of the amplified RF output. The detector's output, after processing by an audio amplifier, excites a loudspeaker.

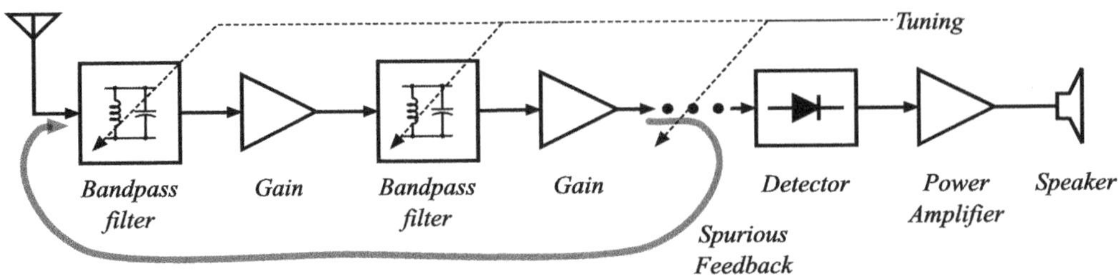

Figure 5.2 Practical problems in a tuned-RF radio receiver.

There are several practical problems with the receiver of Figure 5.1, as explained using Figure 5.2. Since the carrier frequency is very high in relation to the channel bandwidth, the LC resonators that realize the bandpass filters must have extremely large quality factors. Assuming an RF frequency of 800 kHz and a channel bandwidth of 4 kHz, we see that the Q of the LC resonators should be in the range of several hundreds. This is difficult to achieve in practice.

To receive multiple stations (not at the same time!), the receiver must be tunable, which means that the center frequencies of all the bandpass filters must be moved in unison. The resonators should thus be *ganged*. As one might expect, moving all the filters' center frequencies by *exactly* the same amount is a challenge, and almost always needs adjustments of the individual resonators. Further, changing the center frequency of the LC tanks alters the bandwidth if the Q does not change.

As seen from Figure 5.1, most of the gain and bandpass filtering needed in the receiver is realized at (high) RF frequencies. This is the root cause of the most damning of the issues afflicting the tuned-RF receiver. To understand the magnitude of the problem, note that the gain of the receiver must be of the order of several thousand, so that RF signals whose strengths are a few μW can be amplified to the level of a few hundred milliwatts. Also, bear in mind that the multiple resonators in cascade introduce significant phase shift between the input and output of the RF chain. This means that even small amounts of spurious feedback (paths with gains of the order of 10^{-4}) from the output of the RF chain to its input can cause the chain to become unstable. From the discussion above, we see that the primary reason for the maladies of the tuned-RF radio is the fact that gain and selectivity of the receiver are all obtained at the *same frequency*, namely the carrier frequency.

A modern radio receiver [1, 2], whose block diagram is shown in Figure 5.3 avoids the problems posed by the tuned-RF receiver. It consists of a relatively low gain stage at RF (perhaps with a gentle RF bandpass filter), whose output is multiplied by two sinusoids generated by a local oscillator. The sinusoids are simply phase-shifted from one another by 90° and are given by $A_o \cos(2\pi f_o t)$ and $A_o \sin(2\pi f_o t)$. The RF input is given by

$$v_{rf}(t) = m_I(t)\cos(2\pi f_o t) + m_Q(t)\sin(2\pi f_o t), \tag{5.1}$$

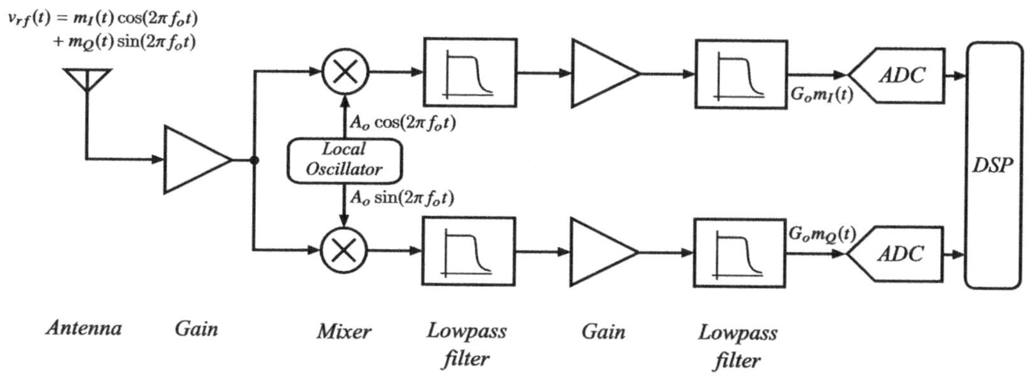

Figure 5.3 Block diagram of a direct-conversion receiver architected for integrated circuit implementation.

where $m_I(t)$ and $m_Q(t)$ are the baseband information-bearing signals. In RF parlance, multiplication is termed mixing (or heterodyning or down-conversion). The outputs of the mixers consist of the low-frequency message signals and components at $2f_o$. The latter are removed by the lowpass filters subsequent to the mixers. The (weak) baseband signals are further amplified and filtered, and digitized by an analog-to-digital converter (ADC) before being processed by the digital signal processor (DSP). The DSP uses sophisticated algorithms to decode information from $m_I(t)$ and $m_Q(t)$, and correct for some practical impairments in the RF signal chain. Since the local oscillator's frequency is the same as that of the RF carrier frequency, such a receiver is called the *direct-conversion receiver*.

How does the direct-conversion architecture solve the problems associated with the tuned-RF receiver? We notice that in a marked departure from the latter, most of the gain and selectivity (filtering) needed are now achieved at baseband. This greatly relaxes the design of the amplifiers. The lowpass filters now needed are a lot easier to design than the high-Q bandpass filters needed in the tuned-RF receiver. They can be realized using active techniques; thus, the use of inductors can be dispensed with. As a result, the filters are compact, and can be integrated on a chip. Tuning the receiver is easy to accomplish, since only the LO frequency needs to change. Further, the bandwidth is independent of the frequency to which the receiver is tuned. Finally, spurious feedback at baseband is much smaller than that at RF, making stability of the baseband chain a nonissue. Parasitic coupling between the baseband output and the RF input still exists; however, this is no longer problematic, as the feedback signal at baseband finds itself translated to f_o, and subsequently eliminated by the subsequent lowpass filters. We thus see that the direct-conversion receiver, in one stroke, eliminates most of the problems of the tuned-RF structure.[1] It is no wonder, then, that

[1] In life and in engineering, there is never a free lunch. While direct conversion does eliminate most issues in a tuned-RF receiver, it introduces problems of its own. Fortunately, these are far easier to address.

the direct-conversion radio is extensively used in practice and that we are probably carrying at least half a dozen of them in our pockets right now.

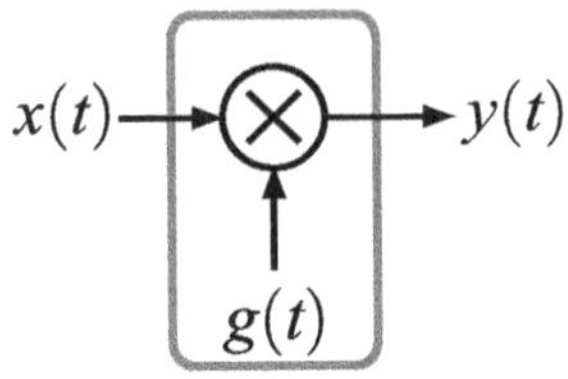

Figure 5.4 Mixing is a linear operation, though not time-invariant.

Staring at Figures 5.2 and 5.3, we see that the key concept that makes all the wonders of direct conversion possible is that of mixing (or multiplication). Figure 5.4 shows a multiplier, one of whose input ports is excited by $g(t)$. Our "system" is the one enclosed in the gray box, whose gain varies with time according to $g(t)$. The input and output are $x(t)$ and $y(t)$, respectively. Contrary to preliminary instincts, it is linear since it obeys superposition, i.e. if input $x_1(t)$ yields an output $y_1(t)$, and $x_2(t)$ yields $y_2(t)$, $\alpha x_1(t) + \beta x_2(t)$ yields $\alpha y_1(t) + \beta y_2(t)$ as shown below.

$$\begin{aligned} x_1(t) &\rightarrow y_1(t) = g(t)x_1(t), \\ x_2(t) &\rightarrow y_2(t) = g(t)x_2(t), \\ \alpha x_1(t) + \beta x_2(t) &\rightarrow y(t) = \alpha \underbrace{g(t)x_1(t)}_{y_1(t)} + \beta \underbrace{g(t)x_2(t)}_{y_2(t)}. \end{aligned}$$

However, it is time-varying, since delaying $x_1(t)$ by t_o does *not* result in $y_1(t - t_o)$.

$$\begin{aligned} x_1(t) &\rightarrow y_1(t) = g(t)x_1(t), \\ x_1(t - t_o) &\rightarrow g(t)x_1(t - t_o) \neq \underbrace{g(t - t_o)x_1(t - t_o)}_{y_1(t-t_o)}. \end{aligned} \tag{5.2}$$

An ADC is another example of a time-varying system. As shown in Figure 5.5a, its role is to convert the continuous-time, continuous-level input $x(t)$ into a discrete-time, discrete-level sequence $v[n]$. From a signal-processing viewpoint, an ADC is equivalent to the following, as illustrated in Figure 5.5b. $x(t)$ is multiplied by a periodic train of continuous-time Dirac impulses. The resulting modulated train, namely $x_s(t)$, is converted into a discrete-time impulse train by the *c2d* block. At this point, the signal is discrete-time but has continuous levels (Figure 5.5c). The output of the *c2d* block is quantized to yield the discrete-level sequence $v[n]$. This process can be thought of as adding a quantization error sequence $q[n]$. In a practical application, an ADC with a sufficiently small $q[n]$ would be chosen so that $v[n]$ is virtually the same as $x(nT_s)$. It is the *multiplication* of $x(t)$ by the impulse train that makes an ADC a time-varying system.

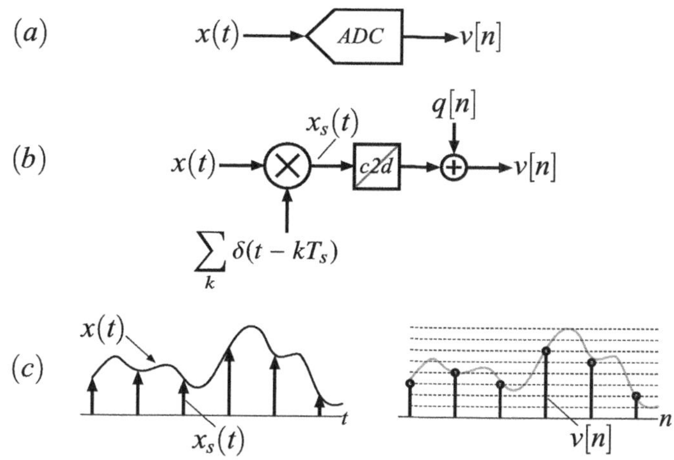

Figure 5.5 (a) An analog-to-digital converter (ADC) converts a continuous-time, continuous-level signal into a discrete-time, discrete-level one. (b) Signal processing equivalent of an ideal ADC. $q[n]$ is the quantization error sequence added by the ADC. (c) $x_s(t)$ is a modulated train of continuous-time Dirac impulses, which is converted into discrete-time and quantized to yield $v[n]$.

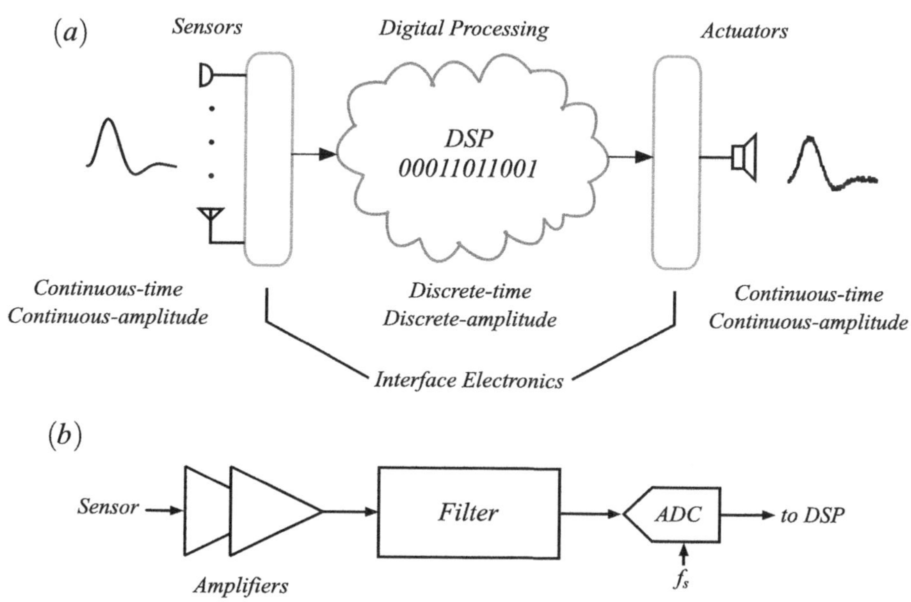

Figure 5.6 (a) A modern electronic system—while processing is performed digitally, real-world signals need to be conditioned and digitized by the interface electronics. (b) Block diagram of an analog front end.

Why are ADCs so important in electronic system design? Computational and signal processing tasks are now performed predominantly by digital means, since digital circuits are robust and can be realized by extremely small and simple structures that can in turn be combined to obtain very complex, accurate, and fast systems. Every year, the speed and density (of transistors) of digital integrated circuits (ICs) increase, thereby enhancing the dominance of digital methods in almost all areas of communications and consumer products. Since the physical world nevertheless remains stubbornly analog (and is likely to remain so), data converters are needed to interface with the digital signal processing (DSP) core. As the speed and capability of DSP cores increases, so too must the speed and accuracy of the converters associated with them. This presents a continual challenge to the engineers dedicated to the design of data converters!

Figure 5.6a illustrates the block diagram of a signal processing system with analog input and output signals, plus a central digital engine. As shown in part (b) of the figure, the analog input signal (usually after some amplification and filtering) enters an ADC that transforms the input signal into a digital data stream. This stream is processed by the DSP core, and the resulting digital output signal is reconverted into analog form by a digital-to-analog converter (DAC). The DAC output is usually also filtered and amplified to obtain the final analog output signal.

The discussion in this section has hopefully convinced the reader about the central role played by linear time-varying systems in modern electronic systems. The question then is not "Why should I care about time-varying circuits?", but rather, "How come I never learned about the analysis of time-varying circuits and systems" in school? Our guess is the following. In the past, most designers dealt with circuits that worked completely with continuous-time signals, like amplifiers and continuous-time filters, which (at least ideally) behaved like linear time-invariant systems. A bit later, when switched-capacitor circuits were in vogue, signals of interest were discrete-time. Thus, barring the front-end sample-and-hold, switched-capacitor amplifiers and filters could be treated as discrete-time linear time-invariant systems. RF designers, who were perhaps the earliest that needed to understand time variance, would learn about the tools of the trade on the job. The absence of simulation support for analysis of time-varying circuits meant that designers often found ad hoc, context-dependent approaches to understand, analyze, and design such circuits.

All this changed with the advent of cellular telephony in the early 1990s. Consumer electronics, in the form of daily-use gadgets like mobile phones, wireless LAN routers, and Bluetooth-enabled peripherals, led to a rapid increase in the demand for low-cost systems-on-chip that integrated complex RF front ends with baseband filters, ADCs, DACs, and DSPs. Needless to say, there was a corresponding massive increase in the community of so-called RF and "mixed-signal" designers. This in turn created a commercial motivation for computer-aided design (CAD) companies to incorporate linear time-varying circuit analyses in their simulation tools. Thanks to this, designers were not only able to simulate time-varying RF circuits, but were also in a position to sharpen their understanding of the underlying mechanisms of operation. Consequently, many improved circuit techniques exploiting these mechanisms were developed. A case-in-point is the recent interest in passive switched-capacitor circuits and N-path filters for RF applications.

Another set of designers needed to understand time variance with sufficient depth. These were ADC designers, whose job was to conceive of and create circuits that efficiently interface between continuous- and discrete-time domains. Unlike RF designers, who are typically interested in the continuous-time output of a time-varying system, an ADC designer is interested in the properties of the system's *sampled* output. The ready availability of "RF simulators," i.e., those CAD tools that included the capabilities of time-varying circuit simulation, helped them understand and verify aspects that had a bearing on data-converter performance. An example of a commercially important ADC family, whose analysis relies heavily on time-varying circuit principles, is the continuous-time delta-sigma modulator (CTΔΣM). It turns out that a CTΔΣM is an example of an ADC with in-built antialiasing. Having motivated our course, we now begin our voyage.

We first review the basics of linear time-invariant systems and the convolution integral. We then derive, from first principles, the convolution integral for a linear time-varying (LTV) system. A linear periodically time-varying (LPTV) system is a special LTV system whose properties change periodically with time. We derive the convolution integral of an LPTV system. Next, we describe the frequency response of an LPTV system by means of what are referred to as harmonic transfer functions. Having covered the basics, we then understand the process of writing the modified nodal analysis (MNA) equations for an LPTV circuit. It turns out that the process is entirely analogous to that followed in a linear time-invariant network. The MNA equations are not only useful in the computation of the harmonic transfer functions but also form a basis for understanding the important concepts of reciprocity and inter-reciprocity as applied to LPTV networks.[2] We then discuss the N-path principle, which is a clever way of configuring N LPTV systems, each varying at a frequency f_s, to realize an LPTV system that appears to vary at Nf_s.

5.2 Linearity and Time (In)variance

Assume that an initially relaxed system (all initial conditions are zero), yields outputs $y_1(t)$ and $y_2(t)$ when excited by $x_1(t)$ and $x_2(t)$, respectively. The system is linear if it obeys superposition; namely, an input $\alpha x_1(t) + \beta x_2(t)$ yields the output $\alpha y_1(t) + \beta y_2(t)$.

$$x_1(t) \to y_1(t),$$
$$x_2(t) \to y_2(t),$$
$$\alpha x_1(t) + \beta x_2(t) \to \alpha y_1(t) + \beta y_2(t).$$

A system is linear *and* time-invariant (LTI) if, in addition to the constraints above, delaying the input by τ results in an output that is also delayed by τ. That is,

$$x_1(t - \tau) \to y_1(t - \tau). \tag{5.3}$$

[2]Reciprocity and inter-reciprocity in LPTV circuits are covered in Chapter 6.

An LTI system is characterized by its impulse response $h(t)$, which is the output of the initially relaxed system for an input $\delta(t)$. Due to time invariance, $h(t)$ can also be interpreted as the response observed at an arbitrary time t_1 due to an impulse applied at a time t before the time of observation, namely, at time $(t_1 - t)$.

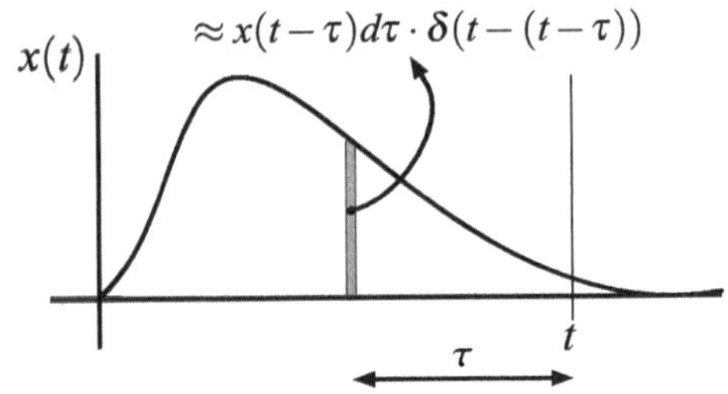

Figure 5.7 Response of an LTI system at time t to an input $x(t)$.

How does the system respond to an arbitrary input $x(t)$? To determine this, we express the input as a sum of slivers with width $d\tau$ and height $x(t-\tau)$, as shown in Figure 5.7. The response at time t due to the sliver is given by $x(t-\tau)\,d\tau \cdot h(\tau)$. The output due to $x(t)$ can thus be obtained by summing the response due to all the slivers, and this is given by the familiar convolution integral

$$y(t) = \int_0^\infty \underbrace{x(t-\tau)\,d\tau}_{\substack{\text{size of impulse}\\ \text{applied at } (t-\tau)}} \quad \underbrace{h(\tau)}_{\substack{\text{response to impulse}\\ \text{applied } \tau \text{ earlier}}} = \int_0^\infty h(\tau)x(t-\tau)\,d\tau. \tag{5.4}$$

A complex exponential is especially significant in the study of linear systems. When the LTI system is excited by $x(t) = e^{j2\pi ft}$, using the convolution integral above yields

$$y(t) = \int_0^\infty h(\tau)e^{j2\pi f(t-\tau)}\,d\tau = e^{j2\pi ft} \underbrace{\int_0^\infty h(\tau)e^{-j2\pi f\tau}\,d\tau}_{H(j2\pi f)}. \tag{5.5}$$

It is thus seen that the output of an LTI system, when excited by a complex exponential, is simply a scaled version of the input. The "gain," which depends on the frequency of the input, is the complex number $H(j2\pi f)$. As seen above, $H(j2\pi f)$ is the Fourier transform of the impulse response $h(t)$, and it can be thought of in the following way:

$$H(j2\pi f) = \frac{\text{Response to } e^{j2\pi ft}}{e^{j2\pi ft}}. \tag{5.6}$$

Thus, a complex exponential at frequency f results in a *steady-state* output that is the input scaled by a complex number $H(j2\pi f)$. Conversely, if the output of an LTI system happens to be a complex sinusoid at frequency f, it must follow that the input is also a sinusoid at f. In a circuit simulator, the frequency response $H(j2\pi f)$ is obtained by running an .AC analysis.

Many networks used in practice are nonlinear and time invariant. Further, they are often used around a time-invariant *operating point*, with the inputs themselves being called *small signals*. When linearized about a time-invariant operating point, the nonlinear network results in a time-invariant linear network. In a circuit simulator, the small-signal frequency response is obtained, therefore, by first running a `.OP` analysis, which yields the operating point and the small-signal LTI network, after which an `.AC` analysis is performed.

5.3 Linear Time-Varying Systems

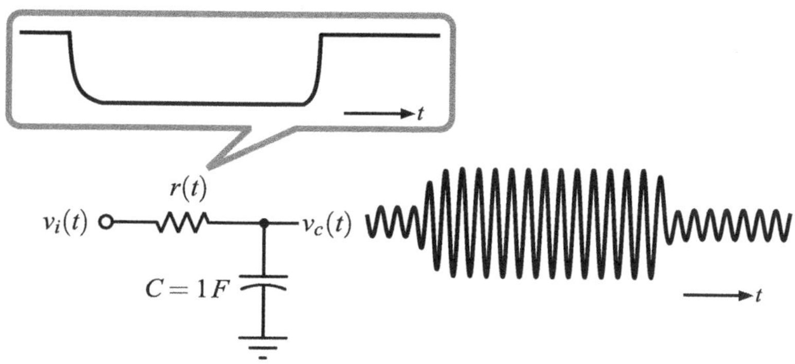

Figure 5.8 First-order RC filter, with the resistance varying with time. $v_i(t)$ is a sinusoidal excitation.

Consider our first-order RC filter once again driven by a sine wave, but with the resistance made time-varying as shown in Figure 5.8. What should we expect to see at the output? When the resistance is high, the amplitude of the output sine wave ought to be small, since the bandwidth of the filter is low. When r is lowered, the bandwidth increases, and we should expect an increased output amplitude. In other words, the gain experienced by the input sine wave is no longer constant but changes with time. This makes sense, since the "properties" of the network are changing with time.

Such systems are best described as being LTV. As in the LTI case, an LTV system is characterized by an impulse response—however, as the reader may well know, this impulse response depends on the time at which the input impulse is applied. The response of an LTV system at a time t, due to an impulse applied at a time $t - \tau$, is denoted by $h(t, \tau)$:[3]

$$\text{Impulse response} = h(\underbrace{t}_{\substack{\text{time of}\\ \text{observation}}}, \underbrace{\tau}_{\substack{\text{impulse applied}\\ \tau \text{ before } t}}). \tag{5.7}$$

[3]Recall that in an LTI system, the output of the system at t when it is excited by an impulse at $t - \tau$ is simply $h(\tau)$.

To reiterate, the first variable in the impulse response denotes the time of observation. The second variable indicates that the system was excited by an impulse launched at a time τ *before* the time of observation. Thus, the response of an LTV system not only depends on the time elapsed between the excitation and observation times but also on the actual time of observation. Contrast this to an LTI system, whose response depends *only* on the time elapsed between the times of excitation and observation by the impulse. An LTI system, therefore, can be thought of as a special case of an LTV system whose impulse response satisfies $h(t,\tau) = h(\tau)$.

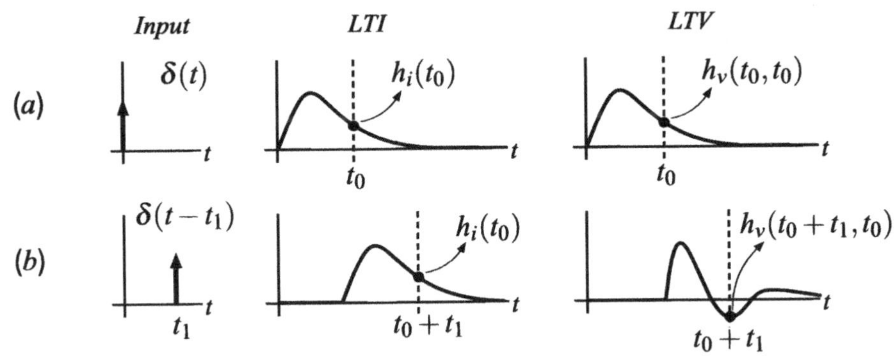

Figure 5.9 Impulse responses of LTI and LTV systems.

The responses of LTI and LTV systems to impulse excitations are illustrated in Figure 5.9. Part (a) of the figure shows the responses of example LTI and LTV systems to an impulse applied at $t = 0$. The output of the former is $h_i(t)$, where h_i is the impulse response. At an observation time t_0, the output is $h_i(t_0)$. The output of the LTV system is given by $h_v(t, t)$, where h denotes the impulse response. When observed at a time t_0, the output is $h_v(t_0, t_0)$. This is to be interpreted as the output at t_0 due to an impulse applied t_0 prior to the time of observation—that is, the input is $\delta(t - (t_0 - t_0)) = \delta(t)$.

When the input is delayed by t_1, as shown in Figure 5.9b, the output of the LTI system is given by $h_i(t - t_1)$. Thus, if the time of observation moves by t_1 to $t_0 + t_1$, the output remains $h_i(t_0 + t_1 - t_1) = h_i(t_0)$. What happens in the LTV case? When observed at $t_0 + t_1$, the output is seen to be $h_v(t_0 + t_1, t_0)$. This is not (necessarily) equal to $h_v(t_0, t_0)$, since the output of an LTV system not only depends on the time elapsed between excitation and observation but also on the actual time at which the output is observed.

How does an LTV system with impulse response $h(t,\tau)$ respond to the complex sinusoid $x(t) = e^{j2\pi ft}$? Proceeding as we did in the LTI case, we obtain

$$y(t) = \int_0^\infty h(t,\tau)e^{j2\pi f(t-\tau)}\,d\tau = e^{j2\pi ft}\underbrace{\int_0^\infty h(t,\tau)e^{-j2\pi f\tau}\,d\tau}_{H(j2\pi f,t)}. \tag{5.8}$$

We see, as in the time-invariant case, that the input sinusoid is scaled by a complex number $H(j2\pi f, t)$. However, the "gain" experienced by the sinusoid is not only a function of its frequency (as in the LTI case) but also a function of time. This makes sense, since the system's properties are varying with time. $H(j2\pi f, t)$, called the time-varying frequency response, is given by

$$\boxed{H(j2\pi f, t) = \int_0^{\infty} h(t,\tau)e^{-j2\pi f\tau}\, d\tau.} \tag{5.9}$$

Further, we note that the (time-varying) frequency response can be interpreted, as in the time-invariant case, as [3]

$$H(j2\pi f, t) = \frac{\text{Response to } e^{j2\pi ft}}{e^{j2\pi ft}}. \tag{5.10}$$

Conjugate Symmetry of $H(j2\pi f, t)$

Consider an LTV system whose impulse response is real. As we have just seen, its frequency response is given by

$$H(j2\pi f, t) = \int_0^{\infty} h(t,\tau)e^{-j2\pi f\tau}\, d\tau.$$

Taking the complex conjugates on both sides yields

$$H^*(j2\pi f, t) = \int_0^{\infty} h(t,\tau)e^{j2\pi f\tau}\, d\tau.$$

Replacing f with $-f$ on both sides of the equation above, we have

$$H^*(-j2\pi f, t) = \int_0^{\infty} h(t,\tau)e^{-j2\pi f\tau}\, d\tau = H(j2\pi f, t)$$

leading us to conclude that

$$\boxed{H(j2\pi f, t) = H^*(-j2\pi f, t).} \tag{5.11}$$

This is consistent with the conjugate symmetry of the frequency response of an LTI system (with a real impulse response), which can be thought of as an LTV system whose frequency response is independent of time.

Suppose one is given an LTV system and asked to determine $H(j2\pi f, t)$. How does one proceed? A way to do this is to somehow measure $h(t,\tau)$ for all t and τ and use (5.9) to determine $H(j2\pi f, t)$. In the time-invariant case, this approach is analogous to finding the frequency response by evaluating the Fourier transform of the impulse response at f. Another approach is to excite the LTV system by $\cos(2\pi ft)$ and $\sin(2\pi ft)$ and record the corresponding responses $w_i(t)$ and $w_q(t)$, respectively,

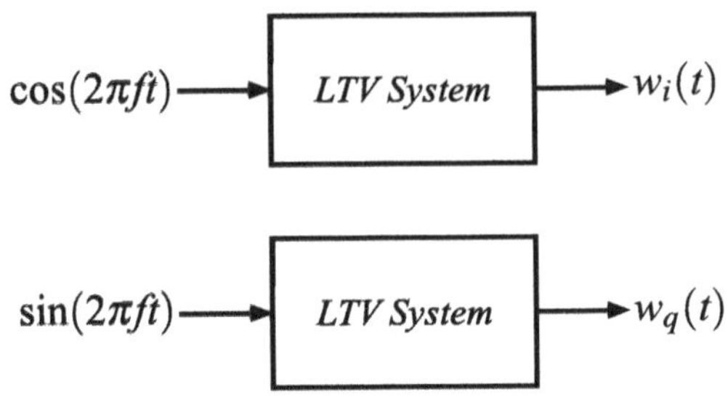

$$\mathrm{Re}(H(j2\pi f, t)) = w_i(t)\cos(2\pi ft) + w_q(t)\sin(2\pi ft)$$
$$\mathrm{Im}(H(j2\pi f, t)) = w_q(t)\cos(2\pi ft) - w_i(t)\sin(2\pi ft)$$

Figure 5.10 Determining $H(j2\pi f, t)$ in the time domain.

as shown in Figure 5.10. Since $\cos(2\pi ft) = (1/2)(e^{j2\pi ft} + e^{-j2\pi ft})$ and $\sin(2\pi ft) = (1/2j)(e^{j2\pi ft} - e^{-j2\pi ft})$, we see that

$$w_i(t) = \frac{1}{2}\{H(j2\pi f, t)e^{j2\pi ft} + H(-j2\pi f, t)e^{-j2\pi ft}\}, \tag{5.12}$$

and

$$w_q(t) = \frac{1}{2j}\{H(j2\pi f, t)e^{j2\pi ft} - H(-j2\pi f, t)e^{-j2\pi ft}\}. \tag{5.13}$$

From these equations, it is straightforward to see that

$$H(j2\pi f, t) = (w_i(t) + jw_q(t))e^{-j2\pi ft}, \tag{5.14}$$

$$Re(H(j2\pi f, t)) = w_i(t)\cos(2\pi ft) + w_q(t)\sin(2\pi ft), \tag{5.15}$$

$$Im(H(j2\pi f, t)) = w_q(t)\cos(2\pi ft) - w_i(t)\sin(2\pi ft).$$

Example: Time-Varying RC Network

Figure 5.11a shows a time-varying first-order RC network, where the resistance $r(t)$ is varied as shown in part (b) of the figure. Let us determine $H(j2\pi f, t)$ for $2\pi f = 4$. To that end, we excite the filter with $\cos(4t)$ and $\sin(4t)$ and record their outputs, denoted by $w_i(t)$ and $w_q(t)$ respectively. $w_i(t)$ (obtained from simulation) is shown in Figure 5.11c. We then determine $H(j2\pi f, t) = (w_i(t) + jw_q(t))e^{-j2\pi ft}$. The magnitude and phase of $H(j2\pi f, t)$ are shown in Figure 5.11d,e, respectively.

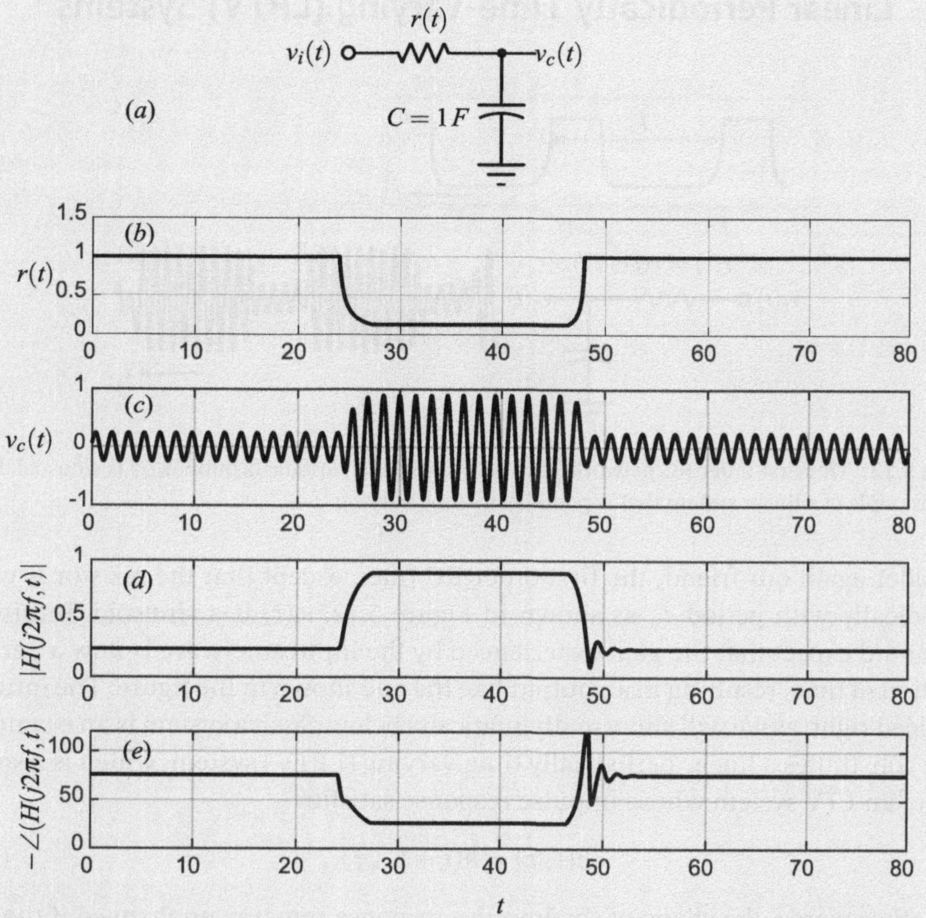

Figure 5.11 (a) RC network with a time-varying resistor, (b) $r(t)$, (c) $v_c(t)$ in response to an input $\cos(8\pi t)$, (d) $|H(j2\pi f, t)|$, and (e) $-\angle H(j2\pi f, t)$.

Let us make sense of $w_i(t)$, which is the response for $v_i(t) = \cos(4t)$. When $r(t) = 1\ \Omega$, the bandwidth of the filter is 1 rad/s. Since the input frequency is well outside the filter's bandwidth, we should expect the amplitude and phase lag of the output sinusoid to be about 0.25 and 75°, respectively.

When $r(t)$ reduces to $1/8\ \Omega$, the filter bandwidth increases to 8 rad/s, and the input lies in band. We should therefore expect the output amplitude to eventually increase to about 0.89 and the phase lag to reduce to $\tan^{-1}(0.5) \approx 25°$. This is confirmed by the simulation results shown in Figure 5.11d,e.

5.4 Linear Periodically Time-Varying (LPTV) Systems

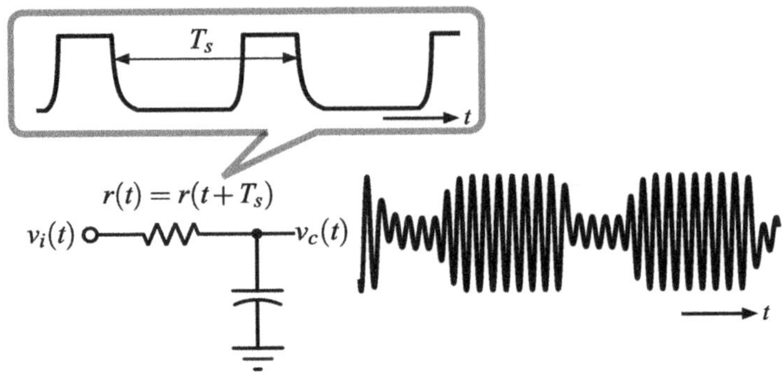

Figure 5.12 Our first-order RC network, with $r(t)$ varied in a periodic fashion. $v_i(t)$ is sinusoidal. This is an example of a linear system that is periodically time-varying.

Consider again our friend, the first-order RC filter, except that the resistor is varied periodically with period T_s as shown in Figure 5.12. $v_i(t)$ is a sinusoid. Intuitively, we should expect that the gain experienced by the input sine-wave is now a periodic function of time, resulting in an output like the one shown in the figure. The intuition is indeed right, as we will show mathematically below. Such a system is an example of what constitutes a linear periodically time-varying (LPTV) system, which is a special case of an LTV system whose impulse response satisfies

$$h(t, \tau) = h(t + T_s, \tau). \tag{5.16}$$

In other words, the shape of the impulse response remains unchanged if the time of observation of the output (t) and the time at which the system is excited (t_1) are *both* shifted by T_s. T_s, which is a characteristic of the system, is called the period of the LPTV system.[4]

Figure 5.13 compares the responses of LTI, LTV, and LPTV systems when excited by impulses. In the LPTV case, we see that the output, when the system is excited by an impulse delayed in time by "this special" T_s, is simply a time-delayed version of the response obtained when the impulsive input occurs at $t = 0$. In fact, if one only considers the first and third rows of Figure 5.13, the LPTV system could be mistaken to be a time-invariant one! In that sense, an LPTV system can be considered to be "closer" to an LTI system than to a time-varying one.

[4] An analogy is the following. Consider taking the stairs to get to your office. The agility with which one does this is perhaps the greatest in the morning. Post lunch, one is most likely slower, and is probably sluggish in the evening. The "response", therefore, is a function of the time of the day. The next morning, however, one is back to his/her sprightly self and is able to conquer the flight with the same speed as one did last morning. The response time is thus periodically time-varying with a period of 24 hours.

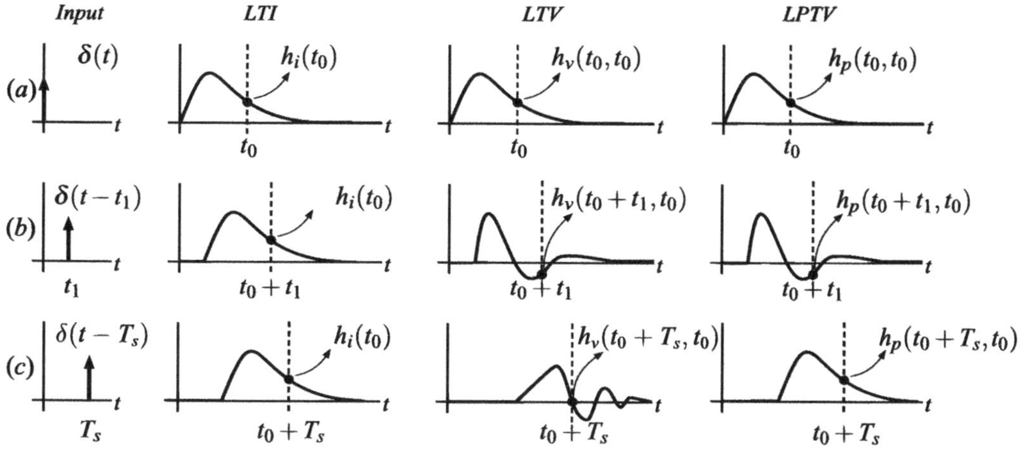

Figure 5.13 Impulse responses of LTI, LTV, and LPTV systems.

How does an LPTV system respond to an input $x(t) = e^{j2\pi ft}$? From (5.8), we see that the output is given by

$$y(t) = H(j2\pi f, t)e^{j2\pi ft}.$$

Since $h(t, \tau) = h(t + T_s, \tau)$, we have

$$\begin{aligned} H(j2\pi f, t) &= \int_0^{\infty} h(t, \tau)e^{-j2\pi f\tau}\, d\tau \\ &= \int_0^{\infty} h(t + T_s, \tau)e^{-j2\pi f\tau}\, d\tau \\ &= H(j2\pi f, t + T_s). \end{aligned} \tag{5.17}$$

Thus, the frequency response $H(j2\pi f, t)$ of an LPTV system is periodic with T_s. What this means is that exciting an LPTV system with $e^{j2\pi ft}$ results in an output that is $e^{j2\pi ft}$ scaled by a periodically changing gain. This is intuitively satisfying, since the properties of the system are varying periodically with time. Note that the "gain" referred to above is a complex quantity.

Example: Periodically Time-Varying RC Network

Figure 5.14a shows a first-order RC filter, now with a periodically time-varying resistor. $r(t)$ has a period T_s, and is shown in part (b) of the figure. Figure 5.14c shows the resulting output for $v_i(t) = \cos(4t)$. As expected, the output sinusoid is higher in amplitude when $r(t)$ goes low and vice versa. Note that this waveform need not necessarily be periodic; it will be so only for special choices of the input frequency.[5] To calculate $H(j2\pi f, t)$, we use (5.15) (which is valid, since an LPTV system is a special case of an LTV system). Figure 5.14d,e shows the magnitude and phase lag of $H(j2\pi f, t)$. As expected, they are periodic with time period T_s.

(Continued)

(Continued)

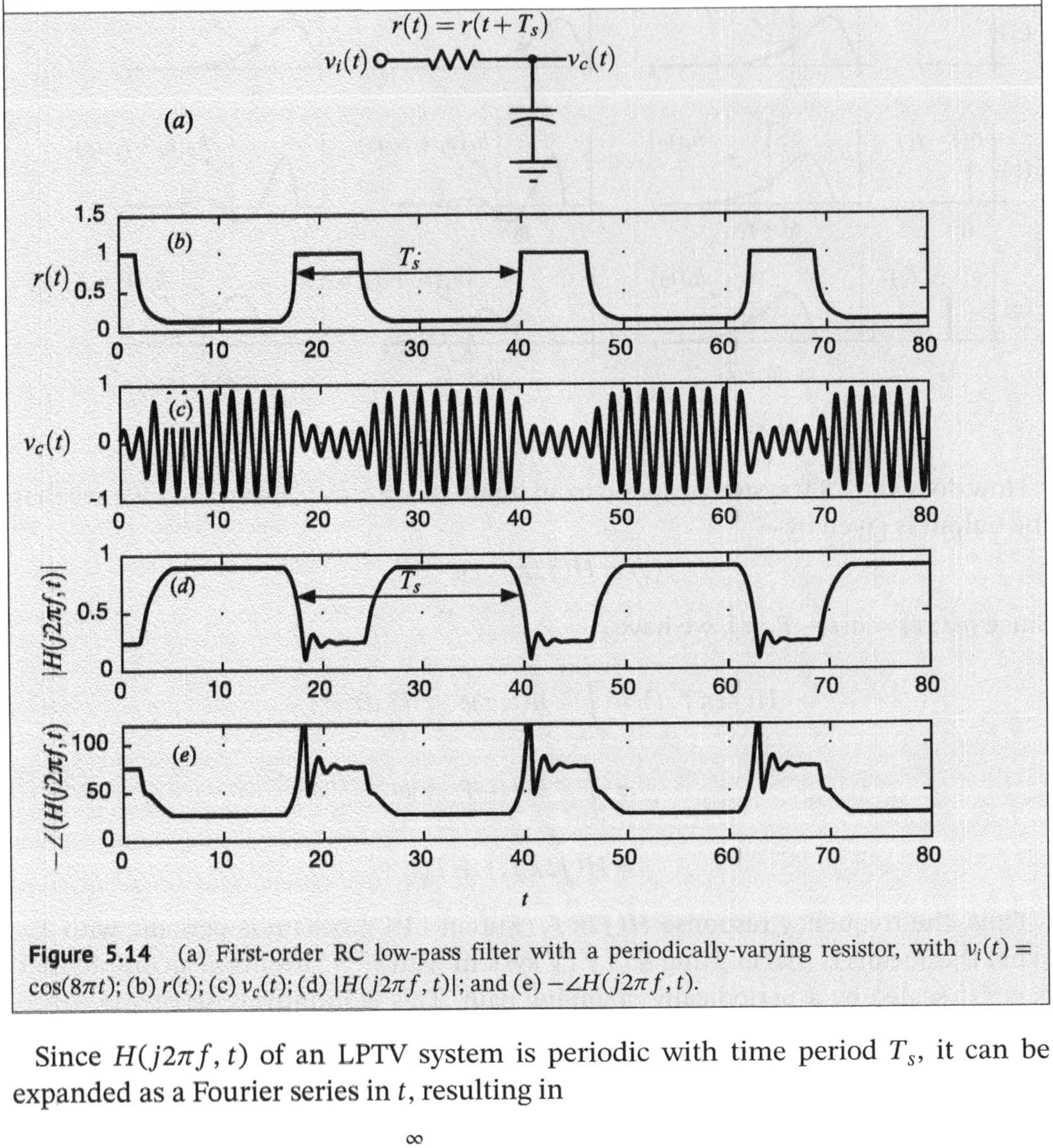

Figure 5.14 (a) First-order RC low-pass filter with a periodically-varying resistor, with $v_i(t) = \cos(8\pi t)$; (b) $r(t)$; (c) $v_c(t)$; (d) $|H(j2\pi f, t)|$; and (e) $-\angle H(j2\pi f, t)$.

Since $H(j2\pi f, t)$ of an LPTV system is periodic with time period T_s, it can be expanded as a Fourier series in t, resulting in

$$H(j2\pi f, t) = \sum_{k=-\infty}^{\infty} H_k(j2\pi f)e^{j2\pi f_s kt}, \qquad f_s = 1/T_s. \tag{5.18}$$

A few comments regarding the equation above are in order. $H(j2\pi f, t)$ is a complex function that is periodic with time. f is simply treated as a parameter, i.e., for each f, $H(j2\pi f, t)$ is in general a *different* periodic function of t. However, for all f, the period with which $H(j2\pi f, t)$ repeats is the same and is equal to T_s. The fundamental

[5]The output will be periodic only if the product of the input frequency f_{in} and T_s is a rational number.

frequency of the Fourier expansion is thus $f_s = 1/T_s$. When $H(j2\pi f, t)$ is expanded in a Fourier series, therefore, the coefficients of the expansion will depend on f. These "f-dependent" Fourier coefficients are what are denoted by $H_k(j2\pi f)$.

How do we determine the $H_k(j2\pi f)$ given $H(j2\pi f, t)$? We proceed as one would do for any Fourier expansion. To determine the (complex) strength of the kth harmonic, multiply $H(j2\pi f, t)$ by $e^{-jk2\pi ft}$ and average the resulting waveform over one period. The $H_k(j2\pi f)$ are called *harmonic transfer functions* and are given by

$$\boxed{H_k(j2\pi f) = \frac{1}{T_s}\int_0^{T_s} H(j2\pi f, t)e^{-j2\pi k f_s t}\,dt.} \tag{5.19}$$

The response of an LPTV system to $e^{j2\pi ft}$ can therefore be expressed as

$$H(j2\pi f, t)e^{j2\pi ft} = \left\{\sum_{k=-\infty}^{\infty} H_k(j2\pi f)e^{j2\pi k f_s t}\right\} e^{j2\pi ft} = \sum_{k=-\infty}^{\infty} \underbrace{H_k(j2\pi f)}_{\substack{\text{Harmonic}\\ \text{Transfer Function}}}\ \underbrace{e^{j2\pi(f+kf_s)t}}_{\substack{\text{frequency}\\ f+kf_s}}. \tag{5.20}$$

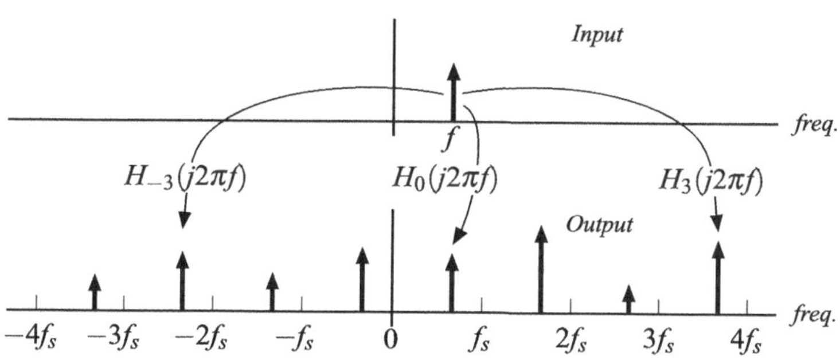

Figure 5.15 In an LPTV network varying at f_s, and excited by $e^{j2\pi ft}$, the output consists of tones at frequencies $f + kf_s$, where k is an integer. $H_k(j2\pi f)$ represents the complex gain from the input (at frequency f) to an output at frequency $f + kf_s$.

From the discussion above, we see that an LPTV system excited by a sinusoid $e^{j2\pi ft}$ results in an output that consists of tones at frequencies of the form $(f + kf_s)$ where k is an integer, as shown in Figure 5.15. $H_k(j2\pi f)$ quantifies the gain experienced by an input at frequency f as it is translated to a frequency $(f + kf_s)$ at the output. Since this gain is with respect to an output frequency around the kth *harmonic* of f_s, $H_k(j2\pi f)$ is called the kth-order *harmonic transfer function*. From the definition (equation 5.19), we see that $H_k(j2\pi f) = H^*_{-k}(-j2\pi f)$. Further, since $H_k(j2\pi f)$ is independent of time, it represents the transfer function of a linear *time-invariant* filter. This means that an LPTV system can be represented as shown in Figure 5.16. The input is processed by a bank of linear time-invariant filters with transfer functions given by $H_k(j2\pi f)$. The outputs of the $H_k(j2\pi f)$ filters are multiplied by $e^{j2\pi k f_s t}$ and added together to yield $y(t)$. This is the **Zadeh expansion** of the LPTV system.

A linear time-invariant (LTI) system may be thought of as an LPTV system with all harmonic transfer functions other than the zeroth-order one being zero.

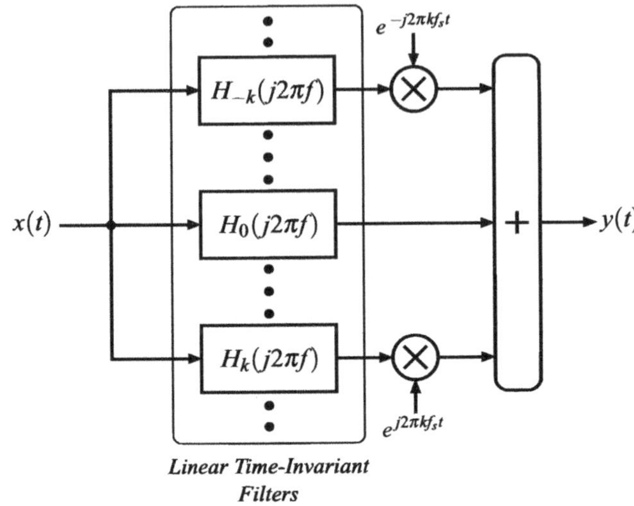

Figure 5.16 Zadeh expansion of an LPTV system. The $H_k(j2\pi f)$ are transfer functions of LTI systems.

An alternative representation of the Zadeh expansion is conceivable, as shown using Figure 5.17. Part (a) of the figure shows the kth arm of the expansion. We place two multipliers in series with the input with their multiplying ports excited by $e^{j2\pi kf_st}$ and $e^{-j2\pi kf_st}$ as shown in Figure 5.17b. Let us examine the contraption enclosed in the gray box. Assuming that the signal at ⓐ is $e^{j2\pi ft}$, let us trace it through the box.

$$\text{ⓐ} \rightarrow e^{j2\pi ft}$$
$$\text{ⓑ} \rightarrow e^{j2\pi(f-kf_s)t}$$
$$\text{ⓒ} \rightarrow H_k(j2\pi(f-kf_s))e^{j2\pi(f-kf_s)t}$$
$$\text{ⓓ} \rightarrow H_k(j2\pi(f-kf_s))e^{j2\pi ft}.$$

We thus see that the signal at ⓓ is $H_k(j2\pi(f-kf_s))e^{j2\pi ft}$. This means that the box is equivalent to a linear *time-invariant* transfer function given by $H_k(j2\pi(f-kf_s))$ as shown in Figure 5.17c.

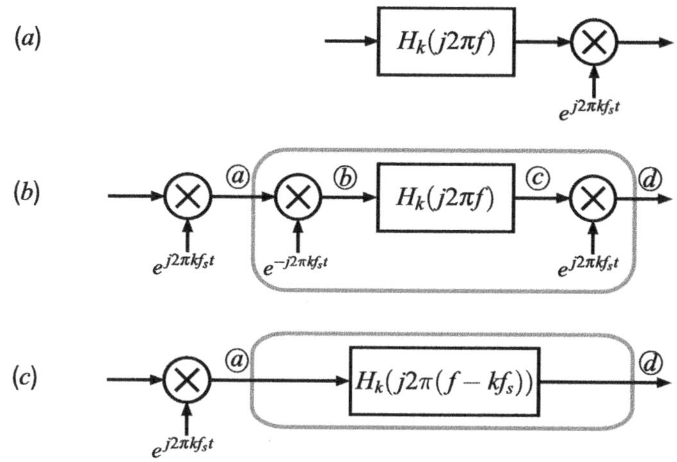

Figure 5.17 (a) The kth arm of the Zadeh expansion. (b) Introducing two premultipliers in the arm. The entity in the gray box is linear and time invariant. (c) Equivalent representation of the box in (a).

The Zadeh expansion of Figure 5.16 can thus be recast in its "mixer-first" form shown in Figure 5.18.

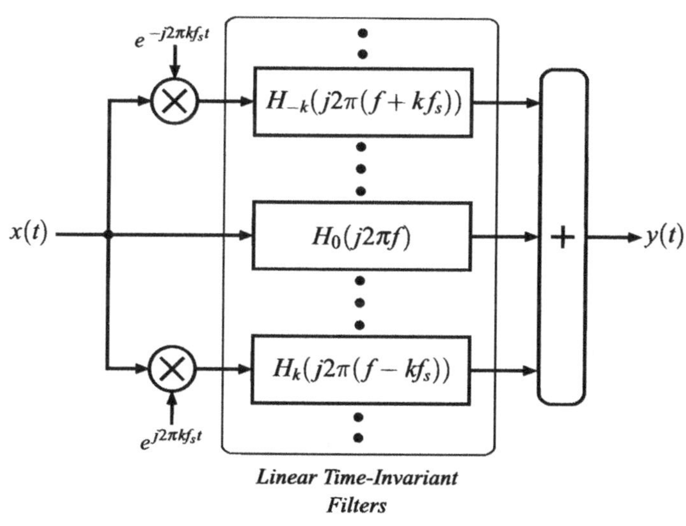

Figure 5.18 The mixer-first view of the Zadeh expansion. The $H_k(2\pi(f-kf_s))$ are transfer functions of LTI systems.

Example

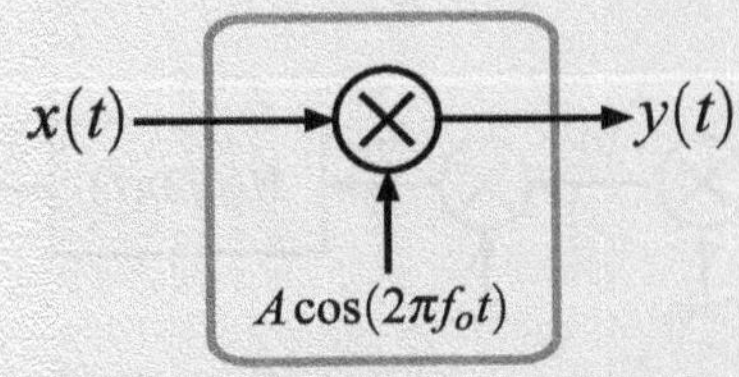

Figure 5.19 An example LPTV system.

What are the harmonic transfer functions of the LPTV system of Figure 5.19? For $x(t) = e^{j2\pi ft}$, we see that

$$y(t) = A\cos(2\pi f_o t)e^{j2\pi ft} = \frac{A}{2}(e^{j2\pi f_o t} + e^{-j2\pi f_o t})e^{j2\pi ft}. \tag{5.21}$$

We see that an input tone with frequency f produces output tones at frequencies $f \pm f_o$. The harmonic transfer functions are thus given by $H_{\pm 1}(f) = (A/2)$, and $H_{\pm k}(f) = 0$ for $k \neq \pm 1$.

We now ask the inverse question: say we observe that the output y is a tone at frequency f. What could x be? Since $H_k(j2\pi f)$ is nonzero only when $k = \pm 1$, it follows that the input must consist of tones at frequencies $(f + f_o)$, or $(f - f_o)$, or both.

Questions to ponder ...

In general, are $|H_k(j2\pi f)|$ and $|H_{-k}(j2\pi f)|$ equal?

What about $|H_k(j2\pi f)|$ and $|H_{-k}(-j2\pi f)|$?

How do we interpret the Zadeh expansion in the time domain? An arbitrary input $x(t)$ can be decomposed as a sum of sinusoids via the Fourier transform as

$$x(t) = \int_{-\infty}^{\infty} X(f)e^{j2\pi ft}\, df. \tag{5.22}$$

The "strength" of the infinitesimal sinusoid with frequency content between f and $f + df$ is given by $X(f)\, df$. The infinitesimal output of the LPTV system to this input is given by

$$dy(t) = (X(f)\, df e^{j2\pi ft}) \sum_{k=-\infty}^{\infty} H_k(j2\pi f)e^{j2\pi k f_s t}. \tag{5.23}$$

The output $y(t)$ is thus seen to be

$$y(t) = \int_{-\infty}^{\infty} X(f)\,df\,e^{j2\pi ft} \sum_{k=-\infty}^{\infty} H_k(j2\pi f)e^{j2\pi kf_st}$$
$$= \sum_{k=-\infty}^{\infty} \underbrace{\int_{-\infty}^{\infty} X(f)H_k(j2\pi f)e^{j2\pi ft}\,df}_{x(t)\,*\,h_k(t)}\; e^{j2\pi kf_st}.$$

where $h_k(t)$ denotes the inverse Fourier transform of $H_k(j2\pi f)$.

Another way of arriving at the result above is the following. Since $h(t,\tau) = h(t + T_s, \tau)$, the time-varying impulse response can be expanded in a Fourier series as follows:

$$h(t,\tau) = \sum_k h_k(\tau)e^{j2\pi kf_st}. \tag{5.24}$$

To arrive at the Fourier series, τ is simply treated as a parameter; the coefficients of the Fourier expansion are therefore functions of τ. The output of the LPTV system for an input $x(t)$ is given by the convolution integral

$$y(t) = \int_0^{\infty} h(t,\tau)x(t-\tau)d\tau = \sum_k \int_0^{\infty} h_k(\tau)e^{j2\pi kf_st}x(t-\tau)d\tau$$
$$= \sum_k \{x(t) * h_k(t)\}\,e^{j2\pi kf_st}.$$

Question to ponder ...

In general, is $h_k(t)$ a real impulse response?

5.5 Calculating Harmonic Transfer Functions

How does one go about calculating harmonic transfer functions in practice? Let us try and relate this to what we know; namely, calculating the transfer function of a time-invariant network [4]. As usual, we work with our familiar first-order RC example shown in Figure 5.20. The input excitation is the complex exponential $e^{j2\pi ft}$. Denoting the node voltages by v_1 and v_2, and the current in the input source by i_s, it is straightforward to write the modified nodal analysis (MNA) equations for the network. The result is the following.

$$\begin{bmatrix} (g + j2\pi fC) & -g & 0 \\ -g & g & 1 \\ 0 & 1 & 0 \end{bmatrix} \begin{bmatrix} V_1 \\ V_2 \\ I_s \end{bmatrix} = \begin{bmatrix} 0 \\ 0 \\ 1\angle 0 \end{bmatrix}. \tag{5.25}$$

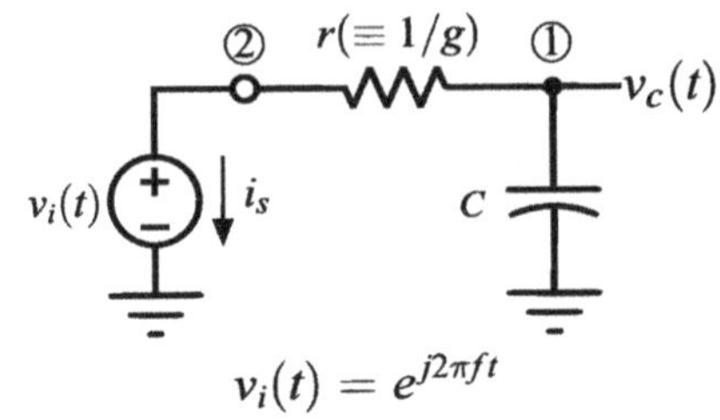

Figure 5.20 LTI network example.

As a refresher, the equation above was derived as given below. The details are described in Chapter 1.

a. The number of unknowns is 3, namely the voltages at the two nodes and the current through the voltage source. The vector of unknowns is thus given by $[V_1 \quad V_2 \quad I_s]^T$. Note that V_1, V_2, I_s are phasors.
b. We form an empty 3 × 3 matrix and add to it the MNA stamp of each element in the network. The resulting matrix premultiplies the vector of unknowns as shown in (5.25).
c. The right-hand side is the vector of excitations. In our example, only the nonzero entry corresponds to the voltage source.

Solving for V_1 yields (the expected)

$$V_1 = \frac{1}{1 + j2\pi frC}. \tag{5.26}$$

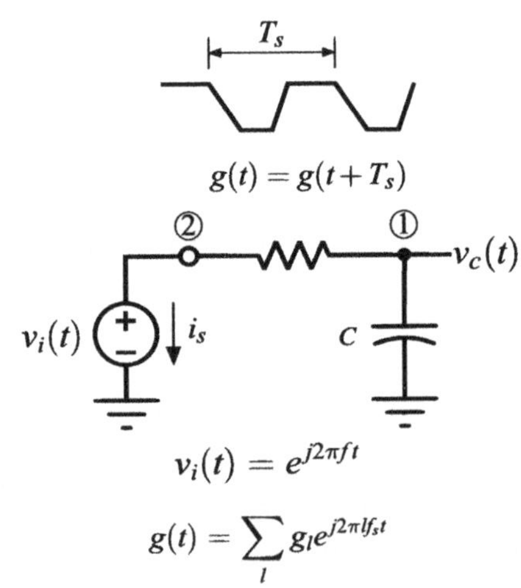

Figure 5.21 First-order RC filter with a periodically-varying conductance.

Let us now attempt to determine the harmonic transfer functions of the first-order RC network, but with the conductance varied periodically according to $g(t) = g(t + T_s)$, with $T_s \equiv 1/f_s$, as shown in Figure 5.21. The voltage $v(t)$ at any node in the LPTV network is the sum of the complex exponentials ..., $e^{j2\pi(f-f_s)t}$, $e^{j2\pi ft}$, $e^{j2\pi(f+f_s)t}$, ...;

i.e., all frequencies of the form $(f + kf_s)$ where k is an integer. The (complex) amplitude of $e^{j2\pi(f+kf_s)t}$ is denoted by V_k. Consequently, $v(t)$ can be expressed as

$$v(t) = \sum_{k=-\infty}^{\infty} V_k e^{j2\pi(f+kf_s)t}. \tag{5.27}$$

V_k therefore represents the phasor corresponding to the frequency $(f + kf_s)$. In practice, k can be restricted to lie in $[-M, M]$, where M is a sufficiently large integer. We denote the vector of phasors by

$$\mathbf{V} = \begin{bmatrix} V_{-M} & \cdots & V_{-1} & V_0 & V_1 & \cdots & V_M \end{bmatrix}^T. \tag{5.28}$$

Observe that the phasor corresponding to $e^{j2\pi ft}$ is the $(M + 1)$th row of $\mathbf{V}$.

A periodically time-varying conductance $g(t)$ can be represented by its Fourier series as follows:

$$g(t) = \sum_l g_l e^{j2\pi l f_s t}. \tag{5.29}$$

Since $g(t)$ is real, $g_{-l} = g_l^*$. The current through the conductance is related to the voltage across it according to

$$i(t) = g(t)v(t) = \left\{\sum_l g_l e^{j2\pi l f_s t}\right\} \sum_k V_k e^{j2\pi(f+kf_s)t}. \tag{5.30}$$

Like $v(t)$, $i(t)$ is also a sum of complex exponentials with frequencies of the form $(f + mf_s)$ where m is an integer. Consequently, $i(t)$ can also be represented in phasor form, which we denote by the column vector $\mathbf{I}$ given by

$$\mathbf{I} = \begin{bmatrix} I_{-M} & \cdots & I_{-1} & I_0 & I_1 & \cdots & I_M \end{bmatrix}^T. \tag{5.31}$$

Since $g(t)$ is periodic, the components of $v(t)$ experience frequency translations to result in the frequency components of $i(t)$. For instance, I_0, which denotes the amplitude of the component of current at a frequency f, is given by

$$I_0 = g_M V_{-M} + g_{M-1} V_{-M+1} + \cdots + g_{-M+1} V_{M-1} + g_{-M} V_M.$$

The amplitudes of the other components of $\mathbf{I}$ can be similarly derived. Expressed in matrix form, $\mathbf{I}$ is related to the Fourier coefficients of the expansion of $g(t)$ and $\mathbf{V}$ by

$$\mathbf{I} = \underbrace{\begin{bmatrix} g_0 & g_{-1} & \cdots & g_{-2M+1} & g_{-2M} \\ g_1 & g_0 & \cdots & g_{-2M+2} & g_{-2M+1} \\ \vdots & \vdots & \ddots & \vdots & \vdots \\ g_{2M-1} & g_{2M-2} & \cdots & g_0 & g_{-1} \\ g_{2M} & g_{2M-1} & \cdots & g_1 & g_0 \end{bmatrix}}_{\mathbf{g}} \underbrace{\begin{bmatrix} V_{-M} \\ V_{-M+1} \\ \vdots \\ V_{M-1} \\ V_M \end{bmatrix}}_{\mathbf{V}} = \mathbf{gV} \tag{5.32}$$

g is what is called a *Toeplitz* matrix formed from the coefficients of the Fourier expansion of $g(t)$. We thus see that the form taken by the equations above is the same as that in the time-invariant case, except that the current and voltage phasor vectors are related to each other through the conductance *matrix* **g**.

An observation regarding **g** is in order. The coefficients g_k of the Fourier expansion of $g(t)$ are related as $g_k = g^*_{-k}$. Consequently, **g** must satisfy $\mathbf{g}^T = \mathbf{g}^*$. We also make the observation (which we will use in Chapter 6 in the context of the adjoint network) that $\mathbf{g}^T$ corresponds to the conductance matrix of the periodically time-varying conductance $g(-t)$.

How is the current $i(t)$ through a linear time-invariant capacitor C related to a voltage of the form $v(t) = \sum_{k=-M}^{M} V_k e^{j2\pi(f+kf_s)t}$ applied across it? $i(t)$ is given by

$$i(t) = C\frac{dv(t)}{dt} = j2\pi C \sum_{k=-M}^{M} (f + kf_s)V_k e^{j2\pi(f+kf_s)t}. \tag{5.33}$$

Consequently, the current phasor can be related to the voltage phasor according to

$$\begin{aligned}\mathbf{I} = j2\pi C &\underbrace{\begin{bmatrix} -Mf_s + f & 0 & \cdots & 0 & 0 \\ 0 & -(M-1)f_s + f & \cdots & 0 & 0 \\ \vdots & \vdots & \ddots & \vdots & \vdots \\ 0 & 0 & \cdots & (M-1)f_s + f & 0 \\ 0 & 0 & \cdots & 0 & Mf_s + f \end{bmatrix}}_{\mathbf{f}} \underbrace{\begin{bmatrix} V_{-M} \\ V_{-M+1} \\ \vdots \\ V_{M-1} \\ V_M \end{bmatrix}}_{\mathbf{V}} \\ &= j2\pi C\mathbf{f}\mathbf{V}. \end{aligned} \tag{5.34}$$

Since the capacitor is time invariant, the **f** matrix is diagonal. Like with the periodically time-varying conductance, we see that the relationship between the current- and voltage-phasors is similar in form to that in the time-invariant case ($I = j2\pi fCV$), except that I, V are vectors and f is replaced by the $(2M + 1) \times (2M + 1)$ matrix **f**.

Referring to our example network of Figure 5.21, Kirchoff's laws must be satisfied at *each* frequency of the form $(kf_s + f)$. Like in the case of a time-invariant network, we can write the (modified) nodal equations, but now with the vectors of the node voltage phasors. The process is entirely analogous to that in a time-invariant network, and the equations can be written down by inspection using the mechanism of MNA stamps. Specifically, the stamp of a periodically time-varying conductance is connected between nodes and is shown in Figure 5.22a. Likewise, that of a capacitor is shown in Figure 5.22b. We write the MNA equations as follows:

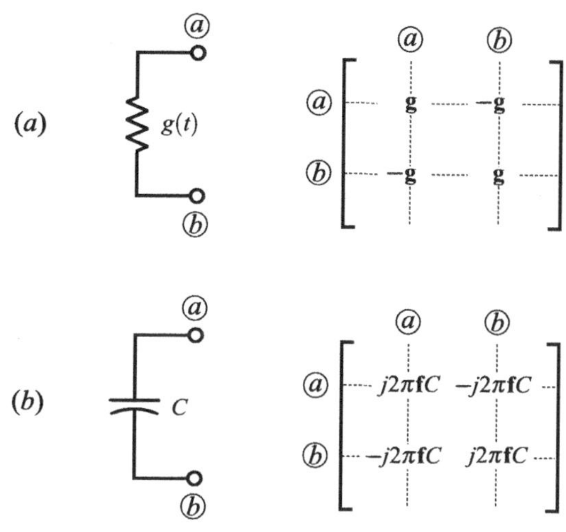

Figure 5.22 MNA stamps of (a) a periodically time-varying conductance and (b) a linear time-invariant capacitor.

a. The number of unknowns is $3 \times (2M + 1)$, namely the voltage phasor vectors at the two nodes, and the current phasor vector through the voltage source, each of which has $(2M + 1)$ unknowns. The vector of unknowns is thus given by $[\mathbf{V}_1 \quad \mathbf{V}_2 \quad \mathbf{I}_1]^T$.
b. We form an empty $3(2M + 1) \times 3(2M + 1)$ matrix, and add to it the MNA stamp of each element in the network. The resulting matrix premultiplies the vector of unknowns as shown in (5.35).
c. The right-hand side is the phasor vector of excitations.

Using this, we see that the MNA equations of the network of Figure 5.21 are given by

$$\begin{bmatrix} (\mathbf{g} + j2\pi\mathbf{f}C) & -\mathbf{g} & \mathbf{0} \\ -\mathbf{g} & \mathbf{g} & \mathbf{1} \\ \mathbf{0} & \mathbf{1} & \mathbf{0} \end{bmatrix} \begin{bmatrix} \mathbf{V}_1 \\ \mathbf{V}_2 \\ \mathbf{I}_s \end{bmatrix} = \begin{bmatrix} \mathbf{0} \\ \mathbf{0} \\ \mathbf{V}_{in} \end{bmatrix} \tag{5.35}$$

where

$$\mathbf{V}_{in} = [\underbrace{0 \quad \cdots \quad 0}_{\text{M zeros}} \quad 1\angle 0 \quad \underbrace{0 \quad \cdots \quad 0}_{\text{M zeros}}]^T. \tag{5.36}$$

Since the excitation is $e^{j2\pi ft}$, $\mathbf{V}_{in}$ is 1 in the $(M + 1)$th row, and zero elsewhere. The **0**s on the right-hand side of (5.35) are $(2M + 1) \times 1$ vectors, while the **0**s on the left-hand side are zero matrices with dimensions $(2M + 1) \times (2M + 1)$. The **1**s on the left-hand side are identity matrices of dimension $(2M + 1)$. When compared to (5.25), we see that the number of equations (and unknowns) in (5.35) is larger by a factor $(2M + 1)$. This makes sense—we are solving KCL/KVL at $(2M + 1)$ frequencies.

From the discussion above, it is easy to appreciate the fact that the analysis of LPTV circuits tends to be *much* more involved than that of LTI networks. For instance, using $M = 10$ for our first-order LPTV network results in 63 equations in as many unknowns!

It is instructive to compare the MNA equations in the LTI and LPTV cases. The summary is shown in Figure 5.23. The key points are as follows:

a. The conductance matrix is 3×3 in the LTI circuit. In the LPTV network, the form of the matrix is identical to that in the LTI case, except that each entry is a $(2M + 1) \times (2M + 1)$ matrix.

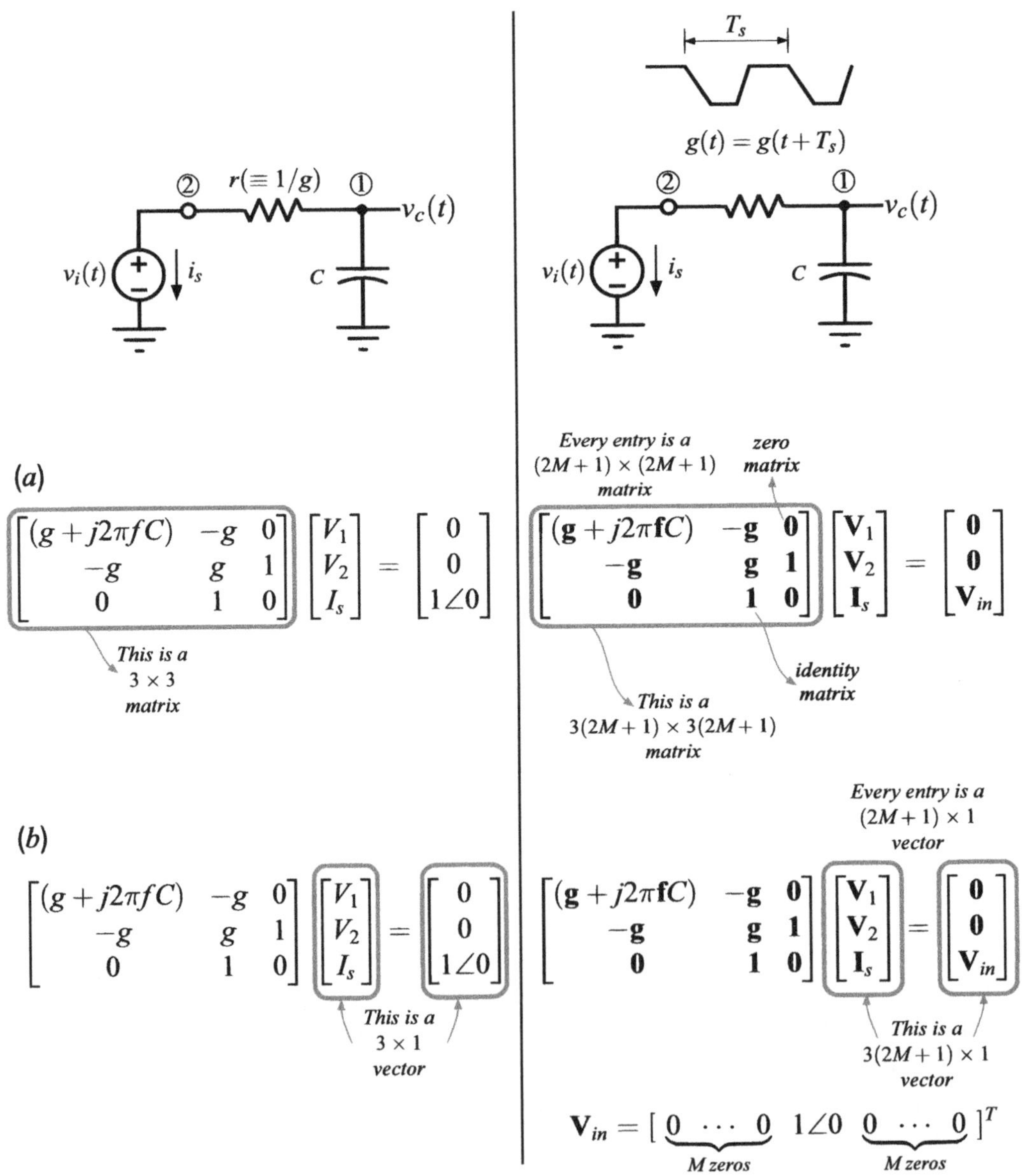

Figure 5.23 Summary and comparison of the MNA equations in the LTI and LPTV RC network.

b. In the conductance matrix of the LPTV network, the **0** and **1** are the zero and identity matrices, respectively. **g** is formed from the coefficients of the Fourier expansion of $g(t)$ according to (5.32). **f** is the diagonal matrix[6] given by $diag(-Mf_s + f, \ldots, f, \ldots, Mf_s + f)$.

c. In the LTI circuit, there are three unknowns, namely V_1, V_2, and I_s that form a 3×1 vector. In the LPTV network, the form of the unknown vector is the same, except that each entry is a $(2M + 1) \times 1$ vector, as seen in Figure 5.23. The number of unknowns is thus seen to be $3(2M + 1)$.

d. The excitation vector on the right-hand side is a 3×1 vector in the LTI circuit. The form of the vector remains identical in the LPTV case, except that each entry is a $(2M + 1) \times 1$ vector.

e. Since $v_i(t) = e^{j2\pi ft}$, the last row is $1\angle 0$ in the excitation vector of the LTI circuit. In the LPTV network, the last entry in the excitation vector is $\mathbf{V}_{in} = [\underbrace{0 \quad \cdots \quad 0}_{M \text{ zeros}} \quad 1\angle 0 \quad \underbrace{0 \quad \cdots \quad 0}_{M \text{ zeros}}]^T$.

Example: LPTV Network with Fixed and Time-Varying Resistors

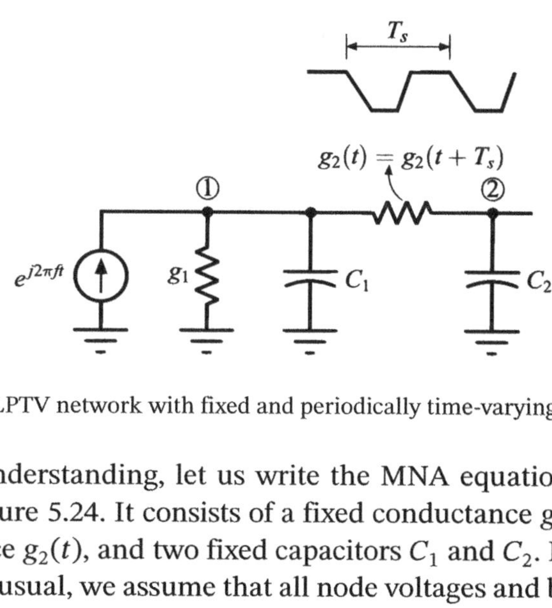

Figure 5.24 LPTV network with fixed and periodically time-varying conductances.

To reinforce our understanding, let us write the MNA equations pertaining to the LPTV circuit of Figure 5.24. It consists of a fixed conductance g_1, and a periodically varying conductance $g_2(t)$, and two fixed capacitors C_1 and C_2. Node ① is driven by a current $e^{j2\pi ft}$. As usual, we assume that all node voltages and branch currents consist of complex exponentials with frequencies ranging from $(-Mf_s + f)$ to $(Mf_s + f)$. We have two unknowns, namely the voltage phasor vectors at nodes ① and ②, each of which has a dimension $(2M + 1) \times 1$. The conductance matrix, formed by summing up the MNA stamps of the components, is of size $2(2M + 1) \times 2(2M + 1)$.

---

[6] $diag(a_1, \ldots, a_N)$ denotes a square matrix of order N with $a_1, \ldots, a_N$ on its diagonal.

The excitation vector is given by $\mathbf{I}_{\text{in}} = [\,\underbrace{0 \quad \cdots \quad 0}_{M\text{ zeros}} \quad 1\angle 0 \quad \underbrace{0 \quad \cdots \quad 0}_{M\text{ zeros}}\,]^T$.

The MNA equations are thus given by

$$\begin{bmatrix} (\mathbf{g}_1 + \mathbf{g}_2 + j2\pi\mathbf{f}C_1) & -\mathbf{g}_2 \\ -\mathbf{g}_2 & \mathbf{g}_2 + j2\pi\mathbf{f}C_2 \end{bmatrix} \begin{bmatrix} \mathbf{V}_1 \\ \mathbf{V}_2 \end{bmatrix} = \begin{bmatrix} \mathbf{I}_{\text{in}} \\ \mathbf{0} \end{bmatrix} \tag{5.37}$$

g_1 does not change with time. The current through it is given by

$$i_1(t) = g_1 \sum_{m=-M}^{M} V_m e^{j2\pi(mf_s+f)t}.$$

Consequently, the current phasor vector $\mathbf{I}_1$ is given by $\mathbf{I}_1 = g_1\mathbf{V}_1 = g_1\mathbf{1}\mathbf{V}_1$ where $\mathbf{1}$ is the identity matrix of size $(2M+1)\times(2M+1)$. It is evident that the MNA stamp of the conductance is $g_1\mathbf{1}$. The key point is that though g_1 is *time invariant*, the MNA stamp is a $(2M+1)\times(2M+1)$ matrix. This makes sense, since it is part of an LPTV network. $\mathbf{g}_2$ is the matrix formed from the coefficients of the Fourier expansion of $g_2(t)$.

5.5.1 MNA Stamps of Periodically Time-Varying Controlled Sources

We now "complete our catalog" of MNA stamps by tabulating those that correspond to the controlled sources. We begin with the VCVS.

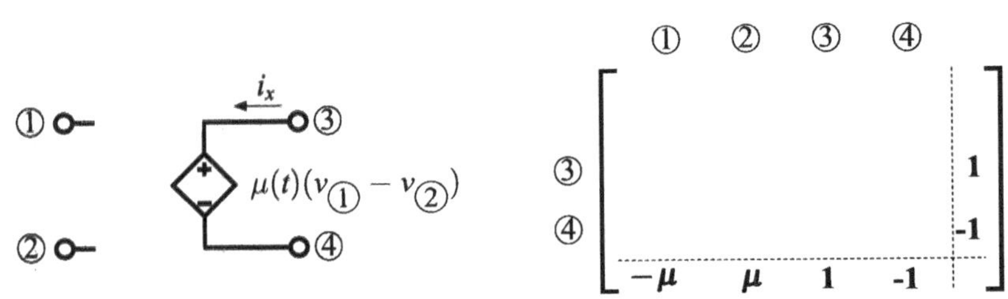

Figure 5.25 A VCVS with a periodically time-varying gain and its MNA stamp.

Figure 5.25 shows a time-varying voltage-controlled voltage source (VCVS), whose gain $\mu(t)$ is varying periodically with time.[7] It can therefore be expanded in a Fourier series that

$$\mu(t) = \sum_{l} \mu_l e^{j2\pi l f_s t}. \tag{5.38}$$

[7]This period must be the same as that with which the network is varying.

Like we did with a time-varying conductance, we form the matrix μ from the Fourier coefficients of $\mu(t)$ according to

$$\mu = \begin{bmatrix} \mu_0 & \mu_{-1} & \cdots & \mu_{-2M+1} & \mu_{-2M} \\ \mu_1 & \mu_0 & \cdots & \mu_{-2M+2} & \mu_{-2M+1} \\ \vdots & \vdots & \ddots & \vdots & \vdots \\ \mu_{2M-1} & \mu_{2M-2} & \cdots & \mu_0 & \mu_{-1} \\ \mu_{2M} & \mu_{2M-1} & \cdots & \mu_1 & \mu_0 \end{bmatrix}. \tag{5.39}$$

The MNA stamp of the VCVS is shown on the right of Figure 5.25. The vector of current phasors flowing into the voltage source is treated as an unknown; consequently, an extra row and column are added to the MNA equations. The **1**'s are identity matrices of order $(2M+1)\times(2M+1)$. If the VCVS is time invariant, μ reduces to a diagonal matrix. The empty rows and columns in the stamp of Figure 5.25 are all zero matrices of order $(2M+1)\times(2M+1)$.

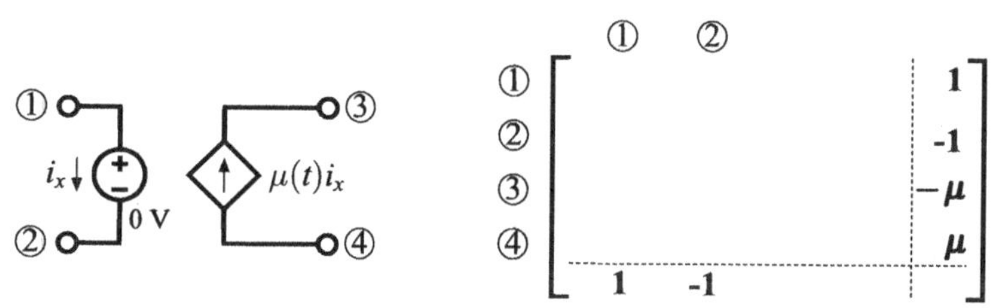

Figure 5.26 A CCCS with a periodically time-varying gain and its MNA stamp.

Figure 5.26 shows a periodically time-varying CCCS and its MNA stamp. μ is given by (5.39). An extra row and column are necessary to accommodate the (unknown) current through the 0 V source.

A periodically time-varying VCCS is shown in Figure 5.27. The transconductance varies according to

$$g_m(t) = \sum_l g_l e^{j2\pi l f_s t}.$$

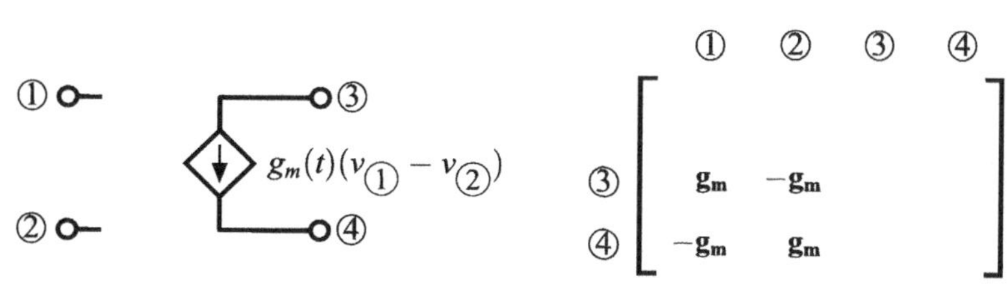

Figure 5.27 A VCCS with a periodically time-varying transconductance and its MNA stamp.

The Fourier coefficients of $g_m(t)$ are arranged in a matrix of size $(2M+1) \times (2M+1)$ according to

$$\boldsymbol{g}_m = \begin{bmatrix} g_0 & g_{-1} & \cdots & g_{-2M+1} & g_{-2M} \\ g_1 & g_0 & \cdots & g_{-2M+2} & g_{-2M+1} \\ \vdots & \vdots & \ddots & \vdots & \vdots \\ g_{2M-1} & g_{2M-2} & \cdots & g_0 & g_{-1} \\ g_{2M} & g_{2M-1} & \cdots & g_1 & g_0 \end{bmatrix}. \tag{5.40}$$

The MNA stamp of the VCCS is shown in Figure 5.27. Notice again the similarity of the stamp's form with that in the case of a time-invariant network.

Figure 5.28 shows a periodically time-varying CCVS. The transresistance varies according to

$$r(t) = \sum_l r_l e^{j2\pi l f_s t}.$$

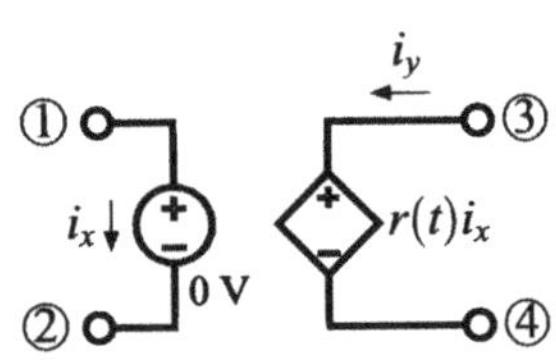

$$\begin{array}{c|cccc:cc} & ① & ② & ③ & ④ & i_x & i_y \\ \hline ① & & & & & 1 & \\ ② & & & & & -1 & \\ ③ & & & & & & 1 \\ ④ & & & & & & -1 \\ \hdashline i_x & 1 & -1 & & & & \\ i_y & & & 1 & -1 & -r & \end{array}$$

Figure 5.28 A CCVS with a periodically time-varying transresistance and its MNA stamp.

The Fourier coefficients of $r(t)$ are arranged in a matrix of size $(2M+1) \times (2M+1)$ according to

$$\boldsymbol{r} = \begin{bmatrix} r_0 & r_{-1} & \cdots & r_{-2M+1} & r_{-2M} \\ r_1 & r_0 & \cdots & r_{-2M+2} & r_{-2M+1} \\ \vdots & \vdots & \ddots & \vdots & \vdots \\ r_{2M-1} & r_{2M-2} & \cdots & r_0 & r_{-1} \\ r_{2M} & r_{2M-1} & \cdots & r_1 & r_0 \end{bmatrix}. \tag{5.41}$$

The MNA stamp of the CCVS is shown in Figure 5.28. There are two additional unknowns, which are the currents through the voltage sources. Two extra columns and rows are thus needed in the MNA stamp. The MNA stamp of an inductor can be derived in a manner analogous to that followed in the case of a time-invariant network. It is left as an exercise for the reader.

To summarize, the MNA equations of an LPTV network can be written in the same form as that for an LTI network, except that the unknowns are now *vectors of phasors*, and the entries in the MNA stamps are matrices of appropriate order. If the maximum order of the harmonic transfer functions (HTFs) is restricted to M, the total number of unknowns is larger by a factor $(2M + 1)$ when compared to an LTI network with the same topology. Consequently, determining the HTFs is a computationally intensive process. How does one choose M? A way to do this is to keep increasing M until the change in the computed HTFs becomes negligibly small as M is increased.

Simulation Corner

Many commercial CAD tools now have what are called "RF simulation" capabilities. This is the capacity to simulate LPTV networks and compute harmonic transfer functions (and other properties of LPTV networks). Let us build on our knowledge of time-invariant circuits to understand what is involved in finding the harmonic transfer functions.

Many practically useful circuits are nonlinear. However, with respect to small-signal excitations, they behave linearly. In "conventional circuits" (those that are time-invariant), frequency response is obtained using an `.AC` analysis. The flow is the following, as illustrated with the circuit of Figure 5.29a. V_{b1}, V_{b2}, and V_{dd} are constant (bias) voltages, and v_i is a small-signal excitation. It is desired to find the response from v_i to the resulting small signals at other points in the circuit.

The simulator assumes that all circuits are nonlinear, and the first order of business is to compute the operating point (the `.OP` analysis). To this end, all small-signal excitations are set to zero, and the nonlinear network is solved. In our example, v_i is set to zero to find the operating point. The resulting network is shown in Figure 5.29b. Once the operating points of the MOSFETs (which are the nonlinear elements in this circuit) are found, their small-signal models can be determined using knowledge of device physics. Next, the small-signal network is formed by setting all bias sources to zero and replacing every element by its small-signal (or incremental) equivalent. In our example, the resulting LTI network is shown in Figure 5.29c, where each MOSFET is replaced by a VCCS. The sources that excite this network are the small-signal excitations (which were turned off during the `.OP` analysis). Finding the frequency response, which is accomplished using the `.AC` analysis involves the solution of the linear network over a specified set of frequencies using the MNA approach we discussed in Chapter 1.

(Continued)

(Continued)

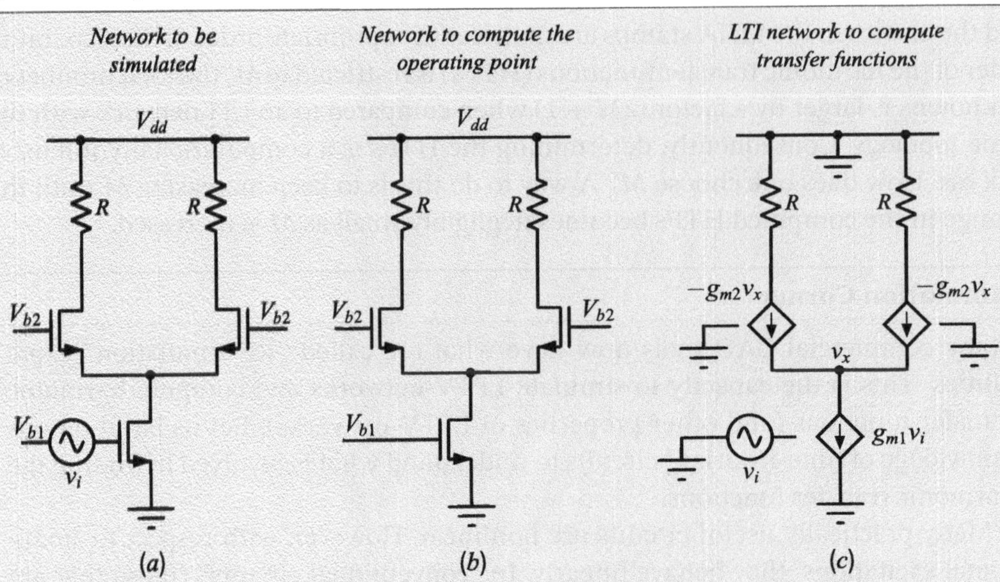

Figure 5.29 (a) The response due to v_i needs to be determined. v_i is a small-signal excitation. (b) Network to determine the operating point. (c) The small-signal equivalent, which is a linear time-invariant network.

Let us now examine the mixer circuit shown in Figure 5.30a. v_i is a small-signal excitation as before. However, the gates of the differential pair are switched by exciting them with a *large* periodic differential signal. Since the "bias" voltages are changing periodically, the operating point of the devices varies in a periodic fashion. The network to determine this periodically varying operating point is shown in Figure 5.30b. This is accomplished by running what is referred to as the *Periodic Steady State* analysis, or `.PSS` analysis in short. Since the operating point of the MOSFETs varies periodically with time, the small-signal model consists of elements that vary periodically. In our simplistic example, the transistors have a transconductance that changes periodically with time.

The resulting small-signal network is shown in Figure 5.30c. It is an LPTV system and is solved using techniques similar to the ones we described in connection with the example circuit of Figure 5.21. This is referred to as the `.PAC` analysis, where the P stands for *periodic*. `.PAC` is analogous to the `.AC`, the latter used when the bias quantities are time-invariant.

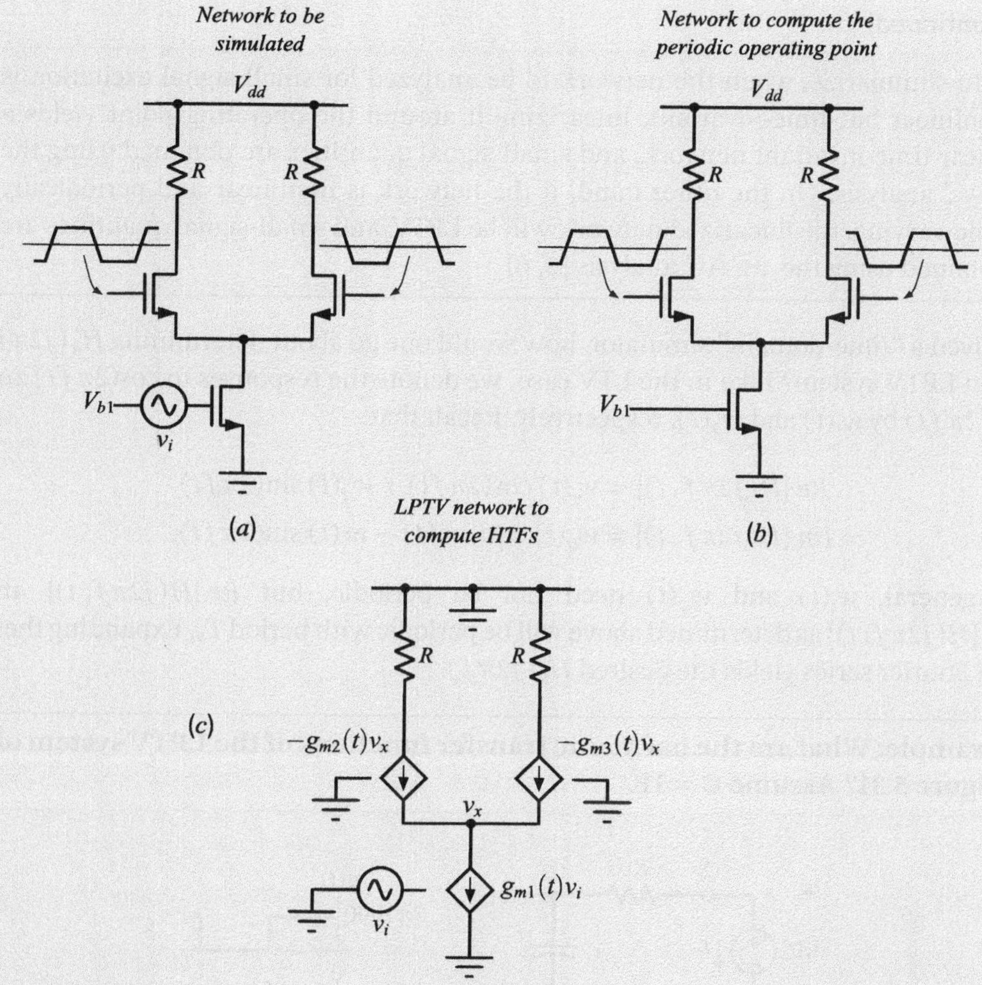

Figure 5.30 (a) The response due to v_i needs to be determined. v_i is a small-signal excitation. The "bias quantities" change periodically. (b) Network to determine the periodic operating point. (c) The small-signal equivalent, which is a linear periodically time-varying network.

In an .AC analysis run over the input frequency range $[f_1, f_2]$, the output frequency range is also $[f_1, f_2]$. However, in a .PAC analysis run over the input frequency range $[f_1, f_2]$, the outputs range over frequencies $kf_s + [f_1, f_2]$ where k is an integer and f_s is the rate at which the network is varying. k, which can range from $[-\infty, \infty]$ is called the *sideband number* and refers to the kth-order harmonic transfer function. Since there are an infinite number of possible sidebands, the desired value(s) of k should be specified before running the .PAC analysis. For instance, when someone refers to the fourth sideband gain for an input frequency of f_1, the frequency at which the output is desired is $(f_1 + 4f_s)$.

(Continued)

(Continued)

To summarize, when the network to be analyzed for small-signal excitation is nonlinear but time-invariant, linearizing it around the operating point yields a linear time-invariant network, and small-signal quantities are obtained using the .AC analysis. On the other hand, if the network is nonlinear and periodically time-varying, the linearized network will be LPTV, and small-signal quantities are obtained using the . PAC analysis [5, 6].

Given a "time-domain" simulator, how would one go about determining $H_k(j2\pi f)$ of an LPTV system? Like in the LTV case, we denote the responses to $\cos(2\pi ft)$ and $\sin(2\pi ft)$ by $w_i(t)$ and $w_q(t)$, respectively. Recall that

$$Re\,\{H(j2\pi f, t)\} = w_i(t)\cos(2\pi ft) + w_q(t)\sin(2\pi ft)$$
$$Im\,\{H(j2\pi f, t)\} = w_q(t)\cos(2\pi ft) - w_i(t)\sin(2\pi ft).$$

In general, $w_i(t)$ and $w_q(t)$ need not be periodic, but $Re\,\{H(j2\pi f, t)\}$ and $Im\,\{H(j2\pi f, t)\}$ as determined above will be periodic with period T_s. Expanding these in a Fourier series yields the desired $H_k(j2\pi f)$.

Example: What are the harmonic transfer functions of the LPTV system of Figure 5.31? Assume C = 1F.

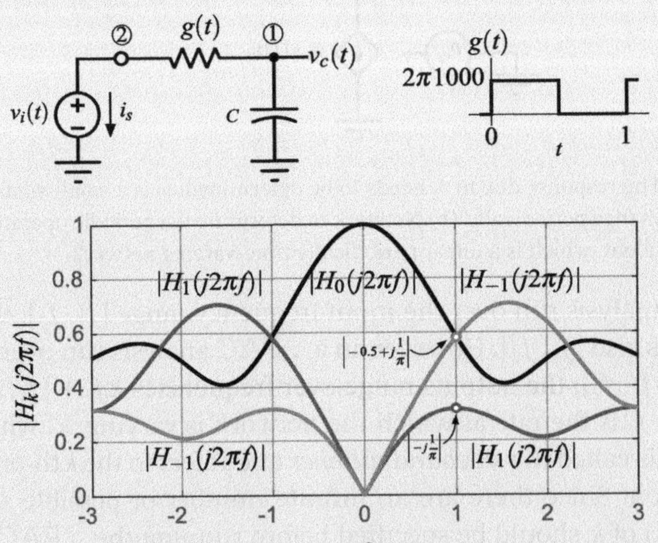

Figure 5.31 Computed harmonic transfer function magnitudes of an RC circuit with a periodically varying conductance. $C = 1$ F. $g(t)$ varies from 0 to $2\pi 1000$ S.

This is an example of an RF sampling mixer. Intuitively, one can think of the time-varying conductance as a periodically operated switch with infinite off-resistance and a small (virtually zero) on-resistance. While this mixer is not particularly useful in practice, it serves as a simple vehicle that gives insight into LPTV circuits. We use (5.35) with $M = 256$ to compute $H_k(j2\pi f)$. The resulting magnitudes for $k = 0, \pm 1$ are plotted in Figure 5.31.

As a sanity check, we see that $|H_0(0)| = 1$ and $H_k(0) = 0$ when $k \neq 0$. This makes sense—with a DC input $v_i = 1$ V, no current flows through the conductance in steady state, and $v_c(t) = 1$ V.

We also see, as expected, that $|H_k(j2\pi f)| = |H_{-k}(-j2\pi f)|$. Let us now verify the values of $H_0(j2\pi)$, $H_{-1}(j2\pi)$ and $H_1(j2\pi)$ obtained by solving (5.35) using "time-domain techniques" for the specific case of $f = f_s = 1$Hz. To do this, we determine $w_i(t)$ and $w_q(t)$ when $v_i(t) = \cos(2\pi t)$ and $v_i(t) = \sin(2\pi t)$ respectively. We then plot $Re\,\{H(j2\pi f, t)\}$ and $Im\,\{H(j2\pi f, t)\}$, as shown in Figure 5.32. It is straightforward to obtain

$$Re\,\{H(j2\pi f, t)\} = \frac{1}{2} + \left\{-\frac{1}{4} - j\frac{1}{\pi}\right\} e^{j2\pi t} + \left\{-\frac{1}{4} + j\frac{1}{\pi}\right\} e^{-j2\pi t} + \cdots$$

$$Im\,\{H(j2\pi f, t)\} = -\frac{1}{\pi} + \left\{-j\frac{1}{4}\right\} e^{j2\pi t} + \left\{j\frac{1}{4}\right\} e^{-j2\pi t} + \cdots$$

From this, we see that

$$H_0(j2\pi) = \frac{1}{2} - j\frac{1}{\pi}$$

$$H_1(j2\pi) = -\frac{1}{4} - j\frac{1}{\pi} + j\left\{-j\frac{1}{4}\right\} = -\frac{j}{\pi}$$

$$H_{-1}(j2\pi) = -\frac{1}{4} + j\frac{1}{\pi} + j\left\{j\frac{1}{4}\right\} = -\frac{1}{2} + \frac{j}{\pi}.$$

These values are in good agreement with computations, as seen in Figure 5.31.

A concluding remark is in order. The "switched"-RC circuit is used in practice as a track-and-hold, where $v_i(t)$ is held on the capacitor when the switch is opened. Though the switch is near ideal (perfect open and almost a short), we see that $H_0(j2\pi f)$ has a 3 dB-bandwidth of about only 0.8 Hz. This *might* give the reader the impression that this circuit cannot track high input frequencies. Clearly, as a simple time-domain argument shows, this is not true. What then is the conundrum here? The answer lies in the fact that the quantity of interest in a track-and-hold is the *sampled* capacitor voltage, while $H_0(j2\pi f)$ calculated above is in relation to the entire voltage waveform (i.e., during the track and hold phases). Sampled quantities, as seen above, can have significantly different behavior when compared to the "entire waveform." We will return to this important aspect later in our journey through the LPTV-system landscape.

(Continued)

(Continued)

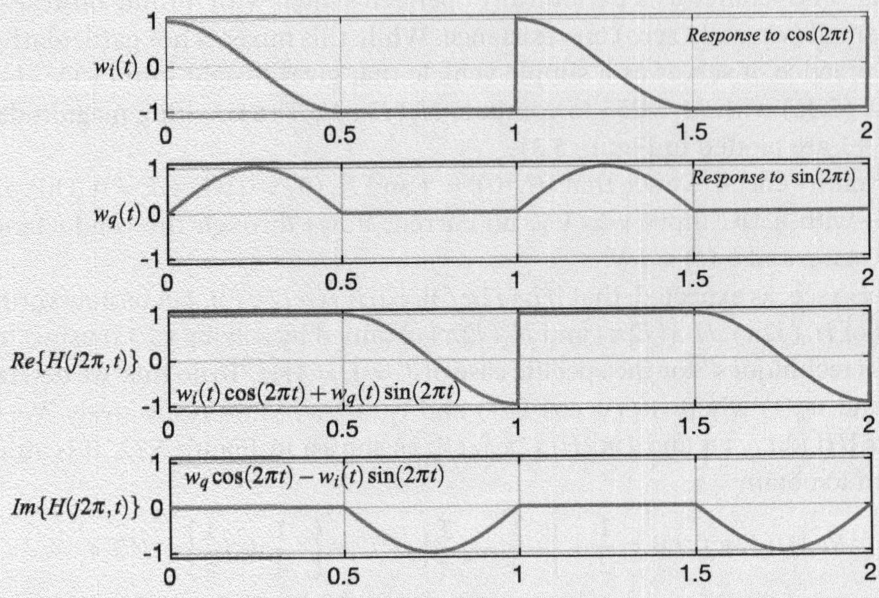

Figure 5.32 $v_c(t)$ in the circuit of Figure 5.31 with $v_i(t) = \cos(2\pi t)$ and $v_i(t) = \sin(2\pi t)$.

5.5.2 Determining $H_{-l}(j2\pi l f_s)$ and $H_l(-j2\pi l f_s)$

In many practical situations,[8] a designer is in the position of having to calculate harmonic transfer functions of the form $H_{-l}(j2\pi l f_s)$, where l is an integer. The input and output frequencies, therefore, are $l f_s$ and 0 respectively. It turns out that in this special (and practically important) case, one does not have to go through the labor of first computing $Re\{H(j2\pi f, t)\}$ and $Im\{H(j2\pi f, t)\}$ and then expanding them in a Fourier series to determine the $H_{-l}(j2\pi l f_s)$ that we seek.

Denoting $H_{-l}(j2\pi l f_s) = a + jb$ and recalling that $H_l(-j2\pi l f_s)$ must equal $(a + jb)^* = a - jb$, we see that the dc components of the responses of the LPTV system to inputs $e^{\pm j2\pi l f_s t}$ are given by

$$e^{j2\pi l f_s t} \rightarrow \underbrace{a + jb}_{H_{-l}(j2\pi l f_s)}$$
$$e^{-j2\pi l f_s t} \rightarrow \underbrace{a - jb}_{H_l(-j2\pi l f_s)}\ .$$

a and b are thus seen to be the average values of the steady-state responses to $\cos(2\pi l f_s t)$ and $\sin(2\pi l f_s t)$, respectively. As a sanity check, consider again the circuit

[8] Determining the conversion gain of RF mixers, and alias rejection of continuous-time delta-sigma modulators are applications that immediately come to mind.

of Figure 5.31. From the discussion above, it is immediately apparent that $H_{-1}(j2\pi) = w_i(t) + jw_q(t) = -0.5 + j/\pi$, in agreement with the frequency-domain computation leading to the plots in Figure 5.31. The reader may verify that $H_{-2}(j4\pi) = 0.5$.

Example: The Switched-RC Circuit

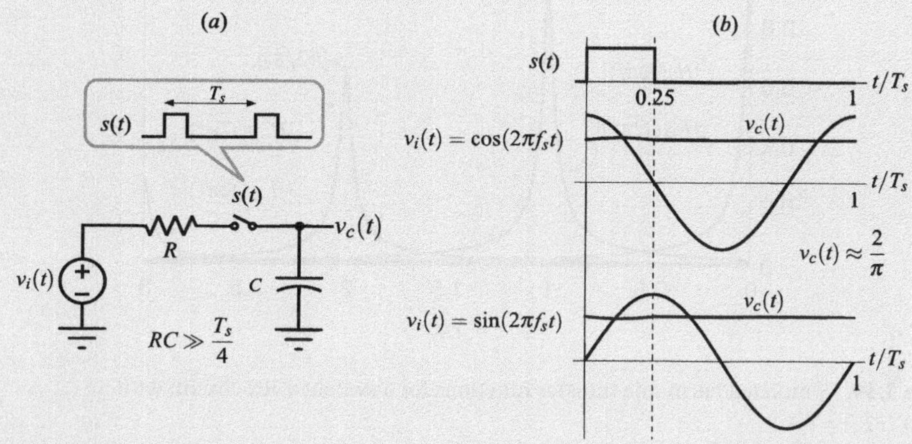

Figure 5.33 Exciting an LPTV system with $\cos(2\pi f_s t)$ and $\sin(2\pi f_s t)$ to determine $H_{-1}(j2\pi f_s)$.

Figure 5.33a shows a switched-RC circuit, where the switch is controlled by the periodic waveform $s(t)$. Its period is denoted by T_s and the duty cycle is 25%. Further, RC $\gg T_s$. Let us determine $H_{-1}(j2\pi f_s)$ and $H_{-2}(j4\pi f_s)$.

RF designers will immediately recognize this as the kernel of a 4-phase filtering mixer, but if you haven't seen this before, don't worry! To determine $H_{-1}(j2\pi f_s)$, we examine the steady-state outputs when $v_i(t) = \cos(2\pi f_s t)$ and $v_i(t) = \sin(2\pi f_s t)$, respectively. When the switch is closed, the source attempts to charge the capacitor, while the latter tries to discharge through the resistor. These two effects must equal each other in a steady state. Further, since $RC \gg T_s$, we expect the capacitor voltage to remain virtually unchanged when the switch is closed. Consequently, $v_c(t)$ in steady state must equal the average value of $v_i(t)$ when the switch is closed. With sine- or cosine-wave excitation, we thus have

$$v_c(t) \approx \frac{1}{0.25\,T_s}\int_0^{0.25T_s} \cos(2\pi f_s t)\,dt = \frac{1}{0.25\,T_s}\int_0^{0.25T_s} \sin(2\pi f_s t)\,dt = \frac{2}{\pi}. \quad (5.42)$$

$H_{-1}(j2\pi f_s)$ is therefore given by $(2/\pi)(1 + j)$. Proceeding along similar lines, it is straightforward to see that $H_{-2}(j4\pi f_s) = 0 + j(2/\pi)$. Our estimates are confirmed

(Continued)

(Continued)

by simulation results shown in Figure 5.34. The reader is invited to verify that $H_{-3}(j6\pi f_s)$ is given by $-\frac{2}{3\pi}(1-j)$.

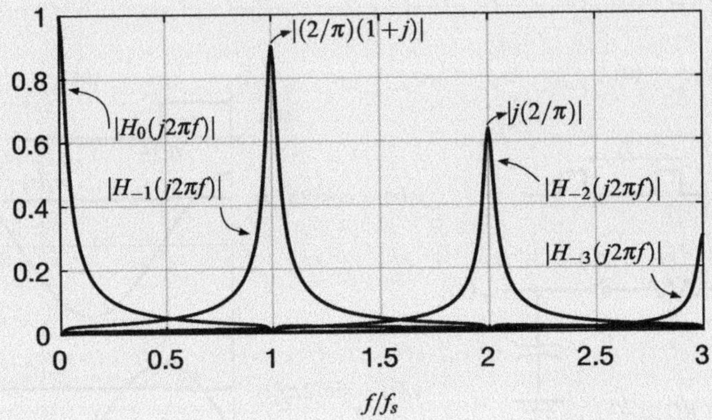

Figure 5.34 Simulated harmonic transfer functions for a switched-RC circuit, with $RC/(0.25T_s) = 6.4$.

The reader might now be tempted to conclude that a similar "trick" can be used to compute $H_{-l}(j2\pi mf_s)$, where l, m are integers with $m \neq l$. That is, excite the system with $\cos(2\pi mf_st)$ and $\sin(2\pi mf_st)$ to obtain $w_i(t)$ and $w_q(t)$, and determine $H_{-l}(j2\pi mf_s)$ from the Fourier components of $w_i(t)$ and $w_q(t)$ at the frequency $(m-l)f_s$. This is true but is somewhat more involved than in the case of $l = m$. The reason is the following. When the system is excited by $\cos(2\pi mf_st)$, which is the sum of $e^{j2\pi mf_st}$ and $e^{-j2\pi mf_st}$, output tones at frequencies $\pm(m-l)f_s$ are the result of the inputs at mf_s and $-mf_s$ according to

$$\cos(2\pi mf_st) \rightarrow \frac{1}{2}\{H_{-l}(j2\pi mf_s) + H_{-l+2m}(-j2\pi mf_s)\}\, e^{j2\pi(m-l)f_st}$$
$$+\frac{1}{2}\{H_l(-j2\pi mf_s) + H_{l-2m}(j2\pi mf_s)\}\, e^{-j2\pi(m-l)f_st}.$$

Similarly, exciting the system at $\sin(2\pi mf_st)$ yields

$$\sin(2\pi mf_st) \rightarrow \frac{1}{2j}\{H_{-l}(j2\pi mf_s) + H_{-l+2m}(-j2\pi mf_s)\}\, e^{j2\pi(m-l)f_st}$$
$$-\frac{1}{2j}\{H_l(-j2\pi mf_s) + H_{l-2m}(j2\pi mf_s)\}\, e^{-j2\pi(m-l)f_st}.$$

Thus, $w_i(t)$ and $w_q(t)$ at $\pm(m-l)f_s$ depend not only on $H_{-l}(j2\pi mf_s)$ (which is what we would like to determine) but also $H_{-l+2m}(-j2\pi mf_s)$. It is only in the special case when $l = m$ that $H_{-l}(j2\pi mf_s)$ and $H_{-l+2m}(-j2\pi mf_s)$ reduce to $H_{-l}(j2\pi lf_s)$ and $H_l(-j2\pi lf_s)$, respectively, which are complex conjugates of each other. The following example should drive home the point.

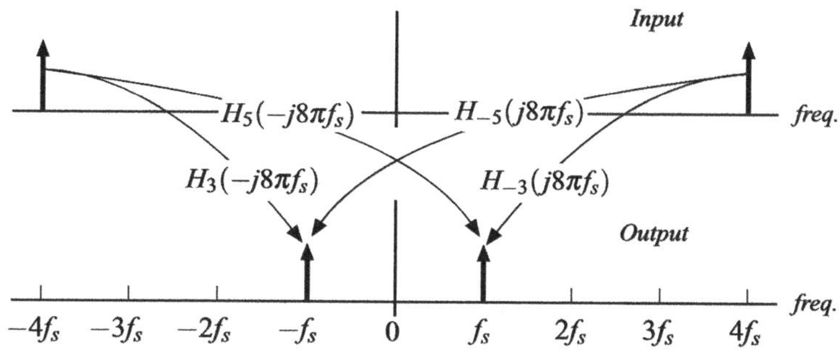

Figure 5.35 Exciting an LPTV system with $\cos(8\pi f_s t)$ to determine $H_{-3}(j8\pi f_s)$.

Example: Determining $H_{-3}(j8\pi f_s)$ of an LPTV system using excitations $\cos(8\pi f_s t)$ and $\sin(8\pi f_s t)$.

The cosine excitation contains complex frequencies $\pm 4f_s$, as shown in Figure 5.35. The strength of the tones in the output of the LPTV system at the frequency f_s, when excited with $\cos(8\pi f_s t)$ is given by

$$\frac{1}{2}\{e^{j8\pi f_s t} + e^{-j8\pi f_s t}\} \rightarrow \frac{1}{2}\{H_{-3}(j8\pi f_s) + H_5(-j8\pi f_s)\}\, e^{j2\pi f_s t} + \frac{1}{2}\{H_3(-j8\pi f_s) + H_{-5}(j8\pi f_s)\}\, e^{-j2\pi f_s t}.$$

An excitation $\sin(8\pi f_s t)$ yields

$$\frac{1}{2j}\{e^{j8\pi f_s t} - e^{-j8\pi f_s t}\} \rightarrow \frac{1}{2j}\{H_{-3}(j8\pi f_s) - H_5(-j8\pi f_s)\}\, e^{j2\pi f_s t} + \frac{1}{2j}\{-H_3(-j8\pi f_s) + H_{-5}(j8\pi f_s)\}\, e^{-j2\pi f_s t}.$$

From the above, we see the strengths of the tones at $\pm f_s$ in the responses to cosine- and sine-wave inputs yield four equations in as many variables, solving which $H_{-3}(j8\pi f_s)$ and $H_5(-j8\pi f_s)$ are both obtained. It seems simpler to calculate the real and imaginary parts of $H(j8\pi f_s)$ (using (5.15)) and look for the components at f_s in them to obtain the desired harmonic transfer function.

5.6 Impedance and Admittance in LPTV Circuits

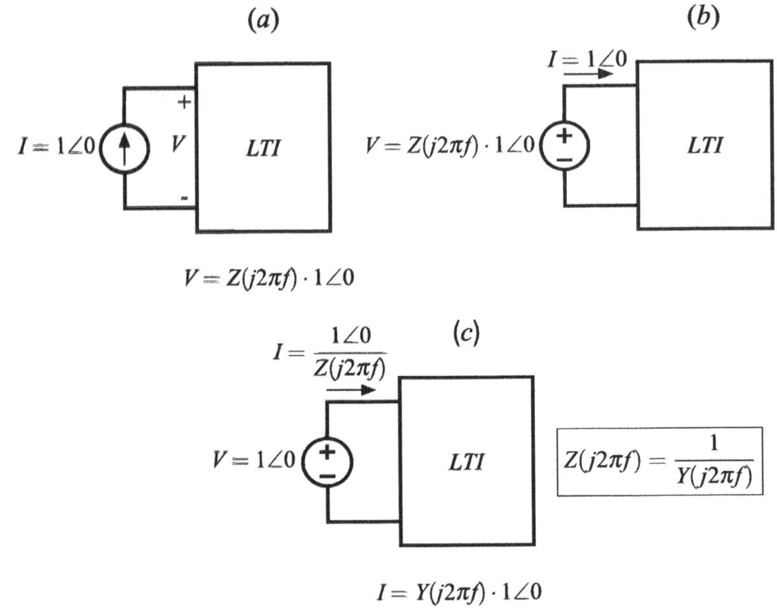

Figure 5.36 In a linear time-invariant network, the driving point impedance and admittance are reciprocals of each other.

When we work with LTI networks, we are used to the notion that the driving-point impedance and admittance of a network are reciprocals of each other. In other words, one can determine $Z(j2\pi f)$ by exciting the network with a current $1\angle 0$ and measuring the voltage developed. Alternatively, one can excite the network with a voltage source $1\angle 0$ and measure the current flowing into it, which yields the admittance $Y(j2\pi f)$. Z can be determined as the reciprocal of Y. This is "common knowledge," but let us deduce it anyway since it gives us a way forward when we consider an LPTV network.

Figure 5.36a shows the network being excited by a current source $1\angle 0$. The voltage developed across the terminals of the network is $Z(j2\pi f) \cdot 1\angle 0$. If the current source is replaced, as shown in Figure 5.36b, by a voltage source $Z(j2\pi f) \cdot 1\angle 0$ the current flowing into the network must be $1\angle 0$. Next, we scale down the voltage source by a number equal to $Z(j2\pi f)$, as shown in Figure 5.36c. Since the network is linear, the current flowing into it should reduce by the same factor. Thus, $I = 1\angle 0/Z(j2\pi f)$. However, by definition, the current that flows into the network when it is excited by a voltage $1\angle 0$ is $Y(j2\pi f) \cdot 1\angle 0$. We thus see that $Z(j2\pi f) = 1/Y(j2\pi f)$.

What happens in the LPTV case? As in the time-invariant case, we excite the network with a current $e^{j2\pi ft}$ (corresponding to the phasor $1\angle 0$) as shown in Figure 5.37a. The voltage developed across it is given by

$$v(t) = Z_0(j2\pi f)e^{j2\pi ft} + \sum_{k\neq 0} Z_k(j2\pi f)e^{j2\pi(f+kf_s)t}. \tag{5.43}$$

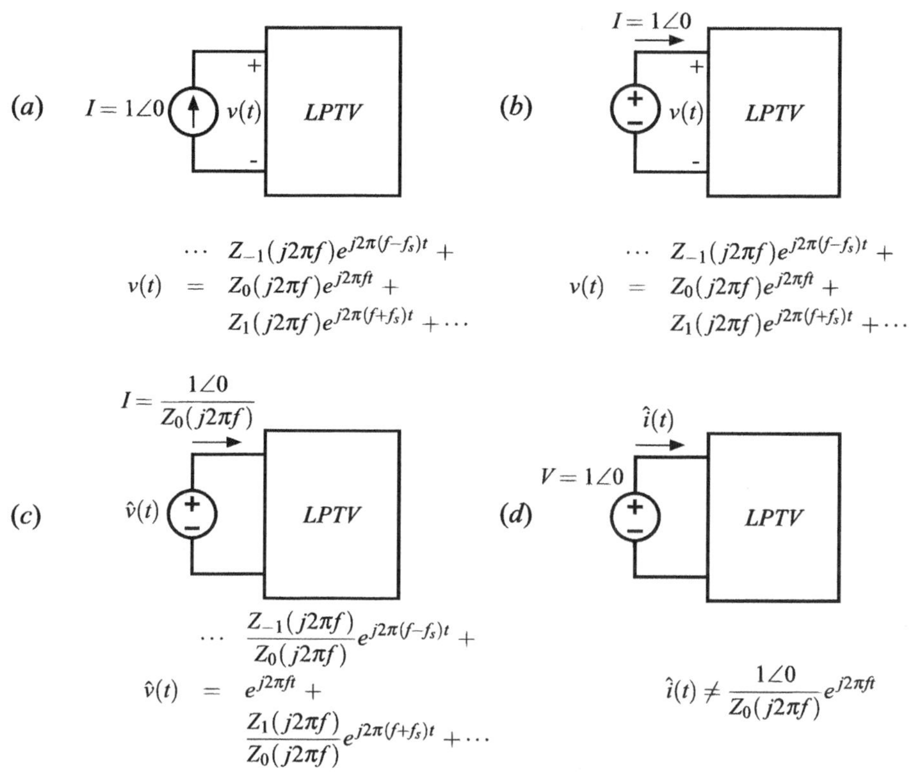

Figure 5.37 In an LPTV network, the driving point impedance and admittance are not reciprocals of each other.

If we excite the network with the voltage source $v(t)$ as shown in Figure 5.37b, the current will be the phasor $1\angle 0$. If we scale down the excitation voltage by the number $Z_0(j2\pi f)$ as shown in Figure 5.37c, the current will reduce by the same factor. From this figure, we observe the following. For I to be $1\angle 0/Z_0(j2\pi f)$, we not only need to apply a voltage at frequency f, but also appropriately chosen voltages at frequencies $(f + kf_s)$. Due to the LPTV nature of the system, voltage excitations at these frequencies get translated to f, and result in a current that is precisely $1\angle 0/Z_0(j2\pi f)$. If we instead excite the network with a voltage source $1\angle 0$, as in Figure 5.37d, the resulting current will no longer be $1\angle 0/Z_0(j2\pi f)$, as it will miss the contribution of the voltages at frequencies $(f + kf_s)$. $\hat{\imath}(t)$, by definition, is given by

$$\hat{\imath}(t) = Y_0(j2\pi f)e^{j2\pi ft} + \sum_{k\neq 0} Y_k(j2\pi f)e^{j2\pi(f+kf_s)t}. \tag{5.44}$$

From Figure 5.37c,d, it is apparent that $Z_0(j2\pi f) \neq 1/Y_0(j2\pi f)$. In terms of phasor vector notation, the voltage phasor vector is related to the current phasor as

$$\mathbf{V} = \mathbf{ZI} \tag{5.45}$$

where **Z** is the impedance matrix given by

$$\mathbf{Z} = \begin{bmatrix} Z_0(j2\pi f) & \cdots & \cdots & Z_{-2K}(j2\pi f) \\ Z_1(j2\pi f) & \cdots & \cdots & Z_{-2K+1}(j2\pi f) \\ \vdots & \vdots & \vdots & \vdots \\ Z_{2K-1}(j2\pi f) & \cdots & \cdots & Z_{-1}(j2\pi f) \\ Z_{2K}(j2\pi f) & \cdots & \cdots & Z_0(j2\pi f) \end{bmatrix}. \tag{5.46}$$

In a similar fashion, $\mathbf{I} = \mathbf{YV}$, where $\mathbf{Y}$ is the admittance matrix that relates the current-phasor vector to the voltage-phasor vector. Clearly, $\mathbf{I}$ must also equal $\mathbf{Z}^{-1}\mathbf{V}$, implying that $\mathbf{Y} = \mathbf{Z}^{-1}$. We thus see that the impedance and admittance matrices are inverses of each other. In general, however, $Y_k(j2\pi f) \neq 1/Z_k(j2\pi f)$.

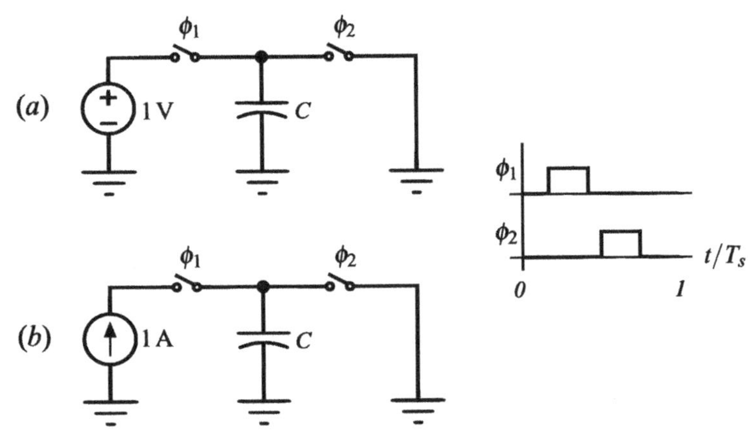

Figure 5.38 The dc "impedance" of a switched-capacitor driven by (a) a voltage source and (b) a current source.

Figure 5.38 shows a simple example that should convince us of this fact. Part (a) of the figure shows a capacitor switched between a 1 V dc source and ground with period T_s. The dc component of the current drawn from the source is seen to be $(C/T_s) \cdot 1\,\text{V}$, resulting in a dc resistance of T_s/C. The resistance obtained when the same network is excited by a current source, as shown in Figure 5.38b, is clearly not the same.

5.7 Thevenin- and Norton-Equivalent Circuits

Consider an LPTV network that varies with frequency f_s, and contains independent voltage and current sources, as shown in Figure 5.39a. These sources *should* be at frequencies of the form $f + kf_s$, where k is an integer. They can therefore be represented as phasor vectors, as we have done in Section 5.6. The lth voltage (current) source phasor vector is denoted by $\mathbf{V}_{sl}$ ($\mathbf{I}_{sl}$). Let the network be excited by

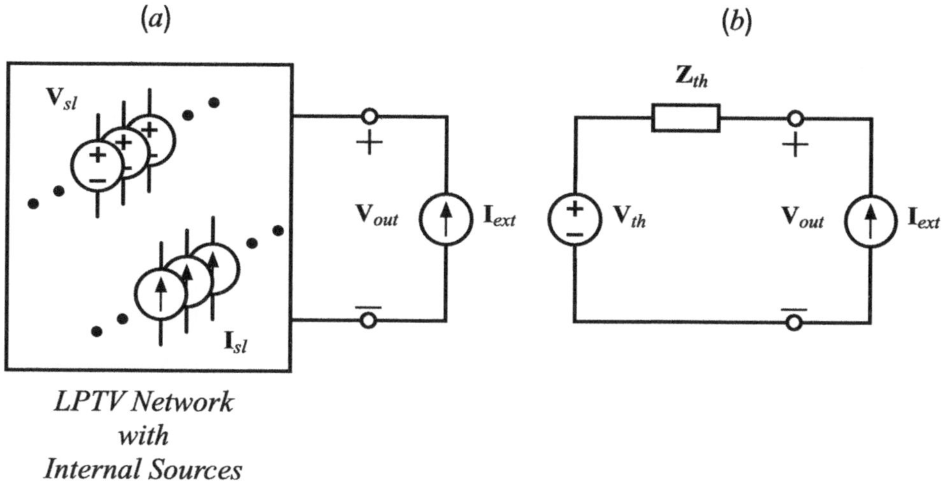

Figure 5.39 (a) An LPTV network with independent voltage and current sources interacting with the external world, modeled by the current source $\mathbf{I}_{\text{ext}}$. (b) Thevenin equivalent.

an external current, whose phasor vector is denoted by $\mathbf{I}_{\text{ext}}$. Note that $\mathbf{I}_{\text{ext}}$ must also contain frequencies of the form $f + kf_s$. We assume that all phasor-vectors are of length $(2K + 1)$, where K is chosen to be sufficiently large.

Since the network is linear, the phasor of the voltage developed across the terminals of the network, namely $\mathbf{V}_{\text{out}}$, is given by

$$\mathbf{V}_{\text{out}} = \underbrace{\sum_l \mathbf{H}_l \mathbf{V}_{sl} + \sum_l \mathbf{Z}_l \mathbf{I}_{sl}}_{\mathbf{V}_{\text{th}}} + \mathbf{Z}_{\text{th}} \mathbf{I}_{\text{ext}} \tag{5.47}$$

where $\mathbf{H}_l$, $\mathbf{Z}_l$, $\mathbf{Z}_{\text{th}}$ are $(2K + 1) \times (2K + 1)$ matrices that quantify the contributions of the various sources to $\mathbf{V}_{\text{out}}$. Denoting $\sum_l \mathbf{H}_l \mathbf{V}_{sl} + \sum_l \mathbf{Z}_l \mathbf{I}_{sl}$ by $\mathbf{V}_{\text{th}}$, we see that $\mathbf{V}_{\text{out}} = \mathbf{V}_{\text{th}} + \mathbf{Z}_{\text{th}} \mathbf{I}_{\text{ext}}$. This relationship can be represented in network form using the Thevenin equivalent circuit, as shown in Figure 5.39b.

$\mathbf{V}_{\text{th}}$ can be interpreted (and found) as the open-circuit voltage *phasor* that is developed across the terminals of the network (i.e., when $\mathbf{I}_{\text{ext}} = \mathbf{0}$), as shown in Figure 5.40a. $\mathbf{Z}_{\text{th}}$ can be determined by exciting the network with $\mathbf{I}_{\text{ext}}$, and all the internal sources being de-energized (i.e., $\mathbf{V}_{sl} = \mathbf{0}$ and $\mathbf{I}_{sl} = \mathbf{0}$), as seen in Figure 5.40b. $\mathbf{V}_{\text{out}}$ and $\mathbf{I}_{\text{ext}}$ are related as $\mathbf{V}_{\text{out}} = \mathbf{Z}_{\text{th}} \mathbf{I}_{\text{ext}}$. Evidently, determining $\mathbf{Z}_{\text{th}}$ needs multiple inputs $\mathbf{I}_{\text{ext}}$.

The discussion above derived the Thevenin equivalent; arriving at the Norton equivalent proceeds similarly and is left as an exercise for the reader.

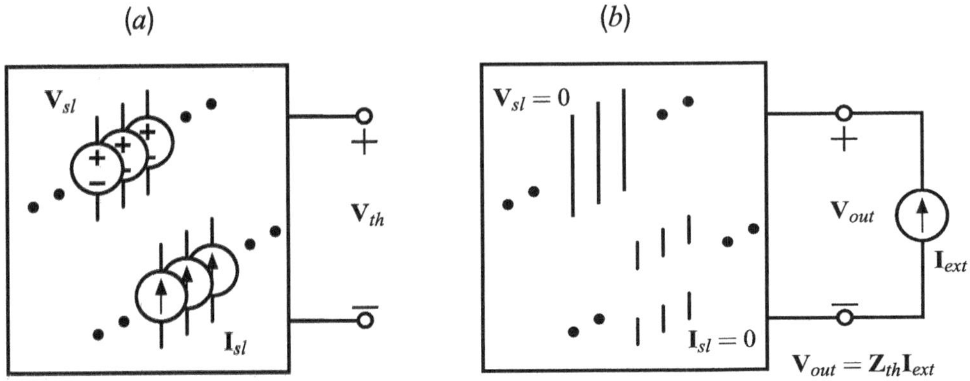

Figure 5.40 Determining (a) the Thevenin equivalent voltage phasor $\mathbf{V}_{\text{th}}$ and (b) the Thevenin impedance matrix.

5.8 The N-Path Principle

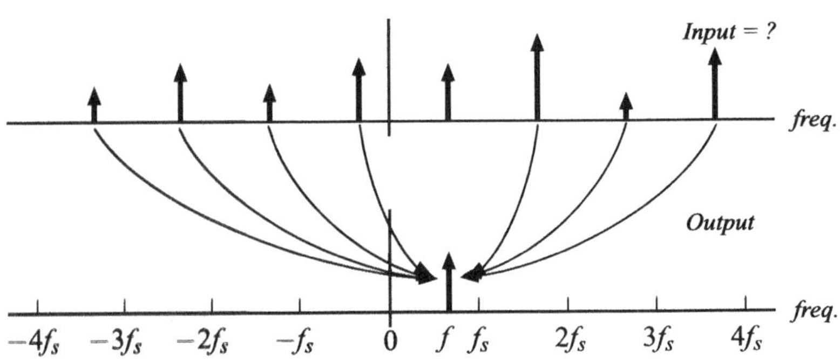

Figure 5.41 Multiple input frequencies can translate to the same output frequency in an LPTV system.

In this section, we motivate and describe the *N-path principle* [7], which is a much-used technique in practice. There are multiple ways of thinking about the need for N-path LPTV systems, and one of them is the following. As we have seen throughout this chapter, if the output of an LPTV system contains a tone at a frequency f, the input could have been any (or all) frequencies of the form $(f + kf_s)$, where k is an integer. There is thus an ambiguity with respect to the frequency content of the input signal as multiple frequencies can fold (or alias) to the same output frequency as shown in Figure 5.41. How do we reduce this ambiguity? A solution is to *somehow* make the "alias transfer functions", namely HTFs of the form $H_{-k}(j2\pi(f + kf_s))$ zero for $k \neq 0$. How this can be accomplished is the crux of the N-path principle.

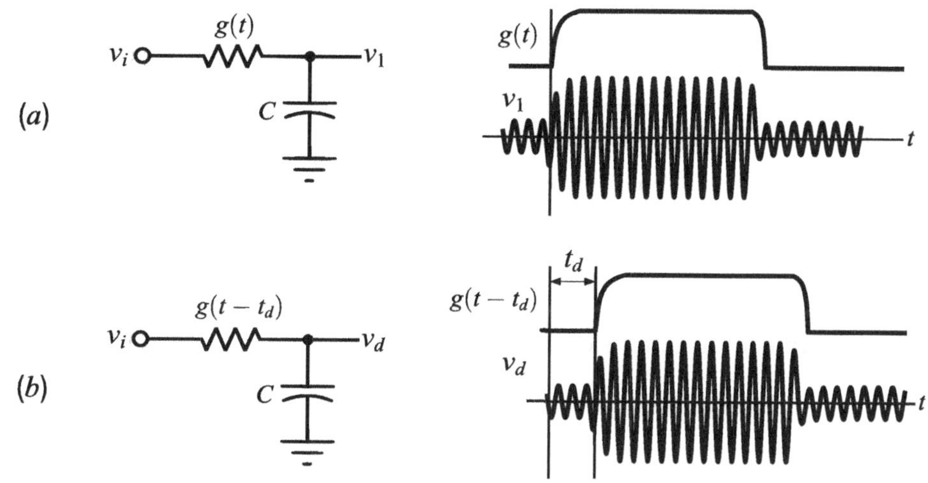

Figure 5.42 (a) An LTV RC circuit. (b) The RC circuit with its timing shifted by t_d.

Consider the simple LTV circuits of Figure 5.42a,b. The conductances in the circuits vary as $g(t)$ and $g(t-t_d)$ respectively, where t_d is a delay. What comment can we make about the time-dependent frequency responses of the two circuits? Let us think through this by way of the impulse responses of the circuits, which we denote by $h(t,\tau)$ and $h_d(t,\tau)$ respectively. Recall that t denotes the time at which the output is observed, while the impulse is applied at $(t-\tau)$. How are $h(t,\tau)$ and $h_d(t,\tau)$ related?

Since the variation of conductance is delayed by t_d in the circuit of Figure 5.42b, delaying both the time of observation *and* the time of application of the impulse by t_d with respect to the circuit of Figure 5.42a should yield the same output. In other words, $h_d(t+t_d,\tau)=h(t,\tau)$. Replacing t with $t-t_d$, we have the equivalent relation

$$h_d(t,\tau)=h(t-t_d,\tau).$$

The time-varying frequency response of the circuit of Figure 5.42b is thus given by

$$H_d(j2\pi f,t)=\int_0^\infty h_d(t,\tau)e^{-j2\pi f\tau}d\tau=\int_0^\infty h(t-t_d,\tau)e^{-j2\pi f\tau}d\tau=H(j2\pi f,t-t_d).$$

The relationship above makes sense due to the following. Recall that $H_d(j2\pi f,t)$ is the (complex) gain experienced by $e^{j2\pi ft}$. Since the variation of conductance in Figure 5.42b is delayed with respect to that in Figure 5.42a, it follows that the gain experienced by $e^{j2\pi ft}$ should also be delayed with respect to that in Figure 5.42a by t_d as shown in the figures. Now consider the case where $g(t)$ is periodic with a period $T_s=1/f_s$, so that

$$H(j2\pi f,t)=\sum_k H_k(j2\pi f)e^{j2\pi kf_st} \tag{5.48}$$

where H_k denotes the kth-order harmonic transfer function of the circuit of Figure 5.42a. From the discussion above, $H_d(j2\pi f, t)$ is seen to be

$$\begin{aligned} H_d(j2\pi f, t) &= \sum_k H_{d,k}(j2\pi f) e^{j2\pi k f_s t} \\ &= H(j2\pi f, t - t_d) = \sum_k H_k(j2\pi f) e^{j2\pi k f_s (t - t_d)} \\ &= \sum_k H_k(j2\pi f) e^{-j2\pi k f_s t_d} e^{j2\pi k f_s t}. \end{aligned}$$

The harmonic transfer functions of the two circuits are seen to be related according to

$$\boxed{H_{d,k} = H_k e^{-j2\pi k f_s t_d},} \tag{5.49}$$

which indicates that the kth-order HTF in Figure 5.42b has a phase lag of $2\pi k f_s t_d$ radians with respect to that of the circuit in Figure 5.42a. This makes sense due to the following. The time-varying gains experienced by the two circuits are shifted in time with respect to each other by a duration t_d. Since the gains are the sum of harmonics of f_s, a time-shift t_d corresponds to a phase shift $2\pi k f_s t_d$ for the kth harmonic. Observe in particular that the zeroth-order HTFs, namely $H_0(j2\pi f)$ and $H_{d,0}(j2\pi f)$, which represent the average gain experienced by $e^{j2\pi f t}$, are equal. This is intuitively satisfying, since the dc value of the "gain" waveform does not change when it is shifted in time.

Now suppose that we form a composite system by taking three such circuits, and stagger their operation in time as shown in Figure 5.43. The successive delay between the three circuits is chosen as $T_s/3$, and their outputs are added with gains (1/3) to yield v_o. What are the HTFs (denoted by $P_k(j2\pi f)$) of this composite system? To determine the HTFs, we excite the composite system by $e^{j2\pi f t}$and examine the components of the output at frequencies $(f + k f_s)$. Referring to Figure 5.43, we can write v_1, v_2, v_3 as

$$\begin{aligned} v_1(t) &= \left\{ \sum_k H_k(j2\pi f) e^{j2\pi k f_s t} \right\} e^{j2\pi f t} \\ v_2(t) &= \left\{ \sum_k H_k(j2\pi f) e^{j2\pi k f_s (t - T_s/3)} \right\} e^{j2\pi f t} \\ v_3(t) &= \left\{ \sum_k H_k(j2\pi f) e^{j2\pi k f_s (t - 2T_s/3)} \right\} e^{j2\pi f t}. \end{aligned}$$

The zeroth-order HTF is the sum of the zeroth-order HTFs of the three subcircuits comprising the system. Though the gains of the three systems are delayed with respect to each other, their zeroth-order HTFs are the same, as discussed earlier. Consequently,

$$P_0(j2\pi f) = \frac{1}{3}\left(H_0(j2\pi f) + H_0(j2\pi f) + H_0(j2\pi f)\right) = H_0(j2\pi f). \tag{5.50}$$

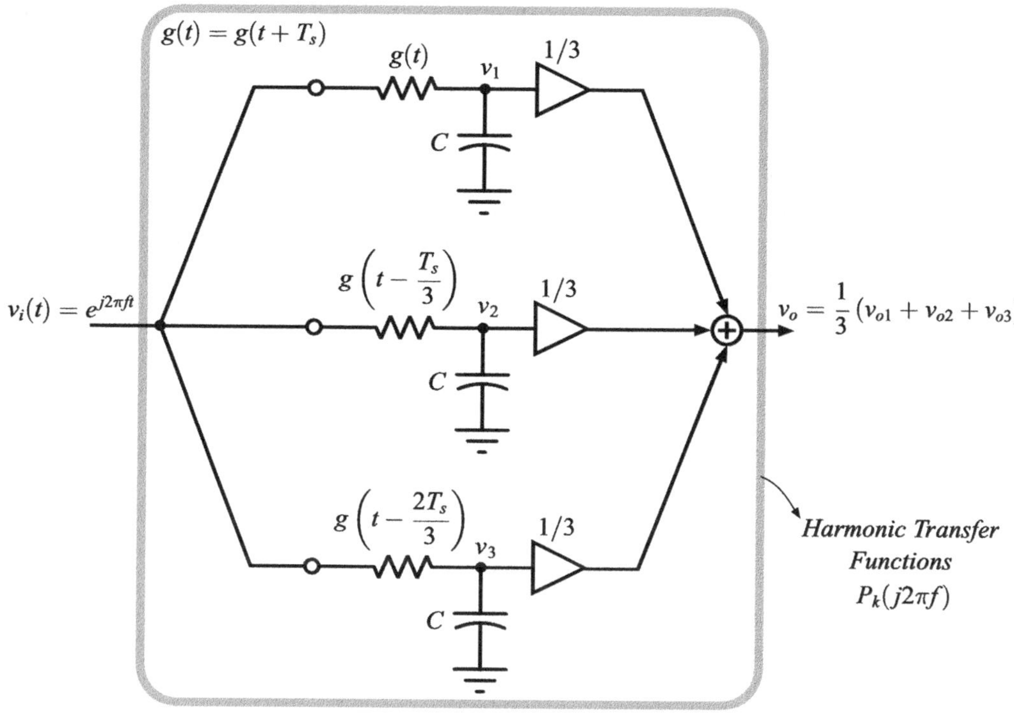

Figure 5.43 An example 3-path system.

What about the first-order HTF, namely $P_1(j2\pi f)$? Since the conductance variations leading to v_2 and v_3 are delayed by $T_s/3$ and $2T_s/3$ respectively, their first-order HTFs in terms of that of the first circuit are given by $H_1(j2\pi f)e^{-j2\pi/3}$ and $H_1(j2\pi f)e^{-j4\pi/3}$. As a result,

$$P_1(j2\pi f) = \frac{1}{3}H_1(j2\pi f)\left(1 + e^{-j\frac{2\pi}{3}} + e^{-j\frac{4\pi}{3}}\right) = 0. \tag{5.51}$$

Similarly, we have

$$P_2(j2\pi f) = \frac{1}{3}H_2(j2\pi f)\left(1 + e^{-j\frac{4\pi}{3}} + e^{-j\frac{8\pi}{3}}\right) = 0. \tag{5.52}$$

However,

$$P_3(j2\pi f) = \frac{1}{3}H_3(j2\pi f)\left(1 + e^{-j\frac{6\pi}{3}} + e^{-j\frac{12\pi}{3}}\right) = H_3(j2\pi f). \tag{5.53}$$

It is straightforward to see that $P_k(j2\pi f) = H_k(j2\pi f)$ when k is a multiple of 3, and zero otherwise. The box in gray is an example of a *3-path* system. A few comments are in order here.

a. Suppose we were given an LPTV system in a box and asked to determine the rate at which the system was varying. What would we do? One way to accomplish this is to excite the system with $e^{j2\pi ft}$ and look at the frequency content of the resulting output. Since we know that the output frequencies would be of the form $(f + kf_s)$, f_s can readily be determined.

If we applied this principle to the gray box of Figure 5.43, we would observe that the output consists of components of the form $(f + 3kf_s)$, where k is an integer. From this we would conclude that the system varies at a rate $3f_s$.

b. Having concluded that the system varies at $3f_s$, we open the box, only to find that all the time-varying components change only at a rate f_s! Three such "slow" systems have been cleverly configured by timing them so as to make it appear as if the overall system is varying at $3f_s$.

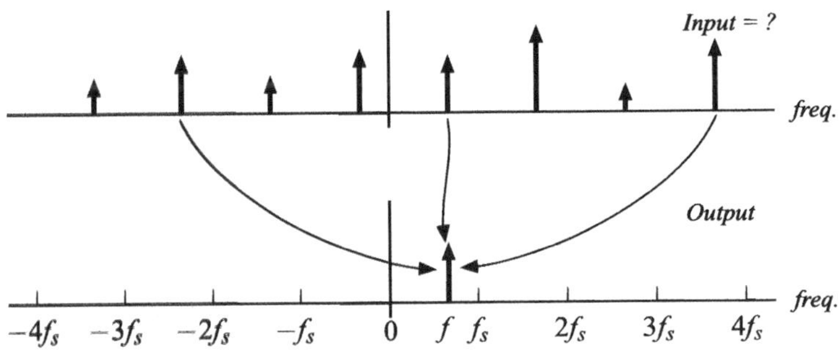

Figure 5.44 $H_{-k}(j2\pi(f + kf_s))$ of the 3-path system is zero for all k that are not a multiple of 3.

c. Since the composite system is varying at a rate $3f_s$, a tone with frequency f at the output can only be due to input tones whose frequencies are of the form $(f + 3kf_s)$ as shown in Figure 5.44. This composite arrangement of LPTV systems, therefore, reduces the number of aliasing (or folding) tones by a factor of 3 when compared to any of its constituent systems.

d. HTFs of the form $P_{3k+1}(j2\pi f)$ and $P_{3k+2}(j2\pi f)$ are zero due to cancellation. This means that mismatch between the three nominally identical LPTV systems that comprise the 3-path system will result in nonzero values for the HTFs due to imperfect cancellation.

Another way of understanding the result above is the following, illustrated in Figure 5.45 using a 3-path example. Consider three nominally identical LPTV systems, labeled A, B, and C in the figure for purposes of identification. They vary with a period $T_s \equiv 1/f_s$, and their outputs are added together. The operation of systems A, B, and C are controlled by the timing waveforms $p(t)$, $p(t - t_d)$ and $p(t - 2t_d)$, respectively, with t_d chosen to be $T_s/3$. Let $x(t) = 0$. If we take a snapshot of the 3-path system at an arbitrary time t_1, the "state" of the individual paths that comprise the system are different and denoted using the symbols shown in the first row of the table in Figure 5.45. At $t = t_1 + t_d$, we notice that B, C, and A assume the states that A, B, and C were in at $t = t_1$. Since the paths are nominally identical (with only their operation shifted in time), it follows that the 3-path system at t_1 cannot be distinguished from that at $t_1 + t_d$. The same argument holds for $t = t_1 + 2t_d$. To summarize, the

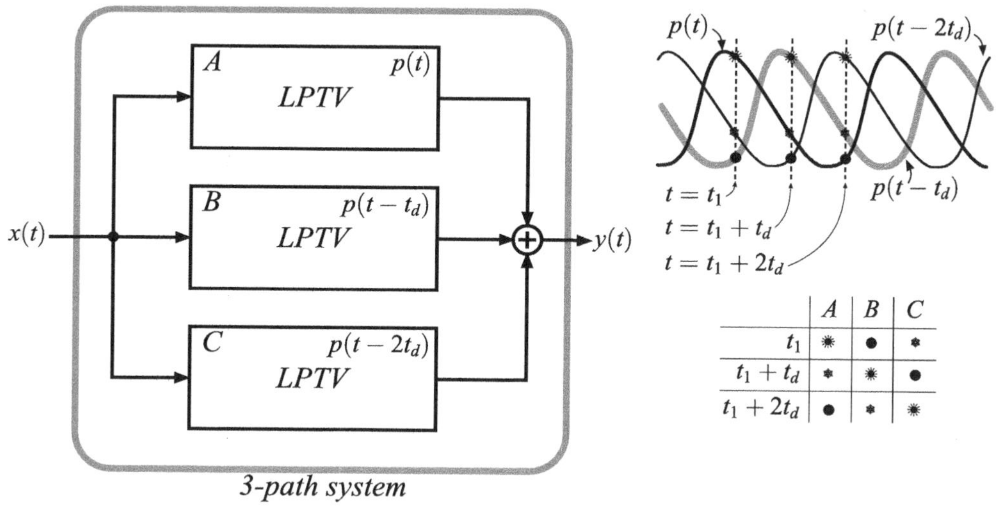

Figure 5.45 A 3-path LPTV system. It "looks identical" at time instants separated by $T_s/3$.

3-path system "looks" identical after every t_d; it is therefore periodic with at a rate $1/t_d = 3f_s$ though each system inside operates at a rate f_s. It is also clear that snapshots of the 3-path system will "look the same" at intervals t_d apart *only if* the individual paths are nominally identical. With path mismatch, the 3-path system will only be periodic at intervals of T_s.

The discussion above used three paths to illustrate the idea. The principle can be generalized to N paths operated in an interleaved fashion, resulting in what are called *N-path* LPTV systems. The basic idea is to stagger the timing between N nominally identical LPTV systems (each varying at f_s) by T_s/N and add their outputs so that all but the HTFs of multiples of N are zero. This way, it appears as if the composite system is varying at Nf_s, though there is no component that varies at that frequency. Consequently, all but the harmonic transfer functions of order $0, \pm N$, $\pm 2N$ etc. are zero. In practice, $N = 4$, or $N = 8$ find common use.

2-Path System Example

As an illustration, consider the LPTV system of Figure 5.46a. The input is multiplied by $\cos(2\pi f_s t)$ and then processed by a lowpass filter with bandwidth B, as depicted in the figure. The filter's output is multiplied by $\cos(2\pi f_s t)$ to result in the LPTV system's output.

(Continued)

(Continued)

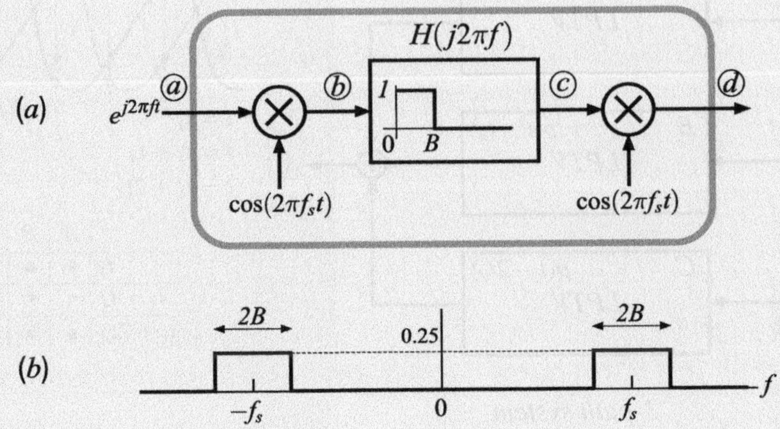

Figure 5.46 (a) LPTV system where $H(j2\pi f)$ is a lowpass transfer function with bandwidth B. (b) $H_0(j2\pi f)$ of the system.

We denote the HTFs of this system by $Q_k(j2\pi f)$. What is the zeroth-order HTF, namely $Q_0(j2\pi f)$ of the system? To determine this, we excite the input by $e^{j2\pi ft}$ and trace the signal through various points in the signal chain.

$$ⓐ \rightarrow e^{j2\pi ft}$$

$$ⓑ \rightarrow \frac{1}{2}e^{j2\pi(f-f_s)t} + \frac{1}{2}e^{j2\pi(f+f_s)t}$$

$$ⓒ \rightarrow \frac{1}{2}H(j2\pi(f-f_s))e^{j2\pi(f-f_s)t} + \frac{1}{2}H(j2\pi(f+f_s))e^{j2\pi(f+f_s)t}$$

$$ⓓ \rightarrow \frac{1}{4}\{H(j2\pi(f-f_s)) + H(j2\pi(f+f_s))\}\, e^{j2\pi ft} +$$
$$\frac{1}{4}H(j2\pi(f+f_s))e^{j2\pi(f+2f_s)t} + \frac{1}{4}H(j2\pi(f-f_s))e^{j2\pi(f-2f_s)t}.$$

We observe the following.

a. The system varies at a rate $2f_s$. This makes sense, since snapshots of the system taken at times t and $(t + T_s/2)$ are equivalent.

b. The zeroth-order HTF, namely $Q_0(j2\pi f)$, which is the ratio of the strength of $e^{j2\pi ft}$ at the output divided by $e^{j2\pi ft}$, is given by

$$Q_0(j2\pi f) = \frac{1}{4}\{H(j2\pi(f-f_s)) + H(j2\pi(f+f_s))\}.$$

$Q_0(j2\pi f)$ is a bandpass transfer function with a peak gain of 0.25, as shown in Figure 5.46b.

c. The only nonzero HTFs (other than that of the zeroth-order) are the ones with order ± 1.

d. A two-path system, formed by adding a second path whose timing is shifted by $t_d = (1/2)\,(T_s/2) = T_s/4$, will result in a system whose first-order HTFs are zero. The resulting 2-path system as shown in Figure 5.47, where the multipliers are controlled by $\cos(2\pi f_s(t - t_d)) = \sin(2\pi f_s t)$. This is a linear *time-invariant* system whose transfer function is $2Q_0(j2\pi f)$.

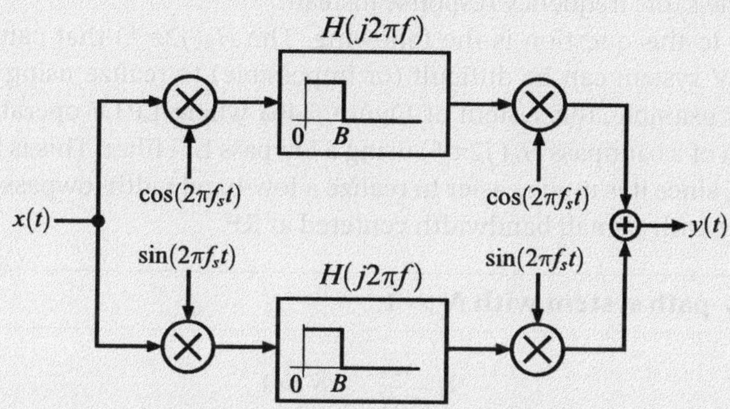

Figure 5.47 Two-path system formed by adding an additional path whose timing is skewed with the system of Figure 5.46.

We thus see that thanks to LPTV operation, a low-pass filter has been converted into one with a bandpass characteristic. Moreover, the center frequency of the bandpass filter is f_s, and can therefore be varied by changing f_s.

What is the impulse response of the N-path LPTV system? Denoting the time-varying impulse response of the single- and N-path systems by $h(t,\tau)$ and $h_N(t,\tau)$, respectively, we see that

$$h_N(t,\tau) = \frac{1}{N}\{h(t,\tau) + h(t - \Delta t,\tau) + \cdots + h(t - (N-1)\Delta t,\tau)\} \tag{5.54}$$

where $\Delta t = T_s/N$. In the limit of $N \to \infty$, the behavior of an N-path system approaches that of a linear time-invariant system since the only nonzero HTF is that of the zeroth order. The impulse response of this ∞-path system, which is LTI, is given by finding the limiting value of $h_N(t,\tau)$ determined above, which we denote by $h_{lti}(\tau)$. It is given by

$$\begin{aligned} h_{lti}(\tau) &= \lim_{N\to\infty} \frac{1}{N}\{h(t,\tau) + h(t - \Delta t,\tau) + \cdots + h(t - (N-1)\Delta t,\tau)\} \\ &= \frac{1}{T_s}\int_{-T_s}^{0} h(t,\tau)\,dt = \frac{1}{T_s}\int_{0}^{T_s} h(t,\tau)\,dt. \end{aligned}$$

The reader might wonder why an N-path system with a large N is useful in practice when its behavior approaches that of an LTI system. Why not simply create an LTI system with an impulse response $h_{lti}(\tau)$ instead, and dispense with the complexity (and hassle) of having to work with N LPTV systems? Recasting this question in the frequency domain, if the only nonzero HTF of the ∞-path system is the zeroth-order one (i.e., $H_0(j2\pi f)$) (which renders it LTI), why should one simply not use an LTI system with the same frequency response instead?

The answer to this question is the following. The $H_0(j2\pi f)$ that can be realized using an LPTV system can be difficult (or impossible) to realize using an LTI network. See, for example, the system of Figure 5.46a where LPTV operation enabled the realization of a bandpass $H_0(j2\pi f)$ using a lowpass LTI filter. This is a significant improvement, since it is much easier to realize a low-bandwidth lowpass filter than a bandpass filter with a small bandwidth centered at RF.

Example: N-path system with $N \gg 1$.

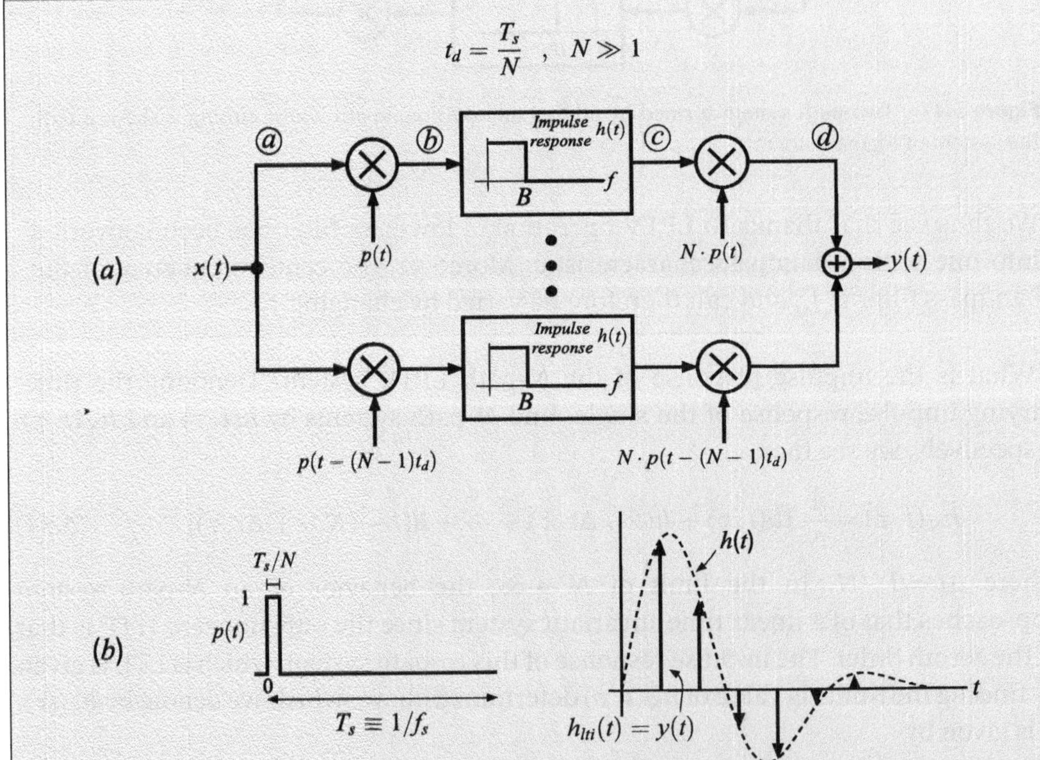

Figure 5.48 (a) N-path system where $N \to \infty$. (b) $p(t)$, and the impulse response of the equivalent LTI system.

Figure 5.48 shows a N-path LPTV system where $N \gg 1$. The input $x(t)$ is multiplied by a periodic pulse $p(t)$ with period T_s and duty cycle $1/N$. The result is processed by a filter with an impulse response denoted by $h(t)$. The output of the filter is multiplied by $N \cdot p(t)$. The output of N such paths, each of whose timing is staggered by T_s/N, are added together to result in the system output, namely $y(t)$. We examine this system in the limit of large N, i.e., $N \to \infty$.

As we discussed earlier, the system becomes LTI as $N \to \infty$. To determine the impulse response of the LTI system, namely $h_{lti}(t)$, we set $x(t) = \delta(t)$ and "measure" $y(t)$. The signals at various points in the topmost path in Figure 5.48 are as follows.

$$\text{ⓐ} \to \delta(t), \qquad \text{ⓑ} \to \delta(t)$$
$$\text{ⓒ} \to h(t), \qquad \text{ⓓ} \to N\,h(t)p(t)$$

As $N \to \infty$, $Np(t)$ tends to the Dirac delta function and

$$y(t) = h_{lti}(t) = h(t) \sum_k \delta(t - kT_s) \tag{5.55}$$

as shown in Figure 5.48b. Assuming $h(t)$ corresponds to a lowpass transfer function with a bandwidth B, the frequency response of the ∞-path LTI system will therefore have passbands centered around all multiples of f_s. The width of each passband will be $2B$. Such a response, (i.e., infinitely many passbands spaced f_s apart) can be obtained using LTI networks only by using ideal transmission lines.

(a) (b) (c)

Figure 5.49 Analogies for (a) single-path system. (b) Three-path system and (c) an ∞-path system. *Source:* Generated with AI using OpenAI.

An analogy to single-, three- and ∞-path LPTV systems is shown in Figure 5.49. Part (a) of the figure shows a single elevator that ferries people from the ground-level of a building to the topmost floor in a periodic fashion (with period denoted by T_s). The time taken for a person arriving at the building to reach the top is a time-varying quantity, depending on when (s)he reaches the lobby. If the person is in luck, the elevator is available right away, else (s)he has to wait for as long as T_s. If one monitors the number of people waiting for the elevator in the lobby, it will have a period T_s.

Figure 5.49b shows three elevators, operating so that each elevator starts off with a time offset $T_s/3$ with respect to the other. This is a three-path system, and a person arriving in the lobby has to wait for a worst-case time of $T_s/3$ before getting on any elevator. Consequently, the number of people waiting in the lobby will have a periodicity of $T_s/3$. The ∞-path system is like having an escalator, which can be thought of as a system of ∞ elevators, each of whose operation is offset in time by an infinitesimal amount. The worst-case waiting time is now zero, consequently there is no "good time" to arrive in the lobby – the time taken to reach the topmost floor will be the same irrespective of when one arrives in the building.

Chapter Summary

- **Time-varying impulse response and convolution integral for linear time-varying and periodically time-varying systems.** The output of a linear time-varying system can be found by convolving the input with the time-varying impulse response.
- **Harmonic transfer functions and the Zadeh expansion.** Exciting an LPTV system at a frequency f results in an output with frequencies of the form $(f + lf_s)$ where l is an integer and f_s is the rate at which the system varies. The lth order harmonic transfer function quantifies the gain from an input at f to an output at $(f + lf_s)$.
- **MNA formulation of circuit equations in LPTV circuits.** This is analogous to the MNA formulation for time-invariant networks, except that the component MNA stamps are matrices that reflect their periodically time-varying nature. "Solving" the network involves significantly more computation, since one has to account for the input frequency f as well as those of the form $(f + lf_s)$ where l is an integer.
- **Periodic operating point, and periodic AC simulations.** These are the LPTV analogs of operating point and AC simulations that one runs on a time-invariant circuit. The periodically-varying nonlinear network has a periodic operating point. Every nonlinear element is linearized around the (periodic) operating point, resulting in an LPTV small-signal network.
- **The N-path principle.** This is a way of configuring N LPTV systems, each varying at a frequency f_s, in a way that the composite system appears as if it is varying at a frequency Nf_s. The advantage of such a realization is that there is no network component that has to vary at Nf_s.

Problems

Problem 1. *In the text, we denoted the impulse response of a linear time-varying system by $h(t, \tau)$, where t is the time at which the output is observed, while the input impulse is applied at an instant $(t - \tau)$. An alternative notation for the impulse response that is also in use, is $\hat{h}(t, \tau)$ where t denotes the time of observation (as before), and τ is the time* *at which the input impulse is applied. The aim of this problem is to develop the results in this chapter using this alternative notation.*

a. *What is/are the constraint(s) that $\hat{h}(t,\tau)$ should satisfy if the LTV system is causal?*

b. *How is $h(t,\tau)$ related to $\hat{h}(t,\tau)$?*

c. *Determine the response of the LTV system to a complex exponential $e^{j2\pi ft}$ in terms of $\hat{h}(t,\tau)$, and thereby express the time-varying frequency response in terms of $\hat{h}(t,\tau)$.*

d. *What is the constraint that must be satisfied by $\hat{h}(t,\tau)$ if the system is also periodically time-varying?*

Problem 2. *This problem involves the use of a SPICE simulator. If you have a favorite, use it. Else, we recommend LTSPICE, which is free and versatile.*

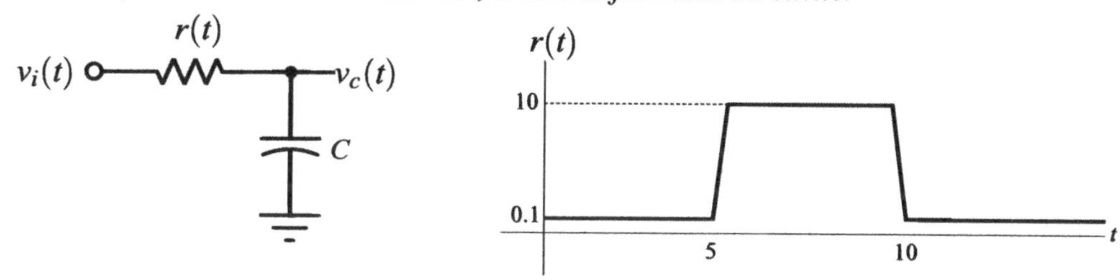

Figure 5.50 A time-varying RC circuit.

Create a first-order time-varying RC filter as shown in Figure 5.50, where $C = 1\,F$. The resistor $r(t)$ varies with time from 0.1 to 10 Ω as shown (the rise- and fall-times are 0.1 s). Plot the magnitude and phase (in degrees) of the time-varying frequency response as a function of time from 0 to 20 s for an input frequency of 1 Hz. Suggest some sanity checks to be convinced of the correctness of your plots.

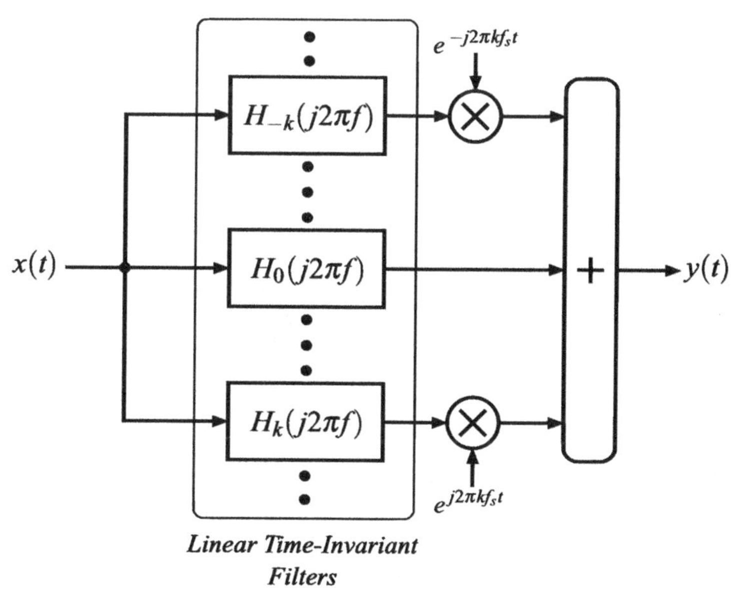

Figure 5.51 Zadeh expansion of an LPTV system.

Problem 3. *Figure 5.51 shows the Zadeh expansion of an LPTV network, where k ranges from $-\infty$ to ∞.*

a. *In general, are the impulse responses corresponding to $H_k(j2\pi f)$ real or complex?*

b. *What is the relationship between the impulse responses of the filters $H_k(j2\pi f)$ and $H_{-k}(j2\pi f)$ for a given k?*

c. *Suppose that $x(t) = \cos(2\pi ft)$. Determine the output of the branches containing $H_k(j2\pi f)$ and $H_{-k}(j2\pi f)$. Also determine their sum. Repeat the exercise above for $x(t) = \sin(2\pi ft)$. What do you conclude?*

Problem 4. *Figure 5.52 shows an alternative representation of an LPTV system. How are the impulse responses of $F_k(j2\pi f)$ related to those corresponding to $H_k(j2\pi f)$ of Figure 5.51?*

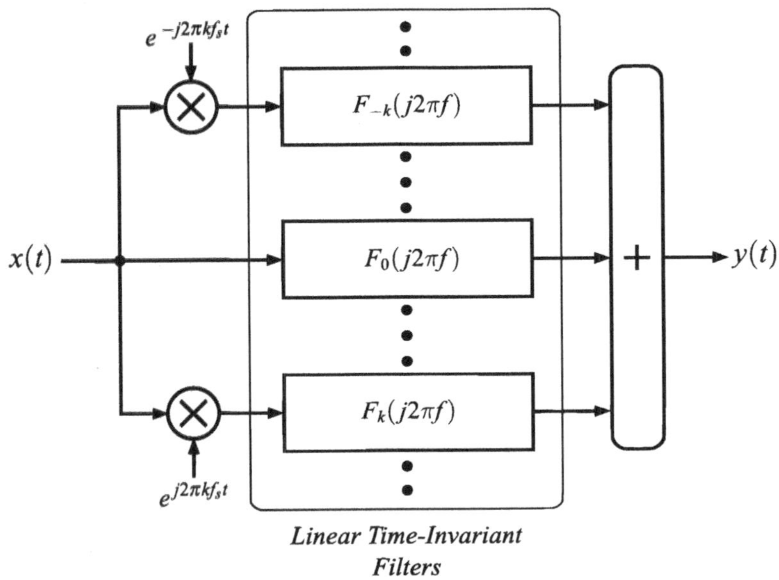

Figure 5.52 Alternative representation of an LPTV system.

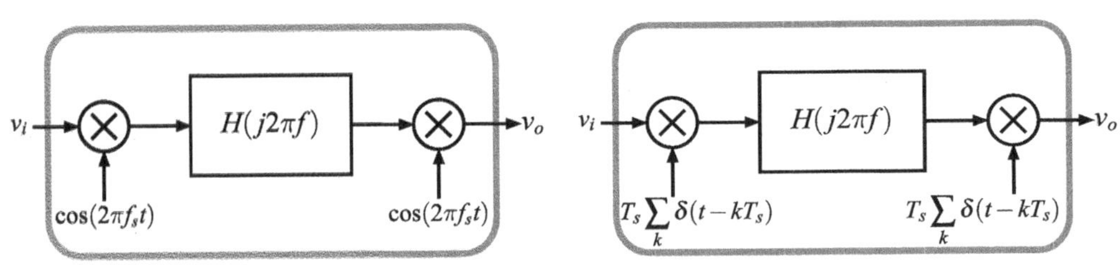

Figure 5.53

Problem 5. *Determine the Zadeh expansion for the systems in Figure 5.53.*

Problem 6. *Two LPTV systems vary at the same frequency f_s. Their HTFs are denoted by $H_k(j2\pi f)$ and $G_k(j2\pi f)$. They are cascaded. Determine the HTFs of the cascade. Does the order in which they are cascaded have a bearing on the result?*

Problem 7. *When an LTI system with transfer function $H(j2\pi f)$ is excited by $\cos(2\pi f t)$, the output is given by $|H(j2\pi f)| \cos(2\pi f t + \angle H(j2\pi f))$. Derive a similar relationship for the output of an LPTV system varying at f_s and HTFs denoted by $H_k(j2\pi f)$.*

Problem 8. *An LPTV system varies at a frequency f_s, and its HTFs are denoted by $H_k(j2\pi f)$. We perform two experiments. In the first, the system is excited by $\cos(2\pi f t)$. The resulting waveform is denoted by $w_i(t)$. In the second experiment, the excitation is $\sin(2\pi f t)$, and the output is denoted by $w_q(t)$. In general, can we relate $w_i(t)$ and $w_q(t)$? How about $\overline{w_i^2(t)}$ and $\overline{w_q^2(t)}$?*

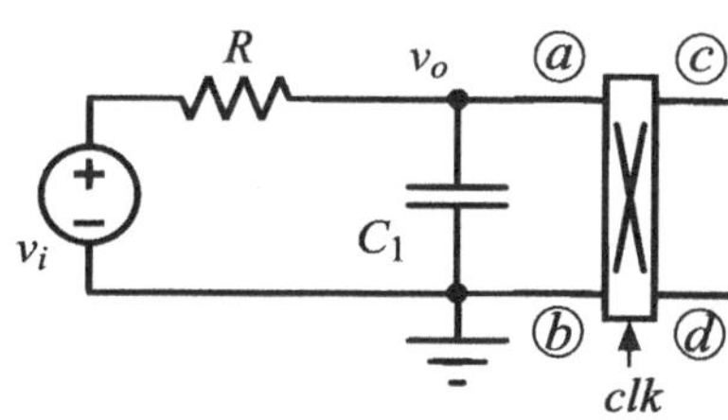

Figure 5.54 LPTV system incorporating a chopper.

Problem 9. *The circuit shown in Figure 5.54 is an LPTV network. The chopper is a network of ideal switches that works as follows. When clk is high, nodes ⓐ and ⓑ are connected to nodes ⓒ and ⓓ respectively. When clk is low, nodes ⓐ and ⓑ are connected to nodes ⓓ and ⓒ respectively. The clk signal has a 50% duty cycle and frequency f_s. Denote the zeroth-order harmonic transfer function from v_i to v_o by $H_0(j2\pi f)$. Determine $H_0(0)$.*

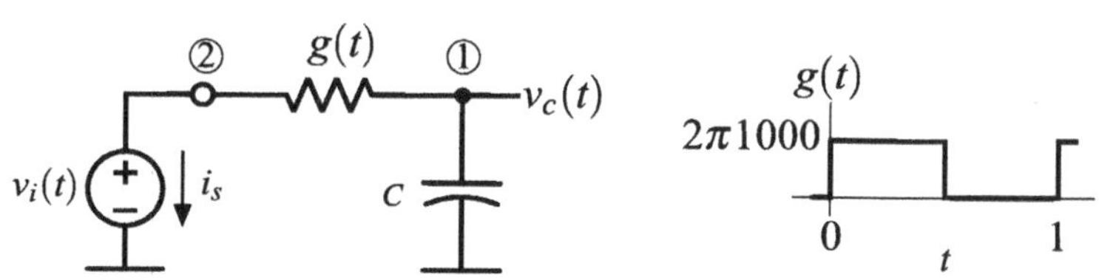

Figure 5.55 Time-varying RC filter for problem 10.

Problem 10. *Figure 5.55 shows an RC network with a periodically time-varying resistor. The capacitor $C = 1\,F$. The conductance varies from 0 to $2\pi 1000\,S$ as shown in the figure. Write a MATLAB program to determine the harmonic transfer functions of this* *network. Assume that $g(t)$ has a rise- and fall-times of 0.1 s and an average value that is half its peak value. Denoting the maximum number of harmonics at any node by M, so that there are $(2M + 1)$ sinusoids at any node, your code should be able to accommodate a user-specified M. Plot $|H_k(j2\pi f)|$ for $k = 0, \pm 1$, for $M = 128, 256, 512$. For uniformity, the range of the y-axis must be 0–1, and the x-axis from −3 to 3 Hz, with increments in f chosen to be 0.05. What do you notice as you change M? Run sanity checks for the values of the harmonic transfer functions at frequencies 1, 2, 3 Hz.*

Suggestion : When you invert the **G** *matrix in MATLAB to solve* **GV = I**, *do not use* `V = inv(G)*I`. *It turns out that matrix inversion is a very computationally intensive process – remember that you are inverting a 3000 × 3000 matrix. Rather, use* `V = G\I`, *which is much quicker. See for yourself how much quicker* `V = G\I` *is with respect to explicitly computing the inverse.*

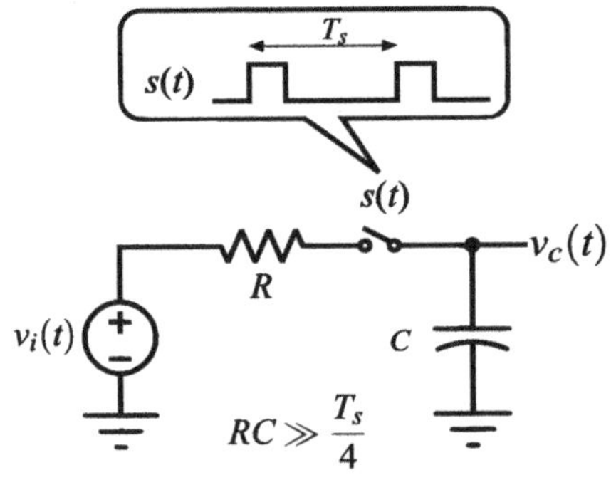

Figure 5.56 A switched RC filter, where $RC \gg T_s/4$.

Problem 11. *Figure 5.56 shows an RC network with a periodically operated ideal switch. The switch is closed when $s(t)$ is high and open when it is low. $s(t)$ has a frequency of 1 Hz and a 25% duty cycle. The capacitor $C = 1$ F. $RC = 10$ s. The output is the voltage at the output of the capacitor $v_c(t)$. Analytically determine $H_{-k}(j2\pi f)$ for 1, 2 (use the fact that $RC \gg T_s/4$ to advantage).*

Next, use the code you developed in the previous problem to compute $H_{-k}(j2\pi f)$ for 1, 2. Use $M = 512$, and a frequency resolution of 0.01 Hz. Plot $H_{-k}(j2\pi f)$ for 1, 2 for f between 0 and 3 Hz at intervals of every 0.01 Hz. What do you notice? For ease of computation, assume that the rise- and fall-times of $s(t)$ are 1% of its period, and that the average of $s(t)$ is 25% of its peak value.

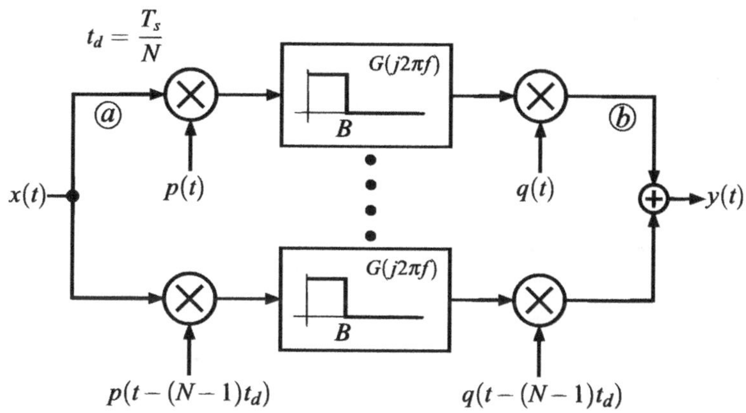

Figure 5.57 N-path system.

Problem 12. *Figure 5.57 shows an N-path system. $G(j2\pi f)$ represents the transfer function of an LTI system with a brickwall lowpass response with bandwidth $B \ll f_s$. $p(t)$ and $q(t)$ are periodic with a period $T_s \equiv 1/f_s$ and are expressed as*

$$p(t) = \sum_l P_l e^{j2\pi l f_s t}$$

and

$$q(t) = \sum_m Q_m e^{j2\pi m f_s t}.$$

a. *Determine an expression for the harmonic transfer functions $H_k(j2\pi f)$ of one arm of the system (between ⓐ and ⓑ).*
b. *Draw a qualitative picture of $|H_0(j2\pi f)|$.*
c. *Determine the harmonic transfer functions $\hat{H}_k(j2\pi f)$ of the N-path system.*
d. *Denoting the impulse response corresponding to $G(j2\pi f)$ by $g(t)$, determine the impulse response of the N-path filter as $N \to \infty$.*

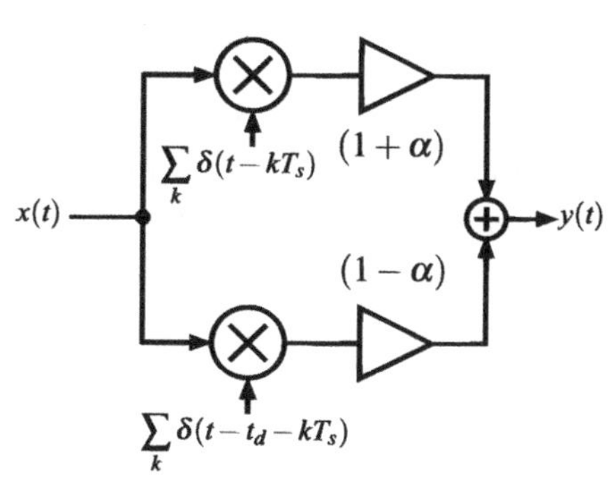

Figure 5.58 A time-interleaved sampler, $t_d = 0.5T_s + t_o$. Under nominal conditions t_o and α are zero.

Problem 13. *In Figure 5.58, $t_d = 0.5T_s + t_o$.*

- *Determine the harmonic transfer functions of the system when α and t_o are 0.*
- *Evaluate the HTFs for small, nonzero α but $t_o = 0$.*
- *Next, repeat for $\alpha = 0$ but small nonzero t_o.*

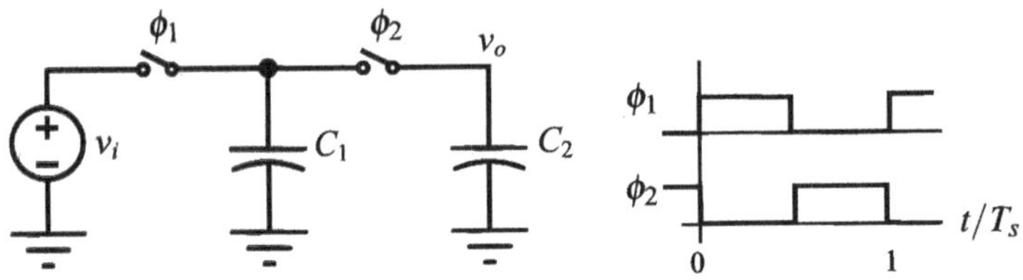

Figure 5.59

Problem 14. *In Figure 5.59, the clock signals ϕ_1 and ϕ_2 do not overlap, and have (almost) 50% duty cycle. v_i is the input, while v_o denotes the output. Determine the HTFs of the circuit $H_k(j2\pi f)$.*

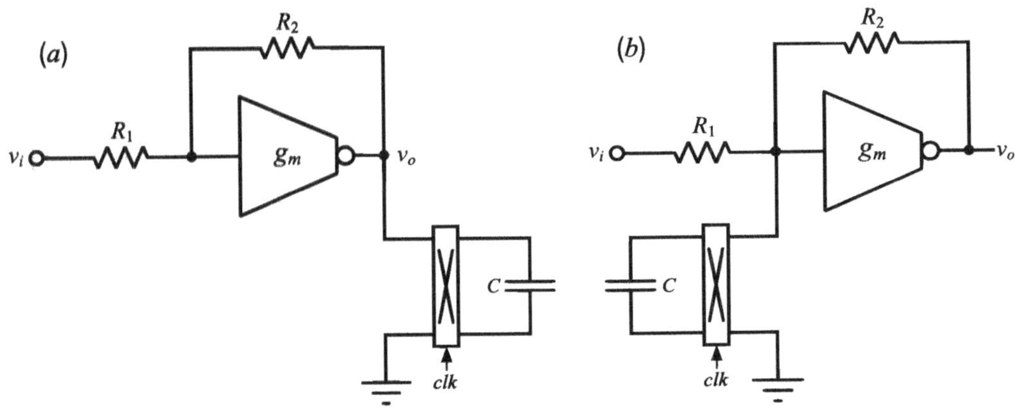

Figure 5.60

Problem 15. *The circuits shown in Figure 5.60 are LPTV networks. The chopper is controlled by a clock signal that has a 50% duty cycle and a frequency f_s. Assume that $C(1 + R_2/R_1)/g_m \ll 1/f_s$ and $g_m R_1, g_m R_2 \gg 1$. For each of the circuits, determine $H_{-k}(j2\pi k f_s)$.*

References

1 B. Razavi, *RF Microelectronics*, 2nd ed. Prentice Hall, 2011.
2 T. H. Lee, *The Design of CMOS Radio-Frequency Integrated Circuits*, 2nd ed. Cambridge University Press, 2004.

3 L. A. Zadeh, "Frequency analysis of variable networks," *Proceedings of the IRE*, vol. 38, no. 3, pp. 291–299, Mar. 1950.

4 P. Vanassche, G. Gielen, and W. Sansen, *Systematic Modeling and Analysis of Telecom Frontends and Their Building Blocks*. Springer US, 2005.

5 K. Kundert, *The Designer's Guide to Spectre RF*. Cadence Design Systems, 2004.

6 K. Kundert, "Introduction to RF simulation and its application," *IEEE Journal of Solid-State Circuits*, vol. 34, no. 9, pp. 1298–1319, Sep. 1999.

7 L. E. Franks and I. W. Sandberg, "An alternative approach to the realization of network transfer functions: The N-path filter," *Bell System Technical Journal*, vol. 39, no. 5, pp. 1321–1350, Sep. 1960.

6

LPTV Circuits and Systems: Advanced Topics

We described the basics of LPTV circuits and systems in Chapter 5. We now build on those foundations. First, we derive the concepts of reciprocal and inter-reciprocal networks. As discussed earlier in the context of linear time-invariant (LTI) circuits, reciprocity and inter-reciprocity are not merely mathematical curiosities—they have practical utility, especially when determining transfer functions from multiple inputs to a single output. This situation arises frequently in noise analysis. If these properties are useful tools in LTI networks, they are even more valuable in linear periodically time-varying (LPTV) circuits. The key reason is that in LPTV circuits, one must also account for frequency translation effects, even in the case of a single input and a single output.

The outputs of many practically useful LPTV circuits and subsystems are sampled. Furthermore, it turns out that the sampling rate is identical to the rate at which the circuit is varying. We discuss this very important class of circuits.

Finally, we study noise in LPTV circuits and systems. As with anything LPTV, the analysis can be expected to be more involved. In the special (and practically relevant) case of networks with inductors, capacitors, resistors, and periodically operated switches, it turns out that inter-reciprocity can be exploited to determine the average noise spectral density and total integrated noise in an efficient manner. These relationships are analogous to the Nyquist- and Bode noise theorems we encountered in the context of RLC networks in the LTI case.

6.1 Reciprocity and Inter-Reciprocity

In this section, we aim to understand reciprocity in LPTV circuits and systems. As in the LTI case, we begin with LPTV networks composed of inductors, capacitors, and periodically time-varying resistors. Notably, the conceptual approach to reciprocity in LPTV networks closely parallels that in LTI systems.

Consider the network of Figure 6.1. The conductances vary periodically with time, and the period is denoted by T_s. The time-variation of one example conductance is sketched below the network. Note that the nature of the variation of the conductance can be different for the various instances; however, the period should be identical

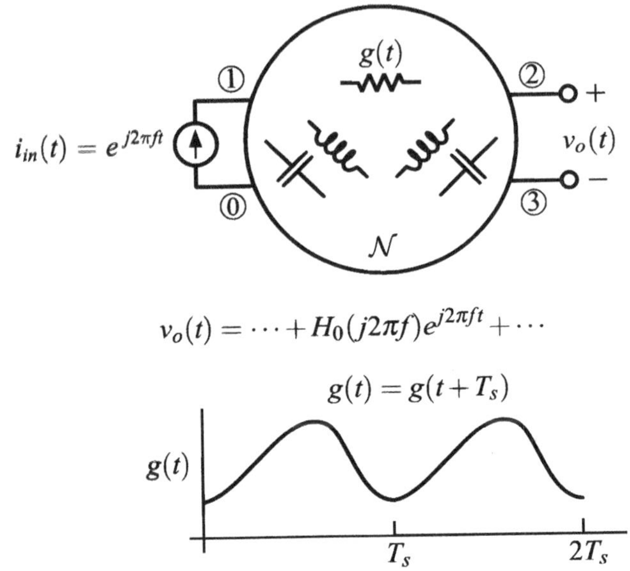

Figure 6.1 Passive LPTV network with periodically varying conductances, inductors and capacitors. The network is excited by a complex exponential with frequency f. The conductances vary with a period denoted by $T_s \equiv 1/f_s$.

for all of them. Clearly, the network is LPTV, varying at a frequency $f_s = 1/T_s$. The network is excited by a current $i_{\text{in}}(t) = e^{j2\pi ft}$, while the "response" we are interested in is the voltage developed across the terminals of the output port. Without loss of generality, we number the nodes as shown in the figure.

In the discussion that follows, we use the notation $\mathbf{1}_r$ to denote a column vector with all entries being 0, except the rth row as shown below.

$$\mathbf{1}_r = [0 \quad \cdots \quad \underbrace{1}_{r^{th}\ term} \quad \cdots \quad 0]^T. \tag{6.1}$$

Like in Chapter 5, we assume that every node voltage and branch current is a sum of complex exponentials with frequencies ranging from $(-Mf_s + f)$ to $(Mf_s + f)$ (i.e., $(2M + 1)$ frequencies in all). The $(2M + 1) \times 1$ current phasor-vector corresponding to the input excitation $e^{j2\pi ft}$ is thus given by $\mathbf{I}_{\text{in}} = \mathbf{1}_{(M+1)}$ (has units of amperes). Using the MNA formulation, the network equations are written as

$$\mathbf{G}\begin{bmatrix} \mathbf{v}_1 \\ \mathbf{v}_2 \\ \mathbf{v}_3 \\ \vdots \end{bmatrix} = \begin{bmatrix} \mathbf{1}_{(M+1)} \\ \mathbf{0} \\ \mathbf{0} \\ \vdots \end{bmatrix}. \tag{6.2}$$

If the network has N unknowns, recall from Section 5.5 that $\mathbf{G}$ is a $(2M+1)N \times (2M+1)N$ matrix formed from the MNA stamps of the individual elements comprising the network. Suppose that we are interested in the transfer function relating the frequency component of $v_o(t)$ at f to the input current. This is the zeroth-order harmonic transfer function, which we denote by $H_0(j2\pi f)$.

$$H_0(j2\pi f) = \begin{bmatrix} \mathbf{0}^T & \mathbf{1}^T_{(M+1)} & -\mathbf{1}^T_{(M+1)} & \mathbf{0}^T & \cdots \end{bmatrix} \begin{bmatrix} \mathbf{v}_1 \\ \mathbf{v}_2 \\ \mathbf{v}_3 \\ \vdots \end{bmatrix}. \tag{6.3}$$

The node-voltage phasor-vectors, obtained by solving the MNA equations, are given by

$$\begin{bmatrix} \mathbf{v}_1 \\ \mathbf{v}_2 \\ \mathbf{v}_3 \\ \vdots \end{bmatrix} = \mathbf{G}^{-1} \begin{bmatrix} \mathbf{1}_{(M+1)} \\ \mathbf{0} \\ \mathbf{0} \\ \vdots \end{bmatrix}. \tag{6.4}$$

From the two equations above, we see that

$$H_0(j2\pi f) = \underbrace{\begin{bmatrix} \mathbf{0}^T & \mathbf{1}^T_{(M+1)} & -\mathbf{1}^T_{(M+1)} & \mathbf{0}^T & \cdots \end{bmatrix}}_{\text{measurement}} \mathbf{G}^{-1} \underbrace{\begin{bmatrix} \mathbf{1}_{(M+1)} \\ \mathbf{0} \\ \mathbf{0} \\ \vdots \end{bmatrix}}_{\text{excitation}}. \tag{6.5}$$

Transposing both sides, we obtain

$$H_0(j2\pi f) = \begin{bmatrix} \mathbf{1}^T_{(M+1)} & \mathbf{0}^T & \mathbf{0}^T & \cdots \end{bmatrix} (\mathbf{G}^T)^{-1} \begin{bmatrix} \mathbf{0} \\ \mathbf{1}_{(M+1)} \\ -\mathbf{1}_{(M+1)} \\ \mathbf{0} \\ \vdots \end{bmatrix}. \tag{6.6}$$

The equation above can be *interpreted* in the following manner, as shown below:

$$H_0(j2\pi f) = \underbrace{\left[\mathbf{1}^T_{(M+1)} \quad \mathbf{0}^T \quad \mathbf{0}^T \quad \cdots\right]}_{\text{measurement}} (\mathbf{G}^T)^{-1} \underbrace{\begin{bmatrix} \mathbf{0} \\ \mathbf{1}_{(M+1)} \\ -\mathbf{1}_{(M+1)} \\ \mathbf{0} \\ \vdots \end{bmatrix}}_{\text{excitation}}. \tag{6.7}$$

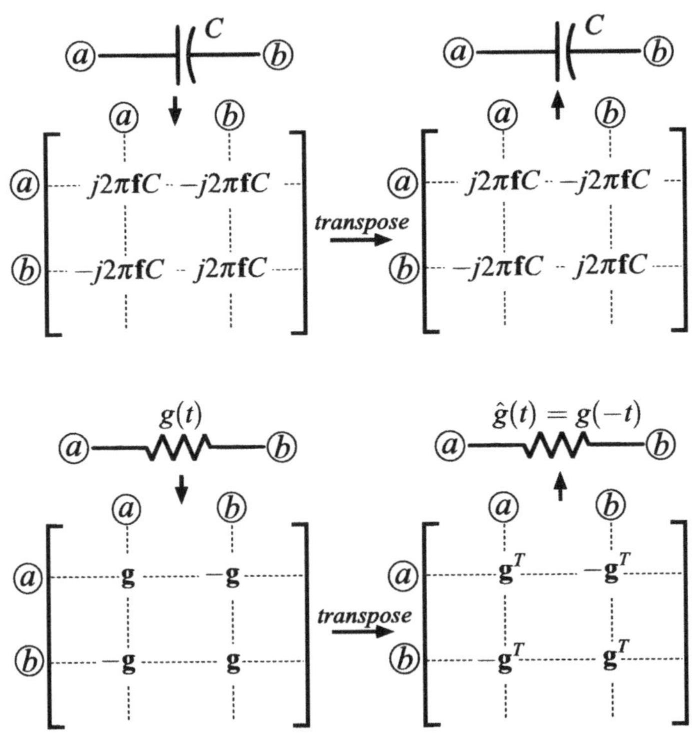

Figure 6.2 A capacitor in $\mathcal{N}$ remains unchanged in $\hat{\mathcal{N}}$, while $g(t)$ in $\mathcal{N}$ is replaced by $g(-t)$ in $\hat{\mathcal{N}}$.

Consider a network whose MNA matrix is $\mathbf{G}^T$. Excite this network with a current source $e^{j2\pi ft}$ placed between nodes ② and ③. The output of interest is the voltage of node ①, at frequency f. The ratio of the output to input is $H_0(j2\pi f)$. Clearly, we have interchanged the location of the excitation and measurement ports. Let us ponder over what the network itself looks like. Evidently, its MNA matrix is the transpose of that of the network of Figure 6.1. Since the MNA matrix of a network is simply the sum of the MNA stamps of the individual elements, the new network can be derived from the original circuit in the following manner. Go element by element in $\mathcal{N}$, replacing each element with one whose MNA stamp is the transpose of that element. We illustrate this process in Figure 6.2.

A capacitor connected between nodes ⓐ and ⓑ has the MNA stamp shown on the left. Recall from Section 5.5 that **f** is a diagonal matrix. Due to this symmetry, the MNA stamp is unchanged when it is transposed, leading us to conclude that capacitors in the original network remain unchanged in the "new" network. Proceeding along the same lines, it is easy to see that inductors in the original network also remain unchanged in the new network.

What about a time-varying conductance? The MNA stamp corresponding to a time-varying conductance is shown in the left of Figure 6.2. $g(t)$ can be expanded into a Fourier series as follows.

$$g(t) = \sum_l g_l e^{j2\pi l f_s t} \tag{6.8}$$

and **g** is formed from the Fourier coefficients of $g(t)$ according to

$$\mathbf{g} = \begin{bmatrix} g_0 & g_{-1} & \cdots & g_{-2M+1} & g_{-2M} \\ g_1 & g_0 & \cdots & g_{-2M+2} & g_{-2M+1} \\ \vdots & \vdots & \ddots & \vdots & \vdots \\ g_{2M-1} & g_{2M-2} & \cdots & g_0 & g_{-1} \\ g_{2M} & g_{2M-1} & \cdots & g_1 & g_0 \end{bmatrix}. \tag{6.9}$$

Since $g(t)$ is real, $g_{-l} = g_l^*$. The transpose of the time-varying conductance's MNA stamp is shown toward the right in Figure 6.2. What conductance $\hat{g}(t)$ does this correspond to? Since $\mathbf{g}^T$ is obtained by replacing every entry of the form g_l in **g** by g_{-l}, it is apparent that

$$\hat{g}(t) = \sum_l g_{-l} e^{j2\pi l f_s t} = \sum_l g_l e^{j2\pi(-l) f_s t} = \sum_l g_l e^{j2\pi l f_s(-t)} = g(-t).$$

Like in the time-invariant case, the "new" network is denoted by $\hat{\mathcal{N}}$, and is called the inter-reciprocal, or adjoint network.

The summary of our discussion so far is as follows [1]. In the original network $\mathcal{N}$, shown in Figure 6.1, the transfer function from an input current at frequency f to an output voltage at the same frequency is denoted by $H_0(j2\pi f)$. The same transfer function can be obtained in a modified network $\hat{\mathcal{N}}$ (called the adjoint). The method of obtaining $\hat{\mathcal{N}}$ and the details of excitation and measurement are as follows.

a. The adjoint $\hat{\mathcal{N}}$ is derived from the original network $\mathcal{N}$ by element replacement as per the following rules, as shown in Figure 6.3.
- A capacitor of value C in $\mathcal{N}$ remains a capacitor with the same value in $\hat{\mathcal{N}}$.
- An inductor of value L in $\mathcal{N}$ remains an inductor with the same value in $\hat{\mathcal{N}}$.
- A time-varying conductance $g(t)$ in $\mathcal{N}$ is replaced by a time-varying conductance $g(-t)$ in $\hat{\mathcal{N}}$. This is often referred to as *time-reversal*.

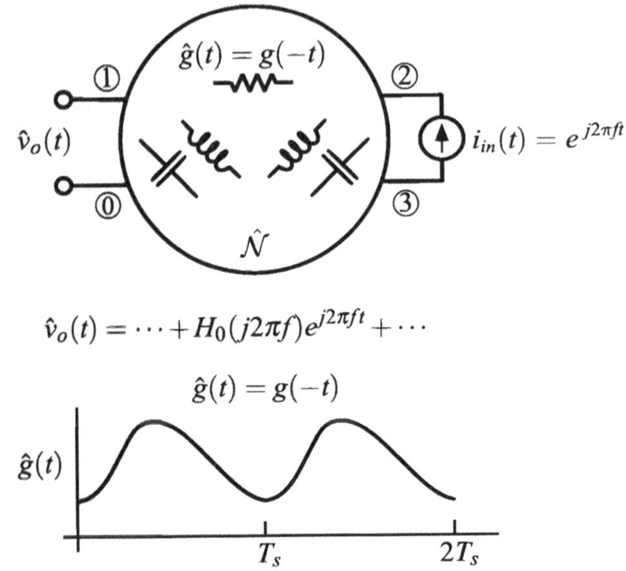

Figure 6.3 The inter-reciprocal (or adjoint) network. Note that the variation of the conductances is reversed in time.

b. Excite the output port of $\hat{\mathcal{N}}$ with a current $e^{j2\pi ft}$.
c. Measure the component of the voltage at the input port of $\hat{\mathcal{N}}$ at the frequency f. It will be $H_0(j2\pi f)e^{j2\pi ft}$.

The result above is called the *transfer-function theorem* (Figure 6.4) [2].

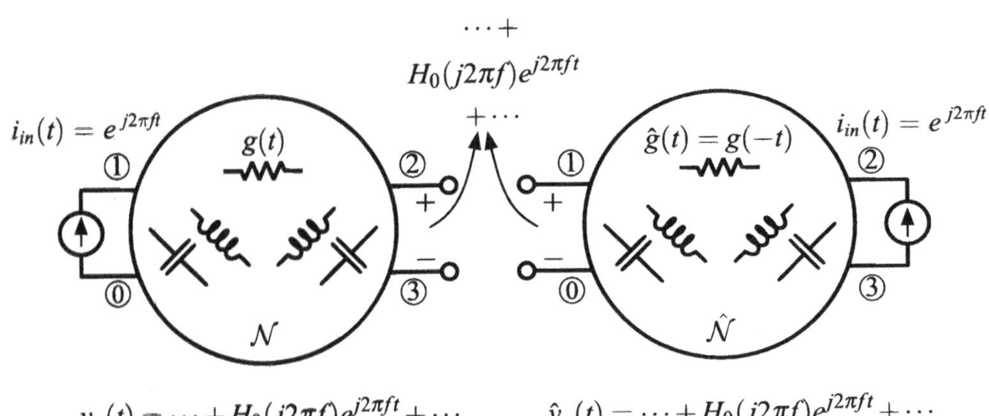

Figure 6.4 The transfer-function theorem.

6.1.1 The Frequency-Reversal Theorem

If the transfer-function theorem in the discussion above was "interesting," what we will see below is positively delicious. Consider again the LPTV network excited by a current source $e^{j2\pi ft}$, as shown in Figure 6.5. The output of interest now is the

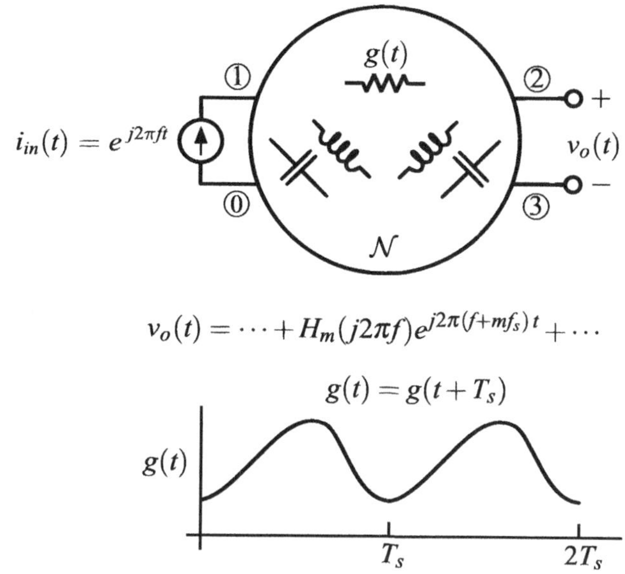

Figure 6.5 LPTV network excited with a current $e^{j2\pi ft}$. The output of interest is the component of $v_o(t)$ at the frequency $(f + mf_s)$.

component of $v_o(t)$ at the frequency $(f + mf_s)$, i.e., the mth-order harmonic transfer function $H_m(j2\pi f)$. As before, the MNA equations can be solved to obtain the node-voltage phasor-vectors. It is straightforward to see that $H_m(j2\pi f)$ can be expressed as

$$H_m(j2\pi f) = \underbrace{\begin{bmatrix}\mathbf{0}^T & \mathbf{1}^T_{(m+M+1)} & -\mathbf{1}^T_{(m+M+1)} & \mathbf{0}^T & \cdots\end{bmatrix}}_{\text{measurement}} \mathbf{G}^{-1} \underbrace{\begin{bmatrix}\mathbf{1}_{(M+1)} \\ \mathbf{0} \\ \mathbf{0} \\ \vdots\end{bmatrix}}_{\text{excitation}}. \tag{6.10}$$

Transposing both sides, we obtain

$$H_m(j2\pi f) = \begin{bmatrix}\mathbf{1}^T_{(M+1)} & \mathbf{0}^T & \mathbf{0}^T & \cdots\end{bmatrix} (\mathbf{G}^T)^{-1} \begin{bmatrix}\mathbf{0} \\ \mathbf{1}_{(m+M+1)} \\ -\mathbf{1}_{(m+M+1)} \\ \mathbf{0} \\ \vdots\end{bmatrix}. \tag{6.11}$$

The equation above can be *interpreted* in the following manner, as shown below.

$$\underbrace{\overbrace{H_m(j2\pi f)}}_{\mathcal{N}} = \underbrace{\underbrace{\left[\mathbf{1}^T_{(M+1)} \quad \mathbf{0}^T \quad \mathbf{0}^T \quad \cdots\right]}_{\text{measurement at frequency f}} (\mathbf{G}^T)^{-1} \underbrace{\begin{bmatrix} \mathbf{0} \\ \mathbf{1}_{(m+M+1)} \\ -\mathbf{1}_{(m+M+1)} \\ \mathbf{0} \\ \vdots \end{bmatrix}}_{\substack{\text{excitation} \\ \text{at frequency} \\ (f+mf_s)}}}_{\hat{H}_{-m}(j2\pi(f+mf_s)) \quad of \quad \hat{\mathcal{N}}} \cdot \tag{6.12}$$

The right-hand-side corresponds to the transfer function of a network whose MNA matrix is $\mathbf{G}^T$. This is the adjoint network $\hat{\mathcal{N}}$. The excitation is a current source $e^{j2\pi(f+mf_s)t}$ placed between nodes ② and ③, while the output of interest is the voltage of node ①, at frequency f. The ratio of output to input, therefore, is $\hat{H}_{-m}(j2\pi(f + mf_s))$. The hat on H serves to remind ourselves that this transfer function pertains to the adjoint network $\hat{\mathcal{N}}$. What Eq. (6.12) is telling us is that $H_m(j2\pi f) = \hat{H}_{-m}(j2\pi(f + mf_s))$. This is the *frequency-reversal theorem*, illustrated in Figure 6.6 [2].

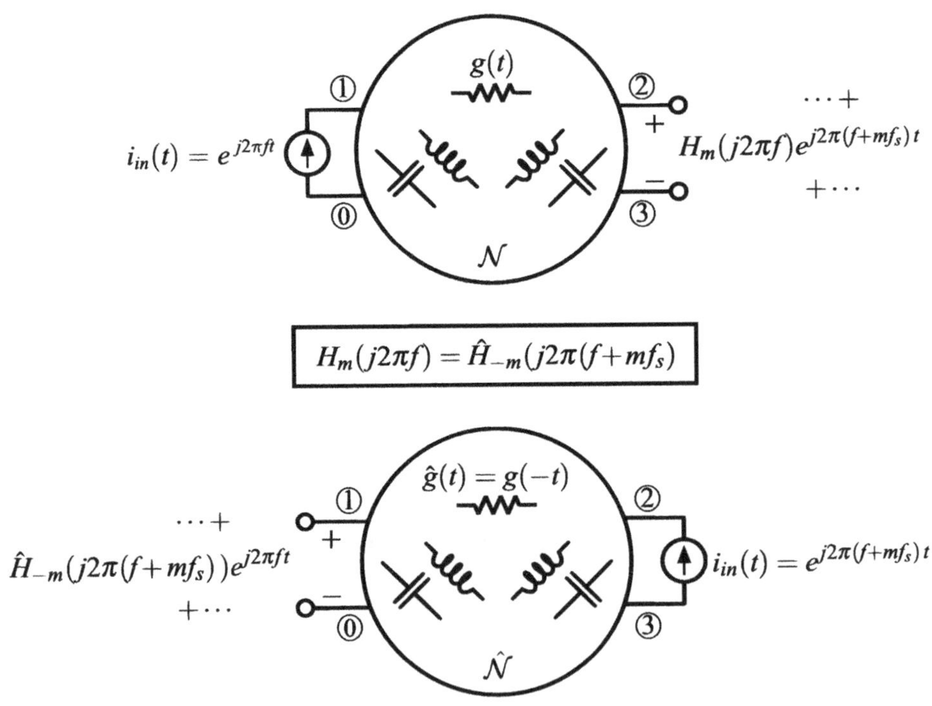

Figure 6.6 The frequency-reversal theorem.

Replacing m with $-m$, and then replacing f with $(f + mf_s)$, we obtain the equivalent relationship

$$H_{-m}(j2\pi(f + mf_s)) = \hat{H}_m(j2\pi f). \tag{6.13}$$

Equivalently, for two frequencies f_1 and f_2, both of which are of the form $f + lf_s$ with l being an integer, the frequency-reversal theorem reads as follows.

> The output at a frequency f_2 at port 2 of an LPTV network $\mathcal{N}$
> excited at port 1 by $e^{j2\pi f_1 t}$
> has the same complex amplitude as
> the output at a frequency f_1 at port 1 of the adjoint network $\hat{\mathcal{N}}$
> excited at port 2 by $e^{j2\pi f_2 t}$.

From the discussion above, the transfer-function theorem is seen to be a special case of the frequency reversal theorem where $f_1 = f_2$, as shown graphically in Figure 6.7.

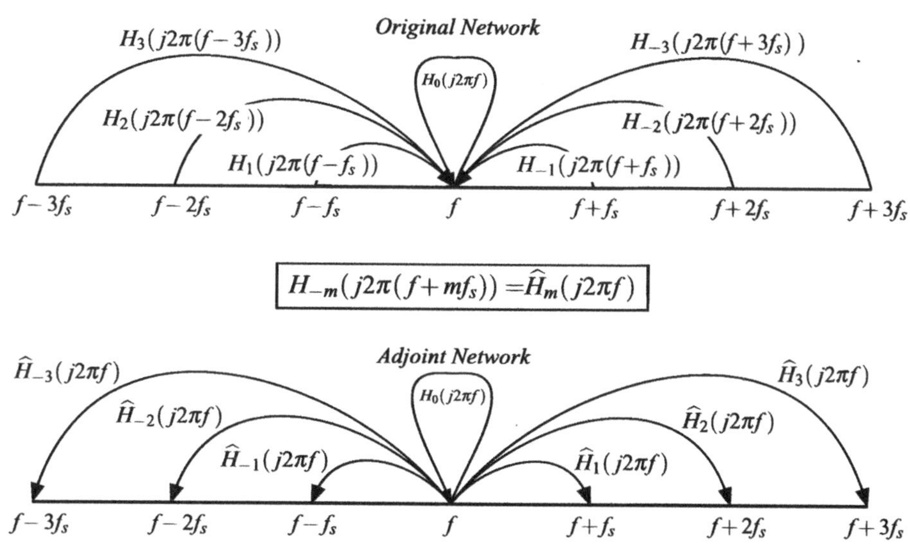

Figure 6.7 The frequency-reversal theorem allows us to calculate the contributions from multiple input frequencies to a single output frequency in one shot.

What is so exciting about the frequency-reversal theorem? It turns out that in many situations that arise in practice (and during noise analysis), we are interested in determining the signal at the output of an LPTV network at a frequency f. We know that a component at f can be due to inputs at frequencies of the form $(f + mf_s)$, where m is an integer. We therefore need to determine, for all m, the harmonic transfer functions

$H_{-m}(j2\pi(f + mf_s))$, which quantify the gain from an input at frequency $(f + mf_s)$ to the output frequency f. How would we go about this if we were not aware of the frequency-reversal theorem? We would excite the system with $e^{j2\pi(f+mf_s)t}$ for all m, *one at a time*, and calculate the output component at f. As one can imagine, this is an extremely laborious process, where we need to solve the MNA equations $(2M + 1)$ times.

The frequency-reversal theorem helps us compute all the $H_{-m}(j2\pi(f + mf_s))$ in *one shot*, as illustrated in Figure 6.8. We do this as follows. We form the adjoint network $\hat{\mathcal{N}}$ using the element-replacement rules discussed earlier in this section. We then excite the output port of the adjoint network with a complex exponential current $e^{j2\pi ft}$. The voltage developed across the input port is given by

$$\hat{v}_1(t) = \sum_m \hat{H}_m(j2\pi f)e^{j2\pi(f+mf_s)t} = \sum_m H_{-m}(j2\pi(f + mf_s))e^{j2\pi(f+mf_s)t}. \quad (6.14)$$

Thanks to the frequency-reversal theorem, the strength of the component of $\hat{v}_1(t)$ at $(f + mf_s)$ yields $H_{-m}(f + mf_s)$.

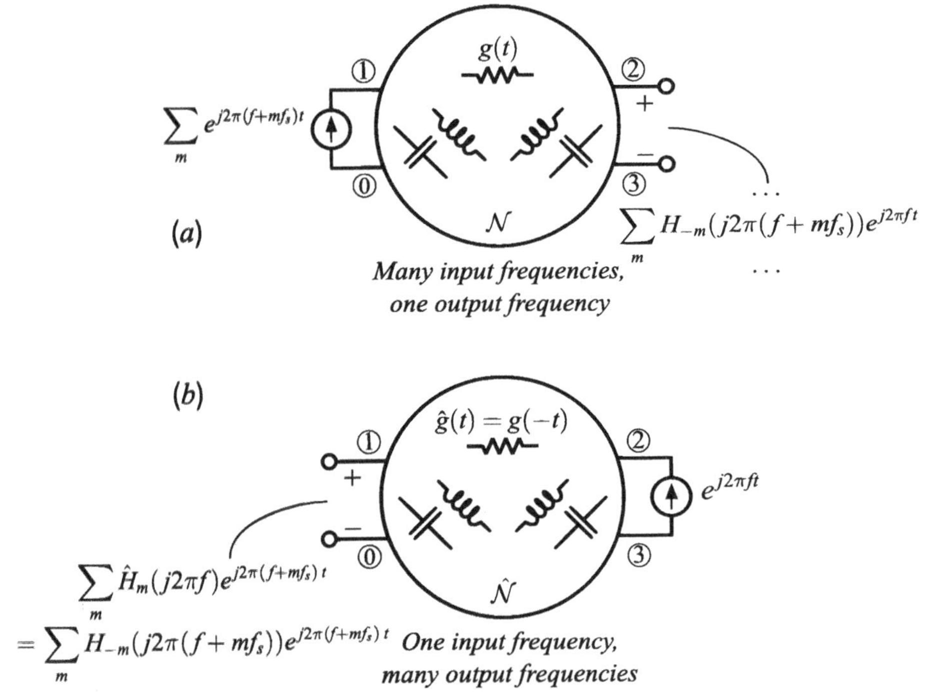

Figure 6.8 Applying the frequency-reversal theorem to calculate gains from multiple input frequencies of the form $e^{j2\pi(f+mf_s)}$ to an output frequency f.

Example: The Switched-RC Network

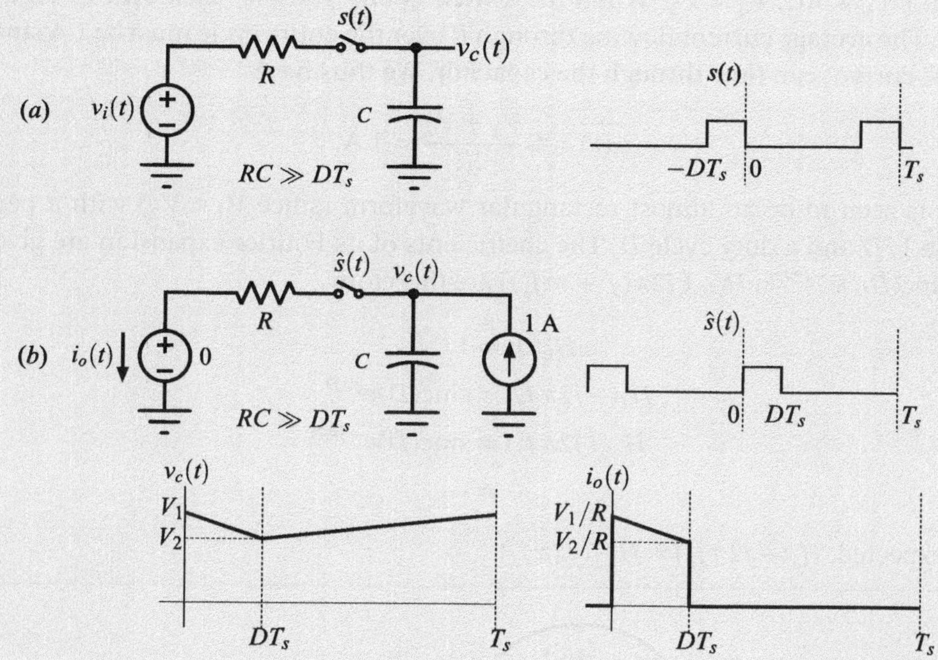

Figure 6.9 Applying the frequency-reversal theorem to calculate gains from multiple input frequencies of the form $e^{j2\pi(f+mf_s)}$ to an output frequency f, for $f = 0$.

Consider the switched-RC network of Figure 6.9a. The switch is controlled by $s(t)$, whose period and duty cycle are denoted by T_s and D respectively. $RC \gg DT_s$. What are $H_{-m}(j2\pi(f + mf_s))$ of this network for $f = 0$?

To determine the desired HTFs in one shot, we exploit the frequency-reversal theorem. The output frequency is f, which is dc. The adjoint network is shown in Figure 6.9b. The switch-control waveform is $\hat{s}(t) = s(-t)$. The input voltage is set to 0, and a 1 A current is injected into the capacitor. Once the steady-state waveform $i_o(t)$ is determined, the coefficients of its Fourier expansion are the desired $H_{-m}(j2\pi(f + mf_s))$.

When $\hat{s}(t)$ is high, the capacitor discharges through R. Since $RC \gg DT_s$, it falls almost linearly from V_1 at $t = 0$ to V_2 at $t = DT_s$. We thus see that

$$V_2 = V_1\left(1 - \frac{DT_s}{RC}\right).$$

(Continued)

(Continued)

Since $DT_s \ll RC$, $V_1 \approx V_2$. When the switch is off, $i_o(t) = 0$, as shown in Figure 6.9b. The average current flowing through R over the entire cycle must be 1 A since no dc current can flow through the capacitor. We thus have

$$\overline{i_o(t)} = D\frac{V_1 + V_2}{2R} = 1\text{ A}.$$

$i_o(t)$ is seen to be an almost rectangular waveform (since $V_1 \approx V_2$) with a peak value $1/D$ and a duty cycle D. The coefficients of its Fourier expansion are given by $\text{sinc}(Dn)e^{-j\pi Dn}$. $H_{-m}(j2\pi(f + mf_s))$ are therefore

$$\begin{aligned} H_0(0) &= 1 \\ H_1(-j2\pi f_s) &= \text{sinc}(D)e^{j\pi D} \\ H_{-1}(j2\pi f_s) &= \text{sinc}(D)e^{-j\pi D} \\ \cdots &= \cdots . \end{aligned}$$

As expected, $H_1(-j2\pi f_s) = H^*_{-1}(j2\pi f_s)$.

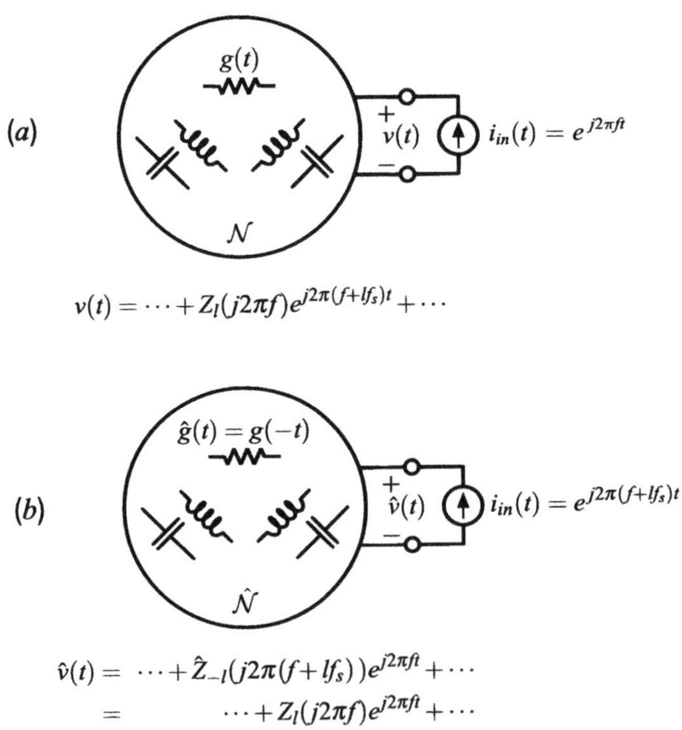

Figure 6.10 Applying the frequency-reversal theorem to determine the relationships between the driving-point impedance of a network and its adjoint.

6.1.2 Implications of the Frequency-Reversal Theorem: Driving Point Impedance of an LPTV Network and Its Adjoint

In this subsection, we discuss some interesting consequences of the frequency-reversal theorem. The first is the relationship between the driving-point impedance of a one-port LPTV network $\mathcal{N}$ and its adjoint $\hat{\mathcal{N}}$. Figure 6.10a shows $\mathcal{N}$ being driven by a current $i(t) = e^{j2\pi ft}$. The voltage developed across the port is comprised of frequencies $(f + mf_s)$. We focus on the component at a frequency $(f + lf_s)$. $Z_l(j2\pi f)$ denotes the lth-*order* driving point impedance, and relates the component of $v(t)$ at $(f + lf_s)$ to the driving current at f. $v(t)$ is expressed as

$$v(t) = \cdots + Z_l(j2\pi f)e^{j2\pi(f+lf_s)t} + \cdots .$$

In the network of Figure 6.10a, where $i(t) = e^{j2\pi ft}$, the component of $v(t)$ at $(f + lf_s)$ is given by $Z_l(j2\pi f)$.

Consider now the network of Figure 6.10b. It shows the adjoint network $\hat{\mathcal{N}}$ excited by a current $i(t) = e^{j2\pi(f+lf_s)t}$. The voltage developed across the port at the frequency f is given by

$$\hat{v}(t) = \cdots + \hat{Z}_{-l}(j2\pi(f + lf_s))e^{j2\pi ft} + \cdots .$$

By the frequency-reversal theorem, interchanging the excitation and measurement ports in $\mathcal{N}$ and $\hat{\mathcal{N}}$, and the input and output frequencies must yield the same transfer function, which is the impedance in this case. Consequently, we have

$$Z_l(j2\pi f) = \hat{Z}_{-l}(j2\pi(f + lf_s)). \tag{6.15}$$

In the special case of $l = 0$,

$$Z_0(j2\pi f) = \hat{Z}_0(j2\pi f)$$

indicating that the zeroth-order driving point impedances of an LPTV network $\mathcal{N}$ and its adjoint $\hat{\mathcal{N}}$ are identical [3].

6.1.3 Implications of the Frequency-Reversal Theorem: Two-Port Parameters

As is evident from our discussion on reciprocity (Chapter 1), an LTI network with resistors, inductors, and capacitors is reciprocal. Consequently, for a two-port network formed using such elements, $z_{21} = z_{12}$ and $y_{21} = y_{12}$ where z and y refer to the impedance and admittance parameters, respectively. Does this hold for LPTV networks consisting of periodically time-varying conductances, inductors, and capacitors?

To understand this, consider the two-port $\mathcal{N}$ shown in Figure 6.11a. To find $z_{21}(j2\pi f)$, we excite port 1 with $i_1 = e^{j2\pi ft}$ and measure the component of the voltage developed across port 2 at the frequency f. In a similar fashion, exciting port 2 of $\hat{\mathcal{N}}$ with a current $e^{j2\pi ft}$ results in a voltage $\hat{v}_1$ with a component $\hat{z}_{12}(j2\pi f)e^{j2\pi ft}$

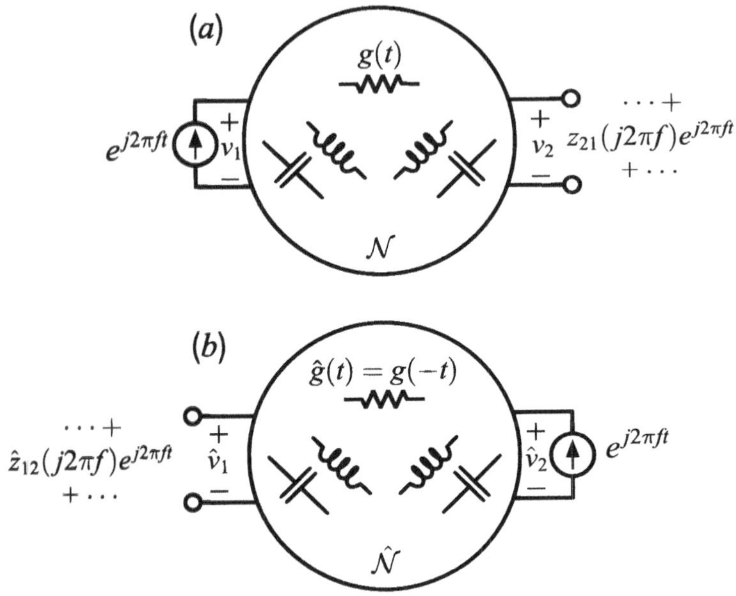

Figure 6.11 Applying the transfer-function theorem to relate (a) z_{21} of a network to (b) $\hat{z}_{12}$ of its adjoint.

as shown in Figure 6.11b. Using the transfer function theorem, we conclude that $z_{21}(j2\pi f) = \hat{z}_{12}(j2\pi f)$. However, $\mathcal{N}$ and $\hat{\mathcal{N}}$ need not be identical. Consequently, $z_{21}(j2\pi f)$ and $z_{12}(j2\pi f)$ of $\mathcal{N}$ are not necessarily equal.

To summarize, LPTV networks with inductors, capacitors and time-varying conductances are capable of achieving non-reciprocal operation, something that is simply not possible with time-invariant RLC circuits.

6.1.4 The Adjoint Network with Periodically-Varying Controlled Sources

How does frequency reversal apply when the network $\mathcal{N}$ periodically contains time-varying controlled sources? It is apparent that, like earlier, we would need to replace each controlled source by a block whose MNA stamp is the transpose of that of the source under consideration. This is best illustrated with an example. Figure 6.12 shows a time-varying voltage-controlled voltage source (VCVS), whose gain $\mu(t)$ is varying periodically with time.[1] It can therefore be expanded into a Fourier series

$$\mu(t) = \sum_{l} \mu_l e^{j2\pi l f_s t}. \tag{6.16}$$

[1]This period must be the same as that with which the network is varying.

Like we did with a time-varying conductance, we form the matrix μ from the Fourier coefficients of $\mu(t)$ according to

$$\mu = \begin{bmatrix} \mu_0 & \mu_{-1} & \cdots & \mu_{-2M+1} & \mu_{-2M} \\ \mu_1 & \mu_0 & \cdots & \mu_{-2M+2} & \mu_{-2M+1} \\ \vdots & \vdots & \ddots & \vdots & \vdots \\ \mu_{2M-1} & \mu_{2M-2} & \cdots & \mu_0 & \mu_{-1} \\ \mu_{2M} & \mu_{2M-1} & \cdots & \mu_1 & \mu_0 \end{bmatrix}. \tag{6.17}$$

Since $\mu(t)$ is real, $\mu_{-l} = \mu_l^*$. The MNA stamp of the VCVS is shown below in Figure 6.12. As usual, the **1**'s are identity matrices with appropriate order. The transpose of this stamp is that of a time-varying current-controlled current source (CCCS) with controlling- and controlled-ports flipped, as shown in the figure. Furthermore, the MNA stamp has in it μ^T. Recall that this corresponds to a gain function that is reversed in time. The CCCS, therefore, has the time-varying gain $\mu(-t)$.

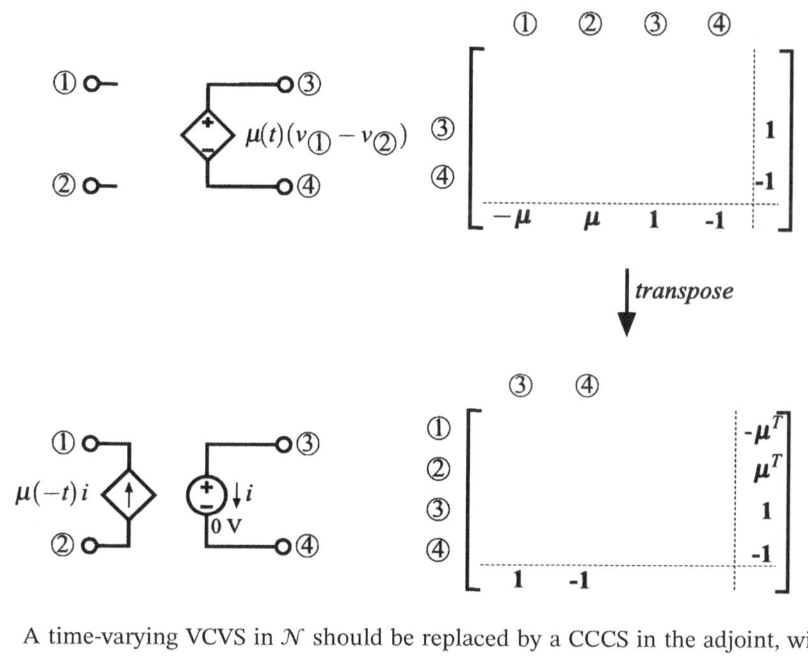

Figure 6.12 A time-varying VCVS in $\mathcal{N}$ should be replaced by a CCCS in the adjoint, with the gain being time reversed. The controlling and controlled ports are interchanged.

Reasoning in a similar manner, we see that a CCCS with gain $\mu(t)$ in $\mathcal{N}$ should be replaced by a VCVS in the adjoint with a gain $\mu(-t)$, with the controlling and controlled ports interchanged. This is shown in Figure 6.13.

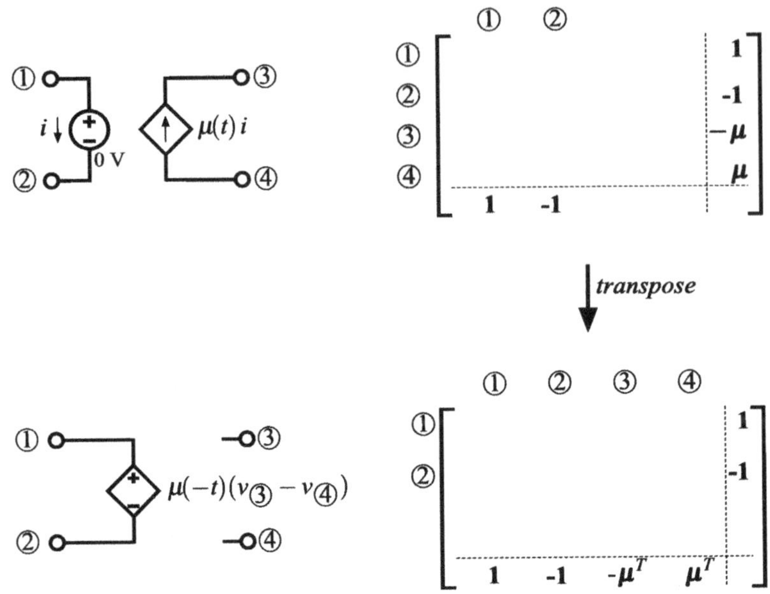

Figure 6.13 A time-varying CCCS in $\mathcal{N}$ should be replaced by a VCVS in the adjoint, with the gain being time reversed. The controlling and controlled ports are interchanged.

Figure 6.14a, b shows the element-replacement rules that should be followed with respect to a voltage-controlled current source (VCCS) and a current-controlled voltage source (CCVS) respectively. Apart from interchanging the controlled and controlling ports, the time-varying transconductance and transimpedance functions are time-reversed in the adjoint.

6.1.5 Multiple Inputs

The concepts of inter-reciprocity and the adjoint network become even more compelling and useful when one is dealing with multiple input sources, as illustrated in Figure 6.15a. The LPTV network $\mathcal{N}$ is excited by p independent voltage sources, and q independent current sources. Each source is a mixture of tones, with every tone's frequency being of the form $(f + lf_s)$ where l is an integer $(|l| \leq M)$. Suppose that we need to determine the component of the output voltage v_o at the frequency f. To accomplish this, we would have to find *all* the harmonic transfer functions from *every* source to the output, needing us to solve the network $(p + q)(2M + 1)$ times.

By exploiting the concept of inter-reciprocity and the frequency-reversal theorem, all the harmonic transfer functions corresponding to all the sources can be obtained in one shot as shown in Figure 6.15b. The "output port" of the adjoint network $\hat{\mathcal{N}}$ is

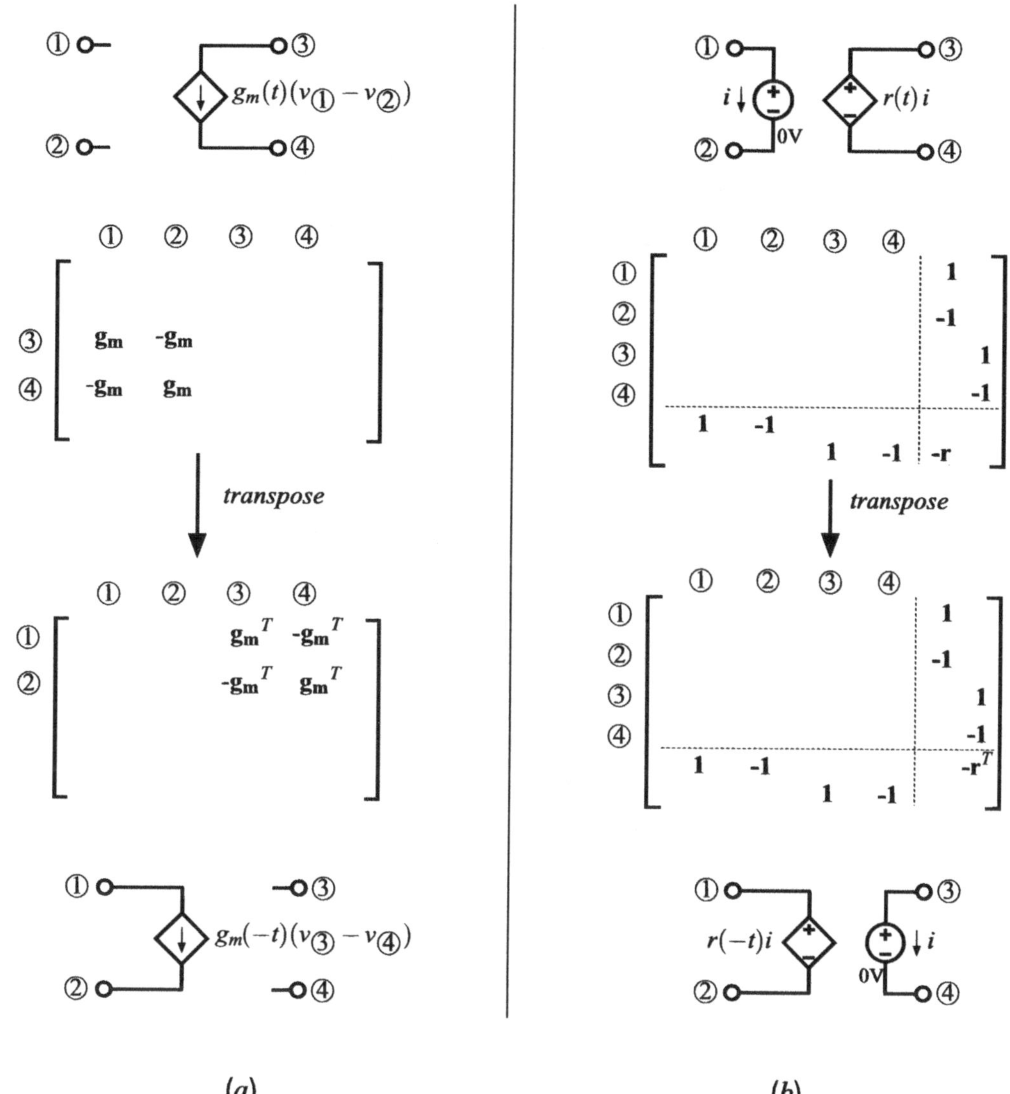

Figure 6.14 Element replacement rules for a (a) time-varying VCCS and (b) time-varying CCVS.

excited by a current $e^{j2\pi ft}$. The voltages developed across the current-input ports and the currents through the voltage-input ports yield all the required harmonic transfer functions, thereby requiring us to solve the network only once.

The reader may wonder if the problem scenario described in Figure 6.15 is relevant in practice. It turns out that such a situation is encountered routinely during noise analysis.

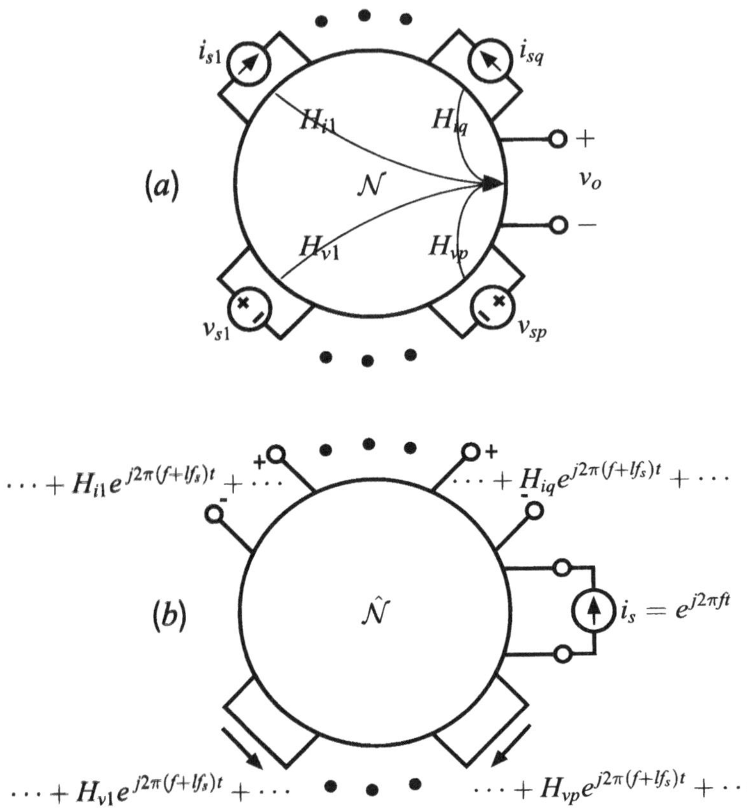

Figure 6.15 (a) LPTV network with multiple input sources and a single output. We are interested in determining the component of $v_o(t)$ at frequency f. (b) The transfer functions from all the sources and at all frequencies of the form $(f + lf_s)$ can be obtained in *one shot* by exciting the "output" port of the adjoint network $\hat{\mathcal{N}}$ with a current $e^{j2\pi ft}$.

Simulation Corner: Periodic Transfer Function (PXF) Analysis

In the previous chapter, we described the idea behind periodic AC analysis (.PAC in short) in an RF circuit simulator. As we saw there, the .PAC analysis computes the harmonic transfer functions from one AC source to any number of designated outputs in an LPTV network. The frequency range of the input source is specified, and so is the desired order of the harmonic transfer functions. The outputs are the $H_l(j2\pi f)$. In other words, the input frequency is f, while the output frequency is $(f + lf_s)$.

The periodic transfer function analysis (.PXF in short) works in the reverse manner, and is useful in the following scenario.

a. There is one output, and a desired output frequency range.
b. There are multiple independent sources.

c. We desire to find the harmonic transfer functions from the (multiple) sources to the output of the form $H_{-l}(f + lf_s)$.

In other words, we seek to answer the question: for each source, how does a component at $(f + lf_s)$ contribute to the output at f? In a . PXF analysis, therefore, the *output* frequency is fixed, while the input frequencies are of the form $(f + lf_s)$ [4].

6.1.6 Signal-Flow Graphs

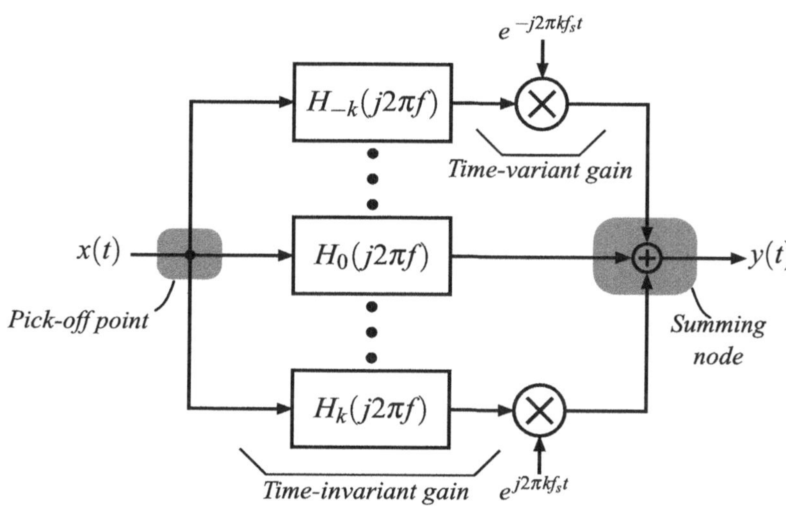

Figure 6.16 The signal-flow graph of an LPTV system (i.e., the Zadeh expansion).

We often encounter signal-flow graphs in our analysis of LPTV systems. A case in point is the flow graph corresponding to the Zadeh expansion of a periodically time-varying system. As seen in Figure 6.16, the graph is comprised of pick-off points, branches with a time-invariant gain (could be frequency dependent), branches with time-varying gain, and summing junctions.

The adjoint signal-flow graph can be derived by element-wise replacement of the individual branches as described below.

a. When a pick-off point with time-invariant weights is encountered, it is replaced by a summing junction. Further, the directions of all arrows are reversed as shown in Figure 6.17a.
b. In a similar vein, a summing junction with weighted inputs is replaced by a pick-off point, again with the directions of all arrows reversed, as seen in Figure 6.17b.
c. A branch with a time-varying gain $w(t)$ is replaced by a branch with direction reversed, and a gain $w(-t)$, as shown in Figure 6.17c.

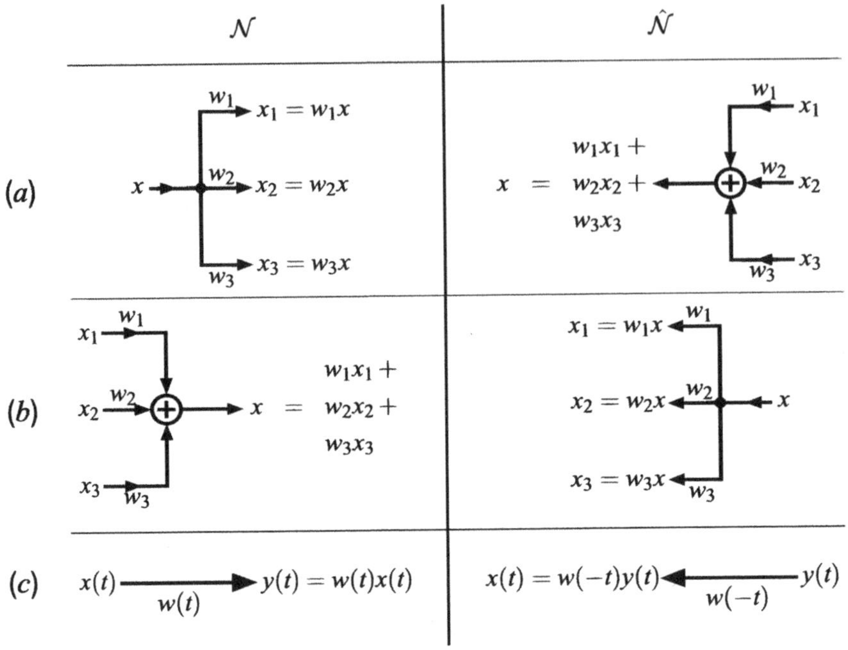

Figure 6.17 (a) A pick-off point with time-invariant weights w_1, w_2, w_3 is replaced by a summing junction and (b) vice-versa. (c) A periodically time-varying gain $w(t)$ is replaced by a periodically time-varying gain $w(-t)$, with the input and output interchanged.

The rules listed above can be derived systematically. However, we should be able to do better by using our experience with controlled sources. Assume all quantities in $\mathcal{N}$ are voltages. The pick-off point of Figure 6.17a reduces to x controlling three VCVS's with gains w_1, w_2, w_3. In the adjoint, the VCVS's will become CCCS's with controlling and controlled ports interchanged. The gains of the CCCS's, whose outputs are shorted, will be w_1, w_2, w_3. It is thus seen that interpreting all quantities in $\hat{\mathcal{N}}$ as currents, a pick-off point in $\mathcal{N}$ will be replaced by a summing junction in $\hat{\mathcal{N}}$. Other rules are similarly derived. Let us now apply what we just learned to the signal-flow graph corresponding to the Zadeh expansion of an LPTV system.

Figure 6.18a shows the Zadeh expansion of an LPTV network $\mathcal{N}$. The adjoint of this signal-flow graph is shown in part (b) of the figure. This corresponds to the Zadeh expansion of the adjoint network $\hat{\mathcal{N}}$. Notice that $y(t)$ is now the input, while $x(t)$ is the output. Further, we see that multiplication by $e^{j2\pi k f_s t}$ in the original graph is replaced with multiplication by $e^{-j2\pi k f_s t}$ in the adjoint graph. We will now confirm the frequency-reversal theorem with respect to these two graphs. If we excite $\mathcal{N}$ with $x(t) = e^{j2\pi(f+kf_s)t}$, the strength of the component of $y(t)$ at the frequency f is given by $H_{-k}(j2\pi(f + kf_s))$.

To see the frequency-reversal theorem in action, we excite the adjoint graph with $y(t) = e^{j2\pi f t}$. This gets up-converted to $(f + kf_s)$ and is processed by the LTI filter H_{-k}. The output component of $x(t)$ at frequency $(f + kf_s)$, therefore, is $H_{-k}(j2\pi(f + kf_s))e^{j2\pi(f+kf_s)t}$. This confirms what we knew already.

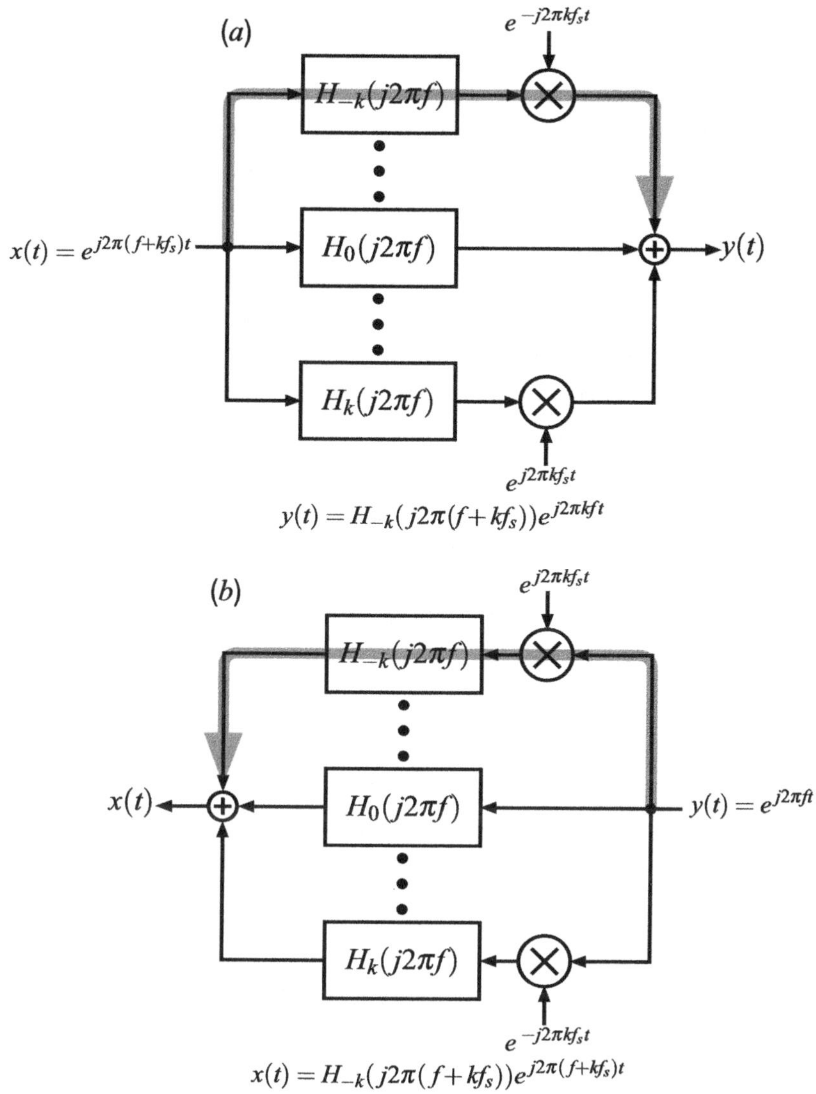

Figure 6.18 (a) Zadeh expansion for an LPTV network and (b) its adjoint.

6.2 A Chopped Amplifier as an LPTV System

Many practical applications need to process low-frequency (and/or dc) inputs. The first step in the signal processing chain is amplification. It turns out that transistors used in these amplifiers not only add broadband noise, but also exhibit another kind of noise that is very strong at low frequencies. This is called flicker noise. It has been empirically found that flicker noise has a spectrum whose density increases inversely with frequency; consequently, it is also referred to as $1/f$ noise. The strength of $1/f$ noise is tied to the manufacturing process, and it suffices to say that the exact mechanisms that govern flicker noise are not "well-understood."

A brute-force way of reducing flicker noise is to use larger device sizes. For instance, increasing the size of a device by 4× reduces $1/f$ noise spectral density by a factor of 2. Brute-force sizing is a common approach, especially when one is dealing with wide signal bandwidths; in such applications low frequencies (where $1/f$ noise is dominant) form only a small fraction of the overall signal bandwidth. Flicker noise, therefore, is a minor irritant that is easily addressed. There are, however, many important applications where the signal is confined to low frequencies. In these cases, flicker noise accounts for almost all the noise added by the transistors. Here, brute-force sizing is not an option for several reasons. Reducing $1/f$ noise by a factor of 10 will need 100× larger devices, which may be impractically large. Increasing device sizes results in larger parasitic capacitances, which in turn degrade amplifier bandwidth. Finally, there can be wide variations (more than 5×) in the strength of flicker noise even with the same manufacturing process. If one sizes devices for the worst case (which may occur in only a small fraction of manufactured ICs), most ICs are terribly oversized, which is wasteful. Circuit designers are, therefore, forced to innovate and come up with circuit solutions that address flicker noise. All these techniques exploit the fact that $1/f$ noise is "slowly varying." One such approach is chopping [5], which is the preferred technique for $1/f$ noise mitigation in continuous-time systems.

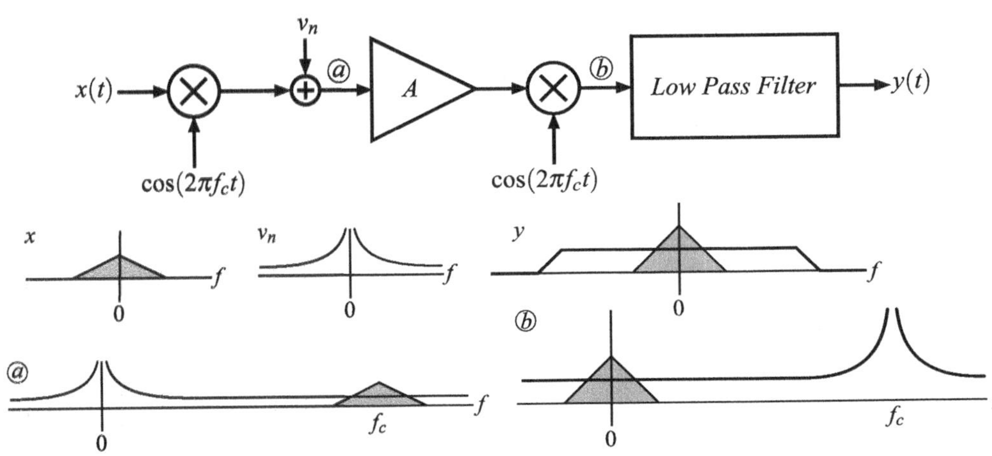

Figure 6.19 Principle behind chopping, and representative spectra at various points in the signal chain.

The basic idea behind chopping is the following. We know that flicker noise becomes very small at high frequencies. Consequently, what if we "move" the spectrum of the signal to a frequency where the noise is white, amplify the "moved" signal, and "move" the amplified spectrum back to dc? This way, the input signal's spectrum does not overlap with that of flicker noise, making the latter inconsequential.

"Moving" the spectrum is nothing but modulation; the system that incorporates the idea above is shown in Figure 6.19, which shows the signal chain and representative spectra at various points in the chain. The baseband input $x(t)$ needs to be processed by the amplifier, whose gain is A. The amplifier's input referred noise is modeled by v_n, which consists of white and $1/f$ components. Directly exciting the amplifier with $x(t)$ would produce an output with a large flicker noise component. Chopping avoids this by first modulating $x(t)$ with a sinusoid at the chopping frequency f_c. The amplifier processes the modulated signal, resulting in an output where the desired amplified input is centered around f_c. The amplifier output is then demodulated by the chopping tone, causing the desired signal to appear around dc. This operation causes v_n to get modulated to around f_c. The low-pass filter eliminates the modulated flicker noise, resulting in an amplified output $Ax(t)$ that is free from $1/f$ noise. Intuitively, we see that the low frequency components of v_n should be eliminated at y, since v_n goes through only one modulator. Clearly, the chopped amplifier is an LPTV system varying at f_c with two "inputs," namely $x(t)$ and v_n. We would be interested in determining how these signals influence the output. Since we are dealing with multiple inputs and the potential for frequency translated versions of the signals to appear at the output, it seems appropriate that we benefit from the simplicity to be had by using the adjoint network.

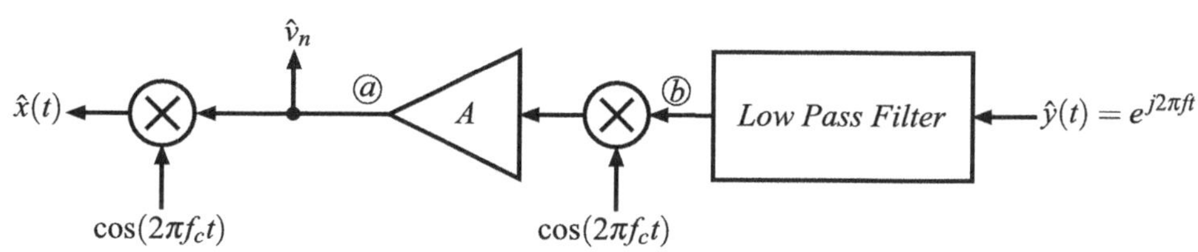

Figure 6.20 The adjoint signal-flow graph of a chopped amplifier.

Figure 6.20 shows the adjoint signal-flow graph of the chopped amplifier of Figure 6.19. The chopping tones remain unchanged, since the cosine is an even function of its argument. The adjoint should help us answer the question – what are all the inputs that can result in an output at a frequency f? In the adjoint, an input $e^{j2\pi ft}$is applied at the $\hat{y}(t)$ "port." Assuming f is within the filter's bandwidth, it goes through to ⓑ, gets modulated by the chopping tone and is then amplified. The output of the amplifier is again modulated by the chopping tone to yield the adjoint network's "output" $\hat{x}(t)$. The harmonic transfer functions from x and v_n to y in the main network are related to the transfer functions from $\hat{y}$ to $\hat{x}$ and $\hat{v}_n$ in the adjoint. It is straightforward to see that

$$\widehat{Hv}_{-1}(j2\pi f) = \widehat{Hv}_1(j2\pi f) = \frac{A}{2}$$
$$\widehat{Hx}_0(j2\pi f) = \frac{A}{2}$$
$$\widehat{Hx}_{-2}(j2\pi f) = \widehat{Hx}_2(j2\pi f) = \frac{A}{4}.$$

All other harmonic transfer functions are zero. From the frequency-reversal theorem, it therefore follows that

$$Hv_1(j2\pi(f - f_c)) = Hv_{-1}(j2\pi(f + f_c)) = \frac{A}{2}$$

$$\widehat{Hx}_0(j2\pi f) = \frac{A}{2}$$

$$Hx_2(j2\pi(f - 2f_c)) = Hx_{-2}(j2\pi(f + 2f_c)) = \frac{A}{4}.$$

All other harmonic transfer functions are zero. As expected, we see that $Hv_0 = 0$, which means that the low-frequency components of v_n (i.e., the amplifier's flicker noise) do not appear at the output y. We also see that the gain from x to y is $A/2$; this is not entirely unexpected, since modulation by a cosine (whose maximum value is 1) should result in a gain loss. Interestingly, $Hx_{-2}(j2\pi(f + 2f_c)) = A/4$, indicating that components in $x(t)$ around the frequency $2f_c$ are down-converted and appear at y with half the gain as that experienced by a low-frequency input! If we know a priori that $x(t)$ is "clean" around $2f_c$, this might not pose a serious problem. There are, however, important applications[2] where this is not true. In such cases, the down-conversion of out-of-band inputs into the signal band is worrisome, and has to be addressed.

The simplistic view of chopping given in Figure 6.19 is a useful aid to understanding. Practical chopped amplifiers deviate significantly from the model of Figure 6.19. These deviations and their ramifications are listed below.

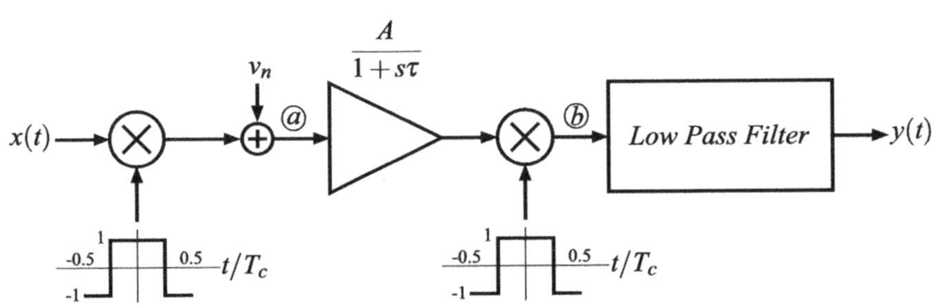

Figure 6.21 A practical chopped-amplifier signal chain.

a. *Square-wave Modulation:* In practice, (de)modulating the input with sinusoidal signals is problematic. Square-wave mixing is the norm, and is particularly easy to implement in fully differential CMOS circuits using switches. Figure 6.21 shows the resulting signal chain with square-wave chopping. Consequently, the input is translated to around all odd multiples of f_c, processed by the amplifier, and then demodulated. Interestingly, if the amplifier's bandwidth is infinite, it is easy to see

[2] A continuous-time $\Delta\Sigma$ modulator employing a chopped input integrator is a case in point.

that $x(t)$ is gained up by a factor A, and multiplied by the square of the chopping waveform (which is 1). Unlike with a sinusoidal chopping tone, there is no down-conversion of out-of-band components into the signal band. The amplifier's offset and flicker noise v_n, which is multiplied by a square wave and subsequently filtered, results in zero offset and flicker noise at y, as in the case with sinusoidal chopping.

b. *Finite Amplifier Bandwidth:* A practical amplifier will have finite bandwidth. While this does not affect the elimination of the dc offset and flicker noise v_n at y, it results in several non-idealities that we will analyze using the adjoint network. In the discussion to follow, the amplifier is modeled as a single-pole system according to

$$A(s) = \frac{A_o}{1 + s\tau}, \tag{6.18}$$

so that its 3 dB bandwidth is $1/(2\pi\tau)$ Hz.

c. *Chopping Embedded in a Feedback Loop:* The amplifier in Figure 6.19 operates in an open-loop fashion. Many practical applications embed the amplifier in a feedback loop. As a result, the signal chain does not lend itself to be modeled in the simple manner of Figure 6.19.

In the discussion that follows, we consider only (a) and (b) above. The combination of a square-wave chopping tone and finite amplifier bandwidth makes the analysis of the signal-chain of Figure 6.21 considerably more involved than that of Figure 6.19. Adjoint-network techniques make our life easy, as described below. Figure 6.22a shows the adjoint signal-flow graph. The direction of signal flow is reversed, and the summing node at the amplifier input is replaced with a pick-off point. The chopping waveform is time-reversed.

As before, we would like to answer the question: what are all the inputs that would result in a low-frequency output at y in Figure 6.21? We are, therefore, interested in determining $H_{-l}(j2\pi l f_c)$. To obtain these transfer functions, $\hat{y}(t)$ in the adjoint signal-flow graph is excited with dc, as shown in Figure 6.22a. The input to the amplifier is a square wave with period T_c. In steady state, the amplifier output (at ⓐ) is an exponential wave, with amplitude A. From the theory of adjoint networks, the Fourier coefficients of this periodic wave must be the harmonic transfer functions from v_n to the output of the original system, evaluated at $j2\pi l f_c$. Since the dc value of the waveform at ⓐ is zero, it follows that dc offset does not appear at the output of the chopped amplifier of Figure 6.21, even when finite bandwidth effects are considered. Moving on, the waveform at $\hat{x}(t)$ is as shown in Figure 6.22b. The Fourier expansion of this waveform not only yields the H_l we are after, but also places in evidence other interesting aspects.

a. The waveform at $\hat{x}(t)$ has a period $T_c/2$. Thus, expanding it in a Fourier series with a fundamental frequency of $f_c = 1/T_c$ will result in all odd coefficients

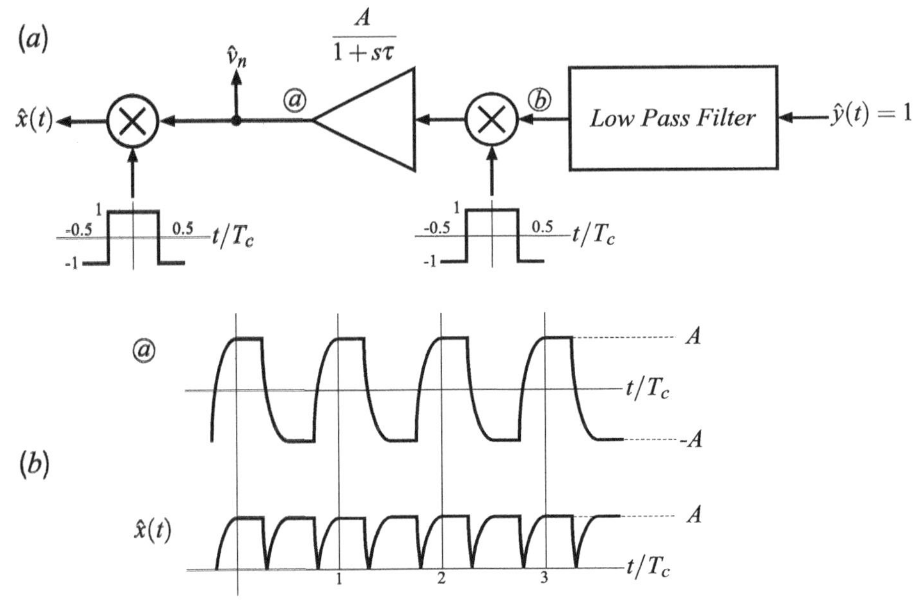

Figure 6.22 (a) The adjoint signal-flow graph and (b) representative waveforms.

being zero. This means that in the signal chain of Figure 6.21, only input components from multiples of *twice* the chopping frequency will appear in the output at dc. Thus, though the system is LPTV with frequency f_c, $H_{-l}(j2\pi l f_c) = 0$ for all odd l.

b. The dc component of $\hat{x}(t)$, which is the dc gain of the chopped amplifier, is given by $A(1 - (4\tau/T_c))$.

c. Observing $\hat{x}(t)$ in Figure 6.22b, we see that it can be written as

$$\hat{x}(t) = A - \sum_k q\left(t - k\frac{T_c}{2}\right), \tag{6.19}$$

where $q(t) = 2A\exp(-t/\tau)$. The Fourier coefficients, obtained by sampling the Fourier transform of $q(t)$, are given by

$$H_{-2l} = -\frac{\tau}{T_c}\frac{4A}{1 - j4\pi l f_c \tau}, \qquad l = 1, 2, \cdots \tag{6.20}$$

Results from periodic transfer function (PXF) simulations of the chopped signal chain of Figure 6.21, for $A = 1$ and $\tau/T_c = 0.05$, confirm the analysis above. Figure 6.23 shows the magnitudes of $H_{-l}(j2\pi l f_c)$ for $l = 0, \cdots, 7$. The calculated magnitudes, from (6.20), fall right on the simulated values. As expected, the conversion gains from odd multiples of the chopping frequency to the output are zero. From (6.20), we see that the aliasing gains from even multiples of f_c are proportional to τ/T_c.

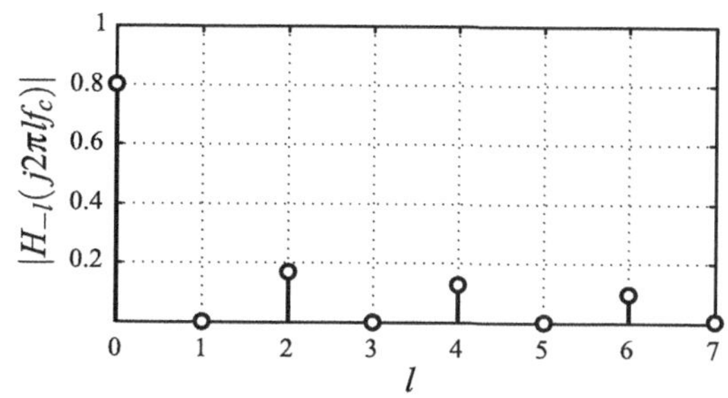

Figure 6.23 Simulated and calculated magnitudes of $H_{-l}(j2\pi l f_c)$ for $l = 0, \cdots, 7$ for the system of Figure 6.21 for $A = 1$ and $\tau/T_c = 0.05$.

6.3 Time Domain Implications of Reciprocity and Inter-Reciprocity

So far, we have seen the transfer function and frequency reversal theorems pertaining to reciprocal and inter-reciprocal networks. Surely, there must be some time-domain relationships that must be equivalent to the frequency-domain truths we arrived at previously. This section is dedicated to examining time-domain aspects. Recall that the impulse response of a general LTV system is denoted by $h(t, \tau)$. The system is excited at $(t - \tau)$, while the output is observed at t. If the system, in addition to being time-varying, is also periodically so, the impulse response satisfies

$$h(t, \tau) = h(t + T_s, \tau) = \sum_k h_k(\tau)e^{j2\pi k f_s t}. \tag{6.21}$$

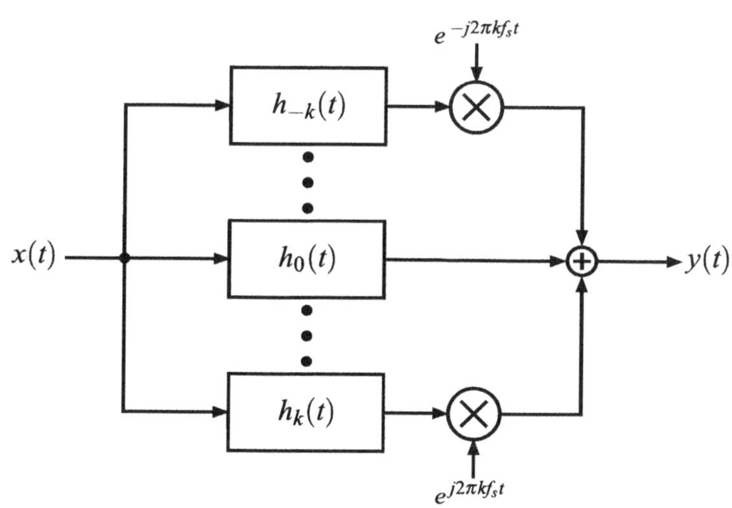

Figure 6.24 Time-domain restatement of the Zadeh expansion.

The response $y(t)$ of an LPTV system to an input $x(t)$ is thus given by the convolution integral

$$y(t) = \int_0^{\infty} h(t,\tau)x(t-\tau)\,d\tau = \int_0^{\infty} \sum_k h_k(\tau)e^{j2\pi k f_s t}x(t-\tau)\,d\tau. \tag{6.22}$$

Since the integral is with respect to τ, the $e^{j2\pi k f_s t}$ can be brought out of the integral. Further, the order of the summation and integration can be interchanged. We then have

$$y(t) = \sum_k \underbrace{\left[\int_0^{\infty} h_k(\tau)x(t-\tau)\,d\tau\right]}_{x(t)*h_k(t)} e^{j2\pi k f_s t}. \tag{6.23}$$

This can be interpreted as follows. The output $y(t)$ of the LPTV system is comprised of contributions from many paths. The output of each path is derived by exciting an LTI system whose impulse response is $h_k(t)$ with the input $x(t)$, and modulating the output of the LTI system by $e^{j2\pi k f_s t}$, as shown in Figure 6.24. The reader will immediately recognize that this is simply the time-domain restatement of the Zadeh expansion, with $h_k(t)$ and $H_k(j2\pi f)$ forming a Fourier-transform pair.

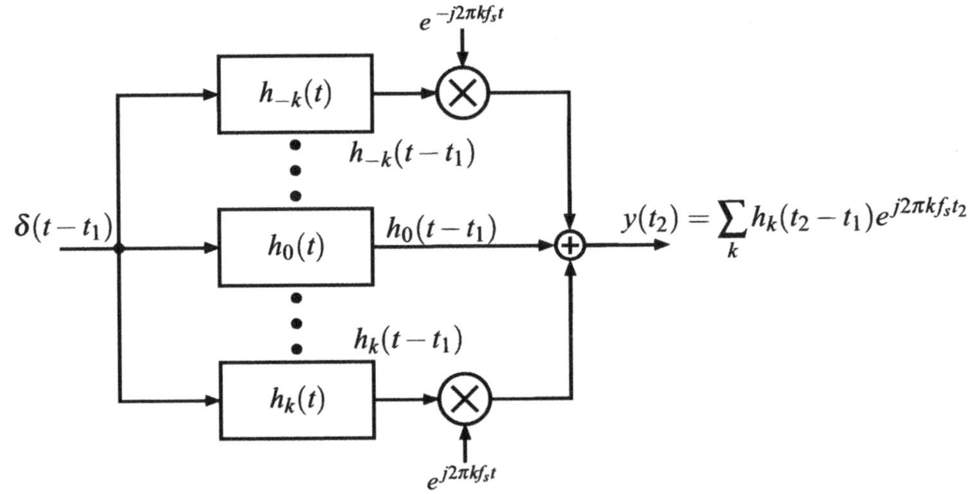

Figure 6.25 Response of an LPTV system measured at t_2, when excited with an impulse at t_1.

Let us now excite the LPTV system with an impulse at t_1 and observe its output at t_2, where $t_2 > t_1$. Referring to Figure 6.25, $x(t) = \delta(t - t_1)$ and the output of the LTI system in the kth path is $h_k(t - t_1)$. The system output at $t = t_2$, namely $y(t_2)$, is thus given by

$$y(t_2) = \sum_k h_k(t_2 - t_1)e^{j2\pi k f_s t_2}. \tag{6.24}$$

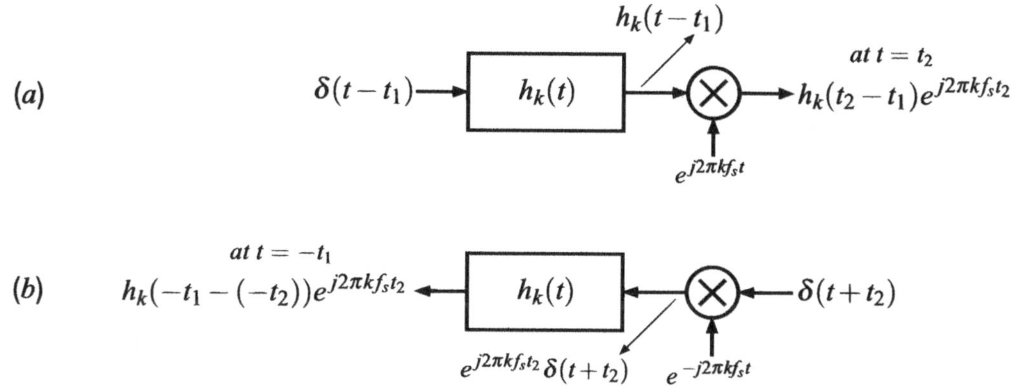

Figure 6.26 Response of the kth arm of the LPTV system measured at t_2, when excited with an impulse at t_1.

Consider the kth path in Figure 6.25, as shown in Figure 6.26a. Denoting the output of this path by $y_k(t)$, we see that

$$y_k(t_2) = h_k\underbrace{(t_2}_{\text{time of observation}} - \underbrace{t_1)}_{\text{time of excitation}} \underbrace{e^{j2\pi k f_s t_2}}_{\text{output scale factor}} \tag{6.25}$$

The right-hand side of the equation above can be *interpreted* in the following manner.

$$y_k(t_2) = h_k(\underbrace{(-t_1)}_{\text{time of observation}} - \underbrace{(-t_2))}_{\text{time of excitation}} \underbrace{e^{-j2\pi k f_s(-t_2)}}_{\text{input scale factor}} = \hat{y}_k(-t_1). \tag{6.26}$$

Suppose there is an LTI system with impulse response $h_k(t)$, whose *input* is multiplied by $e^{-j2\pi k f_s t}$, as shown in Figure 6.26b. If the input to this system is an impulse applied at $t = -t_2$, its output at $-t_1$, denoted by $\hat{y}_k(-t_1)$ is given by $h_k((-t_1) - (-t_2))e^{-j2\pi f_s(-t_2)}$, which is identical to $y_k(t_2) = h_k(t_2 - t_1)e^{j2\pi f_s t_2}$. We thus see that exciting the kth path with an impulse at t_1, and measuring its output $y_k(t)$ at t_2 yields the same response as that obtained by exciting the system of Figure 6.26b with an impulse at $(-t_2)$ and measuring its response $\hat{y}_k(t)$ at $-t_1$. It thus follows that the same relationship must hold for the outputs of the LPTV system of Figure 6.27a,b.

The system of Figure 6.27b can be immediately recognized as being the adjoint of that of Figure 6.27a. We thus have the following [1, 6]:

> The response $y(t_2)$ at port 2 of an LPTV network $\mathcal{N}$ excited at port 1 by an impulse at t_1 is the same as
> the response $\hat{y}(-t_1)$ at port 1 of the adjoint network $\hat{\mathcal{N}}$ excited at port 2 with an impulse at $-t_2$.

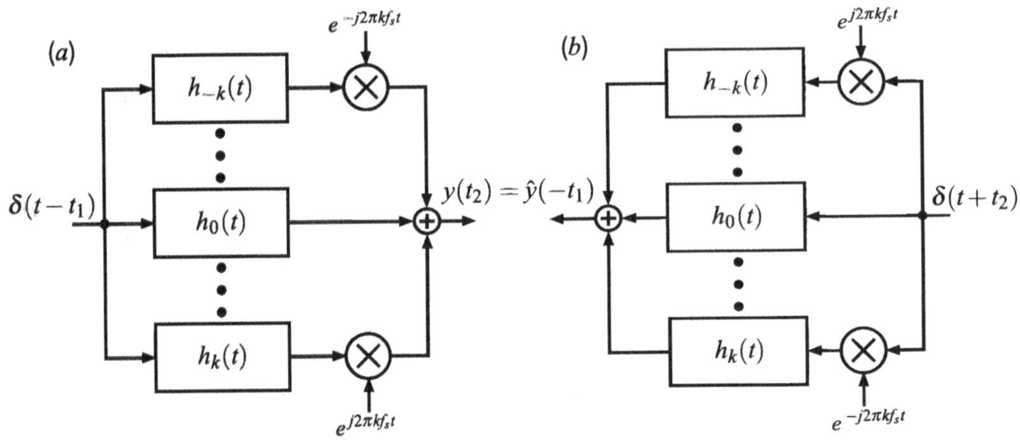

Figure 6.27 (a) Response of an LPTV system measured at t_2, when excited with an impulse at t_1 is identical to (b) the response of the adjoint excited at $-t_2$ and measured at $-t_1$.

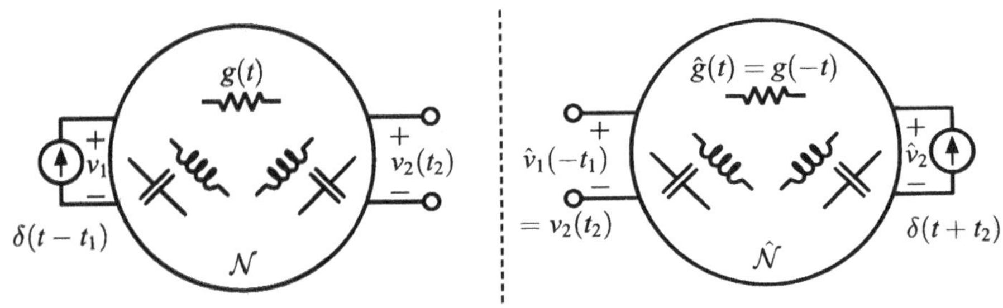

Figure 6.28 The voltage across port 2 of $\mathcal{N}$ at time t_2 when port 1 is excited by a current impulse at t_1 is the same as the voltage developed across port 1 of the adjoint network at time $-t_1$ when its port 2 is excited by a current impulse at $-t_2$.

An example excitation-response pair is shown in Figure 6.28. We have already learned how to form the adjoint network—all time-varying conductances are time-reversed, and controlled sources are replaced appropriately as discussed in the previous section. Some readers might be alarmed (looking at negative times in our discussion) that causality might be violated in the adjoint network. This is definitely not so: since $t_2 > t_1$, it follows that $-t_1 > -t_2$. We can therefore rest in peace, knowing that the response of $\hat{\mathcal{N}}$ occurs only after it is excited.

Another straightforward observation, which is a consequence of the periodically time-varying nature of $\mathcal{N}$, is the following. The response $y(mT_s + t_2)$ at port 2 of an LPTV network $\mathcal{N}$ excited at port 1 by an impulse at $mT_s + t_1$ is the same as the response $\hat{y}(-t_1)$ at port 1 of the adjoint network $\hat{\mathcal{N}}$ excited at port 2 with an impulse at $-t_2$.

We illustrate with an example. Consider again our friend, the periodically-switched RC network, excited with a voltage impulse at $t = t_1$, as shown in Figure 6.29a. The switch control sequence has a period T_s and a duty cycle of 50%. The output voltage

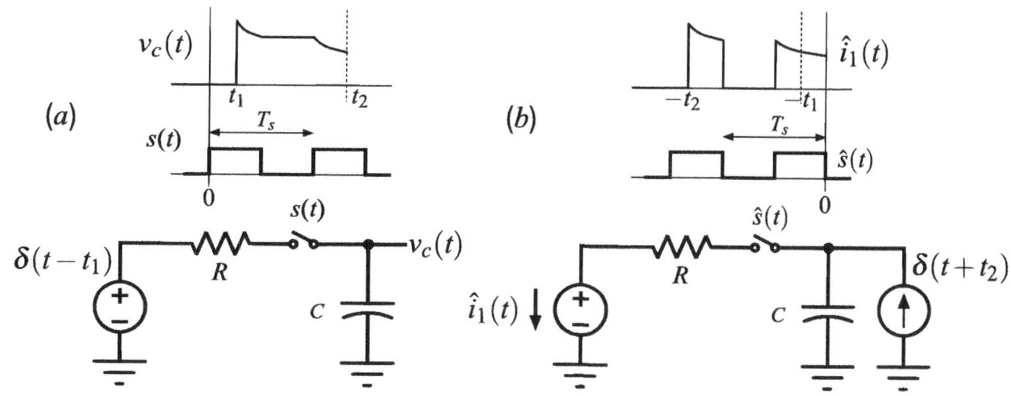

Figure 6.29 The voltage across port 2 of $\mathcal{N}$ at time t_2 when port 1 is excited by a voltage impulse at t_1 is the same as the current through port 1 of the adjoint network at time $-t_1$, when its port 2 is excited by a current impulse at $-t_2$.

$v_c(t_1)$ is $1/RC$, and initially decays exponentially with a time-constant RC. It is easy to see that

$$v_c(0.5\,T_s) = v_c(T_s) = \frac{1}{RC}e^{-\frac{(0.5\,T_s-t_1)}{RC}}, \quad v_c(t_2) = \frac{1}{RC}e^{-\frac{(0.5\,T_s-t_1)}{RC}}e^{-\frac{(t_2-T_s)}{RC}}. \tag{6.27}$$

The response at t_2 due to a voltage impulse excitation at t_1, therefore, is given by

$$\frac{1}{RC}e^{-(0.5\,T_s-t_1)/RC}e^{-(t_2-T_s)/RC}.$$

The adjoint network is shown in Figure 6.29b. The switch-control signal is time reversed, and a current impulse is applied across the capacitor at $t = -t_2$. The current $\hat{i}_1(t)$ through the 0 V input source is the output under observation. $\hat{i}_1(-t_2) = 1/RC$, and initially decays exponentially with a time-constant RC. At $t = -T_s$, when the switch is opened, the current drops to zero. Again, it is straightforward to see that

$$\hat{i}_1(-T_s) = \hat{i}_1(-0.5\,T_s) = \frac{1}{RC}e^{-\frac{(t_2-T_s)}{RC}}, \quad \hat{i}_1(-t_1) = \frac{1}{RC}e^{-\frac{(t_2-T_s)}{RC}}e^{-\frac{(0.5\,T_s-t_1)}{RC}}. \tag{6.28}$$

As expected from the theory, $\hat{i}_1(-t_1)/1\,\text{A} = v_c(t_2)/1\,\text{V}$.

The question that arises is the following. Why is the result discussed in this section useful, or worth knowing? It turns out to be of immense value when dealing with the properties of sampled LPTV systems, which we consider next.

6.4 LPTV Systems with Sampled Outputs

Many practically-important LPTV systems have outputs that are sampled at the same rate as that at which the system is varying. Some examples are shown in Figure 6.30. Part (a) of the figure shows a switched-RC circuit. The switch is controlled by a clock with period T_s and duty cycle d. Depending on the ratio of RC to T_s and the

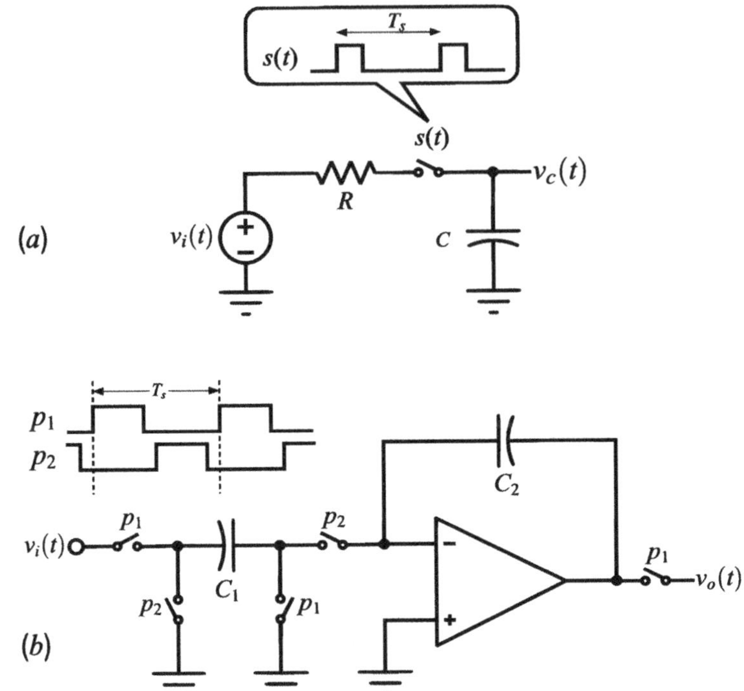

Figure 6.30 (a) A switched-RC network. The output of interest is v_o sampled at the falling edges of $s(t)$. (b) A switched-capacitor integrator. The output of interest is v_o sampled at the falling edges of p_1.

duty cycle d, the network can be used as a track-and-hold or a narrow-band filter. It also forms the basic building block of an N-path filter. It turns out that $v_c(lT_s)$, which is the capacitor voltage sampled at integer multiples of T_s, plays a crucial role in the operation of this circuit.

Figure 6.30b shows a switched-capacitor integrator, where the switches are operated periodically as per the non-overlapping clocks p_1 and p_2. The output of interest is v_o sampled at the falling edges of p_2. To summarize, many practically important LPTV circuits have a continuous-time input and a sampled output. Furthermore, the sampling rate of the output is the same as that at which the circuit is varying.

Consider an LPTV system varying with period T_s, as shown in Figure 6.31a. It is excited by a complex exponential $x(t) = e^{j2\pi ft}$. The output $y(t)$ is sampled at the rate $f_s = 1/T_s$ as shown in the figure to yield a discrete-time sequence $y[n] = y(nT_s)$. Mathematically, sampling implies that $y(t)$ is multiplied by an impulse train to yield a sequence of continuous-time impulses. The c/d converter conceptually converts this sequence of continuous-time impulses into a discrete-time sequence. In system diagrams, it is common to denote the entire contraption in the gray ellipse by a sampling switch as shown in Figure 6.31b.

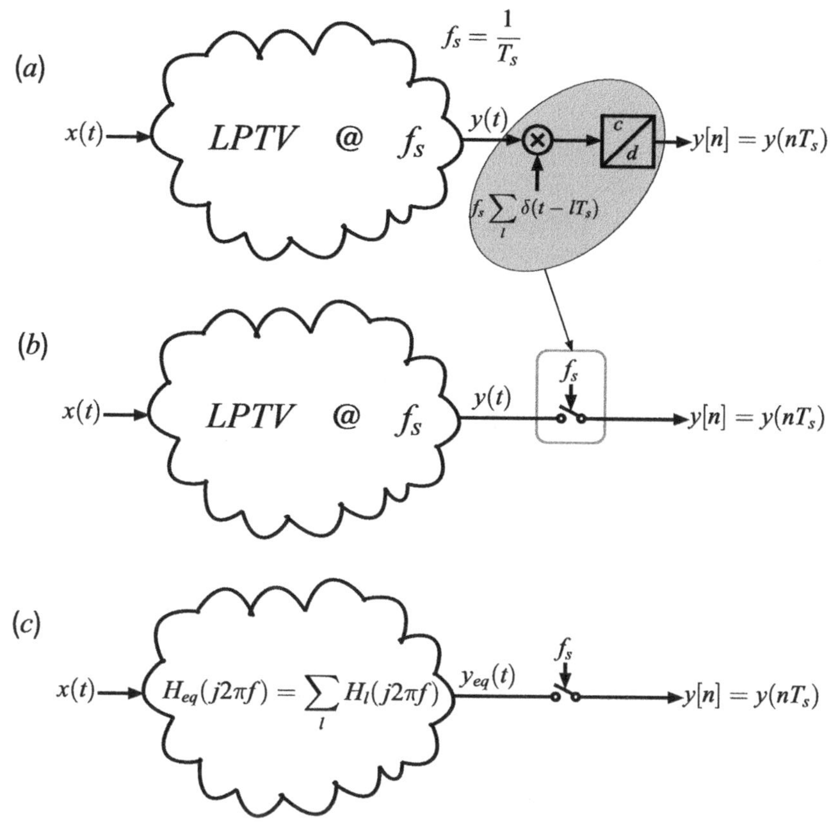

Figure 6.31 (a) An LPTV system (varying at a rate f_s) whose output is sampled at f_s. (b) Equivalent representation of the sampler and continuous-to-discrete time converter. (c) An LTI system whose output sampled at f_s yields the *same* sequence as the system of (a).

Since the system is LPTV, we have

$$y(t) = \sum_{l=-\infty}^{\infty} H_l(j2\pi f)e^{j2\pi(f+lf_s)t}. \tag{6.29}$$

The samples of $y(t)$ are given by

$$\begin{aligned} y(nT_s) &= \sum_{l=-\infty}^{\infty} H_l(j2\pi f)e^{j2\pi(f+lf_s)nT_s} \\ &= e^{j2\pi f nT_s} \sum_{l=-\infty}^{\infty} H_l(j2\pi f). \end{aligned} \tag{6.30}$$

Consider now the system of Figure 6.31c. It is a linear *time-invariant* system, chosen so that its frequency response $H_{\text{eq}}(j2\pi f)$ is *chosen* to be $\sum_l H_l(j2\pi f)$, where the

$H_l(j2\pi f)$ are the harmonic transfer functions of the LPTV system of Figure 6.31a. If this LTI system is excited by $e^{j2\pi ft}$, its output is

$$y_{eq}(t) = e^{j2\pi ft} \sum_{l=-\infty}^{\infty} H_l(j2\pi f). \tag{6.31}$$

When sampled at a rate $f_s = 1/T_s$, we obtain

$$y_{eq}(nT_s) = e^{j2\pi nfT_s} \sum_{l=-\infty}^{\infty} H_l(j2\pi f). \tag{6.32}$$

From Eqs.(6.30) and (6.32), we see that as far as *output samples* are concerned, an LPTV system whose output is sampled at f_s is equivalent to an LTI system whose output sampled at f_s. The equivalent LTI filter has a frequency response [6]

$$H_{eq}(j2\pi f) = \sum_{l=-\infty}^{\infty} H_l(j2\pi f). \tag{6.33}$$

Since an arbitrary input $x(t)$ can be represented as a sum of sinusoids, it follows that the output samples of an LPTV system (when the sampling rate is the same as that at which the system is varying) can be thought of as being obtained by exciting an LTI filter by $x(t)$ and sampling its output at a rate f_s, as shown in Figure 6.31c.

The result derived above makes intuitive sense due to the following. When an LPTV system is excited by a tone at f, the output comprises tones at frequencies $f + lf_s$, where l is an integer. When sampled at f_s, frequency components higher than or equal to f_s are aliased to f. Thus, if one is only interested in the samples of the system's output, they could as well be produced by a properly chosen LTI filter acting on an input tone at a frequency f. Note that the equivalence holds only for samples, and not for the waveforms. Referring to Figure 6.31b,c, we note that $y(nT_s) = y_{eq}(nT_s)$, but $y(t)$ *need* not equal $y_{eq}(t)$.

How do we determine the transfer function of the equivalent LTI system $H_{eq}(j2\pi f)$, given an LPTV system? One way of doing this is to determine the $H_l(j2\pi f)$ of the system, and then use Eq. (6.33). This, however, is inefficient, since we first need to work hard to determine the individual $H_l(j2\pi f)$ for all l, and then throw much of that information away as we are only interested in the sum of the harmonic transfer functions. Fortunately, however, it is easier to proceed in the time domain, as discussed below.

The principle is illustrated using Figure 6.32a. Without loss of generality, we consider the LPTV network $\mathcal{N}$ with a current input and voltage output, namely v_o. The network varies periodically with a frequency f_s, and v_o is sampled at the same rate. As discussed earlier, $v_o(nT_s)$ can be obtained by exciting an LTI system with the same input as $\mathcal{N}$, and sampling its output at $t = nT_s$. We would like to determine the impulse response $h_{eq}(t)$ of the equivalent LTI system. In our example, we note that $h_{eq}(t)$ relates an input current to an output voltage.

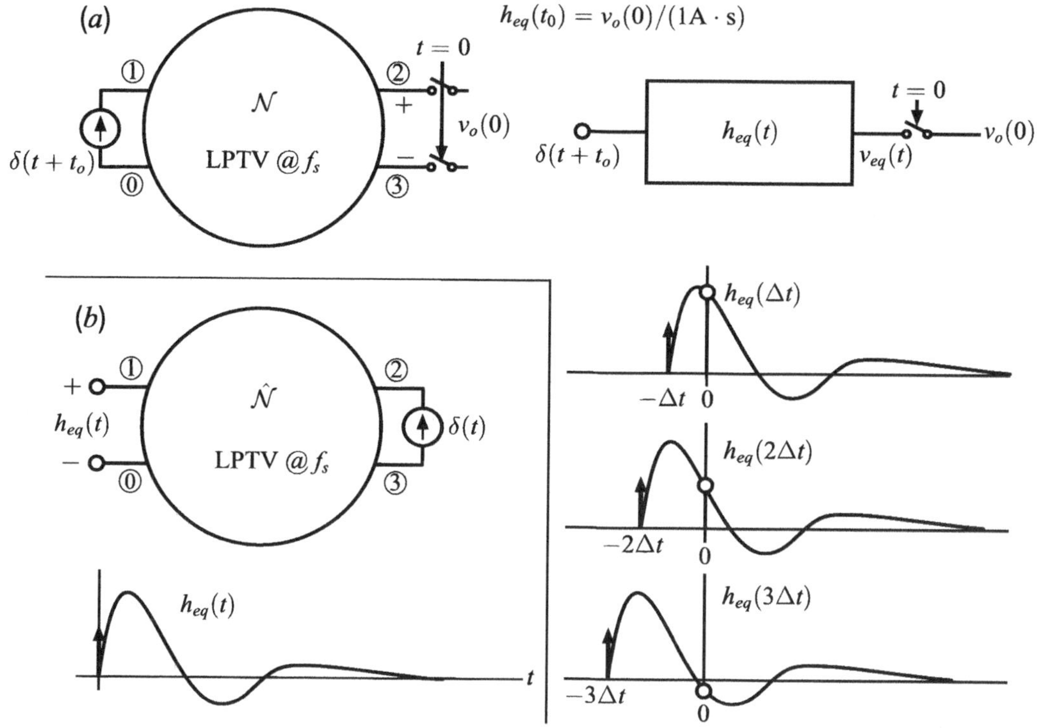

Figure 6.32 (a) An LPTV network excited with an impulse at $-t_o$ and output sampled at $t = 0$. The same output voltage sample is obtained by exciting an equivalent LTI system with an impulse current at $t = -t_o$. (b) Obtaining $v_{\text{eq}}(t)$ in one shot using the adjoint network.

We observe the following:

a. We excite $\mathcal{N}$, as well as the equivalent LTI system with a current impulse at $t = -t_o$.

b. The output of the LTI system is given by

$$v_{\text{eq}}(t) = \delta(t + t_o) \times h_{\text{eq}}(t) = h_{\text{eq}}(t + t_o) 1\text{A} \cdot \text{s} \tag{6.34}$$

We see that $v_{\text{eq}}(t)$ has the same shape as $h_{\text{eq}}(t + t_o)$ except for the units. $h_{\text{eq}}(t + t_o)$, therefore, can be obtained by dividing $v_{\text{eq}}(t)$ by $1\text{A} \cdot \text{s}$.

c. From (b) above, using $t = 0$ we have

$$v_{\text{eq}}(0) = h_{\text{eq}}(t_o) 1\text{A} \cdot \text{s}. \tag{6.35}$$

By sampling the output of the LPTV system at $t = 0$, we obtain $v_o(0)$, which by our theory is identical to $v_{\text{eq}}(0)$. $h_{eq}(t_o)$ is obtained from

$$h_{\text{eq}}(t_o) = v_o(0)/1\text{A} \cdot \text{s}. \tag{6.36}$$

We note that the units of $h_{\text{eq}}(t)$ are Ω/s, which (as expected) means that $H_{\text{eq}}(j2\pi f)$ has units of Ω.

The observations above suggest the following approach to determine $h_{eq}(t)$, as illustrated in Figure 6.32a.

1. Excite $\mathcal{N}$ with a current impulse at $t = 0$, and measure the output voltage $v_o(t)$ of $\mathcal{N}$ at $t = 0$. Calculate $h_{eq}(0) = v_o(0)/(1\text{A} \cdot \text{s})$.
2. Excite $\mathcal{N}$ with a current impulse at $t = -\Delta t$, where $\Delta t \ll T_s$, and measure the output voltage $v_o(t)$ of $\mathcal{N}$ at $t = 0$. Calculate $h_{eq}(\Delta t) = v_o(0)/(1\text{A} \cdot \text{s})$.
3. Excite $\mathcal{N}$ with a current impulse at $t = -2\Delta t$, where $\Delta t \ll T_s$, and measure the output voltage $v_o(t)$ of $\mathcal{N}$ at $t = 0$. Calculate $h_{eq}(2\Delta t) = v_o(0)/(1\text{A} \cdot \text{s})$.
4. $\cdots$

Determining the impulse response of the equivalent LTI filter by applying successively advanced impulses, as shown in Figure 6.32a, is useful but tedious. Fortunately, $v_{eq}(t)$ can be found in one shot by exploiting the time-domain properties of the adjoint network as discussed below.

The basic idea is shown in Figure 6.32b. The adjoint network $\hat{\mathcal{N}}$ is derived from $\mathcal{N}$ using the element-replacement rules discussed earlier. Recall from Section 6.3 that the response of $\mathcal{N}$ at $t = t_2$ when excited by an impulse at $t = t_1$ is identical to the response that would be obtained at the input port of $\hat{\mathcal{N}}$ at $t = -t_1$ when its output port is excited by an impulse at time $t = -t_2$. Applying this to $\mathcal{N}$ of Figure 6.32a, we observe the following.

a. The output of $\mathcal{N}$ at $t = 0$ while applying a current impulse at $t = 0$ is $v_{eq}(0)$. Therefore, exciting the output port of $\hat{\mathcal{N}}$ at $t = 0$ should result in a voltage $v_{eq}(0)$ at the input port, from which we obtain $h_{eq}(0)$.
b. The output of $\mathcal{N}$ at $t = 0$ while applying a current impulse at $t = -\Delta t$ is $v_{eq}(\Delta t)$. Therefore, exciting the output port of $\hat{\mathcal{N}}$ at $t = 0$ should result in a voltage $v_{eq}(\Delta t)$ at the input port, from which we obtain $h_{eq}(\Delta t)$.
c. The output of $\mathcal{N}$ at $t = 0$ while applying a current impulse at $t = -2\Delta t$ is $v_{eq}(2\Delta t)$. Therefore, exciting the output port of $\hat{\mathcal{N}}$ at $t = 0$ should result in a voltage $v_{eq}(2\Delta t)$ at the input port, from which we obtain $h_{eq}(2\Delta t)$.
d. In general, applying a current impulse at the "output port" of the adjoint network at $t = 0$ yields $v_{eq}(t)$ at its input port as shown in Figure 6.32b. Consequently, $h_{eq}(t)$ is simply the voltage waveform developed at the input port of $\hat{\mathcal{N}}$, divided by $1\text{ A} \cdot \text{s}$.

Thanks to the adjoint network, therefore, we are able to obtain the impulse response of the equivalent time-invariant filter in one shot. We illustrate with several practical examples.

6.4.1 The Switched-RC Kernel

The switched-RC circuit of Figure 6.33a is an LPTV network varying at a frequency f_s. The switch is controlled by a clock with period T_s and duty cycle d. The focus of this example is to relate $v_o(lT_s)$ to $v_i(t)$. From the discussion above, $v_o(lT_s)$ can be thought

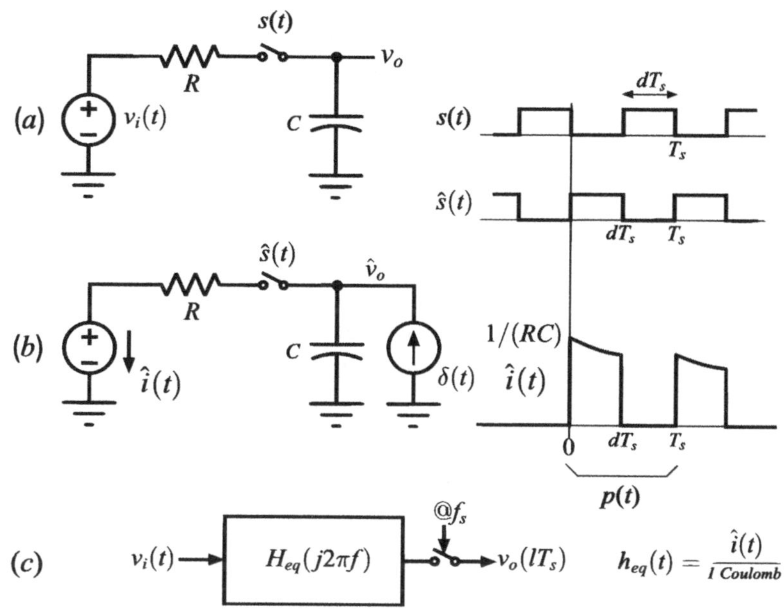

Figure 6.33 (a) The switched-RC kernel. The output of interest is the sampled capacitor voltage at multiples of kT_S. (b) The adjoint network: the current in the input port $\hat{\imath}(t)$ is needed to determine the impulse response of the equivalent time-invariant filter. (c) $h_{\text{eq}}(t) = \hat{\imath}(t)/1$ coulomb.

of as being the sampled output of a linear time-invariant filter driven by $v_i(t)$. The impulse response of this filter is denoted by $h_{\text{eq}}(t)$.

To determine $h_{\text{eq}}(t)$, we first draw the adjoint network. The result is shown in Figure 6.33b. Note that the switch-control signal in the original network is reversed in time. The determination of $h_{\text{eq}}(t)$ proceeds as follows [7, 8]. The excitation and response in the original network are both voltages. Consequently, those in the adjoint network must be currents. The "output" port of the adjoint is excited by an impulse current at $t = 0$, and the current waveform through the input port is recorded. This waveform, divided by 1 A·s, yields $h_{\text{eq}}(t)$.

Referring to Figure 6.33b, the current impulse causes the capacitor voltage to instantly increase to[3] $\hat{v}_o(0+) = 1/C$. In the discussion that follows, we denote $\tau/T_s \equiv d$ and $\exp(-\tau/RC) \equiv \beta$. For $0 < t < \tau$, the capacitor discharges through the resistor with a time-constant RC. $\hat{\imath}(t)$ during this interval, therefore, is $(1/RC)\exp(-t/RC)$. Just before $t = \tau$, the capacitor voltage is $\hat{v}_o(\tau-) = (1/C)\exp(-\tau/RC) \equiv \beta\hat{v}_o(0+)$. For $\tau \le t < T_s$, the switch is open, and $\hat{\imath} = 0$. The voltage across the capacitor does not change during this period.

We denote $\hat{\imath}(t)$ for $0 \le t < T_s$ by $p(t)$. At $t = T_s+$, the switch is closed, and the capacitor begins to discharge again, but now from a value $\beta\hat{v}_o(0+)$. Note that $\hat{\imath}$ is the response to a voltage $\hat{v}_o(0+)$ on the capacitor at $t = 0+$. Therefore, the response to a capacitor

[3] $t-$ and $t+$ denote the time instants just before and just after t respectively. The $1/C$ actually has dimensions of voltage, since the 1 stands for 1 coulomb.

voltage $\beta\hat{v}_o(0+)$ at $t = T_s+$ must be $\beta\hat{\imath}(t - T_s)$. This is a consequence of linearity and the periodically time-varying nature of the network. We thus write

$$\hat{\imath}(t) = p(t) + \beta\hat{\imath}(t - T_s). \tag{6.37}$$

In the frequency domain,

$$\hat{I}(j2\pi f) = \frac{P(j2\pi f)}{1 - \beta e^{-j2\pi f T_s}}. \tag{6.38}$$

Observing Figure 6.33b shows that $p(t)$ can be expressed as the difference between two decaying exponentials as follows, where $h(t) \equiv e^{-t/RC}u(t)$ and $u(t)$ denotes the unit-step function.

$$p(t) = \frac{1}{RC}(h(t) - \beta h(t - \tau)). \tag{6.39}$$

Thus,

$$P(j2\pi f) = \frac{1}{1 + j2\pi f RC}(1 - \beta e^{-j2\pi f\tau}). \tag{6.40}$$

Using Eqs. (6.38) and (6.40), we have

$$\boxed{H_{\text{eq}}(j2\pi f) = \frac{1}{(1 + j2\pi f RC)}\frac{1 - \beta e^{-j2\pi f\tau}}{1 - \beta e^{-j2\pi f T_s}}}. \tag{6.41}$$

When $RC \ll T_s$, $\beta = \exp(-DT_s/RC) \approx 0$, and the switched-RC network is said to be operating in the *sampler mode*. In this mode, $H_{\text{eq}}(j2\pi f)$ approaches $1/(1 + j2\pi f RC)$. Since $RC \ll T_s$, the voltage stored on the capacitor during the hold phase (i.e., when the switch is off) is the input sampled after filtering by a very wide-bandwidth filter.

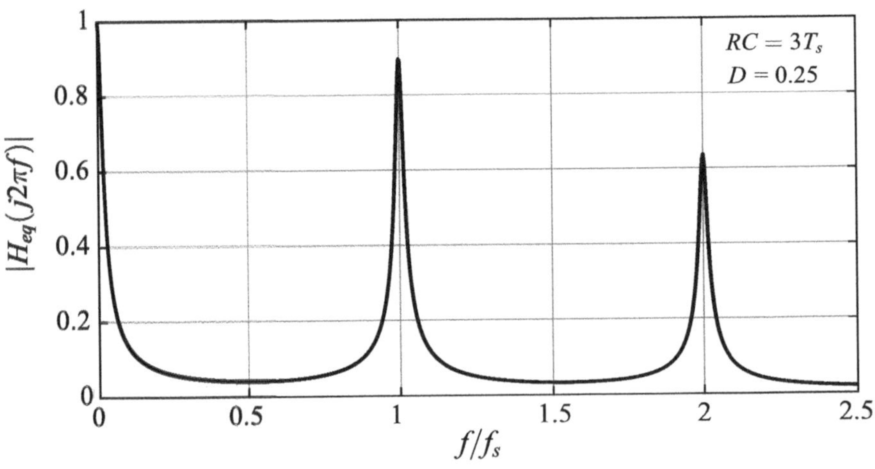

Figure 6.34 $|H_{\text{eq}}(j2\pi f)|$ in the mixer mode, for $RC = 3T_s$, $D = 0.25$.

When $RC \gg T_s$, the switched-RC network operates in what is called the mixer mode. In this mode, the magnitude response of the equivalent filter relating the input to the sampled output exhibits peaks around multiples of f_s as shown in Figure 6.34. This makes intuitive sense due to the following. When the input is a sinusoid at f_s, the waveform in the time window when the switch is closed remains the same in all clock cycles. Consequently, the capacitor "sees" the same portion of the sinewave during every clock cycle and a dc voltage is developed across the capacitor. As the input frequency deviates from f_s, the waveform during the switch's "on-window" keeps changing, reducing the peak voltage on the capacitor.

6.4.2 Integrator with Sampled Feedback

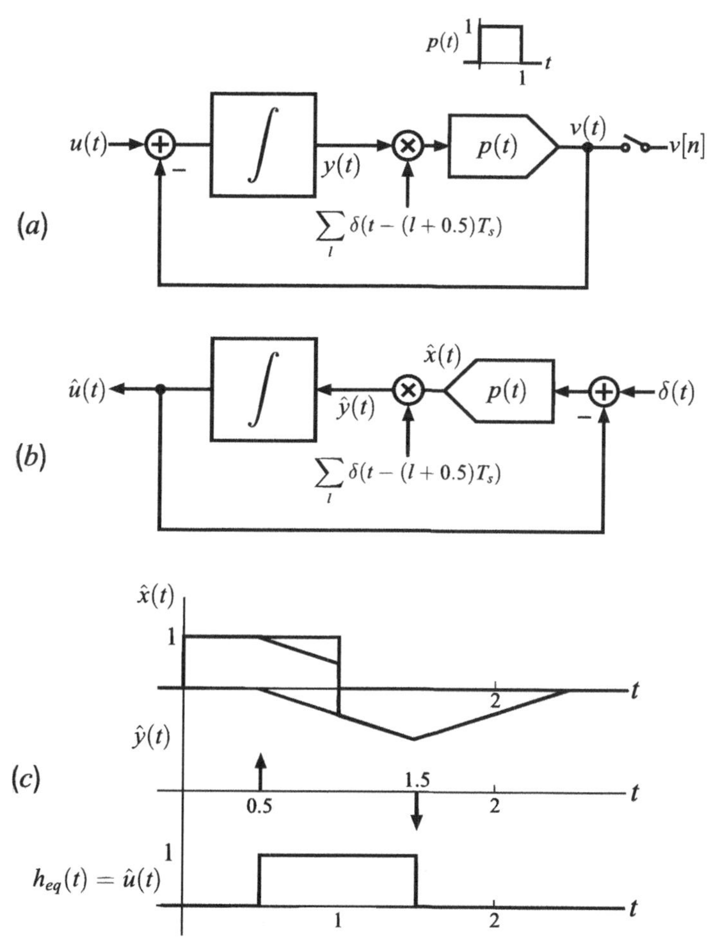

Figure 6.35 (a) Integrator with sampled feedback. (b) The adjoint signal flow graph. (c) Waveforms in the adjoint.

Figure 6.35a shows an LPTV system which consists of an integrator enclosed in a sampled feedback loop.[4] The output of the integrator is multiplied by an impulse train (with period 1 s), which excites an LTI system with a rectangular impulse

---

[4]For those readers familiar with continuous-time delta-sigma data converters, this is the signal processing equivalent of a first-order CT$\Delta\Sigma$M where the quantization noise is set to zero.

response $p(t)$. The output of the filter, namely $v(t)$ is subtracted from the input of the integrator. $v(t)$ is sampled at 1 Hz to yield the sequence $v[n]$. We would like to determine the impulse response of the equivalent linear time-invariant system whose sampled output is identical to $v[n]$ for the same excitation $u(t)$.

As discussed earlier in this section, $h_{\text{eq}}(t)$ of the equivalent LTI system can be obtained using the adjoint signal-flow graph. To form the adjoint, the summing node is replaced by a pick-off point and vice versa. The input and outputs of the other blocks are interchanged, resulting in the signal-flow graph shown in Figure 6.35b. To determine $h_{\text{eq}}(t)$, an impulse is injected into the "output" port of the adjoint. The waveform $\hat{u}(t)$ so developed is the impulse response of the equivalent filter.

Figure 6.35c shows the signals at various points in the adjoint. $\hat{x}(t)$ jumps to 1 at $t = 0+$ and stays that way up to 0.5 s. At $t = 0.5$, $\hat{y}(t)$ is an impulse—consequently, the output of the integrator, namely $\hat{u}(t)$ steps to 1. Since $\hat{u}(t)$ is fed back and inverted at the input of the filter, $\hat{x}(t)$ begins to ramp down at a rate 1 unit per second from its value at $t = 0.5$. $\hat{x}(t)$ is the sum of $p(t)$ and the inverted ramp. At $t = 1.5$, when $\hat{x}(t) = -1$, $\hat{y}(t)$ is a negative impulse which causes $\hat{u}(t)$ to zero. As a result, $\hat{x}(t)$ ramps up to zero at $t = 2.5$ and remains at that value. From the discussion above, we see that $h_{\text{eq}}(t)$ is the unit-rectangular pulse centered at $t = 1$. The frequency response of the equivalent LTI filter is therefore

$$H_{\text{eq}}(j2\pi f) = \text{sinc}(f)e^{-j2\pi f}.$$

The dc gain is seen to be unity. Interestingly, as seen in Figure 6.36, the magnitude response of the equivalent filter has nulls at multiples of the sampling rate, namely 1 Hz. This makes sense due to the following.

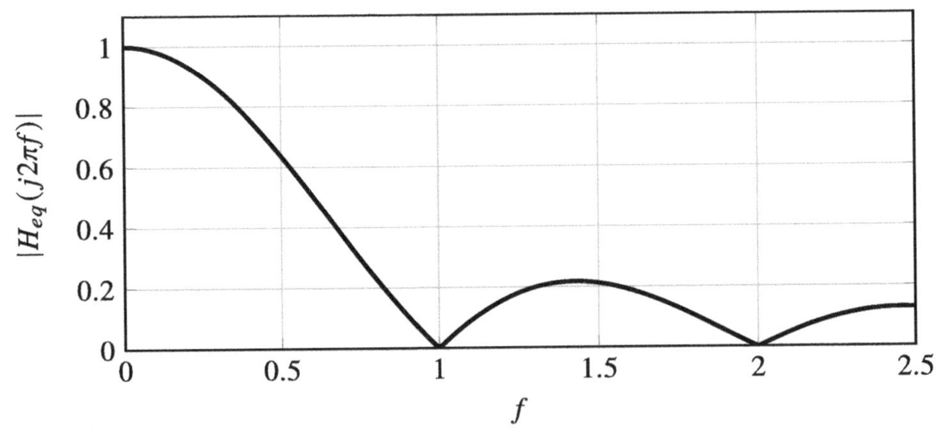

Figure 6.36 Magnitude response of the equivalent LTI filter relating $u(t)$ to $y[n]$ in Figure 6.35.

To determine $H_{\text{eq}}(0)$, we excite the system of Figure 6.35a with a constant input as shown in Figure 6.37a. We proceed as follows.

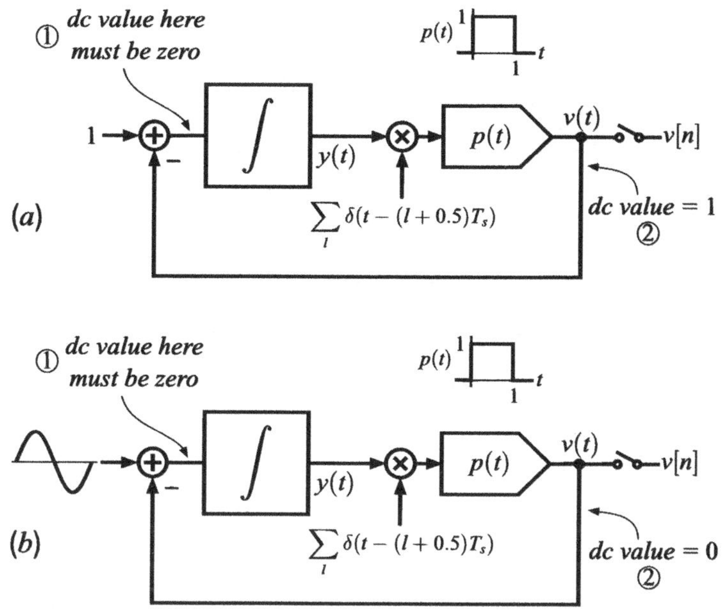

Figure 6.37 Insight to determine $H_{eq}(j2\pi f)$ of the system of Figure 6.35a at (a) dc and (b) at $f_s = 1$ Hz.

- ① Assuming the system is stable, it must follow that the dc value of the integrator's input must be zero. If this was not true, the output of the integrator would go to infinity, contradicting our assumption of stability.
- ② Consequently, the dc value of $v(t)$ must be unity. This means that $H_{eq}(0) = 1$.

Let us now intuitively understand why $H_{eq}(j2\pi f_s) = H_{eq}(j2\pi)$ has to be zero. How would we determine $H_{eq}(j2\pi)$? We should excite the system with a sine wave at the sampling frequency $f_s = 1$ Hz. After being processed by $H_{eq}(j2\pi f)$, the output will be a sine wave scaled by $H_{eq}(j2\pi)$. When sampled at f_s, this sine wave will result in a dc sequence, from which we determine $H_{eq}(j2\pi)$. The application of this idea to our system is shown in Figure 6.37b. We proceed in a step-wise fashion as follows.

- ① Assuming the system is stable, it must follow that the dc value of the integrator's input must be zero, as before.
- ② The dc value of the input is zero. Consequently, the dc value of $v(t)$ must be zero. This means that $H_{eq}(j2\pi)$ should be zero.

6.4.3 Offset Sampling

So far we assumed that the output of the LPTV system is sampled at time instants nT_s, and used the adjoint network to determine $h_{eq}(t)$. In many cases, the sampling instants are offset, in the sense that the LPTV system's output is sampled at $(nT_s + t_x)$ as shown in Figure 6.38a. t_x is called the timing offset. How do we find the impulse response of the equivalent LTI system, namely $h_{eq}(t)$, in this case?

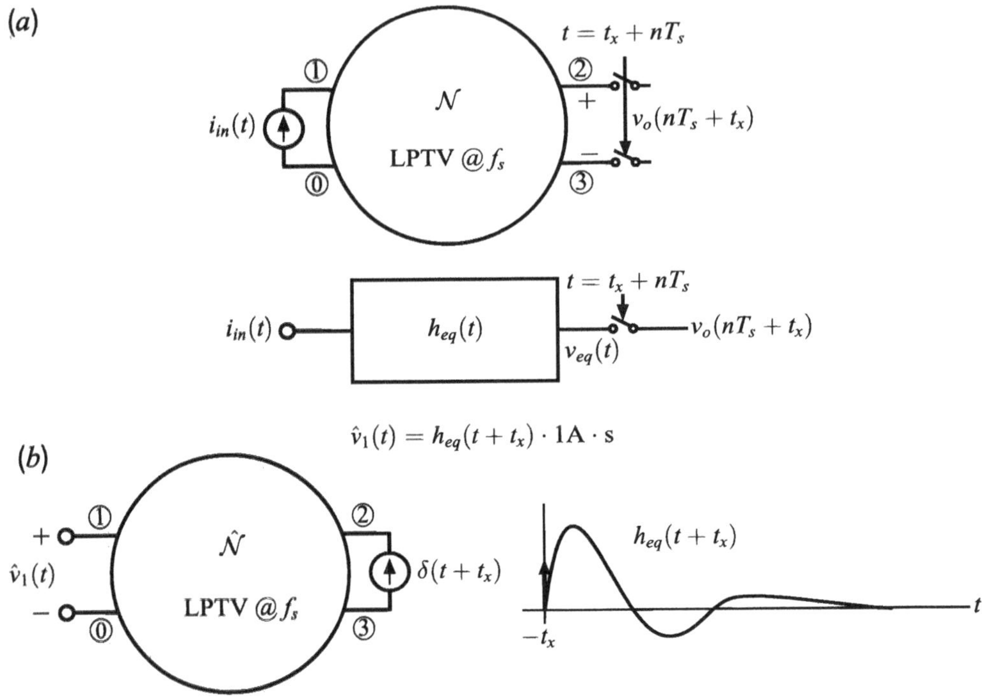

Figure 6.38 (a) LPTV system with offset sampling, and the equivalent model. (b) Determining $h_{\text{eq}}(t)$ using the adjoint network.

The equivalent model of the system is shown in Figure 6.38a. As in the case where $t_x = 0$, we use the adjoint network. Reasoning in the same manner as we did earlier, which we leave to the reader as an exercise, it turns out that $h_{\text{eq}}(t)$ can be obtained as follows. The output port of the adjoint $\hat{\mathcal{N}}$ is excited by an impulse at $t = -t_x$ as shown in Figure 6.38b. The waveform that results at the input port can be shown to be proportional to $h_{\text{eq}}(t)$. In the case of a current input and voltage output, the voltage developed at the input port of the adjoint, denoted by $\hat{v}_1(t)$ in Figure 6.38 is given by $h_{\text{eq}}(t + t_x)\, 1\text{A} \cdot \text{s}$.

Simulation Corner: Determining $H_{\text{eq}}(j2\pi f)$

Suppose that we had an RF simulator capable of running a periodic AC analysis (`.PAC`). How do we determine $H_{\text{eq}}(j2\pi f)$ using such a simulator? Recall that the `.PAC` analysis, in general, is run on the small-signal LPTV network obtained by linearizing a nonlinear circuit about a periodically-varying operating point. As such, the designer has no access to this small-signal network. Creating the adjoint, and exciting it with an impulse to obtain $h_{\text{eq}}(t)$ (from which $H_{\text{eq}}(j2\pi f)$ can be determined) is thus not feasible. Some simulators do have the option to obtain $H_{\text{eq}}(j2\pi f)$ at any timing offset using what is called the *sampled* `.PAC` analysis, but what about a simulator without this feature?

Figure 6.39 Determining $H_{eq}(j2\pi f)$ in an RF simulator without a sampled . PAC option.

One way of accomplishing this is given in Figure 6.39. The output of the LPTV system under consideration is given by $y(t)$, and suppose that we are interested in determining the relationship from the input to the samples $y(nT_s)$. Since the simulator is capable of determining the harmonic transfer functions of the system, $H_l(j2\pi f)$ are known. We have

$$y(t) = \sum_l H_l(j2\pi f) e^{j2\pi(f+lf_s)t}.$$

What we need is to determine the sum of the H_l. To do this, we multiply $y(t)$ with an impulse train to result in $y_c(t)$ as shown in Figure 6.39. (In practice, the impulse train can be approximated by a rectangular wave with a sufficiently small duty cycle). $y_c(t)$ is treated as the output of a new LPTV system. Multiplication with the impulse train will cause tones of the form $(f + lf_s)$ to alias to f. Consequently,

$$y_c(t) = \left\{\sum_k H_k(j2\pi f)\right\} \sum_l e^{j2\pi(f+lf_s)t}.$$

All the harmonic transfer functions of this new system are seen to be the sum of the HTFs of the original system, which is what we wanted to determine.

To summarize, $H_{eq}(j2\pi f)$ is the zeroth-order harmonic transfer function of the composite system of Figure 6.39. $h_{eq}(t)$ can be obtained by computing the inverse Fourier transform of $H_{eq}(j2\pi f)$.

6.5 LPTV Systems Driven by a Class of Modulated Signals

In the previous section, we related the samples of an LPTV system's output to its (continuous-time) input. In this section, we consider a corollary property, where an LPTV system is driven by a class of modulated signals, and we relate the (continuous-time) output waveform to its input.

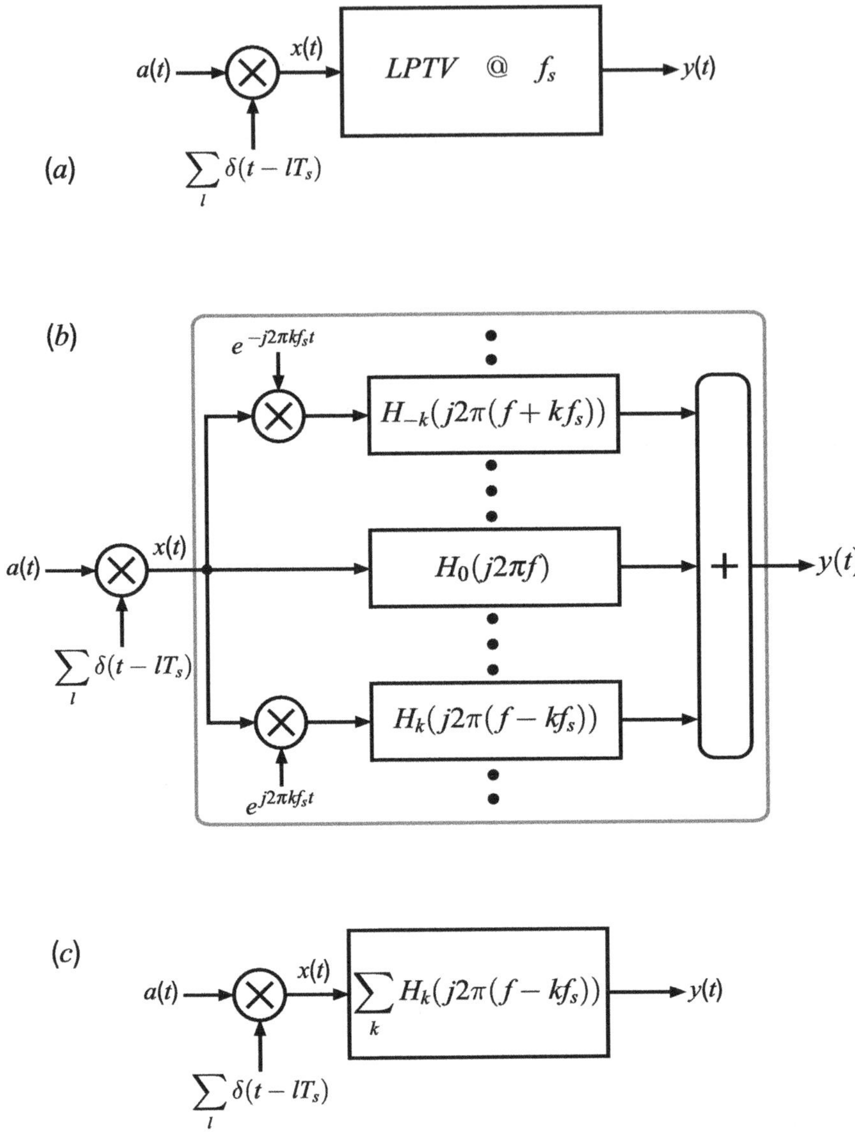

Figure 6.40 (a) An LPTV system varying at frequency f_s. (b) Equivalent representation—the $H_l(j2\pi f)$ represent linear time-invariant filters. (c) Simplified system.

Consider an LPTV system driven by the modulated signal

$$x(t) = \sum_l a(t)\delta(t - lT_s) = \sum_l a(lT_s)\delta(t - lT_s) \tag{6.42}$$

as shown in Figure 6.40a. Notice that the Dirac impulse train is periodic with a fundamental frequency $f_s = 1/T_s$, which is the *same* as that at which the LPTV system is varying. $x(t)$ can be thought of as the result of modulating a continuous-time signal $a(t)$ by the impulse train.

The LPTV system is replaced with the mixer-first form of its Zadeh expansion derived in Chapter 5, resulting in the system of Figure 6.40b. Since $x(t)$ consists of Dirac impulses, it is nonzero only for $t = lT_s$. Thus, $x(t)e^{j2\pi k f_s t} = x(t)e^{j2\pi kl} = x(t)$. As a result, the diagram of Figure 6.40b can be simplified to that shown in Figure 6.40c. Since the H_k represent time-invariant filters, their sum is also time-invariant. It is thus seen that $y(t)$ can be thought of as being the output of a linear *time-invariant* system whose transfer function is

$$H_{\text{eq}}(j2\pi f) = \sum_{k=-\infty}^{\infty} H_k(j2\pi(f - kf_s)). \tag{6.43}$$

Consider now a more general case, where $x(t)$ is of the form $\sum_l a(lT_s)p(t - lT_s)$. The model of this system is given in Figure 6.41a. Conceptually, $x(t)$ can be thought of as being obtained by filtering the impulsive sequence $\sum_l a(lT_s)\delta(t - lT_s)$ with a linear time-invariant filter whose impulse response is $p(t)$. The cascade of the pulse-shaping filter and the LPTV system can be thought of as another LPTV system, whose harmonic transfer functions are given by $P(j2\pi f)H_l(j2\pi f)$, where $P(j2\pi f)$ denotes the Fourier transform of $p(t)$ and $H_l(j2\pi f)$ are the harmonic transfer functions of the LPTV system. $y(t)$ can therefore be thought of as the output of a linear time-invariant filter whose transfer function is

$$H_{\text{eq}}(j2\pi f) = \sum_{l=-\infty}^{\infty} H_l(j2\pi(f - lf_s))P(j2\pi(f - lf_s)). \tag{6.44}$$

The result above can thus be summarized as follows [9].

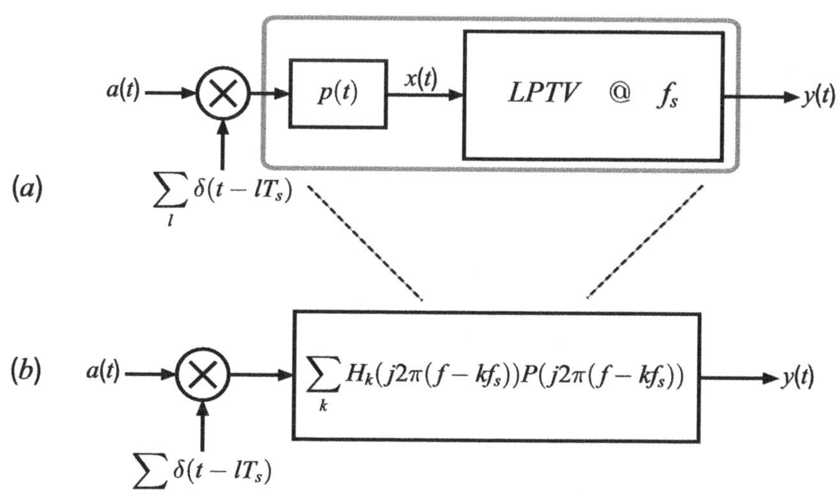

Figure 6.41 (a) An LPTV system varying at frequency f_s. (b) Equivalent representation—the $H_l(j2\pi f)$ represent linear time-invariant filters.

The output $y(t)$ of an LPTV system driven by an input of the form $\sum_k a_k p(t - kT_s)$ is indistinguishable from the output of an appropriately chosen linear time-invariant filter driven by $\sum_k a_k \delta(t - kT_s)$. The frequency response of the filter is given by Equation (6.44) above.

Rather than drive the equivalent LTI filter with $\sum_k a_k \delta(t - kT_s)$, one could drive a filter with transfer function $H_{eq}(j2\pi f)/P(j2\pi f)$ with $\sum_k a_k p(t - kT_s)$. It is straightforward to determine the impulse response of the equivalent LTI filter. As seen in Figure 6.41, applying an impulse at the input results in $y(t) = h_{eq}(t)$.

While the results above are important in their own right, and add to our knowledge of the behavior of LPTV systems, a question that arises is—is this potentially useful in practice, and if so, where?

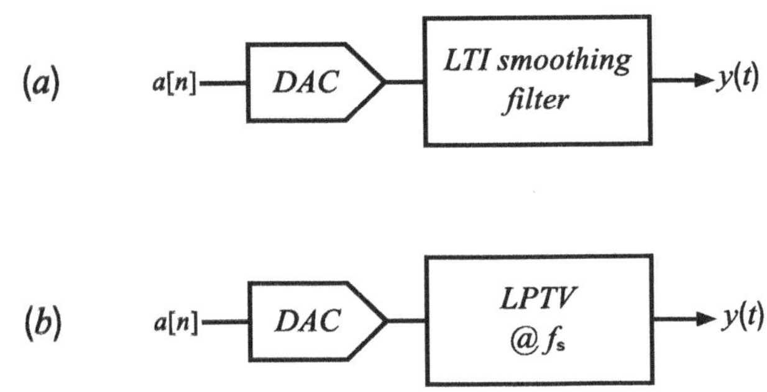

Figure 6.42 (a) A DAC, whose spectral images are attenuated using a time-invariant smoothing filter and (b) the time-invariant filter can be replaced by an LPTV system, potentially simplifying the design of the smoothing filter.

It is well-known that the output of a digital-to-analog converter (DAC) can be expressed as $\sum_k a_k p(t - kT_s)$, where a_k is the discrete-time sequence that needs to be converted into analog form. $p(t)$ denotes the DAC pulse shape. Commonly used DAC pulses are the non-return to zero (NRZ) and return-to-zero (RZ) shapes. From a spectral viewpoint, the DAC output is seen to be the impulsive sequence $\sum a_k \delta(t - kT_s)$ processed by a filter with transfer function $P(j2\pi f)$. Unfortunately, the NRZ and RZ pulses are discontinuous, and therefore, do not adequately attenuate the high-frequency images of the DAC output spectrum. Thus, as shown in Figure 6.42a, a smoothing filter needs to be used. This filter is time-invariant, and needs to have high stop-band attenuation. Further, a common requirement is that the filter should have sufficiently small group-delay distortion (i.e., linear phase). Due to this, high-order Bessel filters are often employed, so as to obtain an adequate stop-band attenuation. The result derived above enables the replacement of the time-invariant smoothing filter with an LPTV system as shown in Figure 6.42b.

Why would the use of an LPTV system in place of an LTI smoothing filter be any better? When compared to an LTI system, an LPTV system with the same number of states has increased degrees of freedom, in the sense that the coefficients of the differential equations describing the system can be varied periodically in a user-defined fashion. As will be shown in the next, this freedom can be exploited to improve filtering characteristics.

6.5.1 Example: LPTV Smoothing Filter

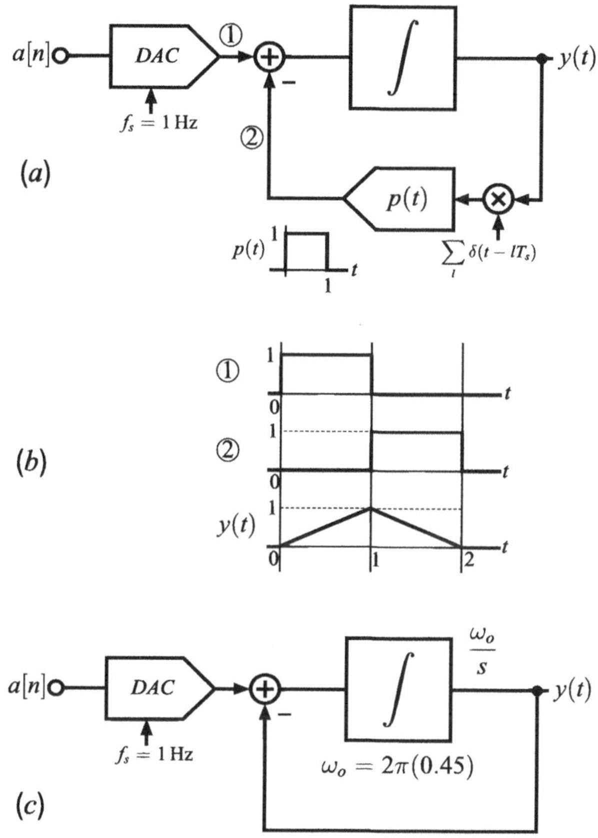

Figure 6.43 (a) NRZ DAC driving an LPTV smoothing filter. (b) Key waveforms. (c) First-order LTI filter with the same 3 dB bandwidth.

In this section, we illustrate the benefits to be had by using an LPTV smoothing filter instead of an LTI one. Figure 6.43a shows an NRZ DAC driving an LPTV system. The sampling rate is assumed to be 1 Hz—this does not lead to a loss of generality. The LPTV "filter" consists of an integrator with sampled-and-held feedback through a zero-order-hold (ZoH). As discussed, even though the network is periodically time-varying, the output cannot be distinguished from that of an equivalent time-invariant filter. $y(t)$ can be written as

$$y(t) = \sum_k a_k \hat{p}(t - kT_s). \tag{6.45}$$

To determine $\hat{p}(t)$, all that one needs to do is to set $a[n] = \delta[n]$. The resulting waveforms at ①, ②, and $y(t)$ are shown in Figure 6.43b. It is seen that the effective DAC pulse-shape at the output of the LPTV system is a triangular pulse. Since the DAC pulse is rectangular, it follows that the LPTV filter effectively behaves like a filter whose impulse response is also a unit rectangular pulse.

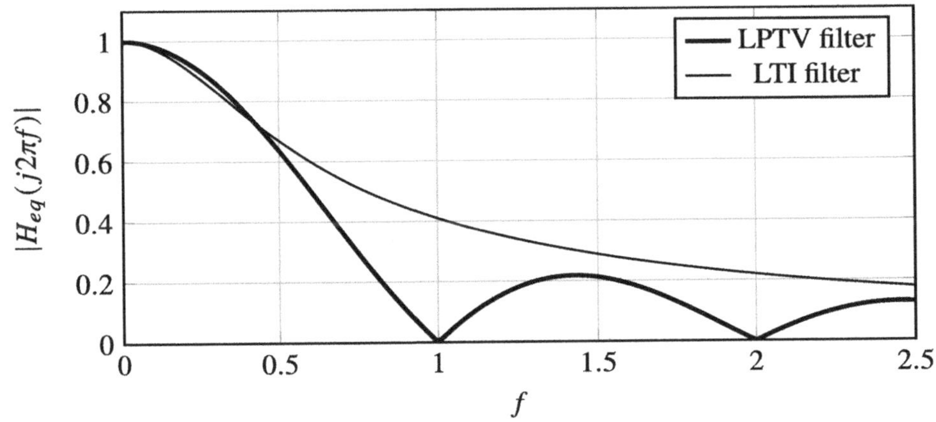

Figure 6.44 Comparison of the magnitude responses of the LPTV filter and a first-order LTI filter with the same 3-dB bandwidth.

Figure 6.44 compares the magnitude response of a first-order linear time-invariant filter with that of the LPTV smoothing filter (which provides a sinc response). The LTI filter is chosen to have the same 3 dB bandwidth as the LPTV structure. The magnitude responses of both filters are almost identical at low frequencies, with the 3 dB bandwidth being 0.45 times the sampling frequency. It is apparent that the LPTV smoothing filter provides a sharper response and nulls at multiples of the sampling frequency. The attenuation of images is, therefore, higher. Furthermore, the phase response of the LPTV filter is *linear*, and no group-delay distortion is incurred. From an implementation viewpoint, the modification from a time-invariant to a time-varying filter is simply the insertion of a sample-and-hold in the feedback path (Figure 6.43a), as opposed to the continuous path that would be necessary in the LTI filter.

It is evident that the LPTV structure performs better than the LTI filter both in terms of the magnitude and phase responses. In the LPTV case, notches are obtained in the frequency response at *all* multiples of the sampling rate by using only one integrator. Such behavior is simply impossible to realize using time-invariant structures.

6.6 Noise in LPTV Networks

LPTV networks are noisy, like their linear time-invariant cousins. Perhaps unsurprisingly, the analysis of noise in LPTV networks is more involved. We illustrate with an example. Consider the system of Figure 6.45a. The RC networks

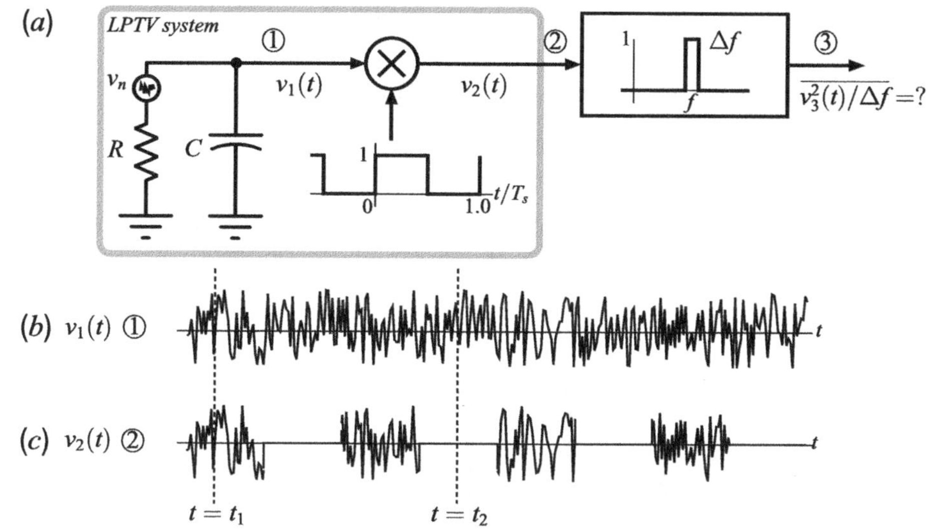

Figure 6.45 (a) Example LPTV network and measurement of the average noise-spectral density. (b) Representative noise waveform at ① and (c) at ②.

produces stationary noise at point ①, with a two-sided noise spectral density given by

$$S_1(f) = \frac{2kTR}{1 + (2\pi fRC)^2}.$$

Figure 6.45b shows a representative noise waveform $v_1(t)$. The mean-square value $\overline{v_1^2(t)}$, as seen earlier, is kT/C. $v_1(t)$ is multiplied by a square wave with 50% duty cycle to result in v_2. The question we wish to answer is "what is the noise spectral density of the waveform at point ②?" Before we go there, however, a few observations are in order.

The waveform at ② is clearly the output of an LPTV system, and is shown in Figure 6.45c. Evidently, there is something "different" about $v_2(t)$ when compared to $v_1(t)$. What is this difference? Suppose we had a million instances of the LPTV system of Figure 6.45a. If we determined the variance at the output of this ensemble at two instants $t = t_1$ and $t = t_2$, the variance at ① would be the same at both time instants. For the noise waveform at ②, however, the variance is clearly varying (pardon the pun). In fact, at $t = t_2$, $v_2(t) = 0$! Furthermore, it is evident that the variance of $v_2(t)$ changes periodically with period T_s. Such random processes, whose statistical properties vary periodically with time, are referred to as being cyclostationary [10]. In an LPTV network, where circuit properties change periodically with time, noise produced is cyclostationary in general. The aim of this section is to give the reader a basic understanding of noise in LPTV circuits adequate for circuit work.

In LTI circuits, we saw that the spectral density at a frequency f was determined as follows.

a. Process the noise waveform by a narrowband filter centered at f and bandwidth $\Delta f \rightarrow 0$.

b. Determine the mean-square value of the filter's output and divide it by Δf to obtain the noise spectral density at the frequency f.

The process described above can also be applied to cyclostationary noise. It turns out that this yields what is termed the "average" noise spectral density at the frequency f. Let us determine this quantity for the system of Figure 6.45a. We proceed as follows.

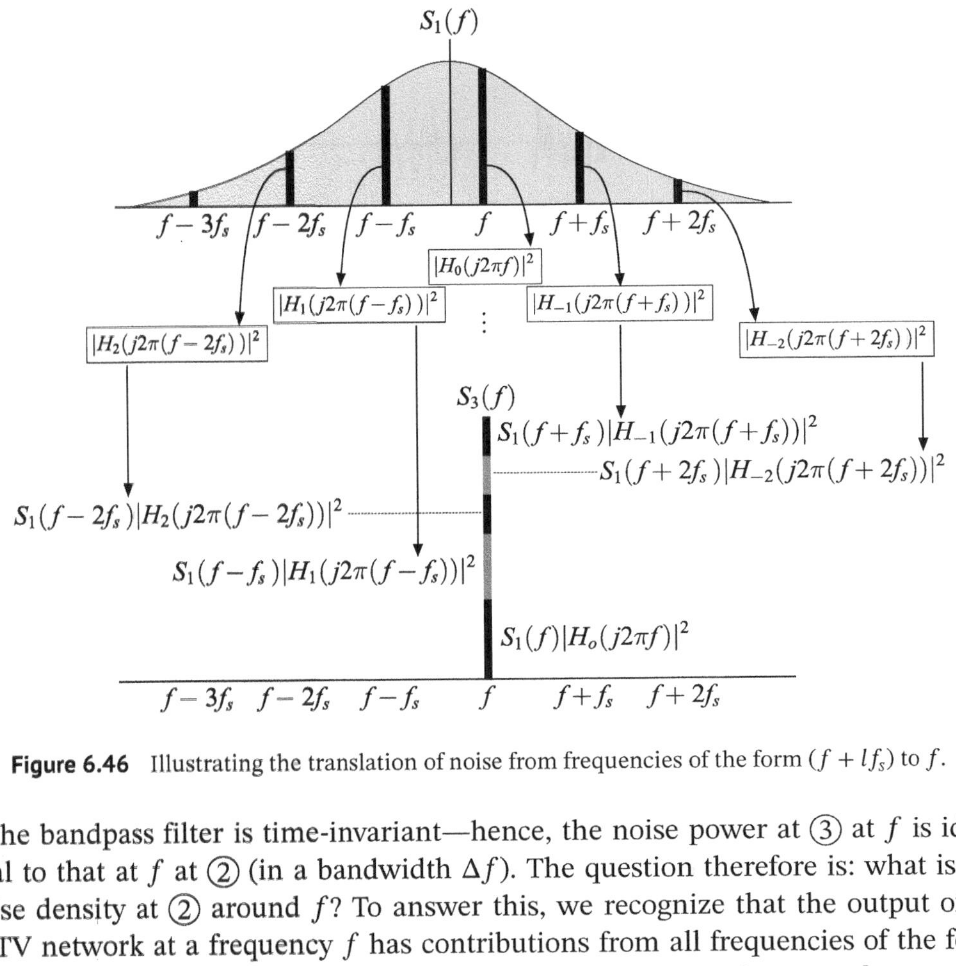

Figure 6.46 Illustrating the translation of noise from frequencies of the form $(f + lf_s)$ to f.

The bandpass filter is time-invariant—hence, the noise power at ③ at f is identical to that at f at ② (in a bandwidth Δf). The question therefore is: what is the noise density at ② around f? To answer this, we recognize that the output of an LPTV network at a frequency f has contributions from all frequencies of the form $(f + lf_s)$. The relative strengths of these contributions depend on two factors. The first is the strength of the input spectral density at $(f + lf_s)$. Second is the magnitude of the harmonic transfer function $H_{-l}(j2\pi(f + lf_s))$ which quantifies the gain from an input at $(f + lf_s)$ to the output at f. In our example, calculation of the NSD at ②, namely $S_2(f)$ proceeds as follows, and is illustrated in Figure 6.46.

a. The noise spectral density (which is the mean-square value of the noise at f normalized to Δf) at ① is $S_1(f)$. Intuitively, one can think of this as a sine-wave at f whose mean-square value is given by $S_1(f)\Delta f$.

b. The NSD at ② at f due to noise at ① at f is given by $S_1(f)|H_0(j2\pi f)|^2$.

c. The NSD at ② at f due to noise at ① at $(f + lf_s)$ is given by $S_1(f + lf_s)|H_{-l}(j2\pi(f + lf_s))|^2$. This is due to noise folding from frequencies of the form $(f + lf_s)$.

d. The power of all the sine waves add; consequently, the NSD at ② at the frequency f is given by

$$S_2(f) = \sum_{l=-\infty}^{\infty} S_1(f + lf_s)|H_{-l}(j2\pi(f + lf_s))|^2. \tag{6.46}$$

In general, therefore, the average NSD at the output of an LPTV system with harmonic transfer functions denoted by $H_l(j2\pi f)$ driven by a stationary noise source with NSD $S_{\text{in}}(f)$ is given by [11]

$$\boxed{S_o(f) = \sum_{l=-\infty}^{\infty} S_{in}(f + lf_s)|H_{-l}(j2\pi(f + lf_s))|^2.} \tag{6.47}$$

6.6.1 Cascaded LPTV Systems

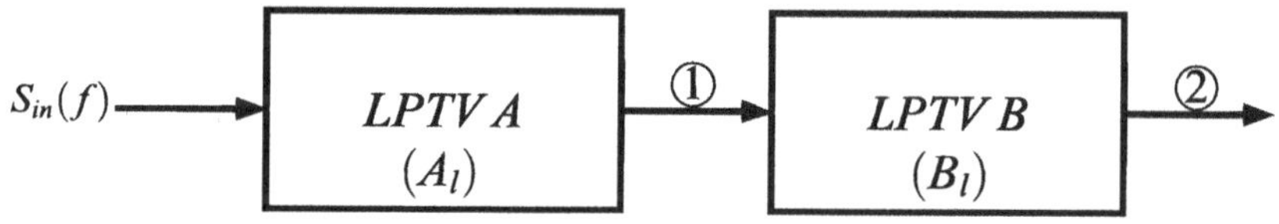

Figure 6.47 Noise in cascaded LPTV systems driven by a stationary noise source.

Consider a cascade of LPTV systems, whose harmonic transfer functions are denoted by $A_l(j2\pi f)$ and $B_l(j2\pi f)$ respectively (Figure 6.47). The input to the cascade is a stationary noise source with NSD given by $S_{\text{in}}(f)$. The average NSD at ① is given by

$$S_1(f) = \sum_{l=-\infty}^{\infty} S_{in}(f + lf_s)|A_{-l}(j2\pi(f + lf_s))|^2.$$

It is tempting, albeit incorrect, to conclude that the average NSD at the output of system B can be obtained using the relation above with S_{in} and A_l replaced by S_1 and B_l respectively. In other words,

$$S_2(f) \neq \sum_{l=-\infty}^{\infty} S_1(f + lf_s)|B_{-l}(j2\pi(f + lf_s))|^2 \quad \text{in general.} \tag{6.48}$$

To see why the relation above is not valid in general, suppose that $S_{\text{in}}(f)$ had a very narrowband spectrum concentrated around f. The NSD at the output of system A would have components around all frequencies of the form $(f + lf_s)$. The origin (or root cause) of all these translated components is the same, namely the input NSD around f. Consequently, these translated components are all *correlated*, which is not accounted for in the relation for S_2 above. The correct approach to obtain $S_2(f)$ is to determine the harmonic transfer functions of the cascade and then use Eq. (6.47).

Example: LPTV Noise Calculations

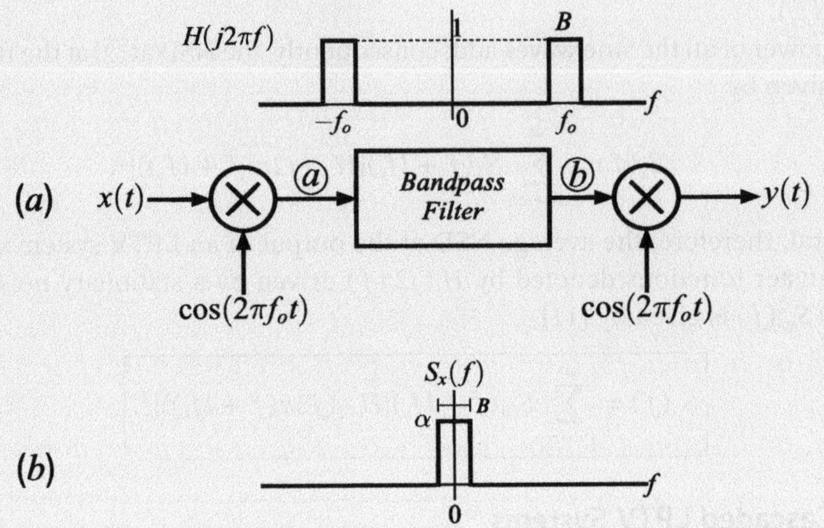

Figure 6.48 (a) What is the mean-square value of $y(t)$? (b) NSD of the stationary noise source x. The mean-square value $\overline{v_x^2}(t) = \alpha B$.

Consider the system of Figure 6.48a. $x(t)$ is a stationary noise source with the spectral density shown in Figure 6.48b. Furthermore, we denote the mean-square value of x by $\overline{v_x^2}$, which equals αB. We seek to answer the following questions.

a. What is the NSD at y?
b. What is the mean-square value of $y(t)$?

Our first order of business is to determine the harmonic transfer functions from x to y. Suppose $x(t) = e^{j2\pi ft}$.

$$
\begin{aligned}
ⓐ &\longrightarrow \frac{1}{2}e^{j2\pi(f+f_o)t} + \frac{1}{2}e^{j2\pi(f-f_o)t} \\
ⓑ &\longrightarrow \frac{1}{2}H(j2\pi(f+f_o))e^{j2\pi(f+f_o)t} + \frac{1}{2}H(j2\pi(f-f_o))e^{j2\pi(f-f_o)t} \\
y &\longrightarrow \frac{1}{4}\{H(j2\pi(f+f_o)) + H(j2\pi(f-f_o))\}\, e^{j2\pi ft} + \\
&\quad \frac{1}{4}H(j2\pi(f+f_o))e^{j2\pi(f+2f_o)t} + \\
&\quad \frac{1}{4}H(j2\pi(f-f_o))e^{j2\pi(f-2f_o)t}.
\end{aligned}
$$

$S_x(f)$ is zero for $|f| > B/2$, i.e., the input frequency content is limited to $|f| \leq B/2$. Consequently, from the equation for y above, we see that for a sine-wave input $e^{j2\pi ft}$where $|f| \leq B/2$,

$$y \to \frac{1}{2}e^{j2\pi ft} + \frac{1}{4}e^{j2\pi(f+2f_o)t} + \frac{1}{4}e^{j2\pi(f-2f_o)t}.$$

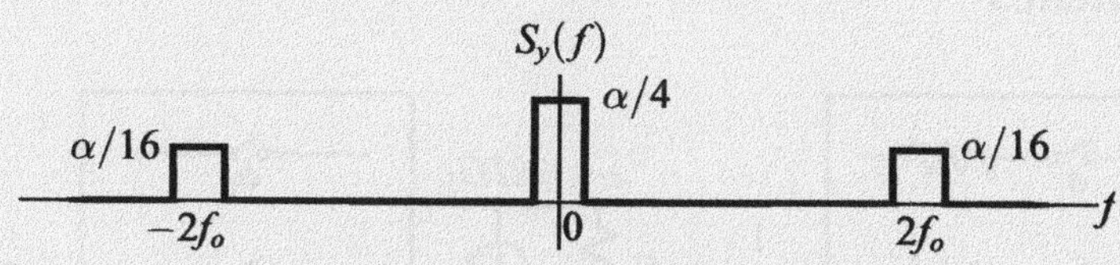

Figure 6.49 Average NSD at y.

Consequently, the NSD of y consists of three parts:

a. $(1/4)S_x(f)$, which is due to the gain $H_0(j2\pi f)$.
b. $(1/16)S_x(f - 2f_o)$, which is due to the up conversion of the input noise to $2f_o$ due to translation by two modulators.
c. $(1/16)S_x(f + 2f_o)$, which is due to the down conversion of the input noise to $-2f_o$, again due to translation by two modulators.

The NSD is shown in Figure 6.49. The mean square value at y, namely $\overline{v_y^2} = (3/8)\alpha B = (3/8)\overline{v_x^2}$.

Let us use this example to illustrate Eq. (6.48). The average NSD at ⓐ is given by $\frac{1}{4}S_x(f - f_o) + \frac{1}{4}S_x(f + f_o)$. Since the spectral width of this noise lies in the passband of the bandpass filter, the average NSD at ⓑ is identical to that at ⓐ. The second multiplier translates this noise to around dc and $2f_o$. The resulting (wrong!) average NSD at y is given by

$$S_y(f) = \frac{1}{8}S_x(f) + \frac{1}{16}S_x(f - 2f_o) + \frac{1}{16}S_x(f + 2f_o)$$

which predicts that $\overline{v_y^2} = \alpha B/4 = (1/4)\overline{v_x^2}$. The correct $\overline{v_y^2}$, as seen earlier, is $(3/8)\overline{v_x^2}$ which is 50% higher than what we incorrectly calculated above [10].

When we discussed noise in linear time-invariant circuits, we found that we could exploit reciprocity to simplify the determination of noise spectral density and total integrated noise is networks comprising of resistors, inductors and capacitors. Specifically, we saw how the Nyquist and Bode noise theorems were derived. What happens when the network consists of periodically operated switches in addition? Perhaps unsurprisingly, it turns out that similar relationships can be found for the average noise spectral density and integrated noise in LPTV networks with resistors, capacitors, inductors, and periodically operated switches. These results are derived next.

6.6.2 Average Noise Spectral Density in RLC Networks with Periodically Operated Switches

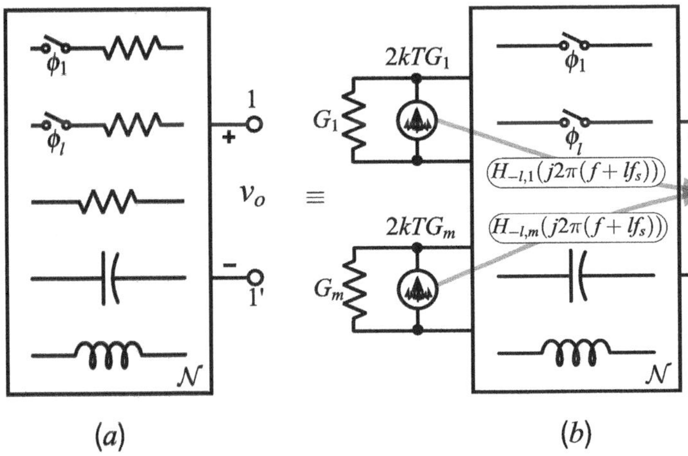

Figure 6.50 (a) LPTV network with inductors, capacitors, resistors, and periodically operated switches. (b) The network with all the resistors pulled out. The pth conductance is associated with a white noise current with spectral density $2kTG_p$. All the elements inside the box are lossless.

We intend to determine the average noise spectral density across terminals 1-1′ of the network $\mathcal{N}$ shown in Figure 6.50a. It consists of inductors, capacitors, resistors, and periodically operated switches. Since every switch is associated with a series resistance it is modeled as an ideal switch in series with a resistance as shown in the figure. $\phi_1, \cdots, \phi_l$ represent the switch-control waveforms, with a switch closed when the corresponding ϕ is high and vice versa. All the switch-control waveforms have the same period T_s, but are otherwise arbitrary. Every resistor adds thermal noise, modeled as a noise current in parallel with it. In the discussion that follows, we will use $G \equiv (1/R)$.

Figure 6.50b shows the same network, but with all the resistors (including those associated with the switches) pulled out. The noise current generated by the pth conductance, which has a spectral density $2kTG_p$, has the harmonic transfer functions $H_{l,p}(j2\pi f)$ to the output. Since the noise generated by the conductances is independent, their noise contributions to the output can simply be added together. Using this along with Eq. (6.47), we see that the double-sided average output noise spectral density is given by

$$S_o(f) = 2kT \sum_{p=1}^{m} \sum_{l=-\infty}^{\infty} G_p |H_{-l,p}(j2\pi(f + lf_s))|^2. \tag{6.49}$$

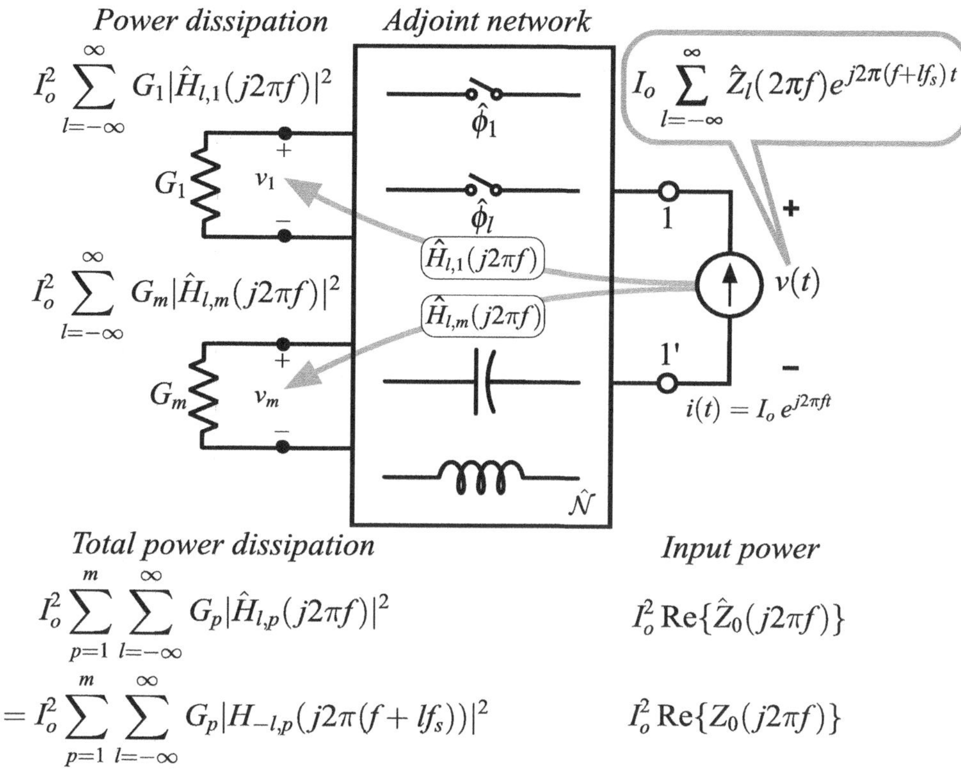

Figure 6.51 Nyquist's noise theorem states that the single-sided voltage noise spectral density across terminals 1-1′ is given by $S_v(f) = 4kT\text{Re}\{Z(j2\pi f)\}$.

Consider now the adjoint network $\hat{\mathcal{N}}$ corresponding to $\mathcal{N}$, as shown in Figure 6.51. The switch control signals in $\hat{\mathcal{N}}$ are reversed in time, i.e., $\hat{\phi}_1(t) = \phi_1(-t)$ etc. We excite the adjoint with a current $i(t) = I_o e^{j2\pi ft}$ at the output port. The voltage across the pth conductance is given by

$$v_p(t) = I_o \sum_{l=-\infty}^{\infty} \hat{H}_{l,p} e^{j2\pi(f+lf_s)t}. \tag{6.50}$$

The power dissipated in G_p is thus seen to be $\sum_l I_o^2 G_p |\hat{H}_{l,p}|^2$. Consequently, the total power dissipated in all the conductances is given by

$$P = \sum_{p=1}^{m} \sum_{l=-\infty}^{\infty} I_o^2 G_p |\hat{H}_{l,p}(j2\pi f)|^2. \tag{6.51}$$

The voltage $v(t)$ developed across the current source $i(t)$ is related to I_o as

$$v(t) = \sum_{l=-\infty}^{\infty} I_o \hat{Z}_l(j2\pi f) e^{j2\pi(f+lf_s)t} \tag{6.52}$$

where $\hat{Z}_l$ is the lth-order harmonic transfer impedance of $\hat{\mathcal{N}}$. Therefore, the power supplied by the current source $i(t)$ is $I_0^2\text{Re}\{\hat{Z}_0(j2\pi f)\}$. Since this must be dissipated in all the conductances, we have

$$P = I_0^2\text{Re}\{\hat{Z}_0(j2\pi f)\} = \sum_{p=1}^{m} \sum_{l=-\infty}^{\infty} I_0^2 G_p |\hat{H}_{l,p}(j2\pi f)|^2. \tag{6.53}$$

Applying the frequency reversal theorem to $\mathcal{N}$ and $\hat{\mathcal{N}}$, we see that $\hat{H}_{l,p}(j2\pi f) = H_{-l,p}(j2\pi(f + lf_s))$. Furthermore, as a consequence of the transfer function theorem, $\hat{Z}_0(j2\pi f) = Z_0(j2\pi f)$. (Recall that $Z_0(j2\pi f)$ denotes the zeroth-order driving point impedance of the original network $\mathcal{N}$). Using the observations above in Eq. (6.53) we obtain

$$\sum_{p=1}^{m} \sum_{l=-\infty}^{\infty} G_p |H_{-l,p}(j2\pi(f + lf_s))|^2 = \text{Re}\{Z_0(j2\pi f)\}. \tag{6.54}$$

Using this in Eq. (6.49), we see that the average double-sided noise voltage spectral density between the terminals 1-1′ of $\mathcal{N}$ is given by $S_o(f) = 2kT\text{Re}\{Z_0(j2\pi f)\}$. Since the real part of $Z_0(j2\pi f)$ is even, we obtain the single-sided voltage NSD as [3]

$$\boxed{S_o(f) = 4kT\ \text{Re}\{Z_0(j2\pi f)\}.} \tag{6.55}$$

Since an LTI network is a special case of an LPTV system, the result above holds for the former, which is the well-known Nyquist noise theorem. The total integrated noise is simply the integral of the average noise spectral density and is given by

$$N_{\text{tot}} = \int_0^{\infty} 4kT\ \text{Re}\{Z_0(j2\pi f)\}\, df. \tag{6.56}$$

It turns out that evaluating this integral can be challenging. An alternative and much simpler way of arriving at N_{tot} uses the concepts of energy and is discussed in the next section.

Using the approach described above, it is easy to show that the average short-circuit noise current spectral density of the one-port network in Figure 6.50 is given by $4kT\ \text{Re}\{Y_0(j2\pi f)\}$. Here $Y_0(j2\pi f)$ denotes the zeroth-order admittance, and is determined by applying a voltage source $e^{j2\pi ft}$ between 1-1′ and measuring the component of current flowing into the network at the frequency f. This is left as an exercise for the reader.

6.6.3 The Switched-RC Kernel

The so called switched-RC kernel is shown in Figure 6.52a. The switch duty cycle is denoted by D. When $RC \ll T_s$, the switched-RC circuit is said to be operating in the sampler mode, while the mixer mode refers to the case where $RC \gg T_s$. Of particular interest in practice is the NSD of the capacitor voltage at very low frequencies. From

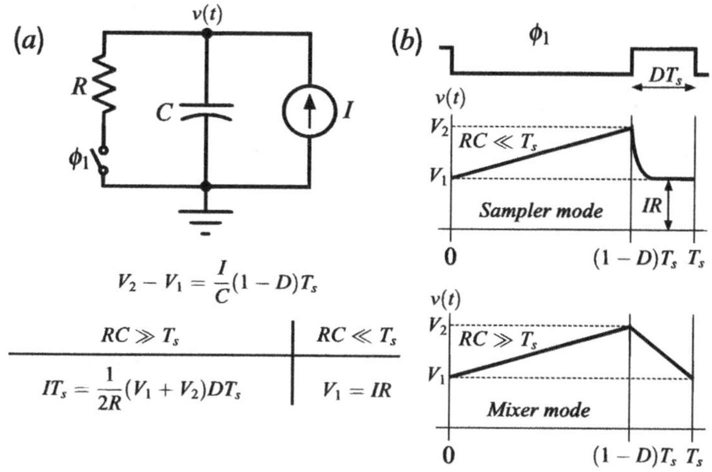

Figure 6.52 (a) The switched-RC kernel. The switch duty cycle is D. (b) The capacitor voltage when $RC \gg T_s$ and $RC \ll T_s$.

our theory, it is necessary to find the real part of the zeroth-order impedance of the network at dc. To this end, a dc current I is injected into the capacitor as shown in blue in Figure 6.52a. The dc value of the resulting $v(t)$ divided by I yields the impedance we are after.

Let V_2 and V_1 denote $v(t)$ at the beginning and end of ϕ_1 respectively. When ϕ_1 is low, the switch is open, and the capacitor is charged by the current source. $V_2 - V_1 = I(1-D)T_s/C$ irrespective of the operating mode.

6.6.3.1 Sampler Mode ($RC \ll T_s$)

When ϕ_1 is high, the capacitor discharges rapidly to IR since $RC \ll T_s$. Consequently, $V_1 = IR$ as shown in Figure 6.52b. The average value of $v(t)$ can easily be calculated as $0.5I(1-D)^2(T_s/C) + IR + I(1-D)R$. The real part of $Z_0(0)$ is thus seen to be $0.5(1-D)^2(T_s/C) + (2-D)R$. The double-sided NSD at low frequencies is thus given by

$$S_v(0) \approx 2kT\left(\frac{(1-D)^2}{2}\frac{T_s}{C} + (2-D)R\right). \tag{6.57}$$

6.6.3.2 Mixer Mode ($RC \gg T_s$)

In this mode, the capacitor discharges (almost) linearly when ϕ_1 is high and attains the voltage V_1 at the end of the phase. The average voltage across R in this phase is $(V_1 + V_2)/2$. Consequently, the charge removed from the capacitor is $DT_s(V_1 + V_2)/(2R)$. This must be balanced by the charge injected by I into the capacitor, namely IT_s. The average value of $v(t)$, which is $(V_1 + V_2)/2$, is thus given by IR/D. The double-sided NSD at low frequencies is therefore seen to be $S_v(0) \approx 2kTR/D$.

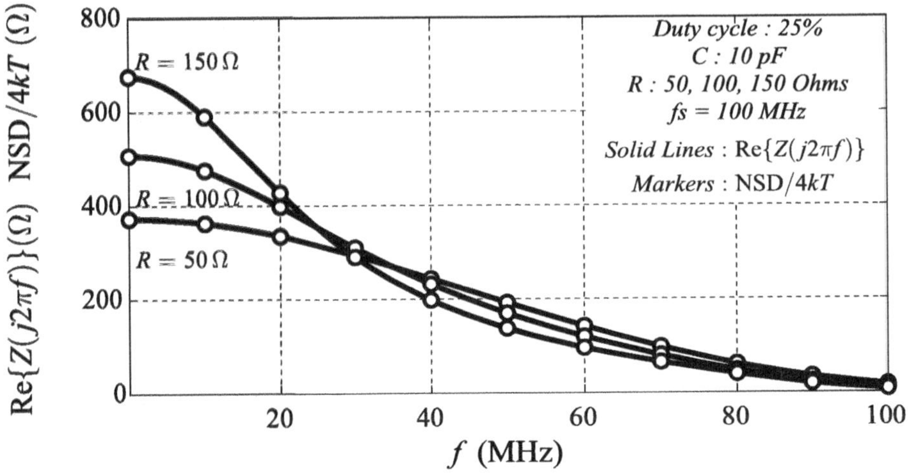

Figure 6.53 Comparing Re$\{Z(j2\pi f)\}$ (solid lines) and single-sided NSD/$4kT$ (markers) for a switched-RC kernel for different values of R.

Figure 6.53 compares the simulated (using Spectre) Re$\{Z(j2\pi f)\}$ (solid lines) and single-sided NSD (normalized to $4kT$) (markers) for the switched-RC kernel as R is varied. Excellent agreement is seen, in accordance with our theory.

6.6.4 First-Order Switched-Capacitor Filter

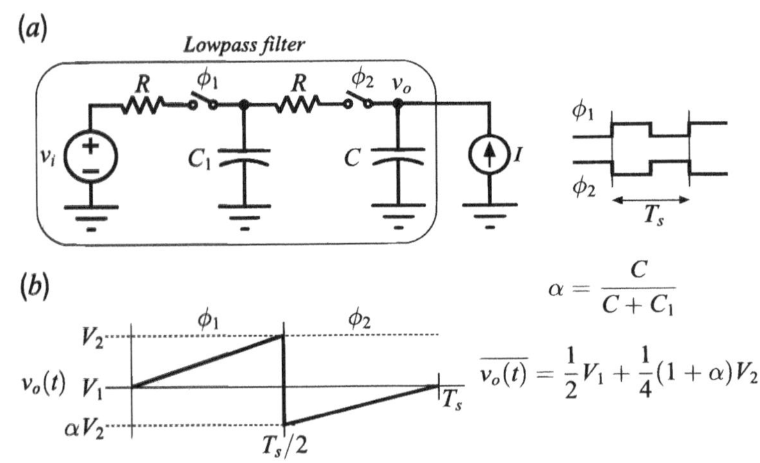

Figure 6.54 (a) A first-order switched-capacitor filter. (b) The capacitor voltage when $RC \ll T_s$ and $RC_1 \ll T_s$.

Consider the first-order switched-capacitor lowpass filter shown in Figure 6.54a. The switches are controlled by two non-overlapping waveforms with period T_s and 50% duty cycle. The resistance of the switches, denoted by R, is assumed to be very small. For $C_1 \ll C$, it is well known that C_1 behaves like a resistor of value T_s/C_1, and the 3 dB bandwidth of the filter is $f_s C_1/C$.

To determine the low-frequency noise spectral density at the output, we set $v_i = 0$ and inject a dc current I into the output node as shown in Figure 6.54a. $Z_0(0)$ is given by $\overline{v_o(t)}/I$ in steady state. To determine $v_o(t)$, we proceed as follows, as shown in Figure 6.54b.

a. During ϕ_1, C_1 is discharged. Denoting v_o at the beginning and end of ϕ_1 by V_1 and V_2 respectively, we see that

$$V_2 - V_1 = \frac{1}{2C} I T_s. \tag{6.58}$$

b. During ϕ_2, C_1 and C are connected together. Due to charge sharing, v_o (almost instantaneously) falls to αV_2 at the beginning of ϕ_2 where $\alpha = C/(C + C_1)$. v_o rises with a slope $I/(C + C_1)$ during this phase, and C_1 and C attain a voltage V_1 at the end of the phase. Consequently, we have

$$V_1 = \alpha V_2 + \frac{1}{2(C + C_1)} I T_s. \tag{6.59}$$

c. From Figure 6.54b, it is straightforward to see that

$$\overline{v_o(t)} = \frac{1}{2} V_1 + \frac{1}{4}(1 + \alpha) V_2. \tag{6.60}$$

Using Eqs. (6.58)–(6.60), we obtain

$$\begin{aligned} V_1 &= \frac{IT_s}{C_1}, \quad V_2 = \frac{IT_s}{C_1} + \frac{IT_s}{2C} \\ \overline{v_o(t)} &= \frac{IT_s}{C_1}\left(1 + \frac{C_1^2}{8C(C + C_1)}\right). \end{aligned}$$

The results make intuitive sense due to the following. In steady state, the average current flowing into C must be zero. Consequently, the current flowing into C (i.e., I) must equal that flowing out, which is $C_1 V_1 / T_s$. V_1 must thus be IT_s/C_1. Since $Z_0(0) = \overline{v_o(t)}/I$, we obtain the single-sided low-frequency noise spectral density as

$$S_v(0) = 4kT \frac{1}{f_s C_1}\left(1 + \frac{C_1^2}{8C(C + C_1)}\right). \tag{6.61}$$

Interestingly, $S_v(0)$ depends on C, reducing to $4kT/(f_s C_1)$ for $C_1 \ll C$.

The circuit of Figure 6.54a was simulated in Spectre for three values of C_1. $f_s = 1$ GHz, $C = 100$ pF and the switch resistance was 3 Ω. Figure 6.55 compares the simulated Re $\{Z_0(j2\pi f)\}$ and the output $NSD/4kT$. Excellent agreement is seen. Simulations indicate that the values of $Z_0(0)$ are 100.7, 27.4, and 18.7 Ω for C_1 =10, 40, and 70 pF respectively. Our analysis (6.61) predicts $Z_0(0)$ of 100.9, 27.9, and 18.5 Ω for these values of C_1, again in good agreement with simulations (given that we assumed switch resistance to be zero).

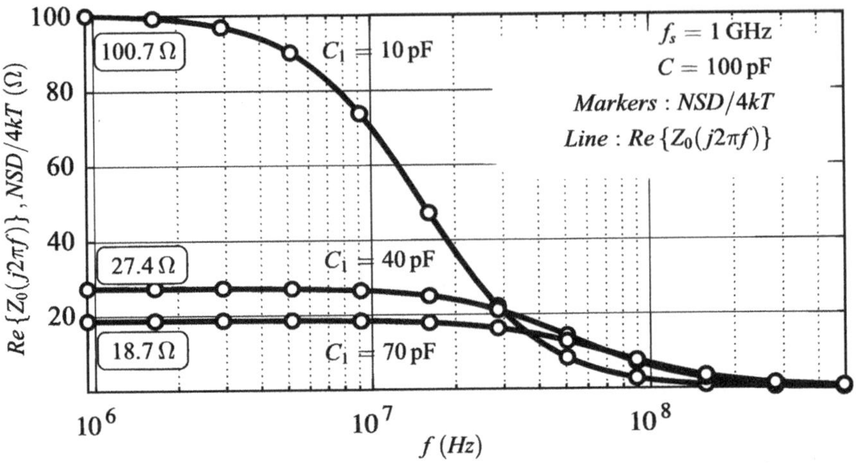

Figure 6.55 Re $\{Z_0(j2\pi f)\}$ and the output $NSD/4kT$. $C = 100$ pF, and the switch resistance is 3 Ω.

6.6.5 Charge-Rotating Switched-Capacitor Filter

Figure 6.56a shows a charge-rotating switched-capacitor filter. The non-overlapping switch-control phases, each with 25% duty cycle, are denoted by $\phi_s, \phi_1, \phi_2, \phi_3$. C_{2a}, C_{2b}, C_{2c} are identical capacitors whose value is C_2. It turns out that v_1, v_o, v_3 represent the outputs of successively higher-order transfer functions.

In the discussion that follows, we apply our theory to evaluate the low-frequency noise voltage spectral density at the output marked v_o. The single-sided density is given by $4kTZ_0(0)$, where $Z_0(0)$ is the zeroth-order impedance seen at the output node at dc. To determine $Z_0(0)$, we inject a dc current I into the output and determine $v_o(t)$ in steady state. $\overline{v_o(t)}/I$ yields $Z_0(0)$. We proceed in the following step-by-step fashion.

a. When ϕ_2 goes high, C_{2b}, which was at a voltage V_c at the end of ϕ_1 shares charge with C_1. Consequently, v_o drops to V_a at the beginning of ϕ_2, and increases at a rate $I/(C_1 + C_2)$. $(V_b - V_a)$ is thus given by ① in Figure 6.56b.

b. When ϕ_2 goes low, I completely flows into C_{2b} and $dv_o/dt = I/C_2$. $(V_c - V_b)$ is thus given ② by in Figure 6.56b.

c. The voltage across C_1, namely $v_x(t)$ tracks $v_o(t)$ in ϕ_2 and remains V_b during ϕ_3. During ϕ_s, v_x decays exponentially with a time constant RC_1, and attains a value V_d at the end of the phase. V_d is related to V_b as shown in ③ in Figure 6.56b.

d. The average value of the current flowing through the resistor R, namely $i(t)$, must be I since the average current through all the capacitors is zero. Charge flows into R only in ϕ_s, and is given by $C_1(V_b - V_d)$. Consequently, $\overline{i(t)}$ is $f_sC_1(V_b - V_d)$ as shown in ④ in Figure 6.56b.

e. In steady state, $v_1(t) = V_d$ and $v_3(t) = V_b$.

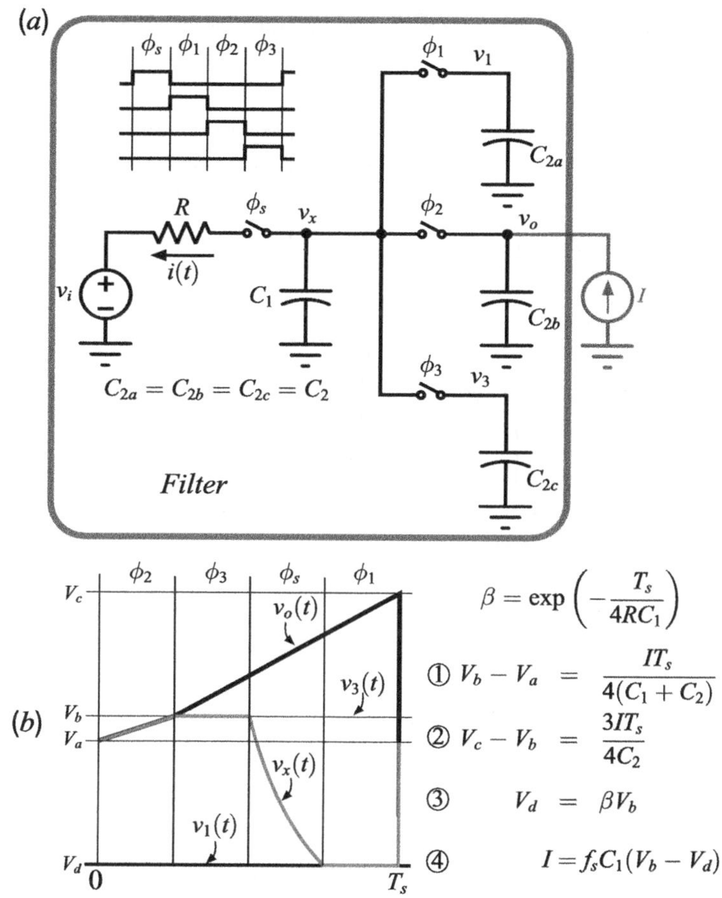

Figure 6.56 (a) A charge-rotating switched-capacitor filter. (b) The capacitor voltage when $RC_2 \ll T_s$ and $RC_1 \ll T_s$.

From Eqs. ①–④ in Figure 6.56b, we obtain

$$
\begin{aligned}
V_b &= \frac{I}{f_s C_1}\frac{1}{1-\beta}, \quad V_d = \beta V_b \\
V_c &= V_b + \frac{3IT_s}{4C_2}, \quad V_a = V_b - \frac{IT_s}{4(C_1 + C_2)}.
\end{aligned}
\tag{6.62}
$$

Furthermore, from Figure 6.56b, it is apparent that

$$\overline{v_o(t)} = \frac{1}{4}\frac{(V_a + V_b)}{2} + \frac{3}{4}\frac{(V_b + V_c)}{2} = \frac{1}{2}V_b + \frac{1}{8}V_a + \frac{3}{8}V_c.$$

From V_a, V_b, and V_c determined above, we obtain $Z_0(0) = \overline{v_o(t)}/I$, and the single-sided low-frequency noise voltage spectral density as

$$S_o(0) = \frac{4kT}{f_s C_1}\left\{\frac{1}{(1-\beta)} + \frac{9C_1}{32C_2} - \frac{C_1}{32(C_1 + C_2)}\right\}.$$

6.6.6 Total Integrated Noise in RLC Networks with Periodically Operated Switches

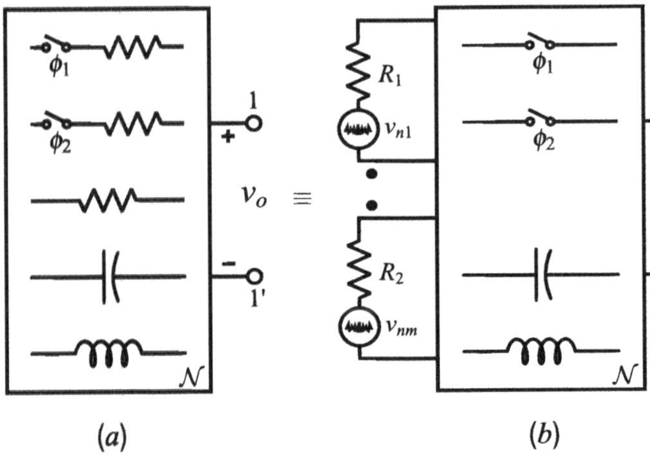

Figure 6.57 (a) The class of periodically-switched networks considered in this paper. (b) Equivalent, with *all* resistors (including the series resistances of the switches) brought out.

Consider a network that consists of linear inductors, capacitors, resistors, and periodically operated switches (all of which are operated at a frequency $f_s = 1/T_s$), as shown in Figure 6.57a. We show, for illustration, switches operated with two phases (ϕ_1 and ϕ_2) – but the number of phases chosen does not result in any loss of generality. Each switch is modeled as a resistor (with its appropriate thermal noise source) in series with an ideal switch, as shown in the figure. Figure 6.57b is the same network, but with *all* resistors (including those corresponding to the series resistance of the switches) "pulled outside the box." The noise source of the pth resistor R_p, labeled v_{np}, has a double-sided spectral density given by $2kTR_p$ V^2/Hz. The problem is to determine the mean square noise voltage $\overline{v_o^2}$. To this end, we first determine the mean square value of the sampled sequence $v_o[lT_s]$, where l is an integer.

As seen earlier, $v_o[lT_s]$ can be related to the noise voltage sources $v_{n1}, \cdots, v_{nm}$ through linear time-invariant filters whose impulse responses are $h_{\text{eq},1}(t), \cdots, h_{\text{eq},m}(t)$, as shown in Figure 6.58a. It is evident that the mean square value of $v_o[lT_s]$ is given by

$$\overline{v_o^2[lT_s]} = 2kT \sum_{p=1}^{m} R_p \int_0^{\infty} h_{\text{eq},p}^2(t)\, dt. \tag{6.63}$$

$h_{\text{eq},1}(t), \cdots, h_{\text{eq},m}(t)$ are obtained as follows. Excite the adjoint of the network of Figure 6.57b with a unit current impulse at $t = 0$, as shown in Figure 6.58b. $h_{\text{eq},p}(t)$ are obtained by dividing the currents $i_1(t), \cdots, i_m(t)$ by 1 coulomb.

Referring to the adjoint network of Figure 6.58b, we denote by E_0 and E_∞ the energies stored in the network at $t = 0+$ and $t = \infty$ respectively for a unit current-impulse

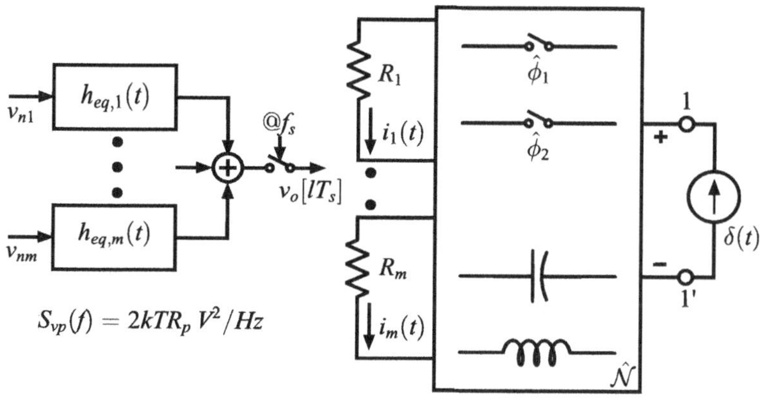

Figure 6.58 (a) Equivalent model to determine the sampled output noise of an LPTV network. (b) Determining $h_{eq,m}(t)$ for the network of Figure 6.57b using the adjoint network.

excitation at the output port. The difference between E_0 and E_∞ must be dissipated in the resistors, i.e., $E_0 - E_\infty = \sum_{p=1}^{m} R_p \int_0^\infty i_p^2(t)\,dt$. Since $i_p(t) = h_{eq,p}(t) \cdot 1$ coulomb, we have [3, 12]

$$E_0 - E_\infty = (1\ \text{coulomb})^2 \sum_{p=1}^{m} R_p \int_0^\infty h_{eq,p}^2(t)\,dt. \tag{6.64}$$

Using Eqs. (6.63) and (6.64), we have

$$\boxed{\overline{v_o^2[lT_s]} = \frac{2kT}{1\ \text{coulomb}^2}(E_0 - E_\infty).} \tag{6.65}$$

If the network's output is sampled at time instants $lT_s - t_o$ where $0 \leq t_o < T_s$, the model that relates the output sequence to the noise sources is still that of Figure 6.58a, except that the impulse responses are different. As earlier, the $h_{eq,m}(t)$ are determined by exciting the output port of the adjoint network of Figure 6.58b with an impulse current at $t = t_o$. Proceeding exactly as we did while deriving (6.65), it is easy to see that $\overline{v_o^2[lT_s - t_o]} = 2kT(E_{to} - E_\infty)$ where E_{to} is the energy delivered to the adjoint network at $t = t_o$ and E_∞ is the energy remaining in the network at $t = \infty$. E_{to} is proportional to the voltage developed across nodes $1 - 1'$ at $t = t_o+$ in Figure 6.58b when an impulse is injected at $t = t_o$. E_0, on the other hand, depends on the voltage developed across nodes $1 - 1'$ at $t = 0+$ when an impulse is injected at $t = 0$. At the instant at which the impulse is injected, no current flows through the resistors – thus, the voltage developed across $1 - 1'$ immediately after application of the impulse is independent of the instant at which the impulse is applied. As a result $E_{to} = E_0$ and $\overline{v_o^2[lT_s - t_o]} = \overline{v_o^2[lT_s]} = 2kT(E_0 - E_\infty)$ for *all* t_0. Since the mean-square value of the sampled voltage sequence is independent t_o, it follows that $\overline{v_o^2(t)} = 2kT(E_0 - E_\infty)$.

Applying (6.64) to determine the mean-square noise voltage across the capacitor C in Figure 6.33 yields, by inspection, $\overline{v_o^2} = kT/C$. (6.64) also applies to time-invariant linear RLC networks, where it is related to Bode's noise theorem.

The total noise in an LPTV network is usually computed in the frequency domain, by considering aliasing effects from many sidebands. This involves considerable algebra and computational effort (see, for instance, the complexity of determining the mean-square noise of the switched-RC kernel). Thanks to the use of reciprocity and the adjoint network, determining the total integrated noise in networks comprising of R,L,C and periodically operated switches is significantly simplified.

6.7 Application Examples

6.7.1 Example 1

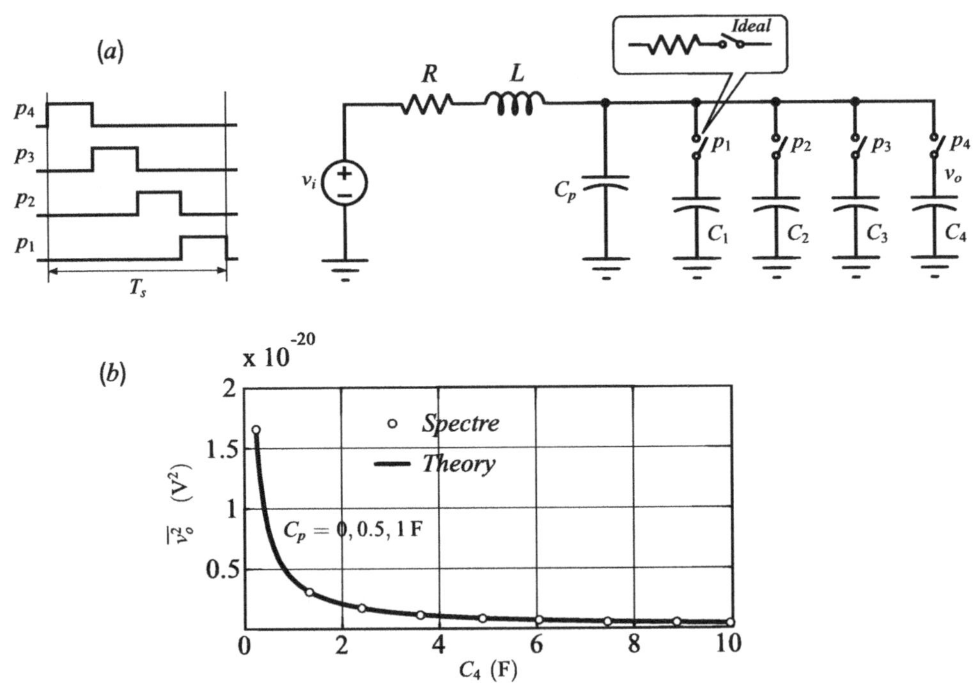

Figure 6.59 (a) Normalized N-path filter with parasitic source inductance and capacitance, with $L = 0.1$ H, $C_1 = 2.5$ F, $C_2 = 3$ F, $C_3 = 3$ F. $T_s = 1$ s, and $\tau = T_s/4$. (b) Comparison of analytical results with periodic noise simulations as C_4 is varied.

Figure 6.59a shows a normalized N-path filter (where the sampling rate is 1 Hz and the source resistance is 1 Ω), with parasitic source capacitance C_p and inductance L. If $C_1, \cdots, C_4$ are equal, and the voltage across C_p is treated as the output, an N-path filter results. We set out to determine the mean square noise across C_4, sampled at lT_s. Without loss of generality, we choose $f_s = 1$ Hz. $L = 0.1$ H, and C_1, C_2, C_3 are deliberately chosen to be unequal. $R_s = 1\ \Omega$ and the switches have a resistance of 40 mΩ. It is seen that $E_0 = 1/(2C_4)$. At $t = \infty$, all the capacitors get discharged and $E_\infty = 0$. We thus conclude that $\overline{v_o^2[lT_s]} = \frac{kT}{C_4}$. Figure 6.59b compares the results obtained using the equation above with that from periodic noise simulations of the circuit of Figure 6.59 as C_4 is swept, for different values of C_p. As expected from our analysis, $\overline{v_o^2[lT_s]}$ is independent of C_p. Determining the mean square noise using conventional techniques would be considerably harder, and involve summation of infinite series to account for noise aliasing from multiple sidebands.

Transient noise simulations were run on the N-path network of Figure 6.59a, with $C_4 = 2$ F. Figure 6.60a shows a representative voltage waveform across C_4. When ϕ_4 is low, the voltage across the capacitor is held, while resistive noise causes changes in v_o when ϕ_4 is high. The histograms in the second row of the figure are for $v_0[lT_s]$ and $v_0[lT_s + 0.9\,T_s]$. The standard deviations are virtually the same, and close to what we expect, namely $\sqrt{kT/C_4} = 45.5$ pV.

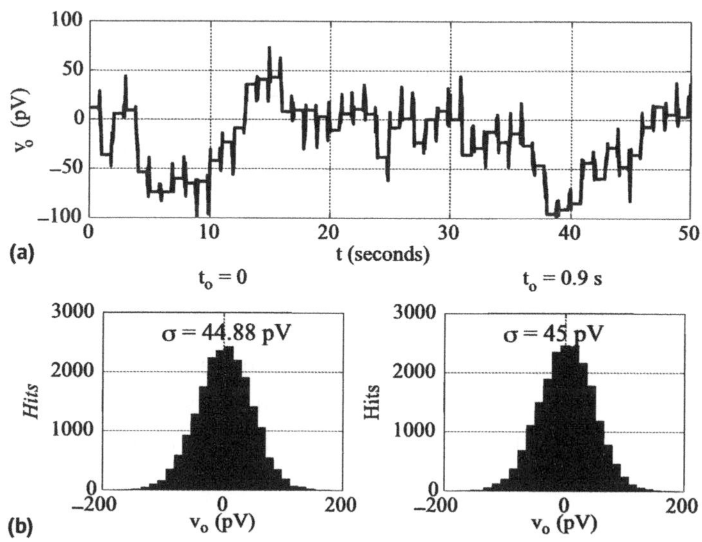

Figure 6.60 Top: v_o from a transient noise simulation of the circuit of Figure 6.59a for $C_4 = 2$ F; Bottom: Histograms of $v_o[lT_s + t_o]$ for two values of t_o.

6.7.2 Example 2

Consider the switched-capacitor low-pass filter shown in Figure 6.61. The switches have a resistance of 4 Ω. The switching frequency is chosen to be 1 Hz. To evaluate the mean square noise at the output node, we form the adjoint, and inject an impulse current. It is apparent that $E_0 = 1/(2C_1)$ and $E_\infty = 0$, resulting in

$$\overline{v_o^2[lT_s]} = \frac{kT}{C_1}. \tag{6.66}$$

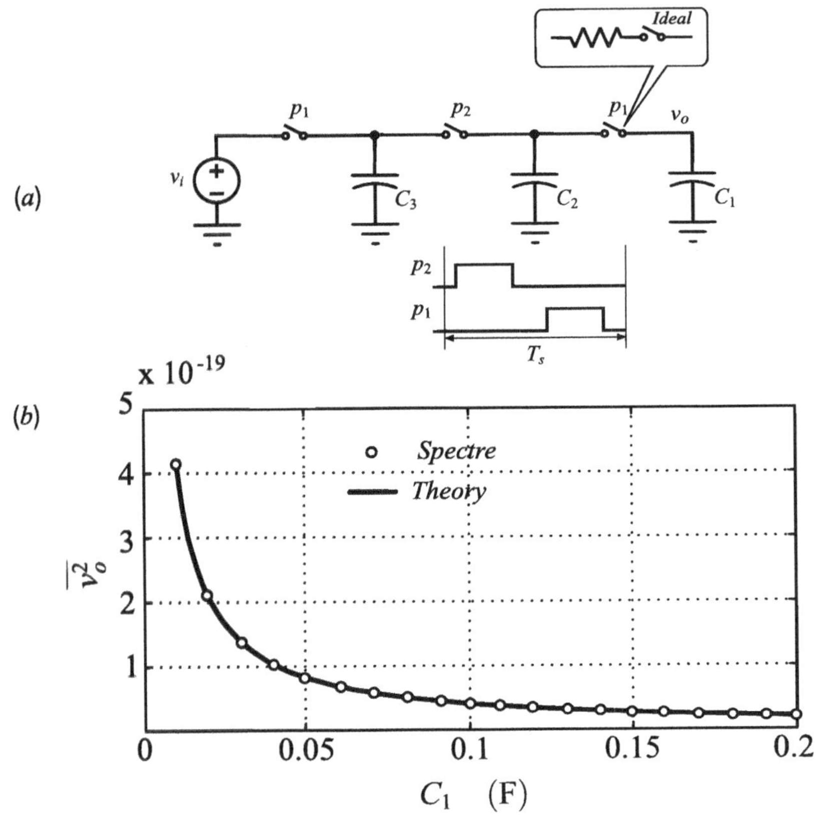

Figure 6.61 (a) Switched-capacitor lowpass filter. The mean-square noise across C_1 is of interest. (b) Comparison of analytical results with periodic noise simulations as C_1 is varied.

Figure 6.61b compares our analytical results with those obtained from periodic noise simulations. Excellent agreement is seen. Though not shown in the figure, simulations indicate (as expected) that the total noise is independent of C_2 and C_3.

6.7.3 Example 3

Lest the reader conclude that E_∞ is always zero, we show a network with a nonzero E_∞. Consider the circuit of Figure 6.62a. p_1 and p_2 are non-overlapping clocks. The switching frequency is 1 Hz. The mean square noise of the voltage across C_1, denoted by v_o, is of interest. Inspecting the circuit, it is apparent that $E_0 = 1/(2C_1)$. At $t = \infty$, steady-state is reached, and no current flows through the switches even when they are on. Thus, the voltages across C_1, C_2, C_3 are equal. The total charge on the top plates of these capacitors must be 1 coulomb—the voltage across them must thus be

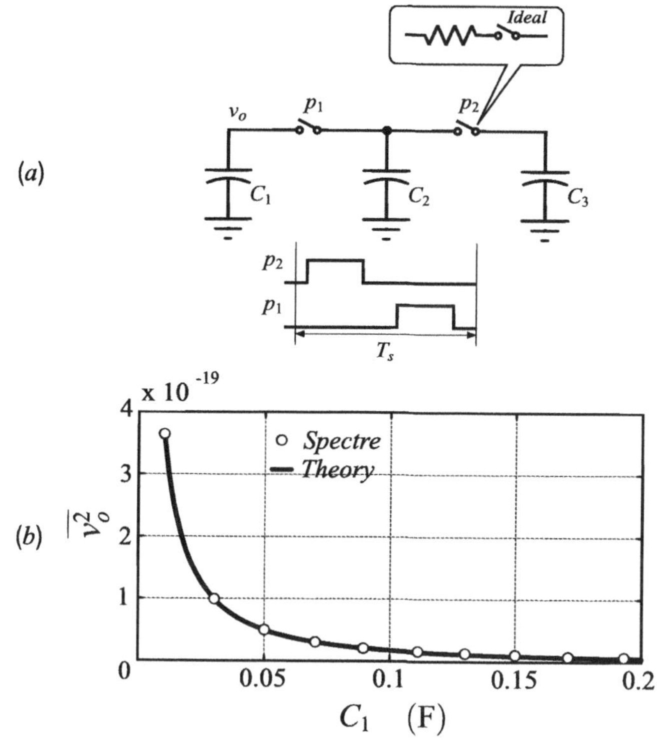

Figure 6.62 (a) Switched-C network. The mean square noise across C_1 is of interest. $C_2 = 25$ mF, $C_3 = 50$ mF. The resistance of the switches is 4 Ω and the switching frequency is 1 Hz. (b) Determining E_0 and $E\infty$.

$v_\infty = 1/(C_1 + C_2 + C_3)$. E_∞, which is $(1/2)v_\infty$ is given by $1/(2(C_1 + C_2 + C_3))$. The mean square output noise is thus given by

$$\overline{v_o^2[lT_s]} = kT\left(\frac{1}{C_1} - \frac{1}{C_1 + C_2 + C_3}\right).$$

Figure 6.62b compares our analytical results with those obtained from periodic noise simulations, for various values of C_1. Excellent agreement is seen. It is easy to see that E_0 and E_∞ remain the same even if p_1 and p_2 overlap, leading us to conclude that the mean-square noise does not change even if the clocks overlap.

Simulation Corner: Periodic Noise Analysis

The average NSD and total integrated noise in an LPTV network can be determined using the periodic noise analysis (referred to in most RF simulators as the `.PNOISE` analysis). The sequence of steps is the following.

a. Determine the periodic operating point using the `.PSS` analysis. This is necessary since the simulator, by default, assumes that the network is nonlinear. The first order of business, therefore, is to determine the periodic operating

(Continued)

(Continued)

point about which the network is linearized. The result is a small-signal network which is LPTV.

b. To find the average NSD at a frequency f, the $H_{-l}(j2\pi(f + lf_s))$ must be found for *all* noise sources in the network. This is accomplished in a computationally-efficient manner using the frequency-reversal theorem.

c. The spectral density of the noise sources is known from the components' noise models. A remark is in order here. The noise added by active components is typically a function of the operating point. For instance, the noise current added by a long-channel MOS transistor in saturation is given by $(8kT/3)g_m$, where g_m denotes the (small-signal) transconductance of the device. g_m, in turn, is given by $\sqrt{2\mu_n C_{ox}(W/L)I_d}$ where I_d denotes the quiescent drain current. If I_d varies periodically in a large-signal fashion, g_m also varies periodically with time. As far as noise is concerned, it is modeled as a stationary noise current multiplied by an appropriately (periodically) varying controlled source.

d. Assuming p uncorrelated stationary noise sources, the average NSD is given by

$$S_o(f) = \sum_{m=1}^{p} \sum_{l=-\infty}^{\infty} S_m(f + lf_s)|H_{-l,m}(j2\pi(f + lf_s))|^2. \tag{6.67}$$

In practice, the maximum number of sidebands that need to be used are an input to be specified by the user. A good strategy is to keep doubling the number of sidebands until the change in $S_o(f)$ is negligible.

e. Integrating the average NSD over the frequency range $0 - \infty$ gives the total integrated noise.

f. Some RF simulators have the option to determine the NSD of the LPTV system's sampled output. This is relevant when, for instance, the LPTV system is followed by an analog-to-digital converter that digitizes the system's sampled output.

Chapter Summary

- **Reciprocity and interreciprocity in LPTV networks.** The inter-reciprocal network corresponding to a given LPTV circuit can be obtained (as in the case of an LTI network) by replacing every element by one whose MNA stamp is the transpose of the element. For a time-varying element, this will correspond to time reversal.
- **The frequency-reversal theorem.** This is an immensely useful relationship that relates harmonic transfer functions in an LPTV network and its adjoint cousin.

- **LPTV networks with sampled outputs.** The sequence obtained by sampling the output of an LPTV network varying at f_s is identical to that obtained by sampling the output of an appropriately chosen linear time-invariant network excited by the same input as the LPTV system. The transfer function of the equivalent LTI system is the sum of the harmonic transfer functions of the LPTV network.
- **Average noise spectral density of an LPTV network.** To determine the average noise power at a frequency f, it is necessary to account for folding of noise from frequencies of the form $(f + lf_s)$ (for all integer l) to f.
- **Average voltage noise spectral density of a periodically-switched RLC network.** It turns out that the average single-sided noise voltage spectral density of a periodically-switched RLC one-port is given by $S_v(f) = 4kT\text{Re}\{Z_0(j2\pi f)\}$ where $Z_0(j2\pi f)$ denotes the zeroth-order harmonic driving-point impedance of the port. This is a generalization of the Nyquist noise theorem for time-invariant RLC networks.
- **Total integrated noise in periodically-switched RLC networks.** The energy-based method applicable to time-invariant RLC networks also works with periodically-switched RLC circuits. The mean-square noise is given by

$$\overline{v_n^2} = 2kT\frac{E_0 - E_\infty}{1\text{ coulomb}^2}$$

where E_0 and E_∞ denote the energy stored in the network at $t = 0+$ and $t = \infty$, respectively when a 1 coulomb charge is injected into the one-port.

Problems

Problem 1. *Derive the MNA stamp of an ideal opamp, which is a part of an LPTV circuit. Also determine the element that corresponds to the transpose of the stamp you just determined.*

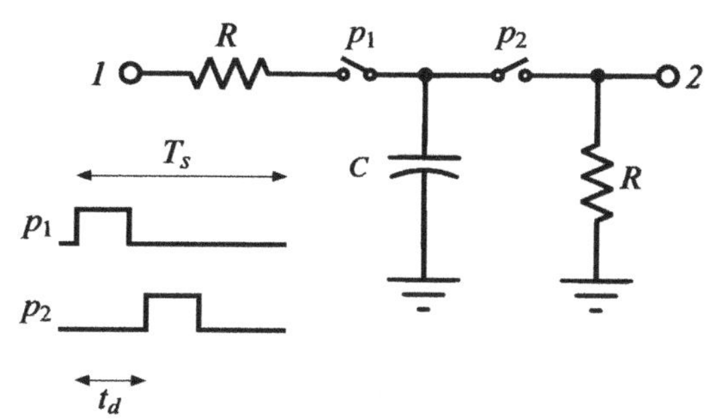

Figure 6.63

Problem 2. *Figure 6.63 shows a two-port network consisting of resistors, ideal periodically operated switches and a capacitor. The switch-control signals have a period*

$T_s = 1$ s, having 25% duty cycle. Choose $R = 1\ \Omega$ and $C = 10$ F. The delay between p_1 and p_2 is t_d. Let $H_0(j2\pi f)$ denote the voltage gain from port 1 to port 2 (zeroth-order HTF) when $v_1 = e^{j2\pi ft}$. Further, let $\hat{H}_0(j2\pi f)$ denote the current gain from port 2 to port 1 (zeroth-order HTF) when $i_2 = e^{j2\pi ft}$. (Obviously, port 1 is shorted for this test). In other words, these are the two standard measurements you would make in a time-invariant network to test for reciprocity. Compute $H_0(j2\pi f)$ and $\hat{H}_0(j2\pi f)$ over the range 0.8–1.2 Hz with a resolution of 0.01 Hz, and for $t_d = 0.3$, 0.4, and 0.5 s. Use $M = 256$. For each t_d, plot on a separate figure the magnitude and phase of the voltage- and current-gains. What do you notice?

Problem 3. *Consider a two-port $\mathcal{N}$. The zeroth-order harmonic admittance matrix (i.e., the y-matrix) is defined as follows.*

$$\begin{bmatrix} I_1 \\ I_2 \end{bmatrix} = \begin{bmatrix} y_{11}(j2\pi f) & y_{12}(j2\pi f) \\ y_{21}(j2\pi f) & y_{22}(j2\pi f) \end{bmatrix} \begin{bmatrix} V_1 \\ V_2 \end{bmatrix}$$

where I_1, I_2 and V_1, V_2 represent the phasors of the port currents and voltages, respectively, at the frequency f. To measure $y_{11}(j2\pi f)$, one would short port 2, apply $v_1(t) = e^{j2\pi ft}$ at port 1 and measure the complex amplitude of $i_1(t)$ at the frequency f. Other parameters are similarly measured.

Suppose $\mathcal{N}$ consists of resistors, capacitors, inductors, and just one periodically operated switch controlled by a rectangular wave with duty cycle D. Prove that $y_{12}(j2\pi f) = y_{21}(j2\pi f)$.

Problem 4. *A one-port network $\mathcal{N}$ consists of resistors, capacitors, inductors, and periodically operated switches. The switches are associated with a series resistance that adds thermal noise. Show that the average noise spectral density of the short circuit current is given by $4kT\,\mathrm{Re}\{Y_0(j2\pi f)\}$, where $Y_0(j2\pi f)$ denotes the zeroth-order admittance. It is determined by applying a voltage source $e^{j2\pi ft}$ across the port and measuring the component of current flowing into the network at the frequency f.*

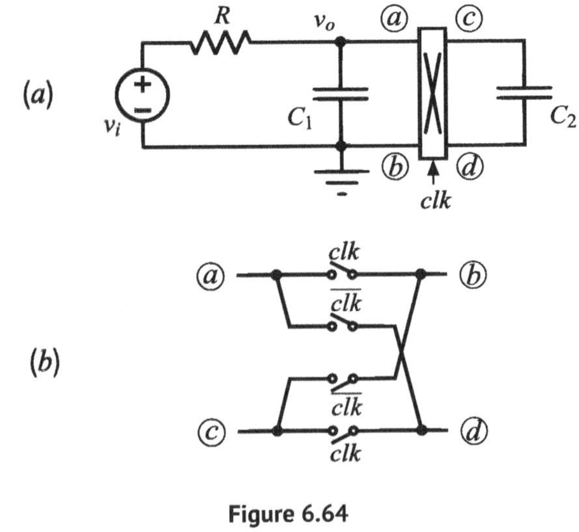

Figure 6.64

Problem 5. *Determine the average noise voltage spectral density at node ⓐ in Figure 6.64a. What is the mean-square noise? Repeat for the voltage at node ⓒ. The equivalent circuit of the chopper is shown in part (b) of the figure. clk has 50% duty cycle, and the switches have a series resistance, which add thermal noise.*

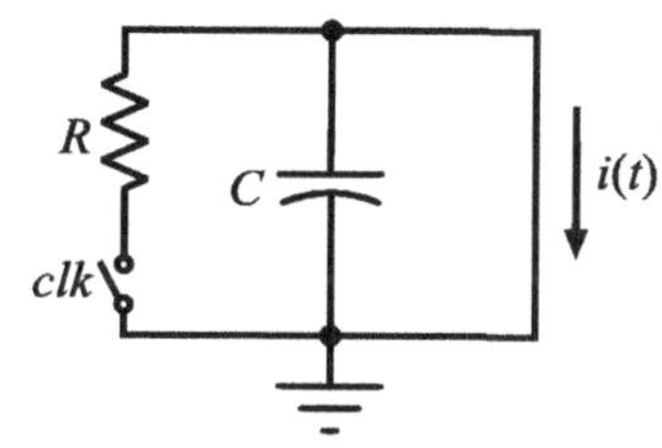

Figure 6.65

Problem 6. *Determine the average noise spectral density of $i(t)$ in Figure 6.65. The switch is ideal, and controlled by clk, which has a period T_s and duty cycle D.*

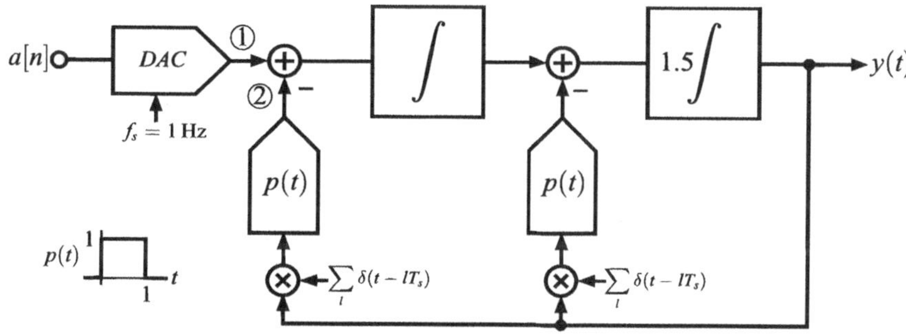

Figure 6.66

Problem 7. *Figure 6.66 shows a DAC driving an LPTV system. Determine the equivalent LTI filter relating the DAC output to $y(t)$. Draw the wave forms at ① and ② for $a[n] = \delta[n]$.*

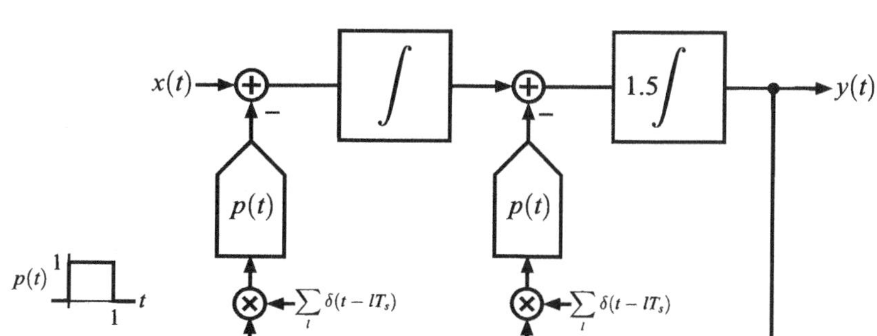

Figure 6.67

Problem 8. *In the LPTV system of Figure 6.67, determine the equivalent LTI filter relating $x(t)$ to the samples of $y(t)$, namely $y[n]$. $T_s = 1\,s$.*

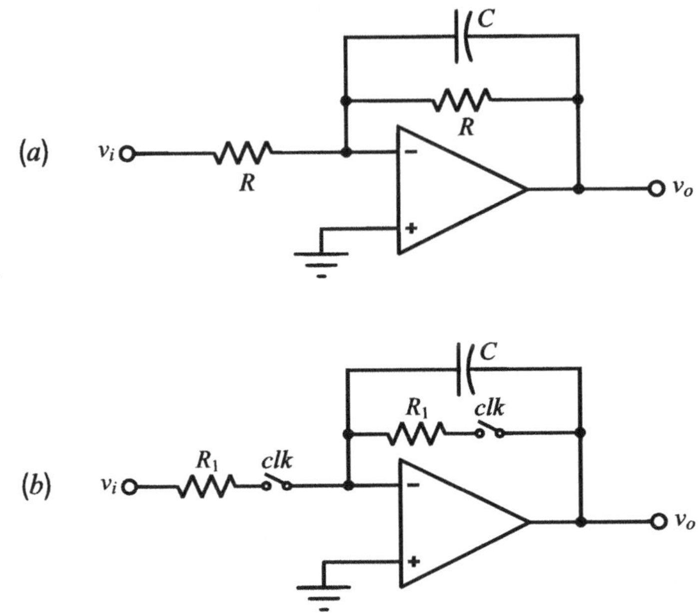

Figure 6.68

Problem 9. *Figure 6.68a shows a lowpass filter with a -3 dB bandwidth of $1/(2\pi RC)$. The opamp is ideal. Suppose C is fixed. If the bandwidth needed is very low, R can become impractically large. A way around this is to duty cycle a smaller (practically realizable) resistor R_1 as shown in Figure 6.68b. The clk signal controlling the switches has a period $T_s \ll RC$, and a duty cycle denoted by D. Estimate the zeroth- and first-order harmonic transfer functions of the switched-RC lowpass filter.*

Determine the low-frequency noise voltage spectral density, and mean-square noise at the outputs of the LTI and LPTV filters.

Write a computer program to calculate the harmonic transfer functions for $C = 10\,pF$, a -3 dB bandwidth of 2 kHz, and $f_s = 5$ MHz. What should R_1 be for $D = 10\%$? Compare your estimated harmonic transfer functions with the computed ones.

Problem 10. *Suppose the output of the switched-RC filter in Figure 6.68b was sampled at the rising edge of clk. Determine the impulse response of the equivalent LTI filter relating v_i to $v_o[nT_s]$. Compare the frequency response of this filter to that of the one in Figure 6.68a.*

References

1 S. Pavan and G. C. Temes, "Reciprocity and inter-reciprocity: A tutorial—part 2: Linear periodically time-varying networks," *IEEE Transactions on Circuits and Systems I: Regular Papers*, vol. 70, no. 9, pp. 3422–3435, 2023.

2 F. Yuan and A. Opal, "An efficient method for the analysis of periodically switched linear networks," *IEEE Transactions on Circuits and Systems I: Fundamental Theory and Applications*, vol. 46, no. 6, pp. 682–694, 1999.

3 S. Pavan, "Noise in periodically-switched RLC networks," *IEEE Transactions on Circuits and Systems II: Express Briefs*, vol. 72, no. 12, pp. 1847–1851, 2025.

4 K. Kundert, *The Designer's Guide to SpectreRF*. Cadence Design Systems, 2004.

5 C. C. Enz and G. C. Temes, "Circuit techniques for reducing the effects of op-amp imperfections: Autozeroing, correlated double sampling, and chopper stabilization," *Proceedings of the IEEE*, vol. 84, no. 11, pp. 1584–1614, 1996.

6 S. Pavan and R. S. Rajan, "Interreciprocity in linear periodically time-varying networks with sampled outputs," *IEEE Transactions on Circuits and Systems II: Express Briefs*, vol. 61, no. 9, pp. 686–690, Sep. 2014.

7 S. Pavan and E. A. M. Klumperink, "Simplified unified analysis of switched-RC passive mixers, samplers, and N-path filters using the adjoint network," *IEEE Transactions on Circuits and Systems I: Regular Papers*, vol. 64, no. 10, pp. 2714–2725, 2017.

8 S. Pavan and E. A. M. Klumperink, "Analysis of the effect of source capacitance and inductance on N-path mixers and filters," *IEEE Transactions on Circuits and Systems I: Regular Papers*, vol. 65, no. 5, pp. 1469–1480, 2018.

9 S. Pavan, "On linear periodically time varying (LPTV) systems with modulated inputs, and their application to smoothing filters," in *IEEE International Symposium on Circuits and Systems (ISCAS)*, 2017, pp.1–4.

10 J. Roychowdhury, D. Long, and P. Feldmann, "Cyclostationary noise analysis of large RF circuits with multitone excitations," *IEEE Journal of Solid-State Circuits*, vol. 33, no. 3, pp. 324–336, 1998.

11 F. Yuan and A. Opal, "Adjoint network of periodically switched linear circuits with applications to noise analysis," *IEEE Transactions on Circuits and Systems I: Fundamental Theory and Applications*, vol. 48, no. 2, pp. 139–151, 2001.

12 S. Pavan, "Simplified analysis of total integrated noise in passive switched-capacitor and N-path filters," in *Proceedings of the IEEE International Symposium on Circuits and Systems (ISCAS)*, 2019, pp.1–5.

7

Continuous-Time Filters

Continuous-time filters are important subsystems that are in wide use in analog signal chains. The area of filter design is vast, and entire books devoted to each aspect of filters and their design have been written. This chapter focuses on those aspects necessary to understand the operation of these analog signal-processing blocks, and develop intuition for the trade-offs involved. We restrict ourselves to topics that are relevant in the implementation of integrated filters. In other words, we hope to address the following questions.

a. How are integrated continuous-time filters most commonly used?
b. How does one derive their transfer functions?
c. What is a reasonable way to implement these filters?
d. What practical nonidealities and trade-offs should one be aware of?

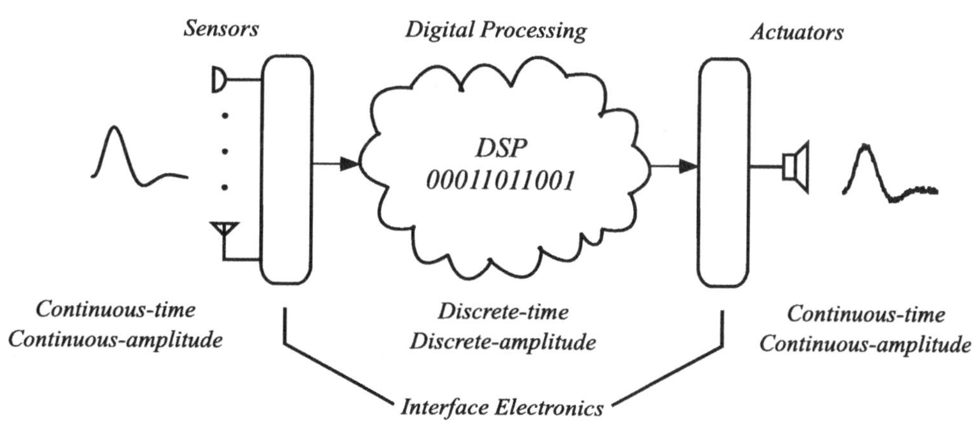

Figure 7.1 Block diagram of a modern signal processing system.

Figure 7.1 shows the block diagram of a modern signal-processing system. It is evident that contemporary signal chains rely heavily on digital processing. The real world, however, remains stubbornly analog (and is likely to remain so in the near future). It is therefore necessary to be able to "translate" signals from the analog world—namely those gathered by sensors like accelerometers, antennas, microphones, etc. into digital form. This is accomplished by analog-to-digital converters (ADC), which convert analog signals (i.e., those that are continuous in level and time) to digital ones, which are discrete-time, discrete-level signals. The ADC, therefore, serves as the link between the analog and digital worlds. It accomplishes this by sampling and quantization. The mathematical basis for sampling is Nyquist's theorem, which states that a continuous-time signal with a bandwidth B can be recovered from its samples if the sampling rate (f_s) exceeds $2B$. Though sensors output signals with a limited bandwidth, direct sampling and digitization of these outputs is impractical due to the following.

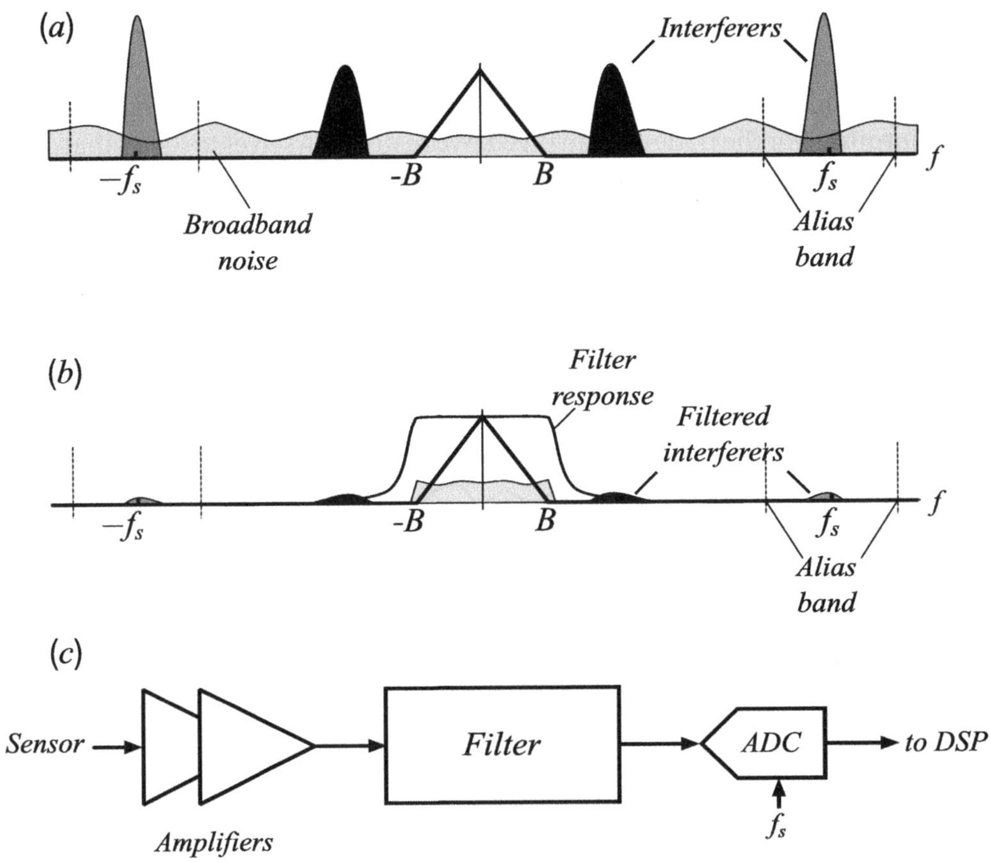

Figure 7.2 (a) Example spectrum at the output of a sensor. (b) Frequency response of the anti-alias filter and the resulting filtered spectrum. (c) Block diagram of the signal chain incorporating an ADC.

The sensor output, besides containing the desired signal, is almost always corrupted by noise and/or interferers. An example spectrum is shown in Figure 7.2a. The signal to be digitized has a bandwidth B. Though the sampling rate f_s satisfies Nyquist's criterion with respect to the desired signal, the same is not true for noise and interference. Specifically, noise and interference in the alias bands, namely signals in the range $mf_s \pm B$ where m is an integer, will be indistinguishable from the desired signal after sampling. To address this issue and enhance signal fidelity, a pre-sampling step is essential. This step is filtering, which aims to reduce/eliminate aliasing problems by adequately attenuating out-of-band noise and interferers. This way, the signal presented to the sampler is almost band limited, thereby satisfying Nyquist's sampling theorem. The filter is termed the *anti-alias filter*. Figure 7.2b shows an example anti-alias filter response and the spectrum after filtering. We see that out-of-band noise is largely eliminated, and interferers are attenuated by a large factor.

The block diagram of the signal chain is shown in Figure 7.2c. The amplifiers are needed to increase the signal level so that the input to the ADC occupies the latter's full scale. The filter aims to eliminate all signals other than the desired one. Ideally, the filter's response should be the so-called brick-wall response, i.e., it should let through frequencies in the range 0-B without any attenuation, while completely eliminating frequencies beyond B. This ideal, of course, can only be approximated in practice.

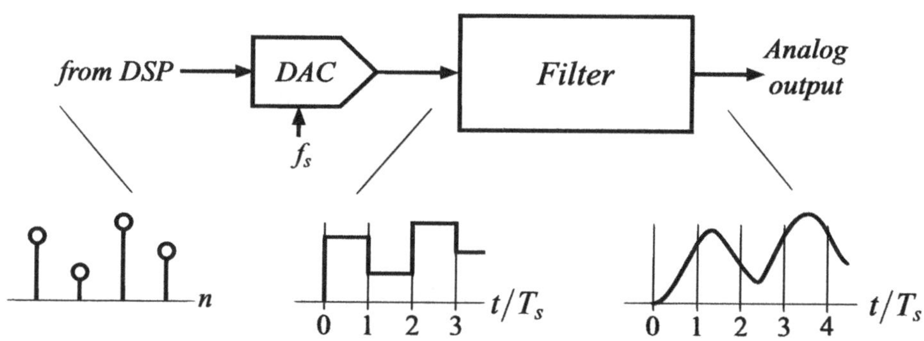

Figure 7.3 Block diagram of the signal chain that converts digital signals to analog form. The smoothing filter removes images from the DAC's output spectrum.

Filtering is also essential when the digital output provided by the DSP has to be converted to analog form. The reason is that the spectrum of the DAC output contains *images* centered around multiples of the sampling rate. These images, which can be problematic in practice, need to be removed. This is accomplished by the so-called *smoothing filter* (also referred to as the reconstruction filter). The block diagram of the digital-to-analog conversion signal path is shown in Figure 7.3. The output of the DAC has step discontinuities (which appear as images in the spectrum). The smoothing filter, as the name indicates, removes these abrupt changes and results in a smooth output waveform. In the frequency domain, this corresponds to attenuation of images.

To summarize, sharp low-pass filters are needed before an ADC (for antialiasing) and after the DAC (for smoothing). Ideally, such filters need to have a brick-wall characteristic. The passband is defined as the frequency range from 0 to ω_o, and the goal is to ensure that the filter's magnitude response is 1 in this range. For $\omega > \omega_o$, the response must be 0. The specification would thus read:

$$\begin{aligned} |H(j\omega)| &= 1, \quad \text{for } \omega < \omega_o \\ &= 0, \quad \text{for } \omega > \omega_o \end{aligned} \tag{7.1}$$

The frequency response $H(j\omega)$, corresponding to the transfer function $H(s)$, is constrained by the fact that the filter consists of resistors, capacitors, and inductors, resulting in a transfer function that is a ratio of polynomials in s:

$$H(s) = \frac{N(s)}{D(s)} = \frac{1 + b_1 s + \cdots b_m s^m}{1 + a_1 s + \cdots a_n s^n}. \tag{7.2}$$

The first order of business, therefore, is to determine the coefficients of $N(s)$ and $D(s)$ so that $|H(j\omega)|$ satisfies the magnitude response specifications. It is, of course, impossible for a ratio of polynomials to result in a response that is an ideal brick wall—we have to be content with an approximation. Intuitively, it seems that using higher-order polynomials will result in a better approximation to the ideal.

As a starting point, simplifying the design by considering basic forms for $N(s)$ and $D(s)$ is a pragmatic approach. Assuming $N(s) = 1$ is a good starting point. In this scenario, the transfer function $H(s)$ is simply given by $H(s) = 1/D(s)$. Such filters are called *all-pole filters*. The term "all-pole" signifies that the transfer function has only poles and no zeros. Recall that poles are complex frequencies where the transfer function goes to infinity, while zeros are complex frequencies where the transfer function goes to zero.

7.1 Magnitude-Approximation Principles

The problem at hand is to determine the transfer function $H(s)$ whose magnitude response has some desired properties. For an all-pole filter

$$H(s) = \frac{1}{D(s)}. \tag{7.3}$$

The denominator $D(s)$ represents an nth-order polynomial, and the condition for a DC gain of 1, implies that $D(s)$ must take the form $1 + a_1 s + a_2 s^2 + \cdots + a_n s^n$. Therefore, mathematically, the problem is to find the coefficients a_1 through a_n such that the magnitude response of the transfer function approximates the ideal brick wall in some sense. How does one relate the magnitude response to the transfer function? Well, that is straightforward:

$$|H(j\omega)|^2 = H(j\omega)H^*(j\omega) = H(j\omega)H(-j\omega) = H(s)H(-s)|_{s=j\omega}. \tag{7.4}$$

To determine $|H(j\omega)|^2$, therefore, we must first form the quantity $H(s)H(-s)$, and evaluate it at $s = j\omega$. The squared magnitude response is a real and even function of frequency [1], so we should expect that $|H(j\omega)|^2$ is a ratio of polynomials in ω^2. Let us indeed see that this is the case. We collect terms in $D(s)$ that contain even powers of s and denote the result as $D_0(s)$. Similarly, terms with odd powers of s are collected as $sD_1(s)$. We see that $D(s)$ can be written as

$$D(s) = 1 + a_1 s + a_2 s^2 + \cdots + a_n s^n = D_0(s) + sD_1(s). \tag{7.5}$$

Note that $D_0(s)$ and $D_1(s)$ are polynomials in s^2. $D(s)D(-s)$ is thus given by

$$D(s)D(-s) = D_0^2(s) - s^2 D_1^2(s). \tag{7.6}$$

To obtain $|D(j\omega)|^2$, we replace s with $j\omega$, which is equivalent to replacing s^2 with $-\omega^2$. Since $D_0(s)$ and $D_1(s)$ are real polynomials in s^2, it follows that $D(s)D(-s)$ must be a real polynomial in s^2.

Example: Magnitude Response of a Second-Order All-Pole Filter

Consider the second-order transfer function

$$H(s) = \frac{1}{1 + a_1 s + a_2 s^2}. \tag{7.7}$$

$D_0(s) = 1 + a_2 s^2$, and $D_1(s) = a_1$. $D(s)D(-s) = (1 + a_2 s^2)^2 - a_1^2 s^2$, which means that

$$|H(j\omega)|^2 = \frac{1}{(1 - a_2\omega^2)^2 + a_1^2\omega^2}. \tag{7.8}$$

So far, we have obtained the squared magnitude response, and concluded that the denominator is a nth-order polynomial in ω^2. The coefficients of this polynomial are functions of the coefficients $a_1, a_2, \cdots, a_n$ of the denominator polynomial $D(s)$. We move to the next (and key) step, namely—to determine the coefficients of the polynomial of $|D(j\omega)|^2$ so that the magnitude of $|H(j\omega)|^2$ achieves the desired specifications. There are many ways of doing this—after all, the ideal brick-wall response cannot be achieved, and there are several ways of approaching this ideal. We will examine this in greater detail going forward—but for the time being, *assume* that we have found the polynomial $|D(j\omega)|^2$ that results in the desired magnitude response. How do we obtain $D(s)$ from $|D(j\omega)|^2$?

Recall that $|D(j\omega)|^2$ was obtained by replacing s^2 with $-\omega^2$ in $D(s)D(-s)$. To determine $D(s)$, therefore, we proceed in the reverse fashion by replacing ω^2 with $-s^2$. The result is a $2n^{th}$-order polynomial in s. The $2n$ roots of this polynomial are found. Since the polynomial can be factored as $D(s)D(-s)$, it follows that the roots exhibit conjugate- and mirror-symmetry. In other words, if p_1 is a root of $D(s)D(-s)$, p_1^*, $-p_1$, and $-p_1^*$ are also roots. To determine $D(s)$ therefore, we collect all poles in the left-half s-plane. Denoting them by $p_1, p_2, \cdots, p_n$, the denominator polynomial is seen to be

$$D(s) = \left(1 - \frac{s}{p_1}\right) \cdots \left(1 - \frac{s}{p_n}\right). \tag{7.9}$$

Example: Going from $|H(j\omega)|^2$ to $H(s)$

Suppose that $|H(j\omega)|^2$ is obtained as

$$|H(j\omega)|^2 = \frac{1}{1+\omega^2}. \tag{7.10}$$

$|D(j\omega)|^2 = 1 + \omega^2$. Replacing ω^2 with $-s^2$, we obtain $D(s)D(-s) = 1 - s^2$. The roots of $(1 - s^2)$ are ± 1. The root of $D(s)$ is the one in the left half s-plane, namely -1. The transfer function is thus given by

$$H(s) = \frac{1}{1+s}. \tag{7.11}$$

To summarize, the process of finding an all-pole transfer function $H(s) = 1/D(s)$ with a desired magnitude response is the following.

a. The squared magnitude response $|H(j\omega)|^2$ is of the form $1/|D(j\omega)|^2$ where $|D(j\omega)|^2$ is a polynomial in ω^2.
b. The polynomial $|D(j\omega)|^2$ is chosen so that $|H(j\omega)|^2$ satisfies the magnitude-response specifications.
c. Once $|D(j\omega)|^2$ is chosen, ω^2 is replaced by $-s^2$ to obtain a polynomial of order $2n$ in s. The left-half plane roots correspond to the roots of $D(s)$. Once the roots of $D(s)$ are known, $D(s)$ is determined using Eq. (7.9).

A significant portion of filter design originated in the context of approximating the ideal "brick wall" response, acknowledging the impossibility of achieving a perfect brick wall and aiming to approach this ideal by increasing the filter order. The ideal brick wall response targeted a magnitude response of 1 up to a cutoff frequency of ω_o rad/s. To avoid carrying terms involving ω_o in all our derivations, we first assume $\omega_o = 1$ rad/s. This choice, however, is not restrictive. Frequency scaling can be applied to extend the design to any desired cutoff frequency. For instance, if a transfer function $H(s)$ achieves a cutoff frequency of 1 rad/s, and one wishes to scale the cutoff frequency to ω_o rad/s, the transformation can be achieved by replacing s in $H(s)$ with s/ω_o. The new transfer function $\hat{H}(s)$ is given by

$$\hat{H}(s) = H\left(\frac{s}{\omega_o}\right). \tag{7.12}$$

In a similar vein, if a network with inductors, capacitors, and resistors yields a cutoff frequency of 1 rad/s, to change the cutoff frequency to ω_o rad/s, inductors and capacitors should be divided by ω_o, while resistors should be left unchanged.

7.2 The Maximally Flat (Butterworth) Approximation

The transfer function $H(s)$ for an all-pole filter is represented as $1/D(s)$, where $D(s)$ is a polynomial. To determine the squared magnitude response $|H(j\omega)|^2$, one evaluates $H(s) \cdot H(-s)$ at $s = j\omega$. This process yields a ratio of polynomials in ω^2, with the numerator being 1, and the denominator taking the form $1 + k_1\omega^2 + k_2\omega^4 + \cdots + k_n\omega^{2n}$. Denoting $k_1\omega^2 + k_2\omega^4 + \cdots + k_n\omega^{2n}$ by $F(\omega^2)$, we have

$$|H(j\omega)|^2 = H(s)H(-s)|_{s=j\omega} = \frac{1}{D(s)D(-s)|_{s=j\omega}} = \frac{1}{1 + \underbrace{k_1\omega^2 + \cdots + k_n\omega^{2n}}_{F(\omega^2)}} \tag{7.13}$$

$$= \frac{1}{1 + F(\omega^2)}.$$

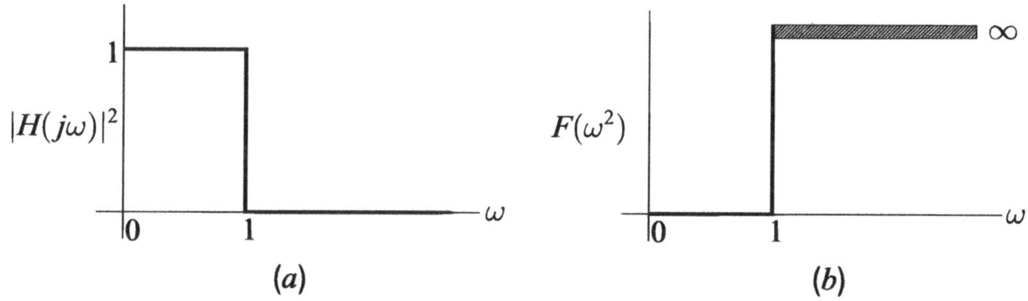

Figure 7.4 (a) Brickwall response to be approximated. (b) Corresponding plot of $F(\omega^2)$.

If $|H(j\omega)|$ has to approximate a brick wall, so does $|H(j\omega)|^2$ (see Figure 7.4). Working with the squared magnitude response saves us the bother of having to deal with square roots. The task at hand, therefore, is to determine $F(\omega^2)$ or equivalently the values of $k_1, \cdots, k_n$ such that the squared magnitude response approximates a brick wall. In mathematical terms, this implies that $1 + F(\omega^2)$ should approximate 1 in the range $[0, 1]$ and go to infinity beyond $\omega = 1$. Equivalently, the goal is to find a polynomial $F(\omega^2)$ such that it is 0 for $\omega < 1$ and approaches infinity for $\omega > 1$. However, finding a polynomial that satisfies these conditions is not possible. The solution involves finding a good approximation. We reason as follows.

At dc ($\omega = 0$), $F(\omega^2) = 0$, which satisfies the requirement. For frequencies slightly away from DC (i.e., $\omega \neq 0$ but $\omega \ll 1$), $F(\omega^2)$ is approximately given by the lowest-order term, namely $k_1\omega^2$. Since the goal is to make $F(\omega^2) = 0$ in the pass band, $k_1\omega^2$ must be set to 0, which is achieved by choosing $k_1 = 0$. Continuing this line of thought, we see that for $\omega \ll 1$, $F(\omega^2)$ is now approximately given by $k_2\omega^4$. Since we want this term to be 0, it follows that $k_2 = 0$. Extending this argument, it is straightforward to see that $k_1, k_2, \cdots, k_{n-1}$ all must be chosen to be zero. What about k_n? Tempting as it may be to set this to zero, we see that this would indeed result in a flat gain in the pass band—but without any frequency selectivity! Therefore, setting k_n to 0 is not feasible. We choose $k_n = 1$ so that $F(\omega^2) = 1$ at $\omega = 1$; this way, $|H(j\omega)|^2$ will be 0.5 at $\omega = 1$, irrespective of n.

We thus see that the function $F(\omega^2) = \omega^{2n}$ (and consequently $|H(j\omega)|^2$) is "as flat as it can possibly be" at $\omega = 0$. The resulting $|H(j\omega)|^2$ is therefore called the maximally flat magnitude response.

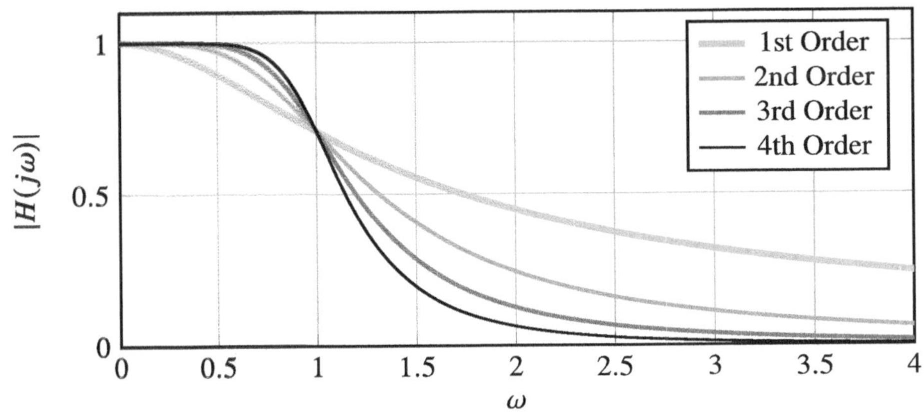

Figure 7.5 Magnitude responses of Butterworth low-pass filters from order 1 to 4. The −3 dB cut-off frequency is 1 rad/s.

Figure 7.5 shows the magnitude responses of first-, second-, third-, and fourth-order maximally-flat low-pass filters with a −3 dB bandwidth of 1 rad/s. As n increases, $|H(j\omega)|^2$ becomes flatter in the pass band and falls more sharply in the stop band, thereby approximating the desired brick-wall response more effectively. In general, the concept of maximal flatness refers to achieving a flat magnitude response at a specific frequency, and higher-order derivatives are made zero to accomplish this flatness. The magnitude response of an nth-order maximally-flat low-pass filter is thus given by

$$\boxed{|H(j\omega)| = \frac{1}{\sqrt{1+\omega^{2n}}}}. \tag{7.14}$$

We have just derived the Butterworth magnitude response, named after the British engineer Stephen Butterworth. The Butterworth filter is designed to achieve a maximally-flat magnitude response around dc, making it a useful approximation for applications where a flat response is desired in the pass band. Let us now address the process of obtaining $H(s)$ from $|H(j\omega)|$.

$$D(s)D(-s) = 1 + (-s^2)^n \tag{7.15}$$

For odd n, $s^{2n} = 1$ and the roots are given by

$$p_k = e^{j\frac{2\pi k}{2n}}, \quad k = 1, \cdots, 2n. \tag{7.16}$$

For even n, $s^{2n} = -1$ and the roots of $D(s)D(-s)$ are given by

$$p_k = e^{j\frac{\pi(2k-1)}{2n}}, \quad k = 1, \cdots, 2n. \tag{7.17}$$

Note that the roots lie on a circle with unit radius.

Example: Denominator Polynomial of a Third-Order Butterworth Filter

For $n = 3$, the roots of $D(s)D(-s)$ are given by

$$e^{j\frac{2\pi}{6}}, e^{j\frac{4\pi}{6}}, e^{j\frac{6\pi}{6}}, e^{j\frac{8\pi}{6}}, e^{j\frac{10\pi}{6}}, e^{j\frac{12\pi}{6}}. \tag{7.18}$$

The roots lie on a circle, as shown in Figure 7.6. Of these, the poles in the left half plane are

$$\begin{aligned} p_2 &= e^{j\frac{4\pi}{6}} = e^{j\frac{2\pi}{3}} \\ p_3 &= e^{j\frac{6\pi}{6}} = -1 \\ p_4 &= e^{j\frac{8\pi}{6}} = e^{-j\frac{2\pi}{3}} \end{aligned} \tag{7.19}$$

$D(s)$ is thus given by $(1 - s/p_2)(1 - s/p_3)(1 - s/p_4)$ which turns out to be

$$D(s) = (1 + s)(1 + s + s^2). \tag{7.20}$$

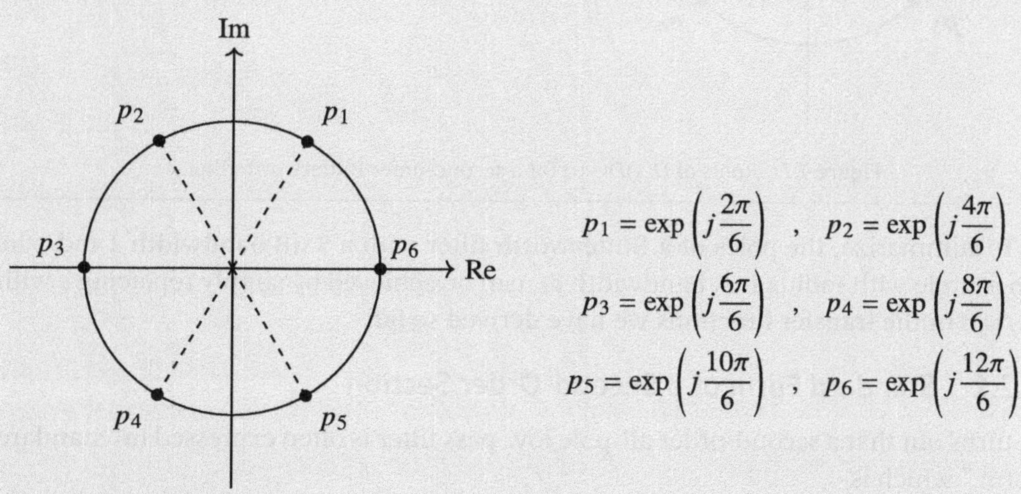

Figure 7.6 Roots of $D(s)D(-s)$ for a third-order Butterworth filter.

Example: Denominator Polynomial of a Second-Order Butterworth Filter

For $n = 2$, the roots of $D(s)D(-s)$ are given by

$$e^{j\frac{\pi}{4}}, e^{j\frac{3\pi}{4}}, e^{j\frac{5\pi}{4}}, e^{j\frac{7\pi}{4}}. \tag{7.21}$$

The roots lie on a circle, as shown in Figure 7.7. Of these, the poles in the left half plane are

$$p_2 = e^{j\frac{3\pi}{4}} \\ p_3 = e^{j\frac{5\pi}{4}} \tag{7.22}$$

$D(s)$ is thus given by $(1 - s/p_2)(1 - s/p_3)$ which turns out to be

$$D(s) = (1 + \sqrt{2}s + s^2). \tag{7.23}$$

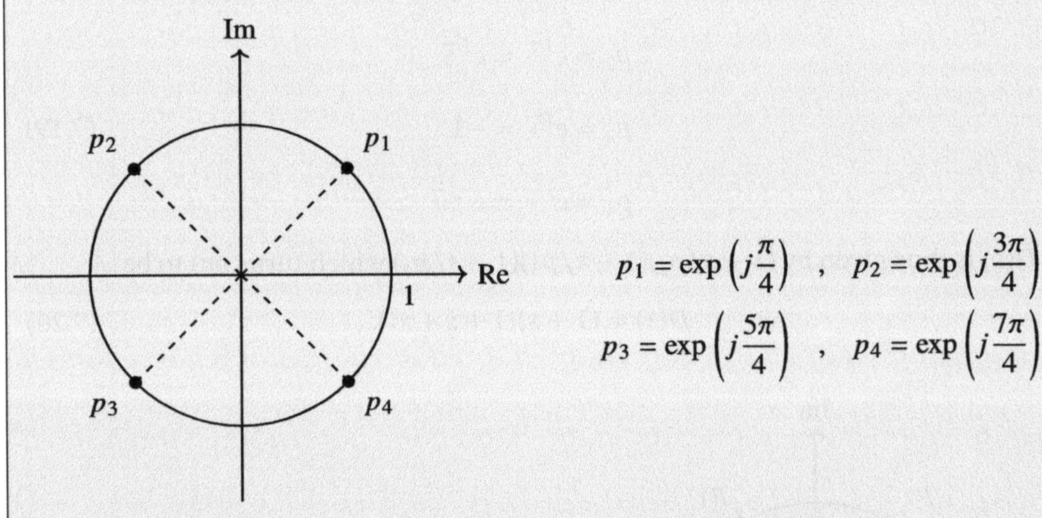

Figure 7.7 Roots of $D(s)D(-s)$ for a second-order Butterworth filter.

To summarize, the poles of a Butterworth filter with a 3 dB bandwidth 1 rad/s lie on a circle with radius 1. A bandwidth ω_o can be achieved by simply replacing s with (s/ω_o) in the transfer functions we have derived so far.

7.2.1 Standard Form of a Second-Order Section

It turns out that a second-order all-pole low-pass filter is often expressed in "standard form," which is

$$H(s) = \frac{G_o}{1 + \frac{s}{\omega_p Q_p} + \frac{s^2}{\omega_p^2}}. \tag{7.24}$$

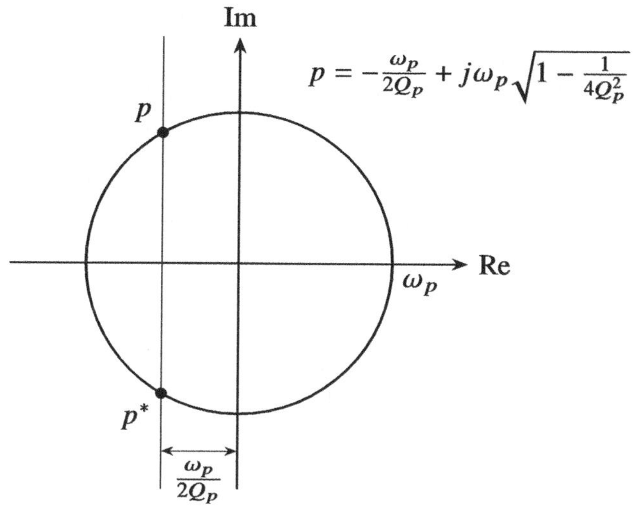

Figure 7.8 Pole positions for the standard form, for $Q_p > \frac{1}{2}$.

Here, G_o is the dc gain, ω_p is termed the "center frequency," and Q_p is called the "quality factor." For $Q_p > 1/2$, the poles are complex conjugates and lie on a circle with radius ω_p, as shown in Figure 7.8. The real part of p is given by $-\omega_p/(2Q_p)$. Such a transfer function is called a *biquadratic transfer function*, and blocks that realize such transfer functions are termed *biquads*.

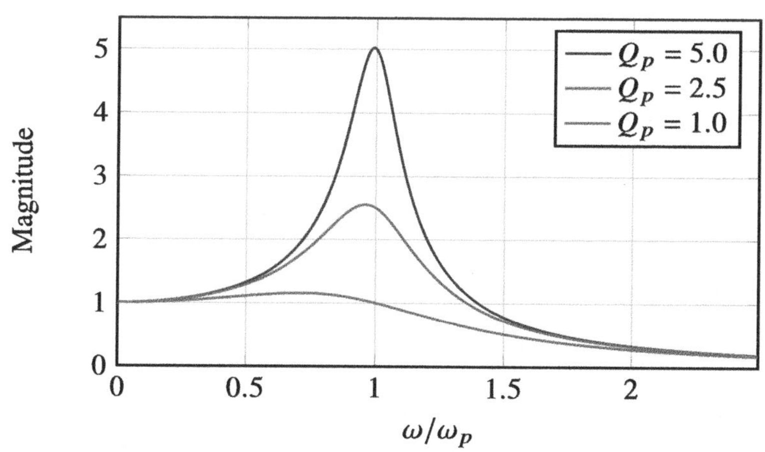

Figure 7.9 Magnitude response of $H(s)$, with $G_o = 1$ and varying Q_p.

Figure 7.9 shows the magnitude response of the standard-form filter for various values of Q_p. A few observations are in order.

a. The gain at $\omega = \omega_p$ is $G_o Q_p$.
b. As Q_p increases, the peak gain approaches $G_o Q_p$ and the frequency at which the peak occurs approaches ω_p.
c. A high-Q_p transfer function is associated with a response with a significant peak—namely, the response rises to and falls from its maximum value of $G_o Q_p$ over a small frequency range.

d. $\angle H(j\omega_p)$ is $-\pi/2$.

e. The group delay, defined as $\tau_g(\omega) = -d\angle H(j\omega)/d\omega$, is $1/(\omega_p Q_p)$ at dc and $2Q_p/\omega_p$ at $\omega = \omega_p$. Figure 7.10 shows the normalized group delay for various values of Q_p. It is evident that a large Q_p entails a significant peaking in the group delay.

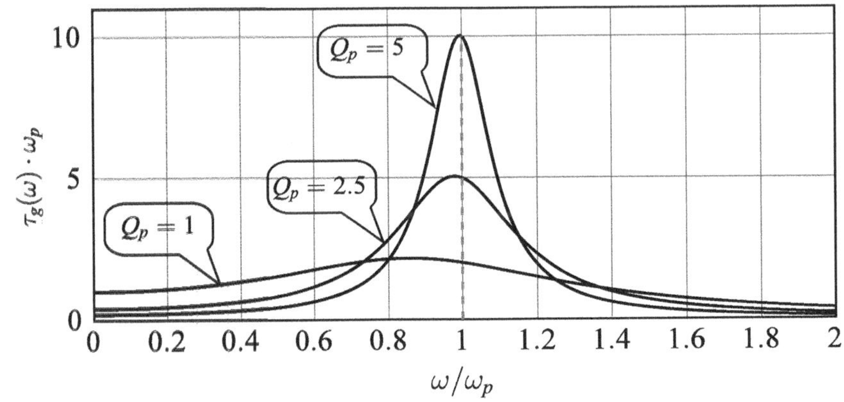

Figure 7.10 Normalized group delay for varying Q_p.

The impulse response corresponding to $H(s)$ is given by

$$h(t) = \frac{G_o \omega_p}{\sqrt{1 - (1/4Q_p^2)}} \cdot e^{-\frac{1}{2Q_p}\omega_p t} \cdot \sin\left(\omega_p \sqrt{1 - \frac{1}{4Q_p^2}} t\right) u(t). \tag{7.25}$$

As Q_p increases, the poles move closer to the $j\omega$ axis. Correspondingly, the impulse response "lasts" longer as the envelope decays at a slower rate. Furthermore, the frequency of ringing in the response approaches ω_p.

Table 7.1 gives the denominator polynomials (in factored form) for Butterworth filters of orders 1 through 5. The third column of the table gives the Q_p's of the second-order sections. Notice that the highest quality factor increases with order.

Table 7.1 Butterworth polynomials.

Order	Butterworth polynomial in factored form	Quality factors
1	$(s + 1)$	
2	$s^2 + \sqrt{2}s + 1$	0.71
3	$(s + 1)(s^2 + s + 1)$	1
4	$(s^2 + 0.77s + 1)(s^2 + 1.85s + 1)$	0.54,1.29
5	$(s + 1)(s^2 + 0.62s + 1)(s^2 + 1.62s + 1)$	0.62,1.62

Highest Pole-Q_p in an nth-Order Butterworth Filter

The highest-Q poles in a Butterworth filter are those closest to the $j\omega$ axis. It is straightforward to see that the angles of these poles are given by $\pm\left(\frac{\pi}{2}+\frac{\pi}{2n}\right)$ radians. It thus follows that

$$\frac{1}{2Q_{p,\max}} = \sin\left(\frac{\pi}{2n}\right)$$

which leads to

$$Q_{p,\max} = \frac{1}{2\sin\left(\frac{\pi}{2n}\right)}.$$

For large n, $Q_{p,\max}$ approaches n/π. This approximate expression predicts that a fifth-order Butterworth transfer function will have a $Q_{p,\max}$ of about 1.59, which is close to the true value, namely 1.62.

Any high-order filter can be realized by cascading second- and/or first-order sections. It is instructive to observe the magnitude responses of individual sections in relation to the response of the whole filter. We use a fifth-order Butterworth transfer function as an example. Figure 7.11 shows the magnitude responses of the individual sections, as well as that of the overall filter. We observe that the sharp transition in the response from the passband to stopband—namely, the sharp selectivity of the Butterworth transfer function is largely due to the highest-Q section. As we have seen earlier, this also results in a sharply increasing magnitude in the passband. The role of the low-Q sections is to tame this peak to result in a flat overall response. The "flatness" in the passband is thus a consequence of the passband-droop of the lower-Q sections' magnitude response compensating for the increasing magnitude of the high-Q biquad. In the stop band, however, the responses of all sections reduce with increasing frequency and aid each other in attenuating out-of-band inputs.

Extending this intuition, it is straightforward to see the following:

a. As the filter order increases, the magnitude response approaches the ideal brick wall. The very sharp drop in the magnitude at the pass-band edge is largely due to the highest-Q section.

b. The sharp peak of the magnitude response of a very high-Q biquadratic section occurs at the center frequency. This means that a high-order filter whose response shape is close to that of a brick wall with cut-off frequency ω_c must contain a high-Q biquad whose center frequency is very close to ω_c.

c. The impulse response of a high-order filter that approximates a brick-wall with cutoff frequency ω_c will have a tail that rings at ω_c rad/s, which is the center frequency of the highest-Q biquad.

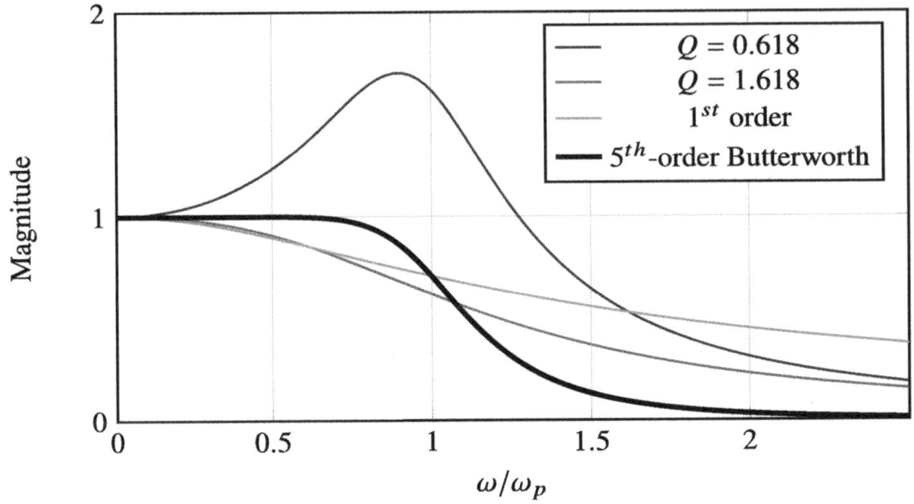

Figure 7.11 Magnitude responses of individual sections in a fifth-order Butterworth lowpass filter.

7.3 The Chebyshev Approximation

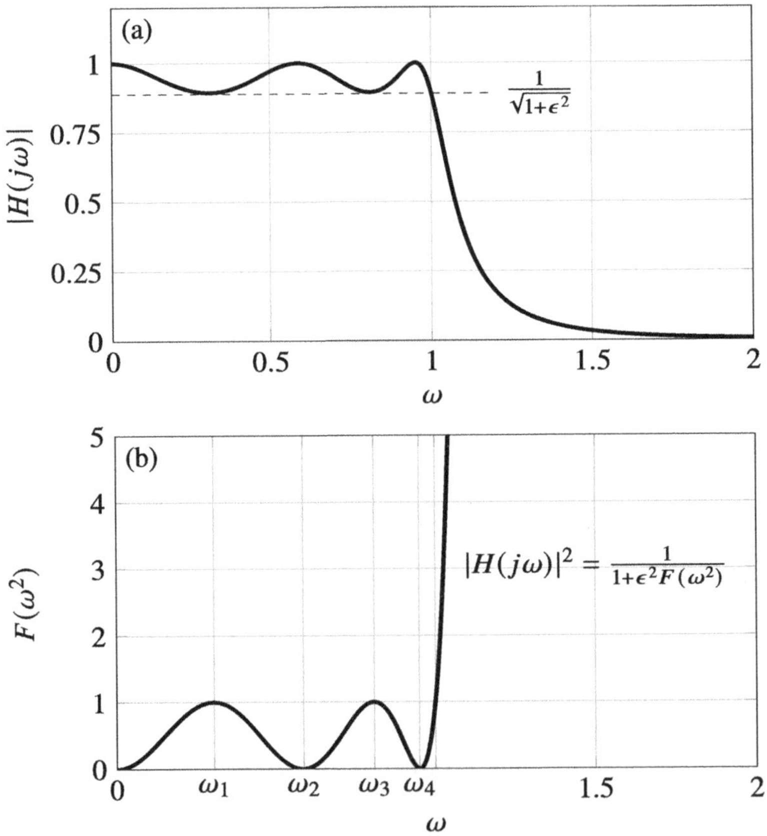

Figure 7.12 (a) A target $|H(j\omega)|$ that approximates the brick wall and (b) the resulting $F(\omega^2)$.

Previous sections discussed the maximally-flat (Butterworth) approximation to the ideal brick-wall low-pass filter. In this section, we will look at another way of approximating the brick wall. The shape of the magnitude response is shown in Figure 7.12a. In the pass band, i.e., $0 < \omega < 1$, the magnitude wiggles between 1 and $1/\sqrt{1+\epsilon^2}$. For $\omega > 1$, the response falls monotonically, approaching zero as $\omega \to \infty$. The pass-band response is referred to as being *equiripple*. Using our experience with Butterworth filters, we write

$$|H(j\omega)|^2 = \frac{1}{1+\epsilon^2 F(\omega^2)}. \tag{7.26}$$

$F(\omega^2)$ is a polynomial in ω^2 that must vary between 0 and 1 in the pass band and monotonically increase in the stop band of the filter, as shown in Figure 7.12b. What is the smallest-order polynomial that can achieve the shape we desire? We observe the following.

a. $F(\omega^2)$ "looks" like a parabola around 0, ω_2, and ω_4. It must therefore have factors of the form ω^2, $(\omega-\omega_2)^2$, and $(\omega-\omega_4)^2$.

b. $F(\omega^2)$ is an even function—it must therefore go to 0 at $-\omega_2$ and $-\omega_4$. It must therefore have factors of the form $(\omega+\omega_2)^2$ and $(\omega+\omega_4)^2$.

c. The minimum-order polynomial that can result in an $F(\omega^2)$ of the form shown in Figure 7.12b is thus of the tenth-order and can be written as

$$F(\omega^2) = k^2\omega^2(\omega^2-\omega_2^2)^2(\omega^2-\omega_4^2)^2 = \{\underbrace{k\omega(\omega^2-\omega_2^2)(\omega^2-\omega_4^2)}_{C_5(\omega)}\}^2 = C_5^2(\omega) \tag{7.27}$$

d. $F(\omega^2)$ is a perfect square. It can therefore be expressed as the square of a fifth-order polynomial denoted by $C_5(\omega)$. For large ω, $C_5^2(\omega) \longrightarrow k^2\omega^{10}$.

We have so far used the fact that $F(\omega^2)$ goes to zero at 0, ω_2, and ω_4 to conclude that it is the square of a fifth-order polynomial in ω. We have not yet exploited the maxima at ω_1 and ω_3, or the fact that the value attained by $F(\omega^2)$ at these frequencies is 1. Since $F(\omega^2) = C_5^2(\omega)$ attains a peak value of 1 in the pass band at ω_1, ω_3, and $F(\omega^2) = 1$ at $\omega = 1$, and $F(\omega^2)$ is an even function, it follows that $C_5^2(\omega) - 1$ can be expressed as

$$C_5^2(\omega) - 1 = k^2(\omega^2-\omega_1^2)^2(\omega^2-\omega_3^2)^2(\omega^2-1). \tag{7.28}$$

As a sanity check, we see that $C_5^2(\omega) - 1$ is also a tenth-degree polynomial which goes as $k^2\omega^{10}$ for large ω. Let us now take a look at the behavior of $C_5(\omega)$, plotted in Figure 7.13. It *must* attain its maxima and minima at $\pm\omega_1$ and $\pm\omega_3$ respectively. $C_5(\omega)$'s derivative must thus have zeros at these frequencies. It can therefore be expressed as

$$\frac{dC_5(\omega)}{d\omega} \propto (\omega^2-\omega_1^2)(\omega^2-\omega_3^2).$$

The constant of proportionality can be obtained as follows. Recall that $C_5(\omega) = k\omega^5$ for $\omega \gg 1$. $\dfrac{dC_5(\omega)}{d\omega}$ must therefore go as $5k\omega^4$ for $\omega \gg 1$.

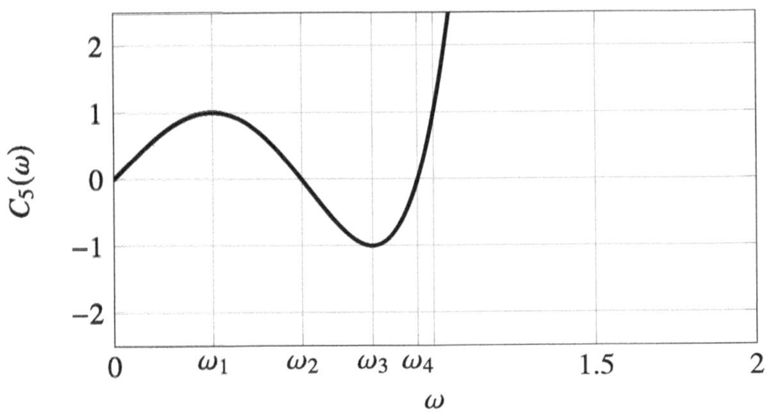

Figure 7.13 Plot of $C_5(\omega)$, which is an odd function of ω.

From the observations above, we see that

$$\frac{dC_5(\omega)}{d\omega} = 5k(\omega^2 - \omega_1^2)(\omega^2 - \omega_3^2). \tag{7.29}$$

From Eqs. (7.28) and (7.29), we have

$$\left(\frac{dC_5(\omega)}{d\omega}\right)^2 = 25\frac{C_5^2(\omega) - 1}{\omega^2 - 1}. \tag{7.30}$$

We therefore obtain the following differential equations governing $C_5(\omega)$, where the terms under the square-root sign are appropriately modified for $\omega < 1$ and $\omega \geq 1$.

$$\begin{aligned} \frac{dC_5(\omega)}{d\omega} &= 5\frac{\sqrt{1 - C_5^2(\omega)}}{\sqrt{1 - \omega^2}}, \quad \omega < 1 \\ \frac{dC_5(\omega)}{d\omega} &= 5\frac{\sqrt{C_5^2(\omega) - 1}}{\sqrt{\omega^2 - 1}}, \quad \omega \geq 1. \end{aligned} \tag{7.31}$$

Solving these differential equations, and using $C_5(0) = 0$, $C_5(1) = 1$ to obtain the constants of integration, we obtain

$$\begin{aligned} C_5(\omega) &= \cos(5\cos^{-1}(\omega)), \quad \omega < 1 \\ &= \cosh(5\cosh^{-1}(\omega)), \quad \omega \geq 1. \end{aligned} \tag{7.32}$$

Unbelievable as it may seem at first sight, $\cos(5\cos^{-1}(\omega))$ is actually a polynomial! It is the fifth-order *Chebyshev polynomial*, named after the brilliant Russian mathematician who discovered the theory of equiripple approximation.

Generalizing to the nth-order case, it is straightforward to see that

$$|H(j\omega)|^2 = \frac{1}{1 + \epsilon^2 C_n^2(\omega)} \tag{7.33}$$

where

$$\begin{aligned} C_n(\omega) &= \cos(n\ \cos^{-1}(\omega)), \quad \omega < 1 \\ &= \cosh(n\ \cosh^{-1}(\omega)), \quad \omega \geq 1. \end{aligned} \tag{7.34}$$

The maxima of transmission occur at frequencies at which $n\cos^{-1}(\omega)$ attains values that are odd multiples of $\pi/2$. For our fifth-order example

$$\omega_{\max} = \cos(\pi/10), \cos(3\pi/10), \cos(5\pi/10). \tag{7.35}$$

Likewise, the pass-band minima occur at frequencies at which $n\cos^{-1}(\omega)$ attains values that are even multiples of $\pi/2$. In our example, we have

$$\omega_{\min} = \cos(2\pi/10), \cos(4\pi/10). \tag{7.36}$$

Sines and Cosines of Complex Numbers

Let $(u + jv)$ be a complex number. What is $\cos(u + jv)$?

$$\cos(u + jv) = \cos(u)\cos(jv) - \sin(u)\sin(jv) \tag{7.37}$$

This brings us to the question: what are $\cos(jv)$ and $\sin(jv)$?

$$\cos(jv) = \frac{e^{j(jv)} + e^{-j(jv)}}{2} = \cosh(v)$$

$$\sin(jv) = \frac{e^{j(jv)} - e^{-j(jv)}}{2j} = j\sinh(v)$$

We therefore have

$$\cos(u + jv) = \cos(u)\cosh(v) - j\sin(u)\sinh(v) \tag{7.38}$$

Having found $|H(j\omega)|^2$, our next order of business is to determine $H(s)$. To find the poles of $H(s)$, we need to find the left-half plane roots of $D(s)D(-s)$ where

$$D(s)D(-s) = 1 + \epsilon^2 C_n^2\left(\frac{s}{j}\right). \tag{7.39}$$

The roots need to satisfy

$$\cos\underbrace{\left\{n\cos^{-1}\left(\frac{s}{j}\right)\right\}}_{u+jv} = \pm\frac{j}{\epsilon} \tag{7.40}$$

Denoting the argument of the cosine function by $u + jv$, we see that

$$\cos(u)\cosh(v) - j\sin(u)\sinh(v) = \pm\frac{j}{\epsilon}. \tag{7.41}$$

Equating real and imaginary parts, we obtain

$$u = (2m+1)\frac{\pi}{2}, \quad v = \pm \sinh^{-1}\left(\frac{1}{\epsilon}\right) \qquad m = 0, 1, 2, \cdots \tag{7.42}$$

Since

$$n \cos^{-1}\left(\frac{s}{j}\right) = u + jv \tag{7.43}$$

we have

$$\frac{s}{j} = \cos\left(\frac{u}{n} + j\frac{v}{n}\right) = \cos\left(\frac{u}{n}\right)\cosh\left(\frac{v}{n}\right) - j\sin\left(\frac{u}{n}\right)\sinh\left(\frac{v}{n}\right) \tag{7.44}$$

which means

$$s = \sin\left(\frac{u}{n}\right)\sinh\left(\frac{v}{n}\right) + j\cos\left(\frac{u}{n}\right)\cosh\left(\frac{v}{n}\right). \tag{7.45}$$

The roots of $D(s)D(-s)$ are therefore given by $\sigma_m + j\omega_m$ where

$$\sigma_m = \pm \sin\left(\frac{(2m+1)\pi}{2n}\right)\sinh\left(\frac{1}{n}\sinh^{-1}\left(\frac{1}{\epsilon}\right)\right), \quad m = 0, 1, 2, \cdots$$

and

$$\omega_m = \cos\left(\frac{(2m+1)\pi}{2n}\right)\cosh\left(\frac{1}{n}\sinh^{-1}\left(\frac{1}{\epsilon}\right)\right), \quad m = 0, 1, 2, \cdots$$

From the equations above, we see that

$$\frac{\sigma_m^2}{\sinh^2\left(\frac{1}{n}\sinh^{-1}\left(\frac{1}{\epsilon}\right)\right)} + \frac{\omega_m^2}{\cosh^2\left(\frac{1}{n}\sinh^{-1}\left(\frac{1}{\epsilon}\right)\right)} = 1$$

which indicates that the roots of $D(s)D(-s)$ lie on an ellipse. Those roots that lie in the left-half s-plane should be chosen to arrive at $D(s)$.

A few observations are in order.

a. For a given ϵ, increasing the order has no effect on the pass-band ripple.

b. For $\omega \gg 1$, $C_n(\omega) = \cosh(n\cosh^{-1}(\omega)) \to 2^{(n-1)}\omega^n$. The stop-band response asymptotically approaches $1/(2^{n-1}\epsilon\omega^n)$. For a fixed order, therefore, choosing a larger pass-band ripple results in a higher stop-band attenuation. There is thus a trade-off between pass-band and stop-band performance.

c. The poles that are closest to the $j\omega$ axis are those that have the least value for $|\sigma_m|$, which corresponds to $m = 0$. The left-half-plane poles are

$$\sigma_0 = -\sin\left(\frac{\pi}{2n}\right)\sinh\left(\frac{1}{n}\sinh^{-1}\left(\frac{1}{\epsilon}\right)\right)$$

$$\omega_0 = \pm\cos\left(\frac{\pi}{2n}\right)\cosh\left(\frac{1}{n}\sinh^{-1}\left(\frac{1}{\epsilon}\right)\right)$$

For a high-order filter (i.e., large n),

$$\sinh\left(\frac{1}{n}\sinh^{-1}\left(\frac{1}{\epsilon}\right)\right) \longrightarrow 0$$

$$\cosh\left(\frac{1}{n}\sinh^{-1}\left(\frac{1}{\epsilon}\right)\right) \longrightarrow 1.$$

Thus, $\sigma_o \ll \omega_o$ which means that $\omega_p = \sqrt{\sigma_0^2 + \omega_0^2} \approx \omega_o \approx 1$. The center frequency of the highest-Q biquad is seen to be close to the band-edge frequency. This makes intuitive sense due to the following.

a. The high-order filter response transitions sharply at the band-edge frequency.
b. This sharp transition in the magnitude response must be due to the high-Q biquad.
c. A high-Q biquad's magnitude response exhibits a sharp transition at its center frequency.
d. The high-Q biquad's center frequency, therefore, must be close to the band-edge frequency.

Highest Pole-Q_p in an nth-Order Chebyshev Filter

The highest-Q poles in a Chebyshev filter are those closest to the $j\omega$ axis. As seen from the discussion above, the ω_p corresponding to these poles is close to the band-edge frequency, namely 1 rad/s. Furthermore, the distance of the poles to the $j\omega$-axis is given by σ_m evaluated for $m = 0$. $Q_{p,\max}$ is thus seen to be $1/(2\sigma_0)$ and is given by

$$Q_{p,\max} = \frac{1}{2\sin\left(\frac{\pi}{2n}\right)\sinh\left(\frac{1}{n}\sinh^{-1}\left(\frac{1}{\epsilon}\right)\right)}.$$

For large n

$$Q_{p,\max} \Longrightarrow \frac{n^2}{\pi\sinh^{-1}\left(\frac{1}{\epsilon}\right)}.$$

This approximate expression predicts that a fifth-order Chebyshev transfer function with 0.5 dB ripple (i.e., $\epsilon = 0.35$) will have a $Q_{\text{p,max}}$ of about 4.5, which is close to the true value, namely 4.47.

It is interesting to see that $Q_{\text{p,max}}$ for a high-order Chebyshev filter goes as n^2, while that of a Butterworth filter is proportional to n. Since the "sharpness" of the filter's response depends on $Q_{\text{p,max}}$, it follows that the Chebyshev filter has a much more abrupt transition than a Butterworth filter. Intuitively, this can be understood as follows. As n is increased, all the energies of the Butterworth approximation are focused on making the passband as flat as possible. In a Chebyshev approximation, on the other hand, the high order is used to achieve a sharp cutoff; the pass-band variation only depends on ϵ and is independent of order.

Example: Fifth-Order Chebyshev Filter, 0.5 dB Ripple

Let us apply the theory above to determine the transfer function of a fifth-order Chebyshev filter with 0.5 dB pass-band ripple (Figure 7.14). The first task is to determine ϵ. Using

$$-20\log\left(\frac{1}{\sqrt{1+\epsilon^2}}\right)=0.5$$

we obtain $\epsilon = 0.35$.

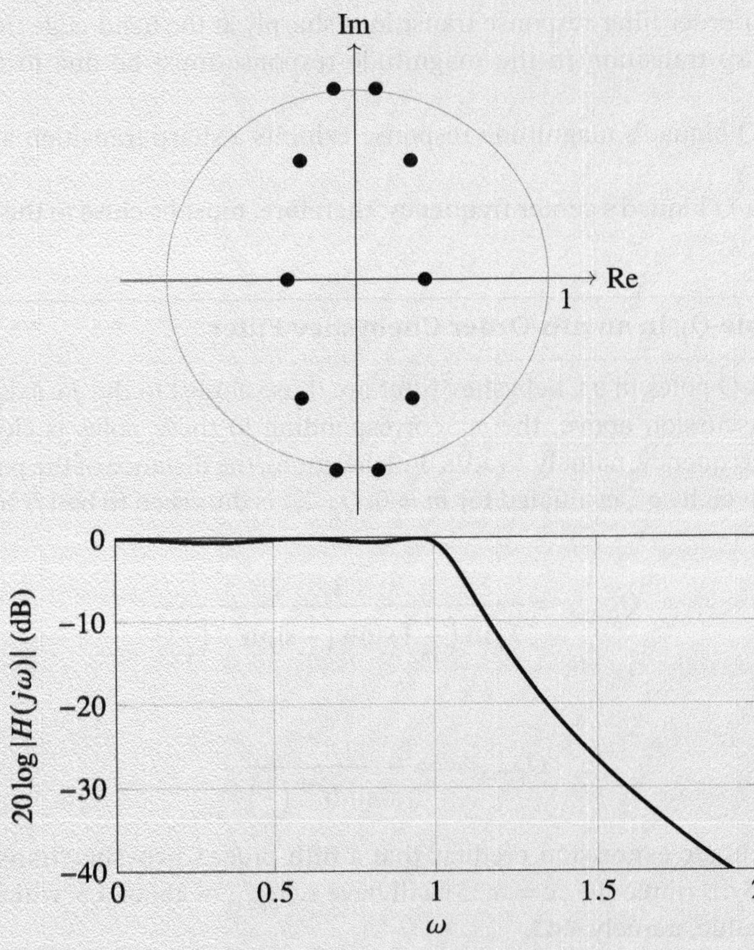

Figure 7.14 Pole constellation of $D(s)D(-s)$ and magnitude response for a fifth-order Chebyshev filter with 0.5 dB pass-band ripple.

The left-half plane poles are obtained as

$$
\begin{aligned}
-\sin(\pi/10)\sinh\left(0.2\sinh^{-1}(1/\epsilon)\right) &\pm j\cos(\pi/10)\cosh\left(0.2\sinh^{-1}(1/\epsilon)\right) \\
&= -0.11 \pm j1.01 \\
-\sin(3\pi/10)\sinh\left(0.2\sinh^{-1}(1/\epsilon)\right) &\pm j\cos(3\pi/10)\cosh\left(0.2\sinh^{-1}(1/\epsilon)\right) \\
&= -0.29 \pm j0.63 \\
-\sinh\left(0.2\sinh^{-1}(2.86)\right) &= -0.3623
\end{aligned}
$$

7.3.1 Even-Order Chebyshev Filters

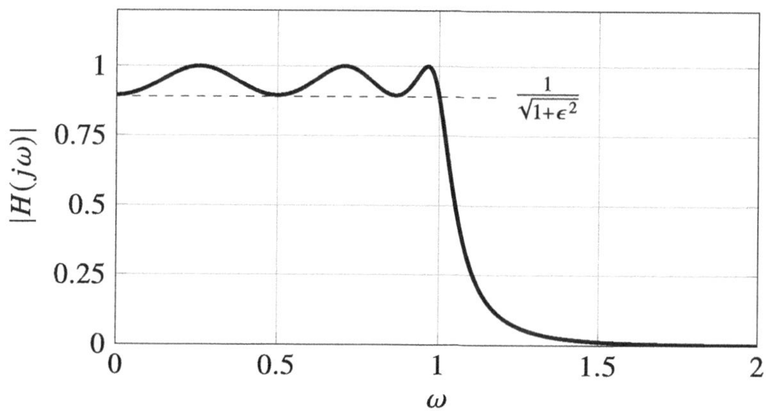

Figure 7.15 Magnitude response of a sixth-order Chebyshev filter.

Even-order Chebyshev filters have the same governing equations as their odd-order cousins. Their response at dc, however, is a pass-band minimum (rather than a maximum in the odd-order case). Figure 7.15 shows the magnitude response of a sixth-order Chebyshev filter. The maxima of transmission occur at frequencies

$$\omega_{\max} = \cos((2m+1)\pi/12), \quad m = 0, 1, 2$$

while the minima occur at

$$\omega_{\min} = \cos(m\pi/6), \quad m = 1, 2, 3.$$

7.4 Butterworth Versus Chebyshev Filters

In system design, one is often faced with the task of choosing an appropriate filter that satisfies a given frequency response mask. For instance, the requirements could read as follows.

a. The dc gain must be unity.
b. The magnitude response must not exceed unity at any frequency.
c. The magnitude response must be greater than 0.95 at $\omega = \omega_1$.
d. The magnitude response must be smaller than 0.01 at $\omega > 5\omega_1$.

The designer has several choices. If (s)he decides to use a Butterworth filter, the order and cutoff frequency are determined as follows.

$$|H(j\omega_1)|^2 = \frac{1}{1 + (\omega_1/\omega_c)^{2n}} > (0.95)^2$$

$$|H(j5\omega_1)|^2 = \frac{1}{1 + (5\omega_1/\omega_c)^{2n}} < (0.01)^2$$

Several combinations of n and ω_c are possible. The lowest-order filter that satisfies this mask is of the fourth-order with $\omega_c = 1.33\omega_1$.

If a Chebyshev filter is chosen, it must be of odd order, since the dc gain needs to be unity. The response at $\omega = \omega_1$ is chosen to be 0.95, which results in $\epsilon = 0.32$ and $\omega_c = \omega_1$. The stop-band constraint results in

$$\frac{1}{\sqrt{1 + \epsilon^2 \cosh^2(n \cosh^{-1}(5\omega_1/\omega_c))}} < 0.01 \tag{7.46}$$

The lowest n that satisfies the inequality above is $n = 3$. We thus see that the Chebyshev approximation results in a lower-order filter. This makes sense, since the highest Q_p of the Chebyshev approximation is considerably higher than that in the Butterworth case.

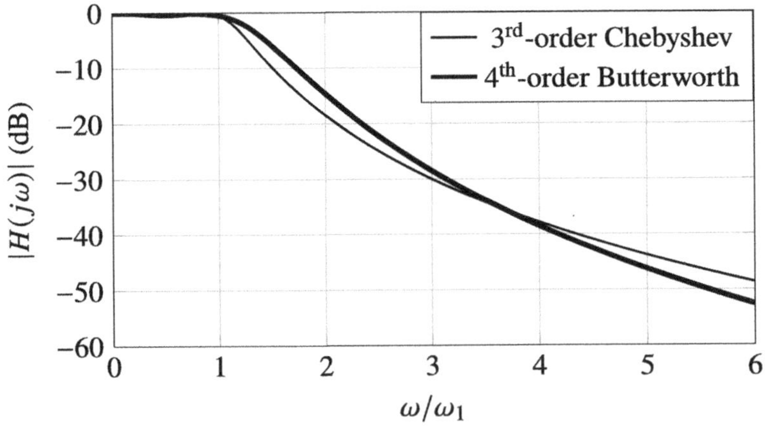

Figure 7.16 Comparing the magnitude responses of Butterworth and Chebyshev filters to achieve the same magnitude-response mask.

Figure 7.16 compares the magnitude responses of the two designs. Though the Chebyshev design is of lower order, its transition band is narrower, thanks to its higher quality factor pole pair. The Butterworth response's roll off at large ω is evidently higher than that of its Chebyshev counterpart—this is to be expected, as its order is higher.

7.5 Realization of Filters

We have seen that the transfer function of a high-order filter can be expressed as a product of second- and/or first-order transfer functions. This means that once we figure out how to implement first-order and biquadratic sections, a filter of arbitrary order can be realized by cascading these building blocks. We begin with the implementation of a first-order section. The transfer function of a first-order all-pole section is given by

$$H(s) = \frac{1}{1 + \frac{s}{\omega_0}}. \tag{7.47}$$

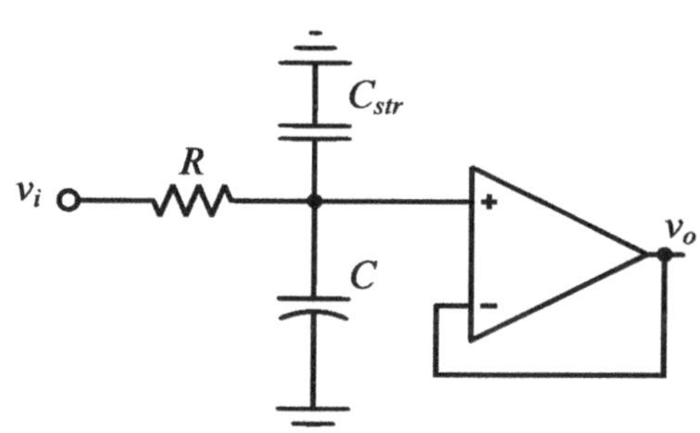

Figure 7.17 Simple-minded realization of a first-order section. The transfer function is sensitive to stray capacitance C_{str}.

This can be implemented using a resistor R and a capacitor C with the condition $RC = 1/\omega_o$. When cascading first- and second-order transfer functions, loading effects can become problematic. To address this issue, a buffer, typically implemented using an operational amplifier (opamp), can be introduced to maintain the integrity of the transfer functions. Figure 7.17 shows an example first-order stage, where the opamp is assumed to be ideal. In practice, however, every node is associated with a stray capacitance. Denoted by C_{str} in the figure, it appears in parallel with the intended capacitance C. The consequence is that the effective bandwidth is altered, and the filter becomes parasitic sensitive. The modified bandwidth is given by $1/R(C + C_{str})$. Stray capacitances at nodes v_i and v_o are not problematic since those are "zero-impedance" nodes.

We now derive an alternative topology that realizes the same transfer function, except for a sign inversion. The end result is well known, but the emphasis here is the *process* by which the topology is derived, so that it can be applied to arrive at the biquad topology in Section 7.5.1.

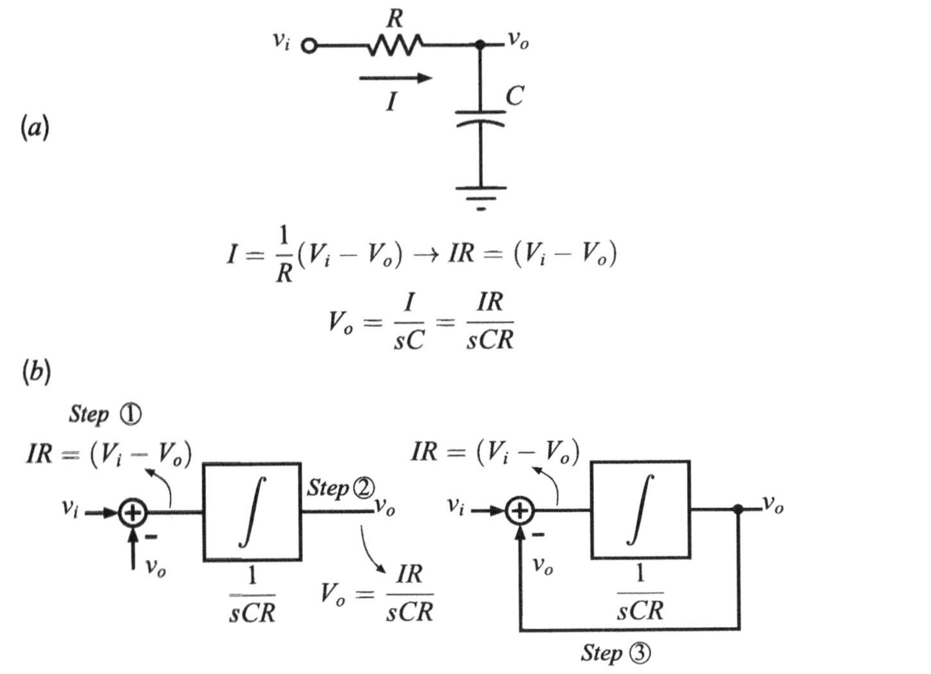

Figure 7.18 (a) Equations governing an RC low-pass filter. (b) Converting the RC low-pass filter into a signal flow graph.

Figure 7.18a shows the first-order RC low-pass filter. We see that[1]

$$I = \frac{1}{R}(V_i - V_o) \rightarrow IR = V_i - V_o$$
$$V_o = \frac{1}{sCR} \cdot IR. \tag{7.48}$$

The equations above suggest the following implementation, as shown in Figure 7.18b.

Step ① Assuming V_o exists, generate the (voltage) quantity IR as $(V_i - V_o)$.
Step ② Obtain V_o by integrating IR. The integrator's unity-gain frequency is $1/RC$ rad/s.
Step ③ Connect V_o obtained in step 2 to the V_o needed in step 1, thereby closing the feedback loop. The resulting block diagram is the first-order low-pass filter with a 3 dB bandwidth given by $1/RC$ rad/s.

The block diagram derived in Figure 7.18b has the advantage that all quantities are voltages. The key building block is the summing integrator, whose function is to add V_i with $-V_o$ and integrate the result. Figure 7.19 shows a circuit realization of such a block. The opamp is ideal. The resistors convert the inputs v_1 and v_2 into currents, which are subsequently summed in the current domain and integrated on

[1]All capitalized quantities refer to the Laplace transforms of the corresponding time-domain quantities.

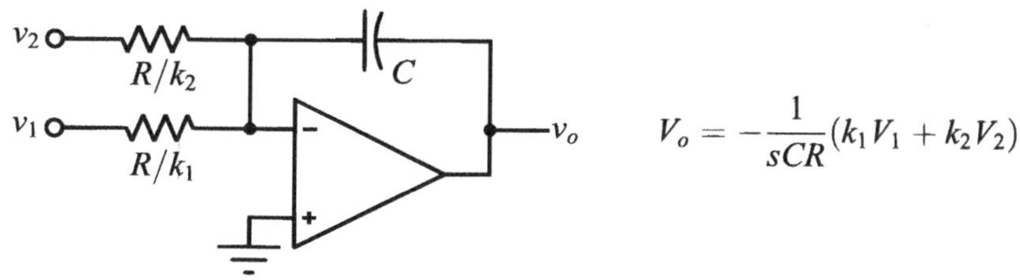

Figure 7.19 A summing inverting integrator.

the feedback capacitor. The result is that v_o, which is the same as the voltage across the capacitor is the weighted and integrated sum of the inputs. A twist, however, is that the output is inverted, so that the output is given by

$$V_o = -\frac{1}{sCR}(k_1 V_1 + k_2 V_2). \tag{7.49}$$

The circuit of Figure 7.19 is called an *active-RC* or an *opamp-RC* integrator. The unity-gain frequency is set by the passives, namely R and C.

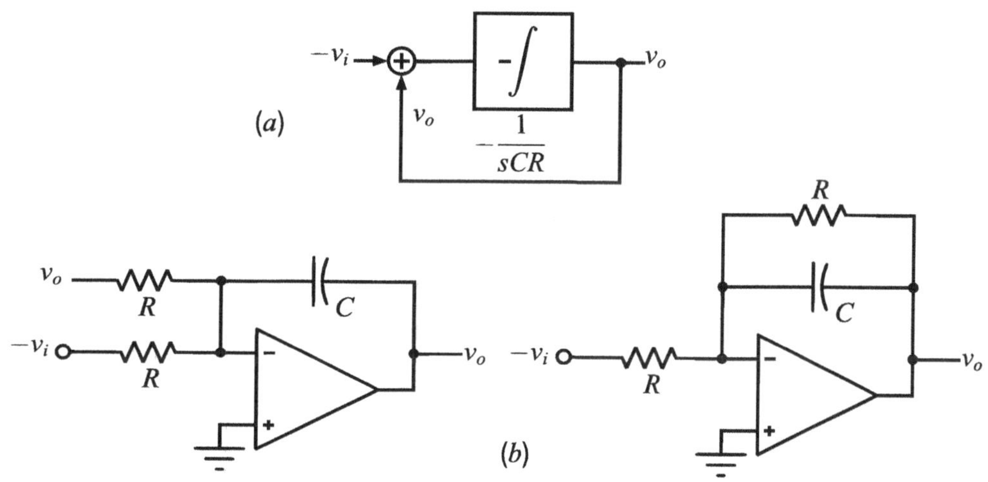

Figure 7.20 (a) Signal flow graph of a first-order low-pass filter using an inverting integrator. (b) Steps in circuit implementation.

To allow for an inverting summing integrator, the block diagram of Figure 7.18b is modified as shown in Figure 7.20a. Since the integrator is inverting, a negative sign is needed at the input. This, when "pushed" through the summer, causes the inversion in the v_o path to be removed and the sign of v_i to be inverted. Figure 7.20b shows the steps in the derivation of the circuit realization. We assume that V_o is available and first form V_o by summing $-V_i$ and V_o, and integrating the result. Since V_o is now available, we connect it to one of the summing integrator's inputs, thereby closing the feedback loop. If the input is $-V_i$, the output is $V_i/(1 + sCR)$. The transfer function realized, therefore, is an inverting first-order one. The inversion in the output is usually not problematic. Assuming an ideal opamp, the output impedance of this stage is zero, making it suitable for cascading with other filter sections. The first-order section derived above is an example of the *active-RC filter* family, so named as the basic building block is the active-RC integrator.

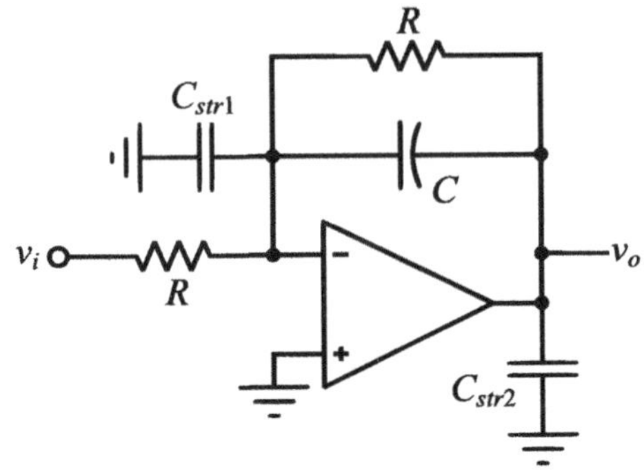

Figure 7.21 The first-order opamp-RC section is parasitic insensitive.

Let us now analyze the impact of stray capacitance on a first-order opamp-RC filter (Figure 7.21). C_{str1} at the virtual ground node does not affect the transfer function. The reasoning is that one terminal of the stray capacitance is connected to real ground, while the other is connected to the virtual ground created by the ideal opamp. As the potential difference across the stray capacitance is effectively zero, no current flows through it. Therefore, C_{str1} is of no consequence. C_{str2} is not important either, even though the voltage across it is v_o. This is due to the following. The output impedance of the stage is zero, thanks to the ideal opamp. The current through C_{str2} will therefore be supplied by the amplifier. v_o, however, only depends on the current through C, which is not altered by that flowing in C_{str2}.

To summarize, the active-RC first-order filter structure not only allows for cascading stages without worrying about loading artifacts, but also results in a design that is insensitive to stray capacitors. Note that a dc gain with a magnitude greater than unity can easily be incorporated into the transfer function by appropriately modifying the input resistor. It also turns out that converting this into a differential structure is straightforward.

7.5.1 Realizing a Second-Order Section

Let us now look at approaches to realize a second-order transfer function. As in the first-order case, we begin with the passive structure. It turns out that an inductor and capacitor are both needed to implement a passive second-order filter with arbitrary ω_p and Q_p. The series RLC network shown in Figure 7.22a realizes a low-pass transfer function given by

$$H(s) = \frac{1}{1 + sRC + s^2LC}. \tag{7.50}$$

In terms of the passive-component values, it is straightforward to see that

$$\omega_p = \frac{1}{\sqrt{LC}}, \quad Q_p = \frac{1}{R}\sqrt{\frac{L}{C}}. \tag{7.51}$$

The problem with this structure, however, is that the large value of inductance needed will result in a bulky realization which is not amenable to integrated-circuit implementation. Recognizing that an inductor, like a capacitor is also an integrator (albeit one where the current is the integral of the voltage), it should be possible to derive a block diagram of the RLC network in a manner similar to that used in Figure 7.18b. We could then use opamp-RC integrators to realize a biquad without using inductors.

Example: Component Values in a Second-Order RLC Low-Pass Filter

Consider the prospect of having to realize a second-order Butterworth low-pass filter with a 3 dB bandwidth of 1 MHz. Assuming $R = 10\ \text{K}\Omega$, calculations show that the values of inductance and capacitance needed are

$$L = 1.13\,\text{mH}, \quad C = 22.5\,\text{pF}. \tag{7.52}$$

Realizing such a large-valued inductor on chip is impractical.

Consider again the RLC low-pass filter of Figure 7.22a. The inductor current is given by

$$I = \frac{V_i - V_o}{R + sL} = \frac{V_i - V_o}{1 + s(L/R)}\frac{1}{R}. \tag{7.53}$$

Like in the first-order case, IR can be expressed as

$$IR = -\frac{V_o - V_i}{1 + s(L/R)}. \tag{7.54}$$

The capacitor voltage, which is also the filter output is

$$V_o = \frac{I}{sC} \Rightarrow \frac{1}{sCR} \cdot (IR) = V_o \tag{7.55}$$

The equations above suggest the following block diagram, as shown in Figure 7.22b.

Step ① Assuming V_o exists, generate the (voltage) quantity IR as $(V_i - V_o)/(1 + s(L/R))$. IR is thus a first-order filtered version of $(V_i - V_o)$, and is realized by enclosing an integrator with unity-gain frequency R/L rad/s in a feedback loop.

Step ② Obtain V_o by integrating IR. The integrator's unity-gain frequency is $1/RC$ rad/s.

Step ③ Connect V_o obtained in step 2 to the V_o needed in step 1, thereby closing the feedback loop. The resulting block diagram is the second-order low-pass filter with $\omega_p = 1/\sqrt{LC}$ rad/s and $Q_p = (1/R)\sqrt{L/C}$, just like the passive prototype it is based on.

(a)

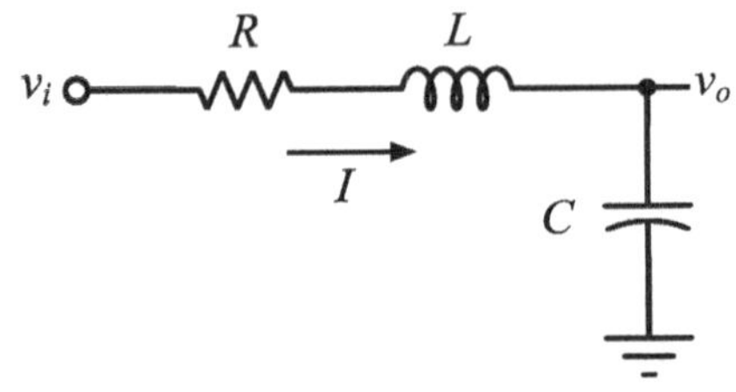

$$I = \frac{1}{R + sL}(V_i - V_o) = (V_i - V_o)\frac{1}{1 + s(L/R)}\frac{1}{R}$$

$$V_o = \frac{I}{sC}$$

(b)

Step ① $IR = \dfrac{1}{1 + s(L/R)}(V_i - V_o)$

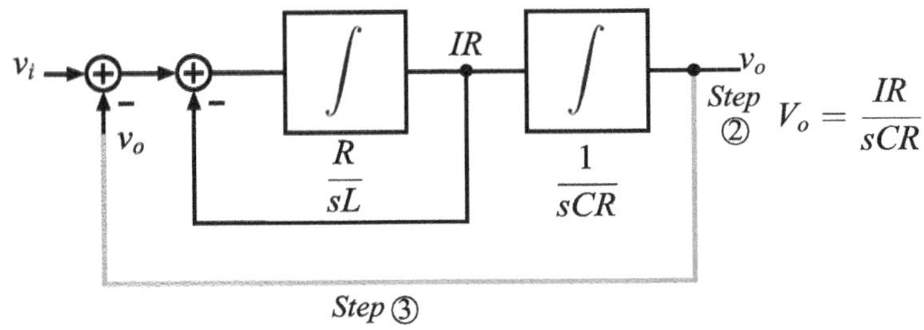

Figure 7.22 (a) Equations governing a second-order RLC filter. (b) Steps to implementing them as a signal flow graph.

Recognizing that we have at hand only inverting active-RC integrators, the block diagram of Figure 7.22b has to be modified by moving inversions around, as shown in Figure 7.23a. The summing low-pass filter with 3 dB bandwidth R/Lrad/s can be implemented as an active-RC structure as shown in Figure 7.23b provided C_1 is chosen so that $C_1R^2 = L$. The gain of -1 is realized using an opamp-based inverting amplifier. The output of the inverting amplifier is integrated and fed back. The result is the three-opamp, six-resistor, two-capacitor structure named the Tow-Thomas biquad [2–4], shown in Figure 7.23b.

As with the first-order section, the transfer function is inverting. Furthermore, no inductors are needed, and the biquad is not affected by stray capacitors, since every node is either a virtual ground or a "zero-impedance" node. There is a direct correspondence between the passive RLC filter of Figures 7.22a and 7.23b. The output of A_1 is proportional to the inductor current I. Since I is a bandpass transfer function, it follows that the output of A_1 is an inverting bandpass filter. The output of A_2 is an inverted version of the same. A_3's output, which represents the capacitor voltage in the network of Figure 7.22a is the low-pass transfer function that we desire.

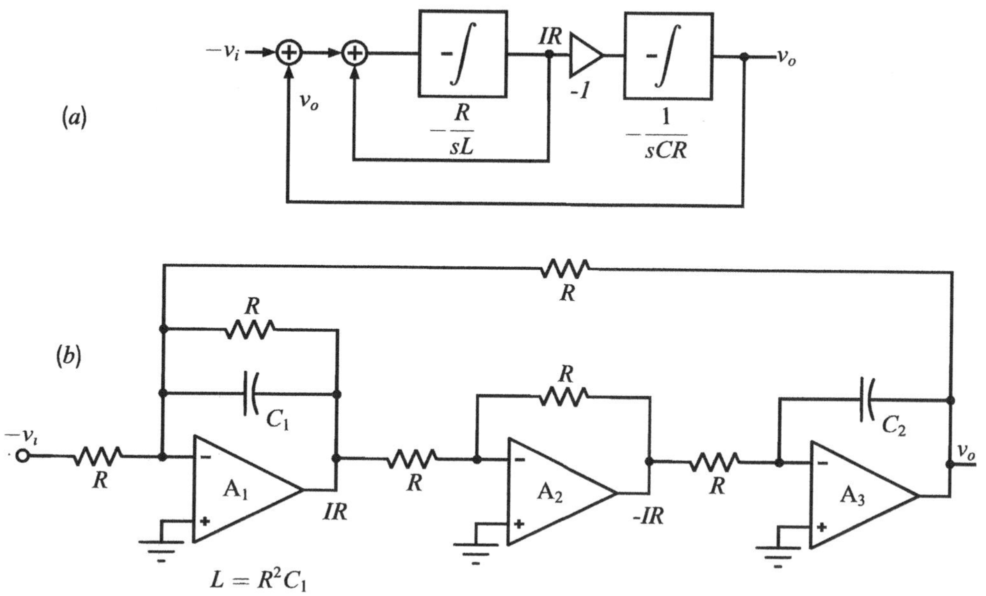

Figure 7.23 (a) Using inverting integrators to realize the signal flow graph of Figure 7.22. (b) Final circuit, known as the Tow-Thomas biquad.

Example: Component Values in a Second-Order Tow-Thomas Low-Pass Filter

Consider the prospect of having to realize a second-order Butterworth low-pass filter with a 3 dB bandwidth of 1 MHz. Assuming $R = 10\ \text{k}\Omega$, calculations show that the values of capacitance needed are

$$C_1 = 11.25\ \text{pF}, \quad C_2 = 22.5\ \text{pF}. \tag{7.56}$$

When compared to the passive network, which needed a large-valued inductor, we see that both these capacitance values are realizable on an integrated circuit.

As we will see later, the Tow-Thomas structure is readily amenable to fully-differential operation, which eliminates the need for the inverting stage. Consequently, only two opamps are required to realize the differential biquad.

Just as there are many paths to Nirvana, the Tow-Thomas biquad can be derived using a completely different approach. We begin with the parallel-RLC bandpass filter network shown in Figure 7.24a. In an attempt to make this amenable to cascading and to incorporate parasitic insensitivity, the RLC network is placed in the feedback path of an ideal opamp (Figure 7.24b). Next, we seek to eliminate the inductors, since we know that they are bulky and not easy to integrate on a chip. To this end, we recognize

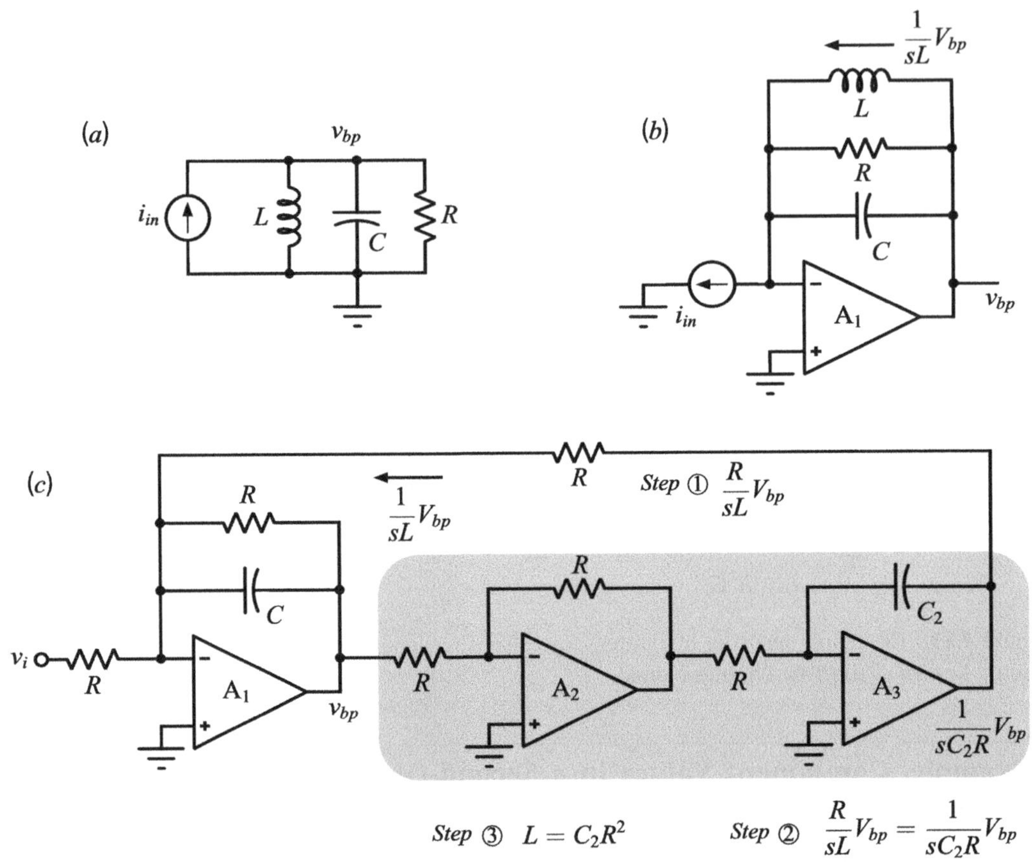

Figure 7.24 (a) Parallel-RLC network. (b) Parasitic-insensitive realization. (c) Eliminating the inductor using a non-inverting opamp-RC integrator.

that the inductor's function is to push a current V_{bp}/sL into the virtual ground node. If we can find a way of injecting the same current without using an inductor, our objective is achieved. We thus ask the question: suppose we replace the inductor in Figure 7.24b with a resistor, what voltage would have to be applied at the other end of the resistor to result in a current V_{bp}/sL? It is easy to see that the resistor must be connected to a voltage $V_{bp}(R/sL)$, which is an integrated version of V_{bp}. Since our building block, namely the integrator, is inverting, we need to cascade it with an inverting amplifier to achieve integration without inversion. This is the circuitry in the gray box in Figure 7.24c. It is apparent that C_2 must be chosen as L/R^2 to achieve the same bandpass transfer function as in Figure 7.24b. Given that the input current source feeds into a virtual-ground node, it can be replaced by a voltage source in series with a resistor, which accomplishes the voltage-to-current conversion (albeit with a sign inversion). The output of A_3, which is the integrated version of the bandpass transfer function, is the low-pass output.

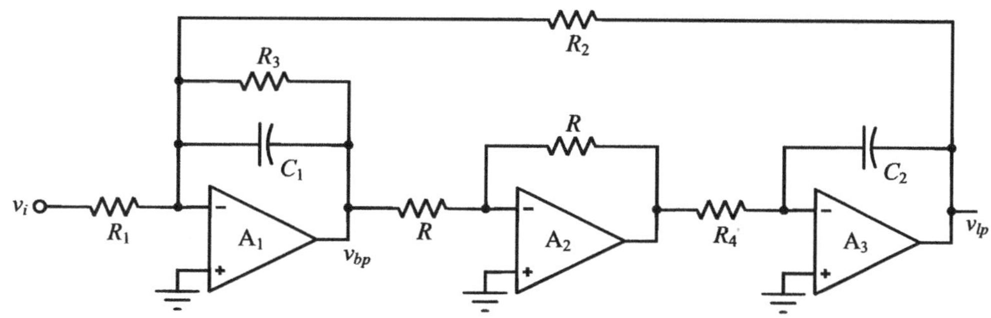

Figure 7.25 The Tow-Thomas biquad with low-pass and band-pass outputs.

The topology for Figure 7.23b was derived from a passive prototype, but there are evidently many more components in the circuit; and one is free to vary these component values. Figure 7.25 shows a more general form of the Tow-Thomas biquad. The gain of the inverting amplifier formed by A_2 is kept at -1 (which is anyway the case in a differential implementation). Straightforward analysis shows that

$$\frac{V_{lp}}{V_i} = -\frac{R_2}{R_1}\frac{1}{1 + sC_2R_2(R_4/R_3) + s^2C_1C_2R_2R_4} \longrightarrow \omega_p = \frac{1}{\sqrt{R_2R_4C_1C_2}},$$
$$Q_p = \frac{R_3}{\sqrt{R_2R_4}}\sqrt{\frac{C_1}{C_2}} \qquad \frac{V_{bp}}{V_i} = -\frac{R_2}{R_1}\frac{sC_2R_4}{1 + sC_2R_2(R_4/R_3) + s^2C_1C_2R_2R_4} \tag{7.57}$$

Let us ponder over the expression for the low-pass transfer function above and see why it makes sense.

a. At dc, C_2 carries no current, which means that the voltage at the output of A_2, and v_{bp} are zero. Consequently, no current flows through R_3. All of the current through R_1 flows through R_2, resulting in an output voltage $-(R_2/R_1)v_i$. The dc gain, therefore, is $-(R_2/R_1)$.

b. The value of the equivalent inductor realized by A_2, A_3, R_4, C_2, R_2 is $L = R_2R_4C_2$. The ω_p of the biquad, therefore, is $1/\sqrt{LC_1} = 1/\sqrt{R_2R_4C_1C_2}$.

c. Q_p is that corresponding to the quality factor of the RLC-parallel network, and is given by $R_3\sqrt{C_1/L} = (R_3/\sqrt{R_2R_4})\sqrt{C_1/C_2}$.

Let us now make sense of the band-pass transfer function. From Figure 7.25, we see that

$$V_{lp} = \frac{1}{sC_2R_4}V_{bp}.$$

Consequently, the V_{bp} is obtained from V_{lp} by multiplying it with sC_2R_4.

7.5.2 The Rauch Filter

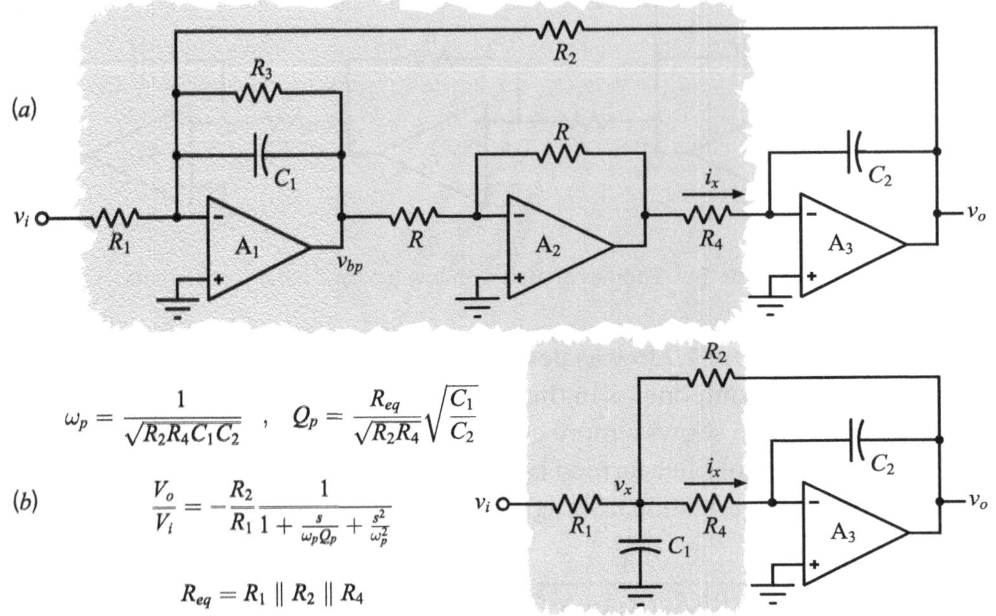

Figure 7.26 (a) The Tow-Thomas biquad—the summing first-order low-pass filter is in gray. (b) Replacing the gray section using passive circuitry to result in the Rauch biquad.

An interesting and widely-used modification of the Tow-Thomas structure is the so-called Rauch biquad. Its evolution is described using Figure 7.26. The portion enclosed in gray in part (a) of the figure injects a current i_x through R_4 given by

$$I_x = \left(\frac{V_i}{R_1} + \frac{V_o}{R_2}\right)\frac{1}{R_4}\frac{R_3}{1 + sC_1R_3}. \tag{7.58}$$

We see that I_x is a first-order low-pass filtered version of a weighted sum of v_i and v_o. The weighted summation and subsequent low-pass filtering is created by A_1, R_1, R_2, R_3, and C_1. Unfortunately, the output of A_1 has a sign inversion, which is remedied by the inverting amplifier created by A_2.

Weighted summation of v_i and v_o, and low-pass filtering the result without inversion can be achieved without opamps by passive summation. This is the idea behind the Rauch structure shown in Figure 7.26b, where the circuitry in the gray box of the Tow-Thomas biquad is replaced by a significantly simpler network. The reader may readily verify that

$$I_x = \left(\frac{V_i}{R_1} + \frac{V_o}{R_2}\right)\frac{1}{R_4}\frac{R_{eq}}{1 + sC_1R_{eq}} \tag{7.59}$$

where $R_{eq} = R_1 \parallel R_2 \parallel R_4$. The expressions for I_x in the Tow-Thomas and Rauch structures are the same form, indicating that the same transfer function can be realized by both topologies with an appropriate choice of component values. The transfer function of the Rauch biquad of Figure 7.26b turns out to be

$$\frac{V_o}{V_i} = -\frac{R_2}{R_1} \frac{1}{1 + \frac{s}{\omega_p Q_p} + \frac{s^2}{\omega_p^2}} \tag{7.60}$$

where

$$\omega_p = \frac{1}{\sqrt{R_2 R_4 C_1 C_2}}, \quad Q_p = \frac{R_{eq}}{\sqrt{R_2 R_4}} \sqrt{\frac{C_1}{C_2}}. \tag{7.61}$$

Why do the expressions above make sense? Let us begin with the dc gain. The current through C_2, and consequently i_x are both zero at dc. As a result, $v_x = 0$. This means that the dc current through R_1, namely v_i/R_1, flows through R_2. The dc value of v_o is thus $-(R_2/R_1)v_i$, indicating that the dc gain of the Rauch biquad is $-(R_2/R_1)$.

Next, let us understand the expression for ω_p. At very high frequencies (i.e., $s \rightarrow \infty$), v_o is vanishingly small and the current through R_2 is close to 0. V_x is thus approximately given by $V_i/(sC_1R_1)$. V_o is therefore seen to be $-V_i/(s^2C_1C_2R_1R_4)$ at high frequencies. From (7.60), the asymptotic behavior of V_o/V_i is seen to be $-(R_2/R_1)(\omega_p^2/s^2)$. Consequently, we have

$$-\frac{R_2}{R_1}\frac{\omega_p^2}{s^2} = -\frac{1}{s^2C_1C_2R_1R_4} \tag{7.62}$$

from which we have $\omega_p = 1/\sqrt{R_2R_4C_1C_2}$.

It is but natural to wonder about the nature of the transfer function V_x/V_i. Since $V_o/V_x = -1/(sC_2R_4)$, it is evident that $V_x/V_i = sC_2R_4(V_o/V_i)$ is a band-pass transfer function.

The Rauch biquad is evidently a lot simpler than the Tow-Thomas structure. The question that then arises is—why is the latter even relevant? As one might expect, there is no free lunch, and each topology comes with its own set of advantages and challenges. With respect to the Tow-Thomas and Rauch biquads, the issues to be considered are the following.

a. The Rauch structure has fewer degrees of freedom since it employs fewer components.

b. The input impedance of the Tow-Thomas biquad is purely resistive (assuming an ideal opamp). This is not true for the Rauch design.

c. A Rauch biquad is sensitive to stray capacitance that appears in parallel with C_1.

d. As we will see going forward, the Tow-Thomas biquad is significantly less noisy than the Rauch filter for the same power dissipation.

e. It turns out that the out-of-band linearity of the Rauch biquad is much better than that of the Tow-Thomas filter. The intuition behind this is that out-of-band signals in the Rauch design first encounter the passive filter formed by R_1, C_1 and are significantly attenuated before they excite the opamp, which is the source of distortion.

f. Digitally programming component values (by the use of switchable resistor and capacitor banks) is problematic in the Rauch biquad. This is due to the following. Switches are realized using MOS transistors, which are nonlinear. Since v_x in the Rauch structure is not a virtual ground node, there can be significant switch-induced distortion when R_1, R_2, R_4, and C_2 are realized as digitally programmable banks.

To summarize, the Rauch biquad realizes a second-order all-pole low-pass response using only one opamp. It accomplishes this by exploiting passive summation and filtering. It is amenable to fully differential operation. While it has challenges as described above, it is a widely used topology on account of its many advantages.

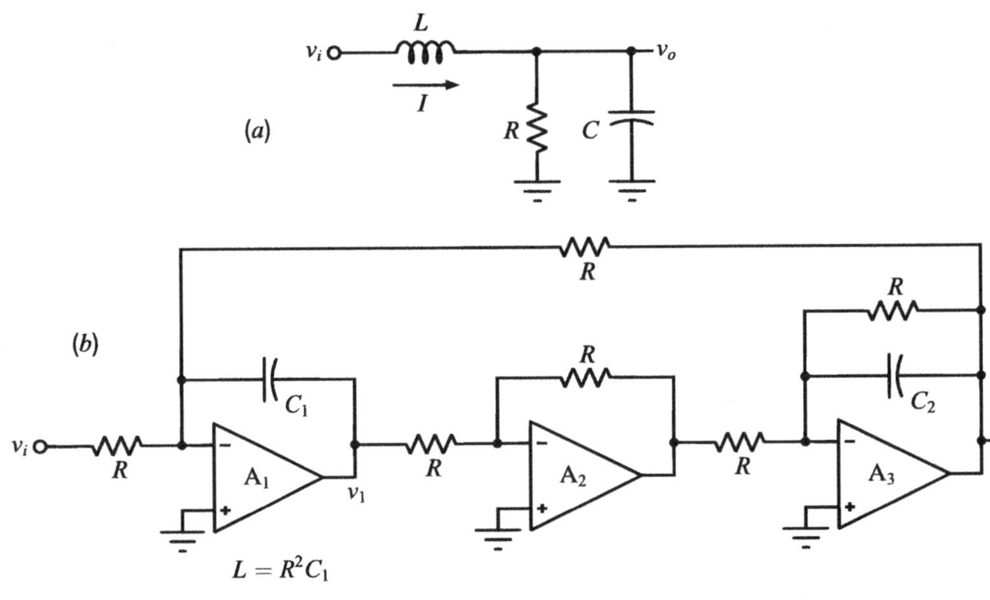

Figure 7.27 (a) Alternative RLC low-pass prototype. (b) Corresponding active-RC prototype.

Other realizations of a biquad are possible. As an example, we show the circuit developed from another passive RLC low-pass structure. Figure 7.27a shows the prototype.[2] Straightforward analysis shows that $\omega_p = 1/\sqrt{LC}$ and $Q_p = R\sqrt{C/L}$. Proceeding in the same manner as we did with the circuit of Figure 7.23a, we arrive at Figure 7.27b. The output of A_3 corresponds to v_o in the passive network. v_1, which is the output of A_1 corresponds to the inductor current (which is no longer a bandpass transfer function). Analysis shows that

[2]The reader would have noticed that this is based on a parallel RLC circuit.

$$\frac{V_o}{V_i} = -\frac{1}{1 + sRC_1 + s^2R^2C_1C_2}$$
$$\frac{V_1}{V_i} = -\frac{1 + sC_2R}{1 + sRC_1 + s^2R^2C_1C_2}. \quad (7.63)$$

which yields

$$\omega_p = \frac{1}{R\sqrt{C_1C_2}}, \quad Q_p = \sqrt{\frac{C_2}{C_1}}. \quad (7.64)$$

A natural question that now arises is: which among the structures Figures 7.27b and 7.23b should one employ in practice? It turns out (as we will see in the subsection on noise) that the modified Tow-Thomas design of Figure 7.27b, being less noisy than the original biquad for the same power dissipation, is the more appropriate choice.

Practical Point: Bottom-Plate Parasitic Capacitance

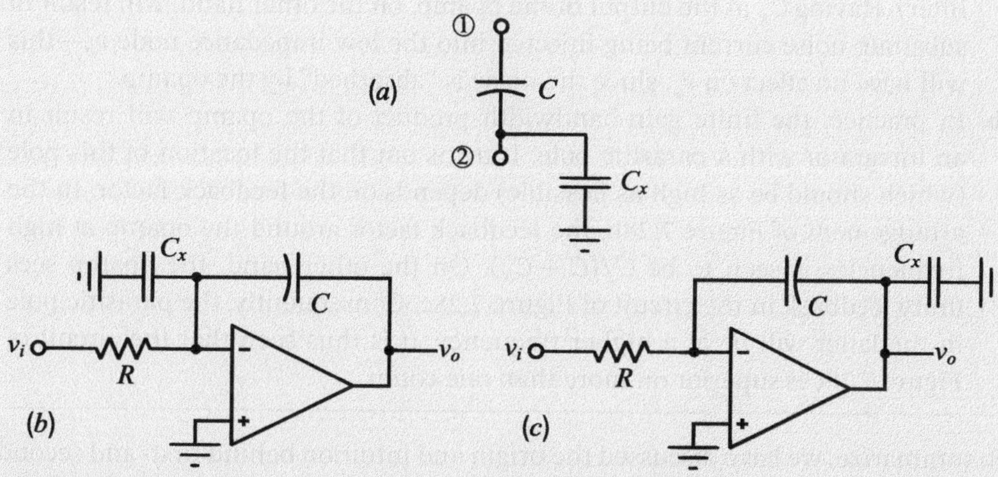

Figure 7.28 (a) Floating capacitor and associated bottom-plate parasitic C_x. (b) Placing C_x at the virtual ground results in a path for noise to couple into the filter. (c) The correct placing of the bottom-plate parasitic.

The workhorse of active filters, namely the active-RC integrator, uses a floating capacitor. It turns out that in practice, one of the terminals of the capacitor has associated with it a *bottom plate parasitic* to ground, as shown in Figure 7.28a. The value of the undesired parasitic is dependent on process technology—it can range from a few percent to well over 20% of the intended capacitance. This then brings up the question: on which side should one place the bottom plate?

(Continued)

(Continued)

Connecting C so that C_x is on the virtual-ground side, as in Figure 7.28b seems like a good idea: since both plates of the parasitic are at the same potential, no current will flow through it, thereby rendering C_x inconsequential. On the other hand, arranging C so that C_x is at the output of the opamp as in Figure 7.28c will need the parasitic capacitor to be charged and discharged like the main capacitor C.

As convincing as the argument above might seem, connecting the bottom plate of C to the virtual-ground node of the opamp is the *wrong* thing to do! The reasons are the following.

a. The "ground" of the parasitic capacitance C_x is the substrate of the integrated circuit. Due to other circuitry on the chip, the substrate node is noisy. Consequently, if C_x is connected to the virtual ground of the opamp, it acts as a source through which substrate noise is injected into the integrator (and the filter). Having C_x at the output of the opamp, on the other hand, will result in substrate noise current being injected into the low-impedance node v_o—this will have no effect on v_o, since the noise is "absorbed" by the opamp.

b. In practice, the finite gain-bandwidth product of the opamp will result in an integrator with a parasitic pole. It turns out that the location of this pole (which should be as high as possible) depends on the feedback factor. In the arrangement of Figure 7.28b, the feedback factor around the opamp at high frequencies is seen to be $C/(C + C_x)$. On the other hand, the opamp sees unity feedback in the circuit of Figure 7.28c. Consequently, the parasitic pole in the latter will be at a higher frequency. It is thus seen that the circuit of Figure 7.28c is superior on more than one count.

To summarize, we have discussed the origin and intuition behind first- and second-order filter building blocks. The topologies we arrived at were based on the opamp-RC integrator, chosen for its parasitic insensitivity. Cascading first- and second-order sections allow us to realize filters of arbitrary order.

7.6 Noise in Opamp-RC Sections

Having seen first- and second-order-section topologies, let us examine the noise properties of these building blocks. We begin with the first-order section.

Figure 7.29 shows the first-order section with noise sources marked. v_{n1} and v_{n2} denote the thermal noise voltage sources associated with the resistors, and v_{n3} denotes the input-referred noise voltage of the opamp. It turns out that in a well-designed opamp, the noise due to the opamp is much smaller than that due to the resistors. The output noise is therefore dominated by that due to the input and feedback resistors.

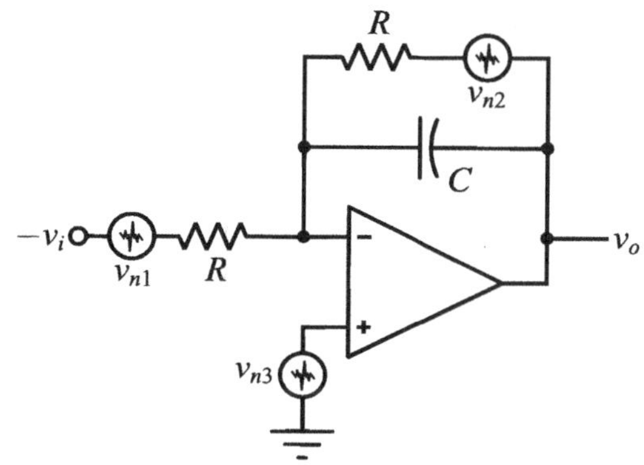

Figure 7.29 Noise sources in a first-order section.

Straightforward analysis shows that the voltage noise-spectral density of the output is given by

$$S_o(f) = 8kTR\frac{1}{1 + 4\pi^2 f^2 R^2 C^2}.$$

The mean-square noise $\overline{v_o^2}$ is seen to be

$$\overline{v_o^2} = 8kTR\int_0^{\infty}\frac{1}{1 + 4\pi^2 f^2 R^2 C^2}df = 2\frac{kT}{C}. \tag{7.65}$$

When compared to a passive-RC filter with the same 3 dB bandwidth, the noise-spectral density and mean-square noise are higher by a factor of 2. This is the price to be paid for the "convenience" of being able to cascade sections without having to worry about loading effects.

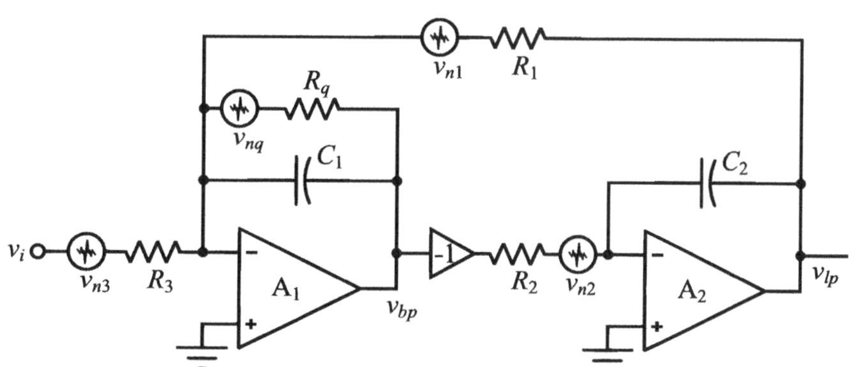

$$\omega_p^2 = \frac{1}{R_1 C_1 R_2 C_2} \qquad Q_p = \frac{R_q}{\sqrt{R_1 R_2}}\sqrt{\frac{C_1}{C_2}}$$

Figure 7.30 Noise sources in a second-order section. The inversion is assumed to be noiseless.

Next, we analyze noise in a Tow-Thomas biquad with low-pass and band-pass outputs. The filter is shown in Figure 7.30. Again, the noise of the opamps is neglected. The low-pass and band-pass transfer functions are given by

$$H_{lp}(s) = -\frac{R_1}{R_3}\frac{1}{1+\frac{s}{\omega_p Q_p}+\frac{s^2}{\omega_p^2}}$$
$$H_{bp}(s) = -\frac{R_q}{R_3}\frac{\frac{s}{\omega_p Q_p}}{1+\frac{s}{\omega_p Q_p}+\frac{s^2}{\omega_p^2}}. \tag{7.66}$$

To determine the noise-spectral density at the filter outputs, we need to find the transfer functions from $v_{n1}, \cdots, v_{nq}$ to v_{lp} and v_{bp}. We make the following observations regarding the low-pass output.

a. The transfer function from v_{n3} to v_{lp} is $H_{lp}(s)$.
b. The transfer functions from v_{n1} and v_{nq} to v_{lp} are $H_{lp}(s)(R_3/R_1)$ and $H_{lp}(s)(R_3/R_q)$, respectively.
c. The transfer function from v_{n2} to v_{lp} is given by $H_{lp}(s)(R_3/R_q) + H_{bp}(s)(R_3C_1/R_2C_2)$.

The noise-spectral density at the low-pass output is thus given by

$$S_{lp}(f) = 4kTR_1\left[1+\frac{R_1}{R_3}+\frac{R_1}{R_q}+\frac{R_1R_2}{R_q^2}\right]\frac{1}{(1-(f/f_p)^2)^2+(f/f_pQ_p)^2}$$
$$+4kTR_1\left(\frac{C_1}{C_2}\right)\frac{(f/f_p)^2}{(1-(f/f_p)^2)^2+(f/f_pQ_p)^2}.$$

where f denotes the frequency in Hz, and $f_p = \omega_p/2\pi$. The noise-spectral density is high in the filter's passband, and peaks around the band-edge frequency. The mean-square noise at the low-pass output is given by

$$\overline{v_{n,lp}^2} = \int_0^{\infty} S_{lp}(f)\,df = 4kTR_1\left[1+\frac{R_1}{R_3}+\frac{R_1}{R_q}+\frac{R_1R_2}{R_q^2}+\frac{C_1}{C_2}\right]\frac{\pi Q_p f_p}{2}. \tag{7.67}$$

Useful Integrals

$$\int_0^{\infty}\frac{1}{(1-f^2)^2+(f/Q_p)^2}\,df = \frac{\pi Q_p}{2} \tag{7.68}$$

$$\int_0^{\infty}\frac{f^2}{(1-f^2)^2+(f/Q_p)^2}\,df = \frac{\pi Q_p}{2}. \tag{7.69}$$

Regarding the bandpass output, we observe the following.

a. The transfer function from v_{n3} to v_{bp} is $H_{bp}(s)$.
b. The transfer functions from v_{n1} and v_{nq} to v_{lp} are $H_{bp}(s)(R_3/R_1)$ and $H_{bp}(s)(R_3/R_q)$, respectively.
c. The transfer function from v_{n2} to v_{bp} is given by $H_{lp}(s)(R_3/R_1)$.

$$S_{bp}(f) = 4kTR_1\frac{R_2C_2}{R_3C_1}\left[1 + \frac{R_3}{R_1} + \frac{R_3}{R_q}\right]\frac{(f/f_p)^2}{(1-(f/f_p)^2)^2 + (f/f_pQ_p)^2}$$
$$+ 4kTR_2\frac{1}{(1-(f/f_p)^2)^2 + (f/f_pQ_p)^2}.$$

The noise-spectral density is seen to be large around the center frequency. The mean-square noise at the band-pass output is given by

$$\overline{v^2_{n,bp}} = \int_0^\infty S_{bp}(f)\, df = 4kTR_1\left\{\frac{R_2C_2}{R_3C_1}\left[1 + \frac{R_3}{R_1} + \frac{R_3}{R_q}\right] + \frac{R_2}{R_1}\right\}\frac{\pi Q_p f_p}{2}. \tag{7.70}$$

A few observations regarding the total integrated noise (at low-pass and band-pass outputs) are in order.

a. The integrated noise is proportional to Q_p. Why does this make sense? For a given f_p, the peak gain from internal noise sources to v_{lp} and v_{bp} is proportional to Q_p. This means that the peak noise-spectral density is proportional to Q_p^2. A higher Q_p, however, renders the response “sharper,” thereby reducing the effective bandwidth. Consequently, the integrated noise is proportional to Q_p.

b. The integrated noise is proportional to f_p. Why does this make sense? For a given Q_p, the bandwidth over which the noise-spectral density is significant is proportional to f_p. The integrated noise must therefore be proportional to f_p.

c. For a given f_p and Q_p, the total noise can be reduced by a factor α by reducing *all resistors* by a factor α. Note that reducing all resistors while keeping f_p unchanged will necessitate increasing all capacitors by the same factor α. Consequently, the opamps will be called upon to supply α times more current, which will in turn increase the power dissipation by a factor α. Doubling the SNR by reducing noise by a factor of 2 therefore entails double the power dissipation. High-frequency, narrowband filters that are designed to achieve low noise should therefore be expected to dissipate a lot of power.

Noise in a Butterworth Biquad

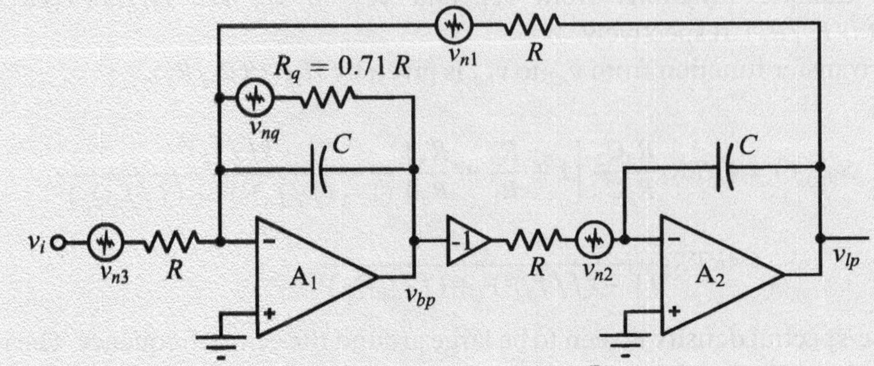

$$\omega_p = \frac{1}{RC} \qquad Q_p = \frac{R_q}{R}$$

Figure 7.31 Butterworth biquad with a dc gain of unity and equal capacitors.

Figure 7.31 shows a Butterworth biquad with a dc gain of unity. It uses equal capacitors. $\omega_p = 1/RC$ and the damping resistor is given by $R_q = Q_pR$. The figure below shows the noise spectral densities at the low-pass and band-pass outputs (normalized to $4kTR$).

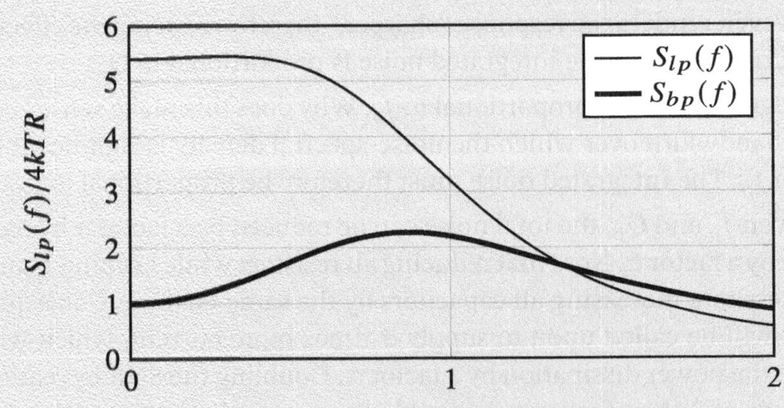

The Tow-Thomas structure of Figure 7.30 provides low-pass and band-pass outputs. The alternate structure, where the damping resistor is placed across C_2, yields only a low-pass output. It turns out, as we show below, that it is the lower-noise alternative—useful if one is not interested in obtaining a bandpass output. Figure 7.32 shows the filter. As usual, the noise of the opamps is neglected.

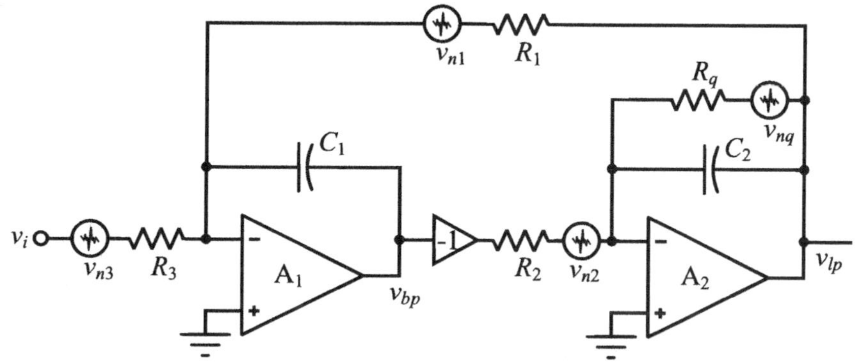

$$\omega_p^2 = \frac{1}{R_1 C_1 R_2 C_2} \qquad Q_p = \frac{R_q}{\sqrt{R_1 R_2}}\sqrt{\frac{C_2}{C_1}}$$

Figure 7.32 An alternative low-noise biquad, useful if one is only interested in a low-pass transfer function.

The low-pass transfer function is given by

$$H_{lp}(s) = -\frac{R_1}{R_3}\frac{1}{1+\frac{s}{\omega_p Q_p}+\frac{s^2}{\omega_p^2}}.$$

To determine the noise-spectral density at the filter output, we need to find the transfer functions from $v_{n1}, \cdots, v_{nq}$ to v_{lp}. We make the following observations.

a. The transfer functions from v_{n3} and v_{n1} to v_{lp} are $H_{lp}(s)$ and $(R_3/R_1)H_{lp}(s)$. respectively.

b. The transfer functions from v_{n2} and v_{nq} to v_{lp} are given by $H_{bp}(s)$ and $H_{bp}(s)(R_2/R_q)$ where

$$H_{bp}(s) = -\frac{\frac{s}{\omega_p Q_p}}{1+\frac{s}{\omega_p Q_p}+\frac{s^2}{\omega_p^2}}.$$

The noise-spectral density at the low-pass output is thus given by

$$S_{lp}(f) = 4kTR_1\left[1+\frac{R_1}{R_3}\right]\frac{1}{(1-(f/f_p)^2)^2+(f/f_pQ_p)^2}$$
$$+\,4kTR_1\left(\frac{C_1}{C_2}\right)\left[1+\frac{R_2}{R_q}\right]\frac{(f/f_p)^2}{(1-(f/f_p)^2)^2+(f/f_pQ_p)^2}.$$

where f and f_p have their usual meanings. The mean-square noise at the low-pass output is given by

$$\overline{v_{n,lp}^2} = \int_0^\infty S_{lp}(f)\,df = 4kTR_1\left[1+\frac{R_1}{R_3}+\frac{C_1(1+R_2/R_q)}{C_2}\right]\frac{\pi Q_p f_p}{2}. \qquad (7.71)$$

Noise in the Alternative Butterworth Biquad

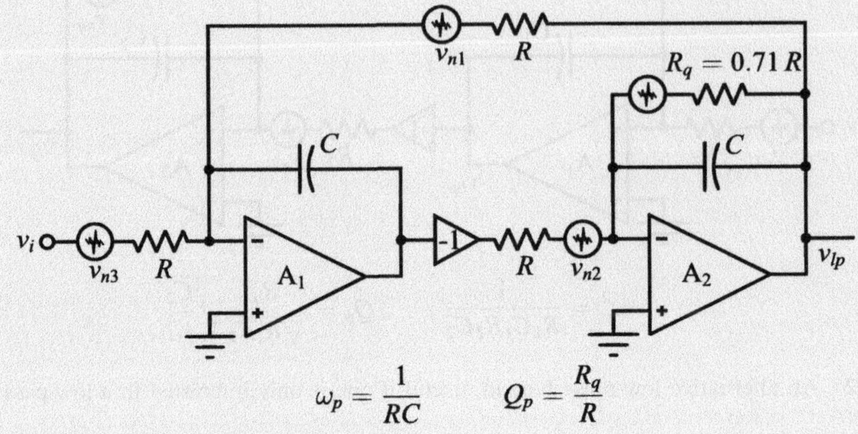

Figure 7.33 Alternative Butterworth biquad with a dc gain of unity and equal capacitors.

Figure 7.33 shows the Butterworth biquad with the damping resistor moved across the second integrating capacitor. It uses equal capacitors. As before, $\omega_p = 1/RC$ and the damping resistor is given by $R_q = Q_pR$. The figure below shows the noise spectral density at the low-pass output (normalized to $4kTR$). It is evident that the noise is considerably smaller than that in the structure of Figure 7.31. The reduction is significant from the power-efficiency viewpoint—remember that to obtain a 2× lower noise using the original structure would need twice the power dissipation!

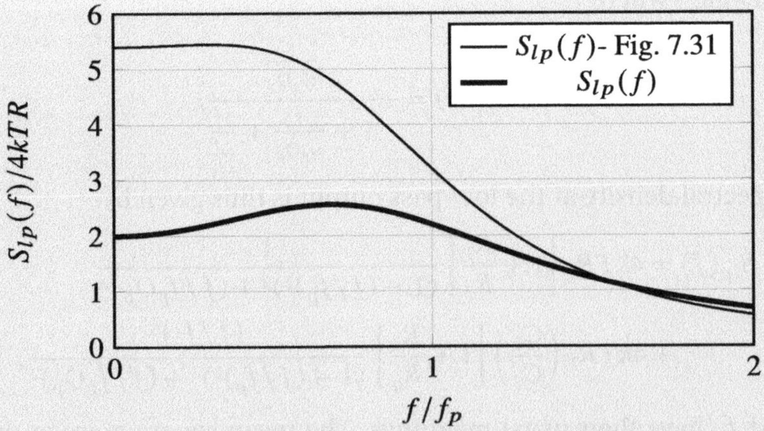

7.7 Dynamic-Range Scaling

We now discuss the rationale behind *dynamic-range scaling*, which is a way to improve the dynamic range of an analog filter for the same power dissipation. Before we go there, however, we first talk about a concept called node scaling.

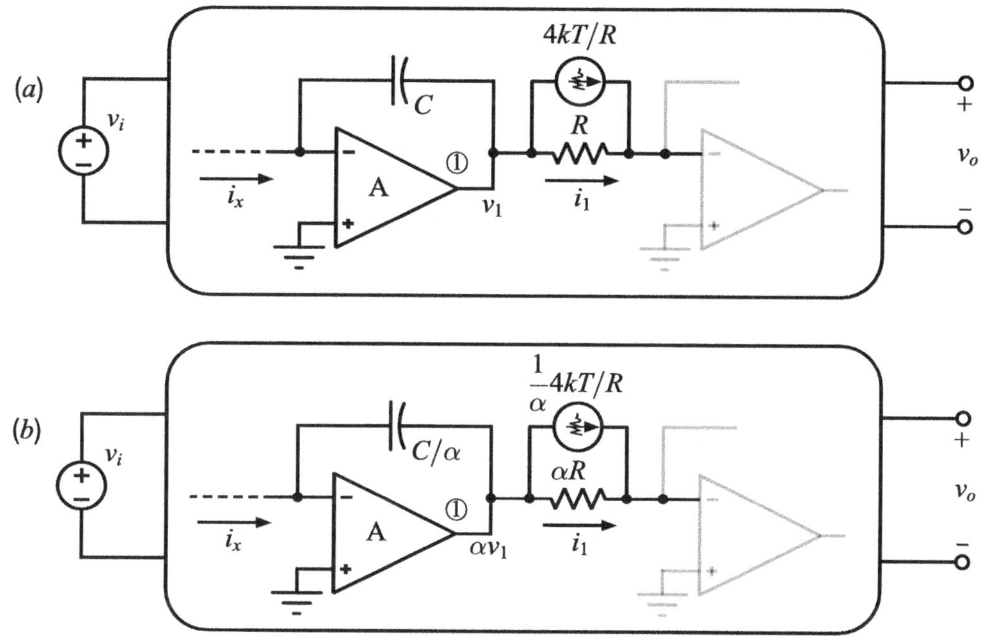

Figure 7.34 (a) Section of an active-RC filter. The transfer function from v_i to v_1 is denoted by $H_1(s)$. (b) Node scaling the transfer function by a factor α reduces due to the resistor.

Consider a portion of an active-RC filter, as shown in Figure 7.34a. The integrating capacitor is denoted by C, and the opamp A drives a resistor of value R, whose other terminal is connected to the virtual ground of an opamp in the filter. The signal current in R is denoted by $i_1 = v_1/R$, and its thermal noise current source is shown. The frequency response at node ① is denoted by $H_1(s) = V_1/V_i$. The filter's transfer function V_o/V_i is denoted by $H(s)$.

What happens to $H(s)$ if C is reduced by a factor α? Clearly, the transfer function will change since i_1 will increase by a factor α. If one wanted the filter transfer function to remain $H(s)$ even when C is reduced by a factor α, R must be increased to αR as shown in Figure 7.34b. This way, I_1 remains the same (i.e., $I_x/(sCR)$) as in Figure 7.34a, and the rest of the filter "will not know" that the integrating capacitor has reduced to C/α. Increasing the resistor to αR, however, reduces the noise power at v_o. This is due to the following. The transfer function from the noise current to v_o is unchanged from part (a) to (b) in Figure 7.34. The noise spectral density in the latter figure, however, is lower by a factor α.

To summarize, reducing C and increasing R by α as in Figure 7.34b has the following consequences:

a. The frequency response at node ① will be *scaled* by α to $\alpha H_1(s)$. This is therefore called *node scaling*. In the time domain, it means that for the same input, the waveform at will be α times larger. The voltage transfer functions to *all nodes other than will remain unchanged.*

b. The noise-power at the filter output reduces, since the spectral density of the noise current injected by αR is lower by a factor α.

From the discussion above, it seems as if using a large α is beneficial on multiple counts—the area of the integrating capacitor reduces, and so does the noise at the filter's output. So what prevents one from using an arbitrarily large α? The answer is opamp saturation, as explained next using an example.

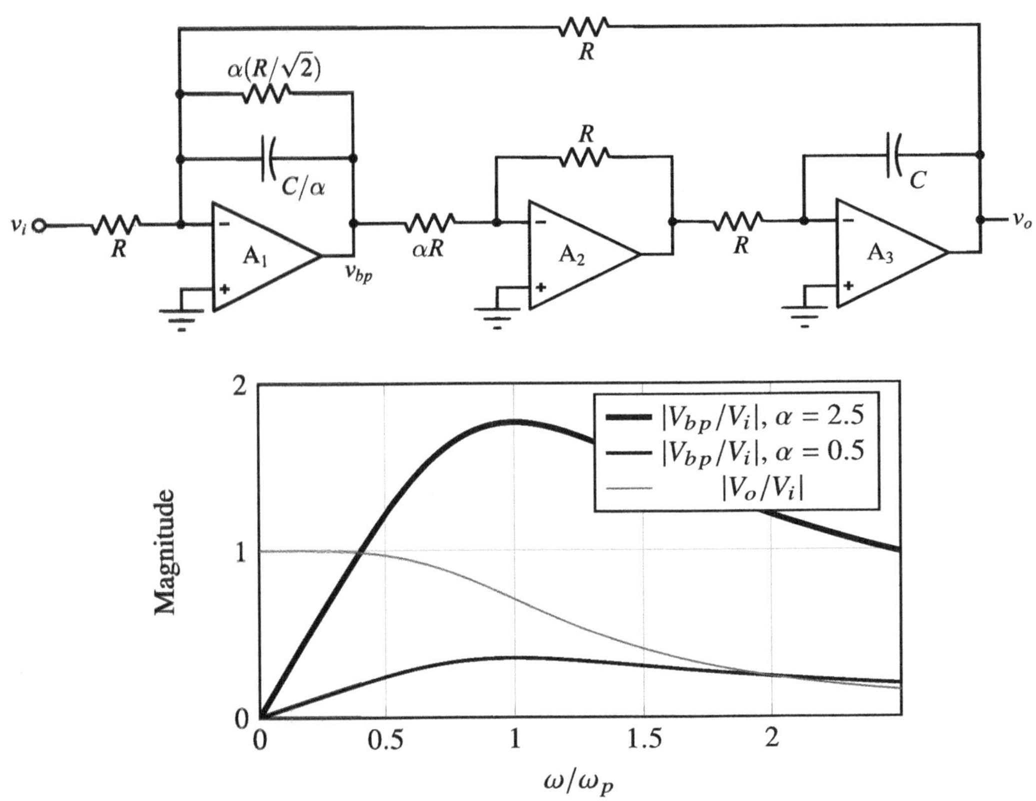

Figure 7.35 Magnitude responses at the low-pass and band-pass nodes of a Butterworth biquad, for $\alpha = 0.5$ and $\alpha = 2.5$.

Figure 7.35 shows the schematic of a Tow-Thomas Butterworth biquad. The impedance in the feedback path of A_1 is scaled by α. To ensure that V_o/V_i remains Butterworth for arbitrary values of α, the resistor that senses v_{bp} should be modified to αR as shown in the figure. The magnitude responses of V_o/V_i and V_{bp}/V_i are plotted in Figure 7.35 for $\alpha = 0.5$ and $\alpha = 2.5$. While the low-pass transfer function does not change with α, we see that V_{bp}/V_i is scaled by a factor of 5.

The key point crucial to understanding the necessity for dynamic-range scaling is the concept of opamp saturation. An opamp operates off of supply rails. The peak-to-peak swing at the output of an opamp is constrained by the supplies, and is usually a large fraction of the potential difference between the positive and negative supply voltages. Peak-to-peak outputs beyond this are simply not possible as the output of the opamp will saturate. We denote the peak-to-peak output swing of an opamp by $2V_{\max}$.

Coming back to our Butterworth example of Figure 7.35, let us assume that $\alpha = 2.5$, and that the biquad is excited by a sine-wave input whose amplitude is $V_{\max}$ and frequency is ω_p. If the opamp were ideal, and its output stage did not saturate, the peak swing at the band-pass node would be about $1.8V_{\max}$. This, however, would cause the real opamp A_1 to saturate (Note that A_2 would still be operating in its linear region). To prevent saturation, the input amplitude should be reduced by a factor of 1.8. This means that the maximum signal power at the filter's output would be smaller by a factor of ($1.8^2 \approx 3.2$) when compared to what it would have been if A_1 did not saturate. A_1, therefore, limits the peak SNR that can be achieved at the filter's output.

We thus see that the act of using a large α (like $\alpha = 2.5$ in our example) to reduce noise at the filter's output has had the deleterious effect of limiting the maximum signal amplitude that the filter can handle. What happens when α is made very small? Using too small an α (say α=0.5 in our example) certainly prevents A_1 from saturating, but it is apparent that the full swing at A_1's output is "not being used." What is the largest α that can be used before A_1 saturates? We see that α can be increased to about 1.41, a choice that results in the peak gain at A_1's output of 1. This in turn will result in lower noise due to the αR resistor.

The discussion above indicates that the "best" value to use for α is the one that results in the largest swing at A_1's output without letting it saturate. Equivalently, the peak of the magnitude response should be unity (like that for A_3). This way, when the filter is excited by a single-tone input of *any* frequency, no single opamp prematurely limits the signal level that the filter can handle. This process of node scaling so that the peak gains of the frequency responses at the outputs of all opamps are unity is termed *dynamic-range scaling*. In our example, using $\alpha = 1.4$ will result in the peak gain of the band-pass transfer function to be unity. This choice will maximize the peak input sine-wave amplitude that the filter can handle while reducing noise due to αR. A smaller α would result in more noise, while a larger α would cause the signal power to reduce.

Example: Dynamic-Range Scaling in a Fourth-Order Butterworth Filter

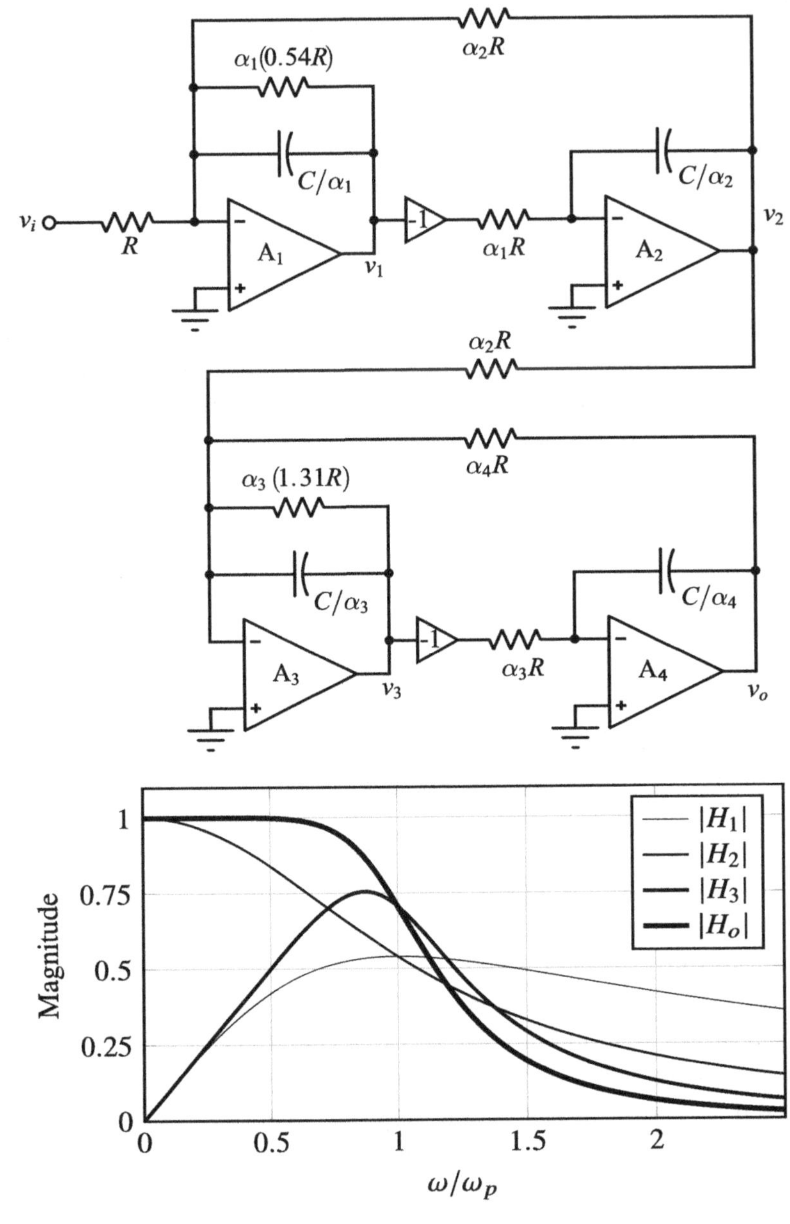

Figure 7.36 Magnitude responses of internal nodes in the fourth-order Butterworth lowpass filter.

Let us apply the techniques we discussed above to a fourth-order Butterworth low-pass filter. Two Tow-Thomas biquads are cascaded to realize the fourth-order transfer function, as shown in Figure 7.36. The quality factors of the biquads are 0.54 and 1.31. The low-Q biquad is placed first. The node scaling coefficients of the integrating capacitors are denoted by $\alpha_1, \cdots, \alpha_4$. We start with all α's being unity. The transfer functions from v_i to $v_1, \cdots, v_o$ are denoted by $H_1(s), \cdots, H_o(s)$ respectively. Figure 7.36

plots the magnitude responses of these transfer functions. We see that the peaks of $|H_2|$ and $|H_o|$ are 1, and occur at dc. The maximum value of $|H_1|$ is 0.54, while that of $|H_3|$ is 0.75. To set the peak gains of all responses to unity, node scaling should be as per $\alpha_1 = 1.85$, $\alpha_2 = 1$, $\alpha_3 = 1.33$, and $\alpha_4 = 1$. The frequency responses at the opamp outputs after dynamic range scaling are shown in Figure 7.37.

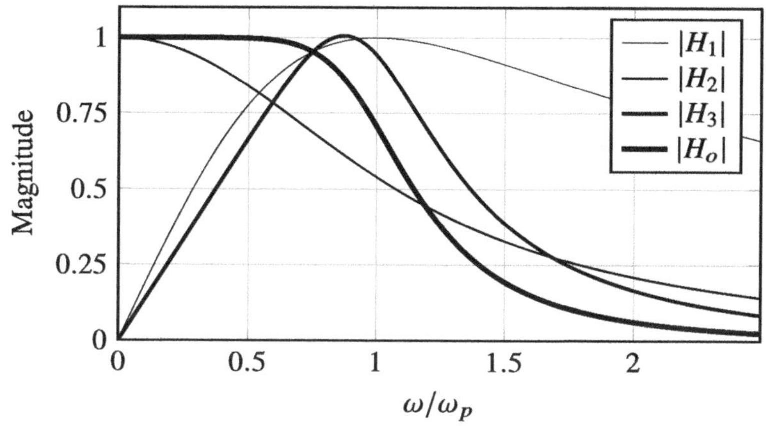

Figure 7.37 Magnitude responses of internal nodes after dynamic-range scaling.

7.8 Fully-Differential Signal Processing, Integrators and Filters

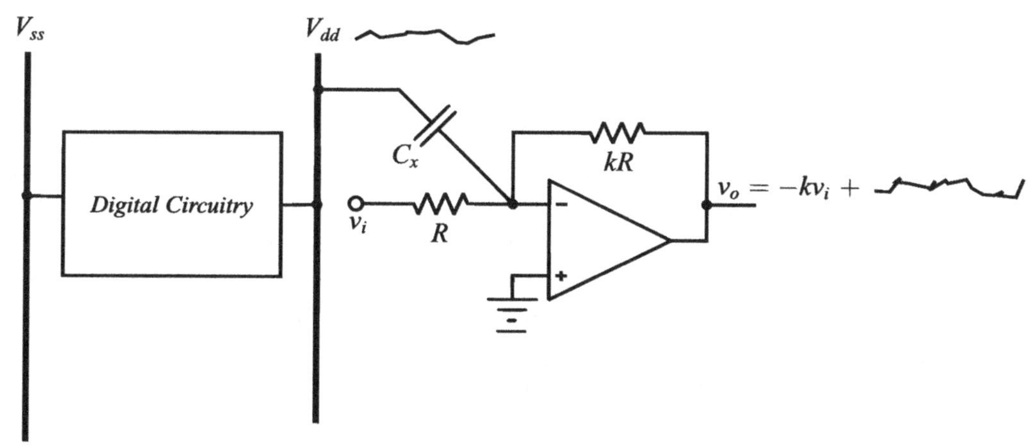

Figure 7.38 Problems with single-ended operation. C_x is a parasitic capacitance to the opamp's virtual ground node.

We have so far worked with so-called single-ended circuits, which means that input and output voltages are referred to ground. On an integrated circuit, this approach is problematic due to the following. Digital circuits on the chip draw spikes of current from the power supply. Consequently, there is a noise-like waveform superimposed on the V_{dd} node, as shown in Figure 7.38. On the same chip, assume that we are trying to build an inverting amplifier with a gain of $-k$. Due to unavoidable reasons, there is a parasitic coupling capacitance C_x between V_{dd} and the opamp's virtual ground. A good design would attempt to make C_x as small as possible, but the key point is to recognize that it cannot be totally eliminated. The result of a nonzero C_x is that noise on the supply will appear at the output of the amplifier as shown in Figure 7.38, thereby corrupting the signals in the analog signal chain.

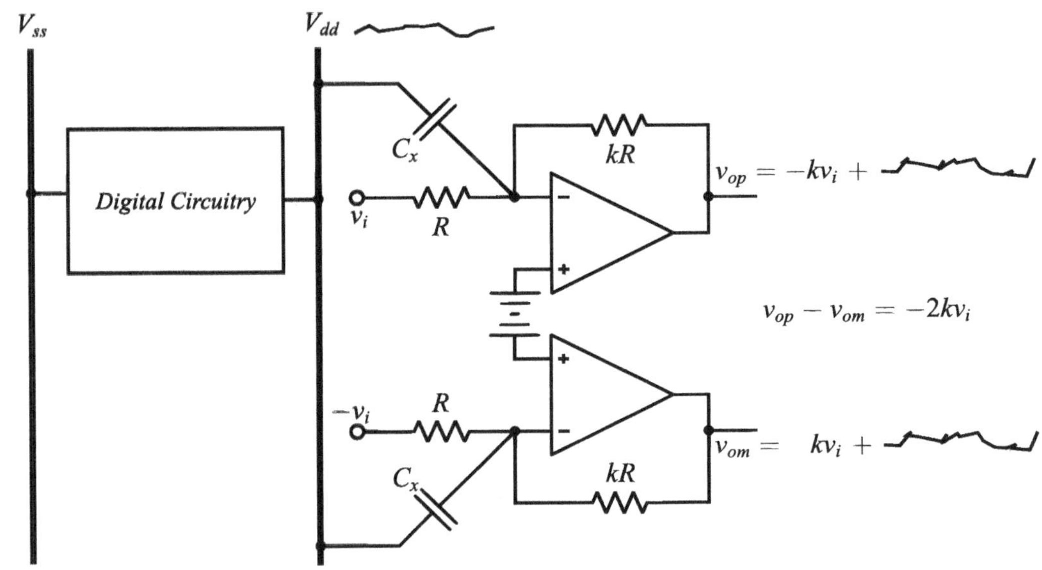

Figure 7.39 Using two identical amplifiers with a differential input. The output of interest is the difference between v_{op} and v_{om}.

A solution to the problem is shown in Figure 7.39. Two *identical* amplifiers are built. By symmetry, the parasitic capacitances coupling the supply to the virtual-ground nodes of the opamps are identical. One amplifier is excited with v_i, while the other is excited with $-v_i$. The output of the former, namely v_{op} is the sum of $-v_i$ and a noise waveform. The second amplifier's output v_{om} is the sum of v_i and the *same* noise waveform that corrupts v_{op}. If however, one observed the difference between v_{op} and v_{om}, it is apparent that the noise would go away, since it is common to both outputs. This way, supply noise coupling does not degrade the signals in the analog path. This mode of operation, where information of interest is the difference between two node voltages (or branch currents), is termed *differential operation*. A few observations are in order.

a. Differential operation achieves robustness to all sources of noise that are common to both copies of the circuit. Such noise is called "common-mode" noise.

b. Common-mode noise is eliminated only if both copies of the circuit are exactly identical. In practice, there will be mismatch between the copies. Consequently, a portion of common-mode noise leaks into the differential output. Referring to Figure 7.39, if the C_x's were not identical the noise waveforms at v_{op} and v_{om}

would not be exactly equal. The difference would appear in the differential output $(v_{op} - v_{om})$.

c. It might seem that differential operation entails a power penalty since two amplifiers are being used. Note, however, that the SNR at the output is increased by a factor of 2 when compared to the single-ended amplifier. This is due to the following. Assuming that $C_x = 0$, so that the noise at the output of each amplifier is due to the resistors and opamp, we see that output noise at v_{op} and v_{om} are independent. The signals, however, are v_i and $-v_i$. Consequently, the noise and signal powers in the differential output $(v_{op} - v_{om})$ will be twice and four times that in the single-ended case respectively. Consequently, the SNR of the differential implementation is higher by a factor of 2 while consuming twice the power, which means that there is no fundamental SNR improvement (or penalty) due to differential operation. The main advantage is robustness with respect to common-mode noise sources (like supply noise discussed earlier).

d. In practice, the amplifiers will be weakly nonlinear. v_{op} and v_{om} will both consist of even- and odd-order distortion components. The differential output, however, is free from even harmonics as they are common to v_{op} and v_{om}. Differential operation, therefore, also improves linearity by eliminating even harmonics.

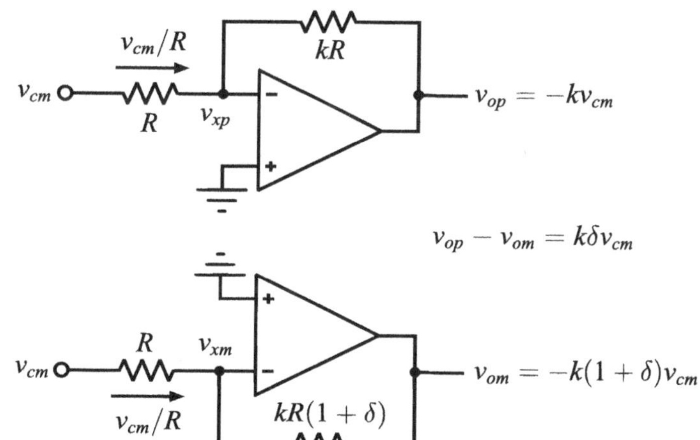

Figure 7.40 Effect of mismatch when two identical inputs are used.

As alluded to earlier, using two nominally identical copies and working with differential signals results in an output that is sensitive to component mismatch. This can become particularly problematic when the *difference* between the two inputs is small, but both inputs vary by a large amount (though they are identical). This is illustrated in Figure 7.40. The inputs are identical, and equal to v_{cm}. The feedback resistor in one of the amplifiers deviates from its nominal value by δR. As a consequence, the differential output voltage which is ideally supposed to be zero, is now seen to be $\delta k v_{cm}$. There is a differential output even though the differential input is zero. How do we address this problem?

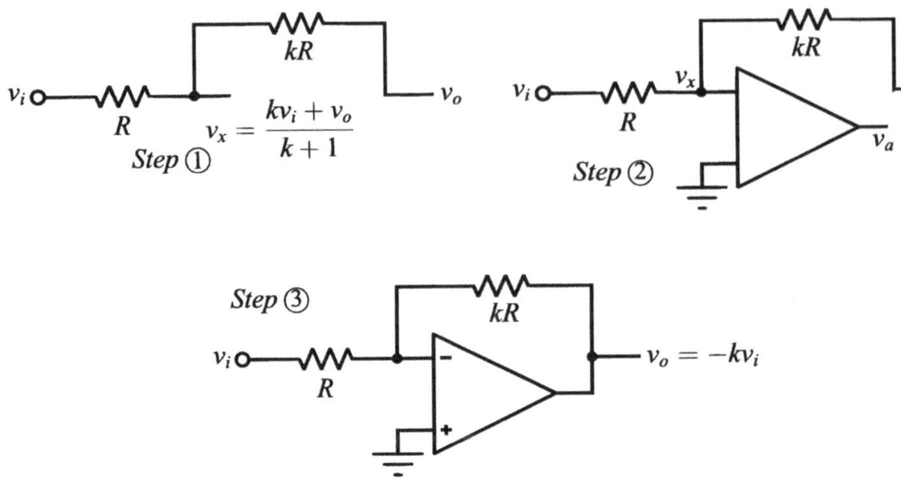

Figure 7.41 Evolution of the inverting amplifier.

The root cause of the problem seen above is the following. Ideally, we would like to amplify only the difference between the two inputs. Each of our amplifiers, however, operate independently and are unable to distinguish whether their inputs are common to both of them, or if they are differential. This suggests the following approach. If the feedback loop leading to the inverting amplifier operated *only for differential signals*, we should hope to see improved performance. To understand this better, consider the evolution of the "standard" inverting amplifier, as shown in Figure 7.41.

Step ① We need to generate $v_o = -kv_i$. Assuming v_o exists, how do we figure out if it is "correct?" To do this we first generate the voltage $v_x = (kv_i + v_o)/(k+1)$ using a resistor divider.

Step ② We compare v_x with 0 using an amplifier with infinite gain. If $v_x > 0$, it means that v_o is not negative enough and must be reduced (and vice versa). Therefore, if $v_x > 0$, the signs of the opamp must be chosen so that v_a goes more negative.

Step ③ The signs of the opamp are chosen appropriately and v_a and v_o are shorted. In steady state, $v_x = 0$ and $v_o = -kv_i$.

Applying the steps above to differential signals, we arrive at the amplifier of Figure 7.42a. The idea is to ensure that $v_{op} - v_{om} = -k(v_{ip} - v_{im})$. To this end, we form the differential error

$$v_x = \frac{1}{k+1}\left[(kv_{ip} + v_{op}) - (kv_{im} + v_{om})\right].$$

As in the single-ended case, $(v_{op} - v_{om})$ is not sufficiently negative if $v_x > 0$. The *fully-differential* opamp senses v_x and varies the *difference* between v_{op} and v_{om} to make v_x go to zero. The opamp, whose equivalent is shown in Figure 7.42b only responds to the differential voltage v_x. v_{op} and v_{om} are balanced about ground, with $(v_{op} - v_{om}) = A(v_{xp} - v_{xm})$ where $A \rightarrow \infty$.

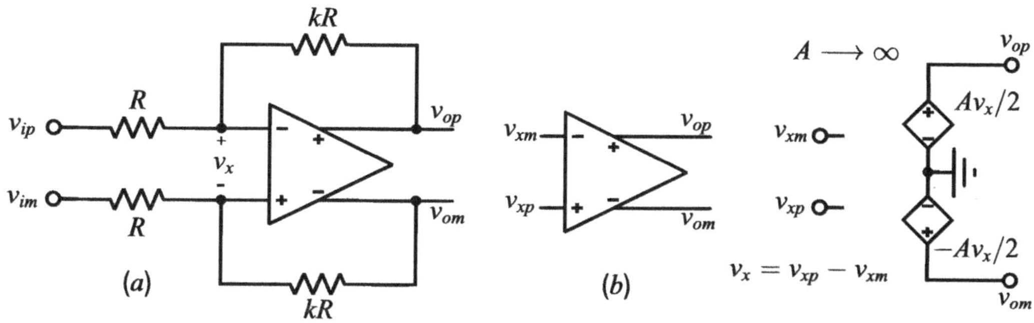

Figure 7.42 (a) Fully-differential inverting amplifier. (b) Opamp equivalent circuit—the outputs are balanced with respect to ground.

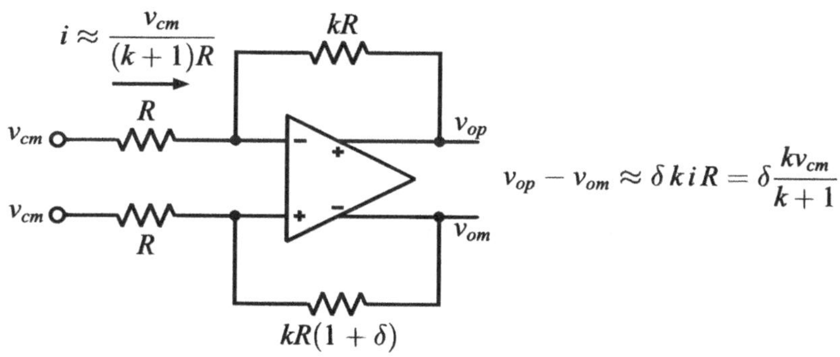

Figure 7.43 Effect of mismatched resistors on the differential output.

Let us now examine the differential output when both inputs are identical, with mismatch in the feedback resistors of the amplifier (Figure 7.43). Ideally, the differential output should be zero. Straightforward analysis shows that the input terminals of the amplifier are at a potential $v_{cm}k/(k+1)$, thereby resulting in a current $i = v_{cm}/(k+1)R$ in both input resistors. The differential output is thus given by $(k/k+1)\delta v_{cm}$. When compared to using two independent single-ended amplifiers, we see that the differential output is $(k+1)$ times smaller.

7.8.1 Fully-Differential Integrators and Filters

We have so far discussed integrators and filters that are single-ended, in the sense that the voltages of interest are referred to ground. As discussed above, there are many advantages to operating circuits in a fully differential fashion. Accordingly, Figure 7.44 compares single-ended and differential integrators.

Figure 7.44b shows a fully differential integrator. The "input" to the integrator is the *difference* between the node voltages marked v_i and $-v_i$. The opamp is also differential, meaning that the gain from its differential inputs to its *differential* outputs is (ideally) infinite. The output of interest is the *difference* between the potentials of the nodes

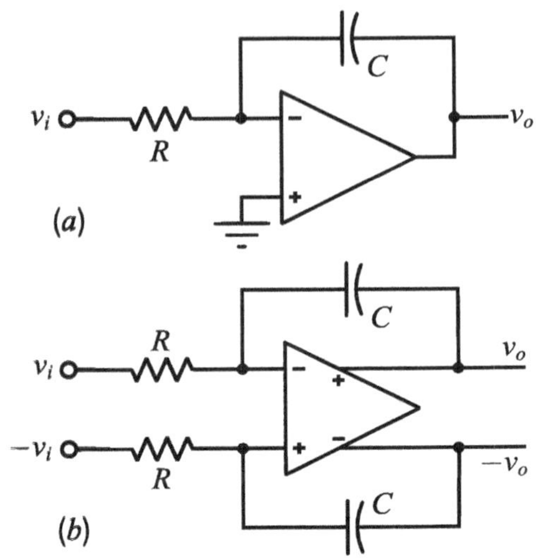

Figure 7.44 (a) Single-ended and (b) fully-differential integrators.

marked v_o and $-v_o$. To reiterate, the signals of interest are *not* v_i or $-v_i$, but their difference. Another useful consequence of fully-differential circuits is that inversion can be had for free, simply by crossing wires.

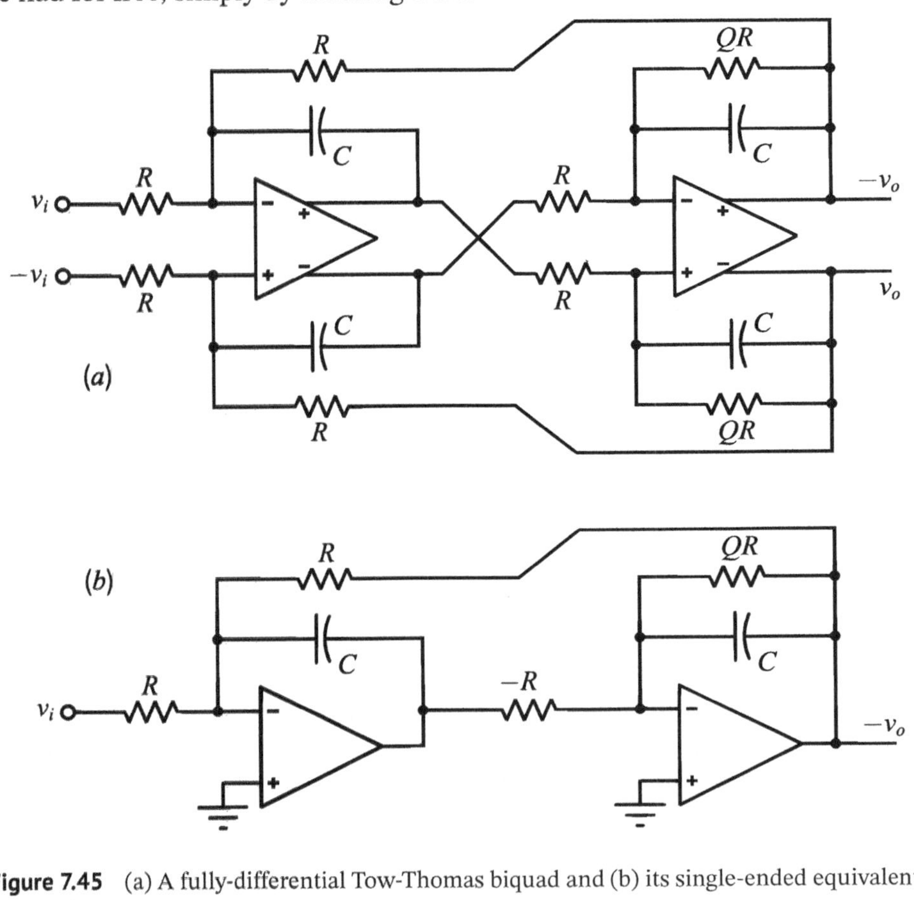

Figure 7.45 (a) A fully-differential Tow-Thomas biquad and (b) its single-ended equivalent.

Figure 7.45 shows a fully-differential biquad that employs equal resistors, and with a damped second integrator. When compared to a single-ended biquad design, we see that the inverting amplifier stage is eliminated. This is because inversion in a differential version can be obtained by simply crossing wires. Fully differential circuits are cumbersome to draw, and are hence often drawn as "single-ended equivalents." Such a representation is shown in Figure 7.45b. The inversion is shown as the negative resistance $-R$ between the two integrators, but should be interpreted as crossing wires of the first integrator's output in the actual circuit.

7.9 Effect of Opamp Finite Gain-Bandwidth Product

Our analysis so far assumed ideal opamps, meaning they have infinite dc gain and bandwidth. In reality, this is not true of course, and it is important to understand what these nonidealities do to the response of a filter.

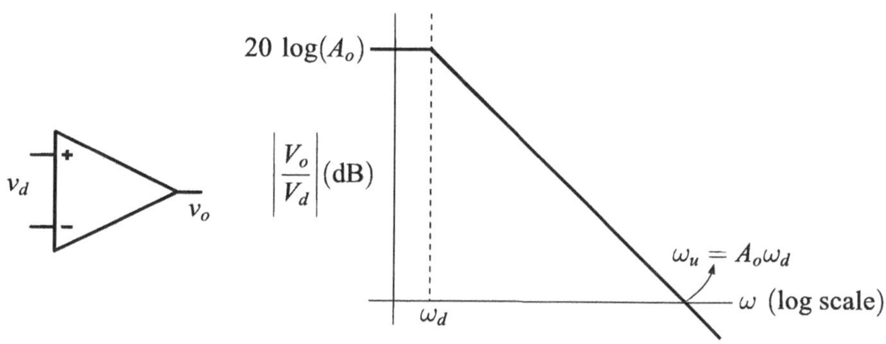

Figure 7.46 Magnitude response of a dominant-pole compensated opamp.

A useful approach to account for the finite gain and bandwidth of a practical opamp is to model it as a single-pole system according to

$$\frac{V_o(s)}{V_d(s)} = A(s) = \frac{A_o}{1 + \frac{s}{\omega_d}} \tag{7.72}$$

where A_o and ω_d denote the dc gain and 3 dB bandwidth respectively. The Bode plot of the opamp's frequency response is given in Figure 7.46. A_o is usually large, often in the range of 10^3 to 10^4. For frequencies larger than ω_d, the opamp's gain is approximately given by

$$\frac{V_o(s)}{V_d(s)} = A(s) \approx \frac{A_o\omega_d}{s} = \frac{\omega_u}{s} \tag{7.73}$$

where ω_u is called the gain-bandwidth product. ω_u is also the frequency at which the magnitude of the opamp's gain drops to unity, and is therefore termed the *unity-gain bandwidth*.

The question we wish to understand and answer is: How does the finite unity-gain bandwidth of the opamp affect the filter transfer function? To address this, we proceed as follows. We will first determine the effect of opamp nonidealities on the properties of an integrator, which is the building block of an opamp-RC filter. We will then figure out how a modification of the integrator's transfer function affects that of the filter.

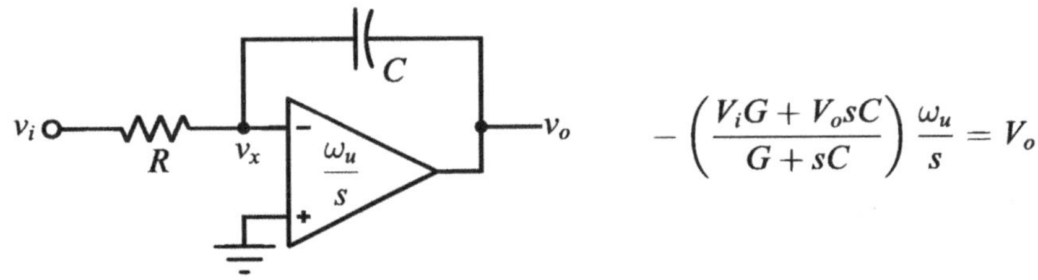

Figure 7.47 An active-RC integrator with a finite-bandwidth opamp.

Figure 7.47 shows an integrator built using an opamp with a finite gain-bandwidth product ω_u. Denoting $G = 1/R$, the relationship between V_o and V_i is given by

$$\underbrace{-\left(\frac{V_iG + V_osC}{G + sC}\right)}_{V_x(s)}\frac{\omega_u}{s} = V_o. \tag{7.74}$$

Rearranging, and using $\omega_o = 1/(RC)$ we obtain

$$\frac{V_o(s)}{V_i(s)} = -\frac{\omega_o}{s}\frac{1}{\left(1 + \frac{\omega_o}{\omega_u}\right) + \frac{s}{\omega_u}} = -\frac{\omega_o}{s\left(1 + \frac{\omega_o}{\omega_u}\right)}\frac{1}{\left(1 + \frac{s}{\omega_u\left(1+\frac{\omega_o}{\omega_u}\right)}\right)}. \tag{7.75}$$

We see that the effect of a finite opamp unity-gain bandwidth is twofold:

a. The integrator's transfer function is now of second-order. The two-amplifier Tow-Thomas biquad (assuming inversion is ideal, which is true in a fully-differential implementation) will be a fourth-order system. The hope is that the frequency response of this network is close to that of the second-order system it is intended to realize.

b. The modified transfer function can be thought of as the cascade of an ideal integrator (albeit with a modified unity-gain frequency) and a first-order low-pass filter with 3 dB bandwidth given by $\omega_u(1 + (\omega_o/\omega_u))$.

Let us make some approximations that are valid in a practical integrator. To achieve a performance close to the ideal, the negative feedback loop should have a high gain so that the virtual-ground node potential is small. This is of course not possible at all frequencies, since the opamp's gain falls with frequency; however, it is clear that its gain must be large at least up to the unity-gain frequency of the integrator, namely ω_o. This means that the quantity $\omega_u/\omega_0 \gg 1$. Using this practical relationship, the integrator's transfer function can be simplified as

$$\frac{V_0(s)}{V_i(s)} \approx -\frac{\omega_o}{s}\frac{1}{\left(1+\frac{s}{\omega_u}\right)}. \tag{7.76}$$

The equation above indicates that the effect of the opamp's finite gain-bandwidth product can be modeled approximately as a high-frequency pole at ω_u. In the time domain, this corresponds to a delay of $1/\omega_u$ with respect to the output of an ideal integrator. Figure 7.48 shows the unit-step responses of ideal and practical integrators.

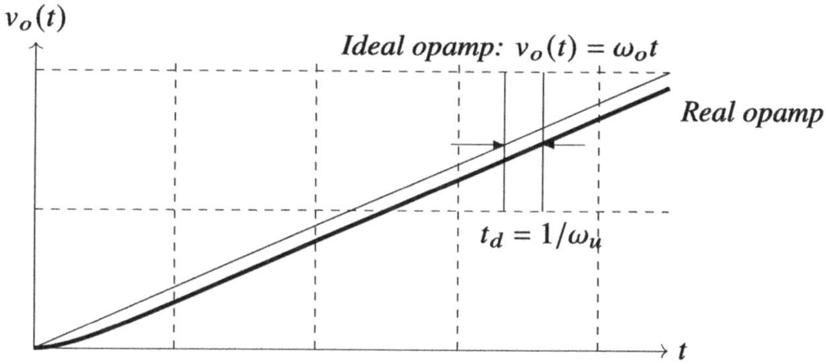

Figure 7.48 Unit-step responses of ideal and practical integrators when $\omega_u \gg \omega_o$. The finite ω_u introduces a delay $1/\omega_u$.

Clearly, as ω_u tends to infinity, the transfer function of a biquad that is built using non-ideal integrators should approach a second-order one. In practice, the high-frequency parasitic pole due to finite opamp bandwidth is far away from the unity-gain frequency of the integrator. This allows us to use concepts like ω_p and Q_p, originally associated with a second-order system, even though the transfer function of the biquad is of a higher order.

Before we dive into an analysis of an integrator's parasitic poles on the frequency response, let us take a step back and ponder over what we should broadly expect. A biquad consists of two integrators enclosed in a negative feedback loop. If there is excess delay in the integrators (due to the finite speed of the opamp), we are in effect introducing delay into a feedback loop. The closed-loop system will therefore tend to move closer to the edge of stability.

Geometric Interpretation of a Transfer Function

Consider the standard form of a second-order all-pole transfer function with complex conjugate poles at p and p^*. It can be expressed as

$$H(s) = \frac{1}{\frac{s^2}{\omega_p^2}+\frac{s}{\omega_p Q_p}+1} = \frac{\omega_p^2}{(s-p)(s-p^*)}. \tag{7.77}$$

(Continued)

(Continued)

To determine $H(s)$ at any complex frequency s_1, we simply substitute $s = s_1$ in the expression above. It is useful to interpret this graphically. Figure 7.49 shows p and p^*. They lie on a circle of radius ω_p. s_1 is an arbitrary complex frequency. To find $H(s_1)$, we first draw arrows from p and p^* to s_1. These arrows represent the complex numbers $(s_1 - p)$ and $(s_1 - p^*)$, and are denoted by a and b respectively. $H(s_1)$ is given by $\omega_p^2/(ab)$. It is thus seen that the magnitude response $|H(s_1)|$ is inversely proportional to the length of the arrows a and b.

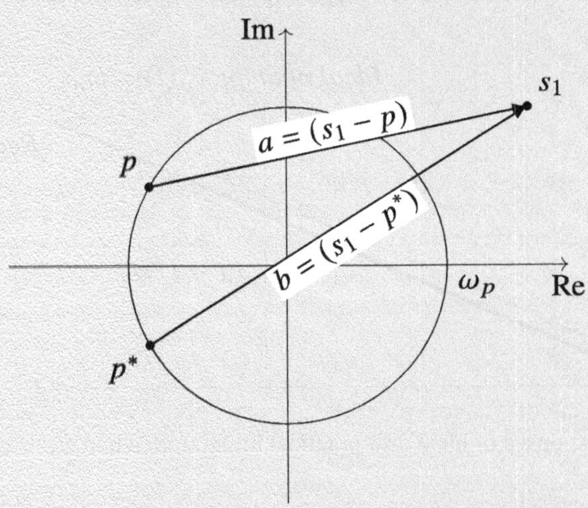

$$H(s_1) = \frac{pp^*}{(s_1 - p)(s_1 - p^*)} = \frac{\omega_p^2}{(s_1 - p)(s_1 - p^*)} = \frac{\omega_p^2}{ab}$$

Figure 7.49 Geometric interpretation of a transfer function.

The frequency response is obtained by evaluating $H(s)$ on the $j\omega$ axis.

Useful Result: $|H(s_1)|$ When s_1 Lies on a Circle of Radius ω_p

We derive below a useful result to evaluate $|H(s_1)|$ when s_1 is any complex frequency that lies on the circle of radius ω_p, where $|p| = \omega_p$. Since $|H(s_1)|$ is given by $\omega_p^2/|ab|$, we need to find $|ab|$. We proceed in the following step-by-step manner.

(a) Referring to Figure 7.50, we recognize that the angle θ subtended by p-p^* at s_1 on the circle is half of that it subtends at the circle's center.

(b) The area of the triangle formed by p-s_1-p^* is given by $(1/2)|ab|\sin\theta$. $\sin\theta$ is seen to be x/ω_p, where x is the Im(p).

(c) The area of the triangle is also given by $x \cdot y$, where $y = |\text{Re}(p) - \text{Re}(s_1)|$.

(d) Using (b) and (c), we have

$$|H(s_1)| = \frac{\omega_p}{2|\text{Re}(p) - \text{Re}(s_1)|} \tag{7.78}$$

As a sanity check, let us evaluate $|H(j\omega_p)|$ using our formula above yields. $\text{Re}(p) = -\omega_p/(2Q_p)$ and $\text{Re}(s_1) = 0$. $|H(j\omega_p)|$ is thus seen to be Q_p, which is what we expect. Turning this result upside down, we can *interpret* Q_p as the magnitude of $H(j\omega_p)$.

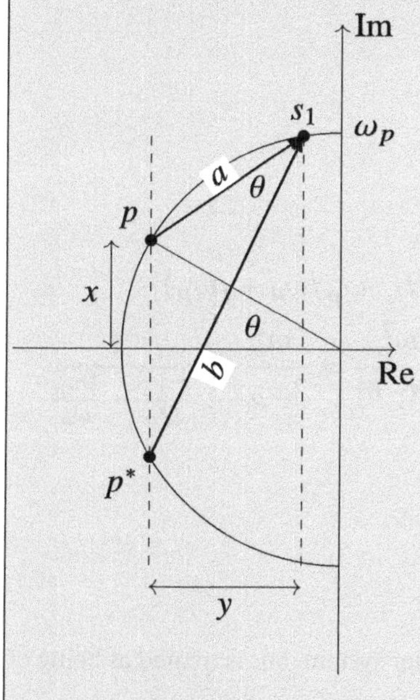

$$\begin{aligned} \text{Area of triangle } p\text{-}s_1\text{-}p^* &= \frac{1}{2}|ab|\sin\theta = xy \\ \sin\theta &= \frac{x}{\omega_p} \\ \Rightarrow |ab| = 2\omega_p y &= 2\omega_p|\text{Re}(p) - \text{Re}(s_1)| \end{aligned}$$

$$\boxed{|H(s_1)| = \frac{\omega_p}{2|\text{Re}(p) - \text{Re}(s_1)|}}$$

Figure 7.50 A useful result concerning $|H(s_1)|$ when s_1 lies on the circle of radius ω_p.

From (7.76), we see that the transfer function of a practical integrator can be obtained by replacing s in that of an ideal integrator by $s(1 + s/\omega_u)$. Denoting the transfer functions of the nonideal and ideal biquads by $\hat{H}(s)$ and $H(s)$, respectively, it is apparent that they are related as

$$\hat{H}(s) = H\left(s + \frac{s^2}{\omega_u}\right). \tag{7.79}$$

The frequency response, which is obtained by evaluating the $\hat{H}(s)$ for $s = j\omega$, is given by

$$\hat{H}(j\omega) = H\left(j\omega - \frac{\omega^2}{\omega_u}\right) \tag{7.80}$$

which can be graphically interpreted as shown in Figure 7.51.

a. Draw the poles of $H(s)$. These are located at $-\frac{\omega_p}{2Q_p} \pm j\omega_x$ where

$$\omega_x = \omega_p\sqrt{1 - \frac{1}{4Q_p^2}}.$$

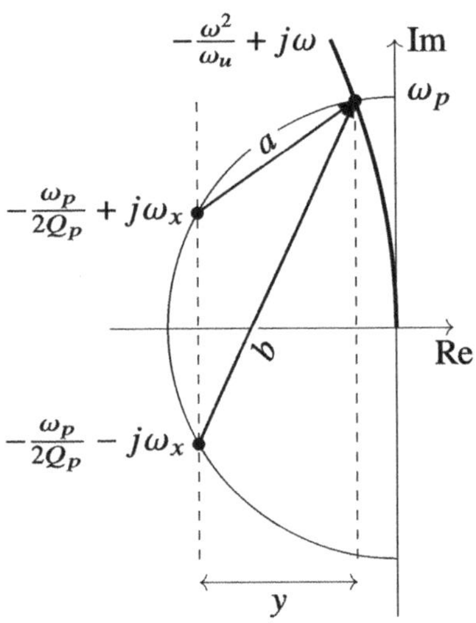

$$\begin{aligned} |\hat{H}(j\omega_p)| &= |H(-\omega_p^2/\omega_u + j\omega_p)| \\ &= \frac{\omega_p^2}{|a \cdot b|} = \frac{\omega_p^2}{2\omega_p y} = \frac{1}{\frac{1}{Q_p} - \frac{2\omega_p}{\omega_u}} \end{aligned}$$

Figure 7.51 $\hat{H}(s) = H(s(1 + s/\omega_u))$. The biquad is now a fourth-order system, but is treated as being of approximately second-order with enhanced Q_p.

The poles lie on circle of radius ω_p. To determine $H(j\omega)$, we go along the imaginary axis. At each point $j\omega$, we draw the vectors to the poles and determine the product of their magnitudes from which $|H(j\omega)|$ can be calculated.

b. Since $\hat{H}(j\omega) = H\left(j\omega - \frac{\omega^2}{\omega_u}\right)$, it follows that $|\hat{H}(j\omega)|$ can be evaluated by using the pole-plot of $H(s)$, but traveling along the curved path $j\omega - \frac{\omega^2}{\omega_u}$ as shown in Figure 7.51. From the figure, it is apparent that $\hat{H}(0) = H(0)$. This makes sense, since the dc gain of the integrator has not been modified by the parasitic pole at ω_u.

c. As ω increases, $|\hat{H}(j\omega)| = \omega_p^2/|ab|$ in Figure 7.51 will be larger than $|H(j\omega)|$. This makes sense due to the following. For a given ω, the corresponding point on the curved path is "closer" to the poles than $j\omega$.

d. $|\hat{H}(j\omega_p)|$ can be evaluated as follows. For $\omega_u \gg \omega_p$, the point $j\omega_p - \omega_p^2/\omega_u$ almost lies on the circle with radius ω_p. Consequently,

$$|\hat{H}(j\omega_p)| = |H(-\omega_p^2/\omega_u + j\omega_p)| = \frac{\omega_p^2}{|ab|} = \frac{\omega_p^2}{2\omega_p y}$$

where

$$y = \frac{\omega_p}{2Q_p} - \frac{\omega_p^2}{\omega_u}.$$

From the two equations above, we have

$$|\hat{H}(j\omega_p)| = Q_p \frac{1}{1 - \frac{2\omega_p Q_p}{\omega_u}}. \tag{7.81}$$

Recall that we interpreted Q_p for a biquad as being the magnitude of the transfer function evaluated at $s = j\omega_p$, namely $|H(j\omega_p)|$. If we *assumed* that our nonideal biquad (a fourth-order system in reality), is a second-order one and determined its quality factor $\hat{Q}_p$ by measuring its gain at ω_p, we would obtain

$$\boxed{\hat{Q}_p = Q_p \frac{1}{1 - \frac{2\omega_p Q_p}{\omega_u}}.} \tag{7.82}$$

Since $\hat{Q}_p > Q_p$, finite opamp bandwidth is said to cause *Q enhancement* [5]. As $\omega_u \to \infty$, i.e., the opamp tends to an ideal one, $\hat{Q}_p \to Q_p$, which makes sense. Furthermore, the factor by which Q_p is enhanced depends on $1 - (2\omega_p Q_p/\omega_u)$. For the same ω_p, a biquad with a higher Q_p will be affected more. Therefore, when one chooses an opamp to realize an active-RC filter, its gain-bandwidth product must be chosen to be much larger than the $\omega_p Q_p$ *product* that needs to be realized. The key point is that the Q_p plays a key role in the choice of the opamp.

Q-Enhancement Example

A student is designing a fully-differential active-RC biquad with a 3 dB bandwidth of 20 kHz and $Q_p = 10$, and decides to use an opamp that has a unity-gain bandwidth of 1 MHz. What Q_p will the biquad actually have?

Using the relationship derived previously, we see that the quality factor due to finite opamp speed is enhanced to

$$\hat{Q}_p = 10 \frac{1}{1 - \frac{2 \cdot 20 \times 10^3 \cdot 10}{10^6}} = 16.67.$$

(Continued)

(Continued)

We see that the Q_p has increased significantly from the desired value. If ω_u reduces by 20% due to a change in ambient temperature, Q_p would increase to 20, indicating that the design is very sensitive to environmental variations.

Having seen the effect of the opamp's finite gain-bandwidth product on the performance of a biquad, it is natural to wonder what happens to a high-order low-pass filter. Such a transfer function is the product of second- and first-order transfer functions. Also, as seen earlier, a high-Q_p biquad whose ω_p is close to the band-edge frequency is needed to achieve the sharp roll-off of the magnitude response. Since this biquad has the highest Q_p among all the biquads in the design, its quality factor is enhanced to the largest extent. One should therefore expect the magnitude response to deviate from the ideal to the largest extent at the band edge.

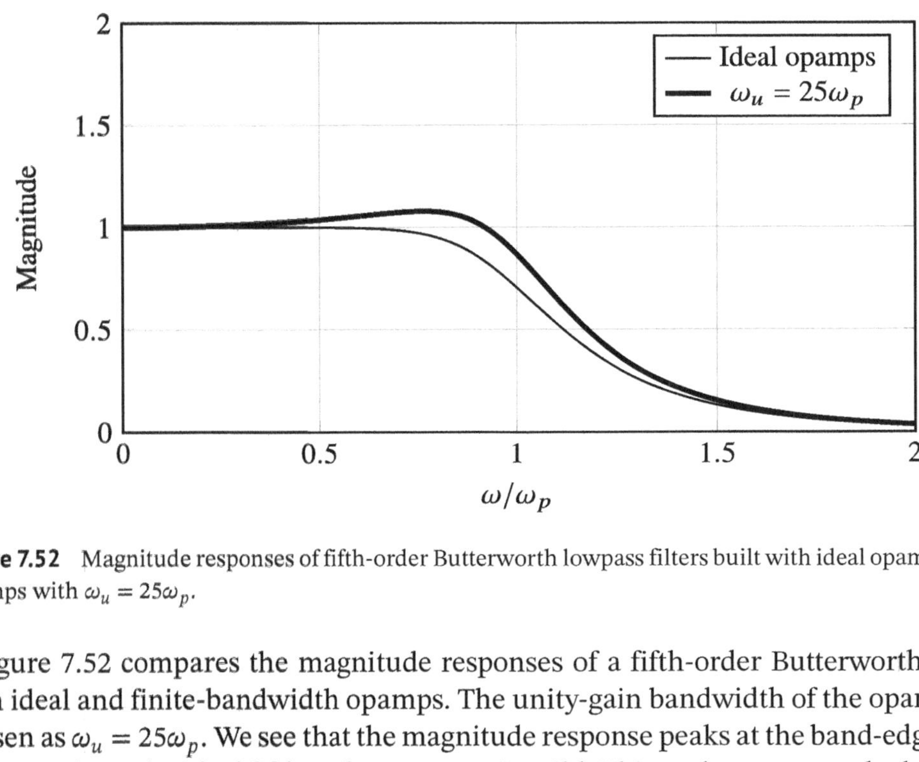

Figure 7.52 Magnitude responses of fifth-order Butterworth lowpass filters built with ideal opamps and opamps with $\omega_u = 25\omega_p$.

Figure 7.52 compares the magnitude responses of a fifth-order Butterworth filter with ideal and finite-bandwidth opamps. The unity-gain bandwidth of the opamps is chosen as $\omega_u = 25\omega_p$. We see that the magnitude response peaks at the band-edge; the deviation from the ideal fifth-order response is mild. This makes sense as the highest quality-factor poles have $Q_p = 1.62$.

A fifth-order Chebyshev filter presents a very different picture (for the same opamp ω_u), as seen in Figure 7.53. The magnitude has increased by a factor of 1.7 at the band edge. This is because the highest-Q_p biquad in the filter is 5.55.

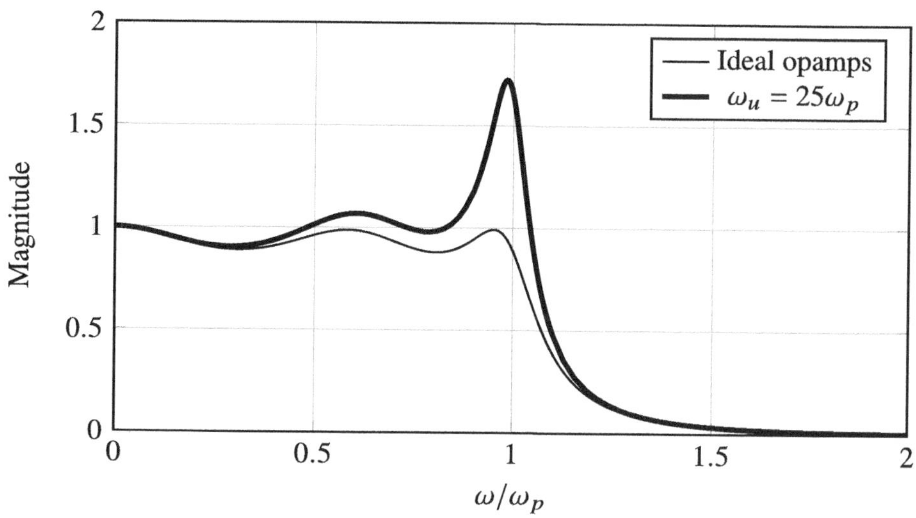

Figure 7.53 Magnitude responses of fifth-order Chebyshev lowpass filters built with ideal opamps and opamps with $\omega_u = 25\omega_p$.

7.9.1 Addressing Finite Opamp Bandwidth

We have just seen how finite opamp bandwidth causes Q-enhancement, which in turn results in a significant deviation of the magnitude response from the one that is desired. How do we fix this issue? One can think of two straightforward ways of addressing the problem.

a. *Wider-bandwidth opamps:* This is the most obvious solution. For example, using $\omega_u = 250\omega_p$ in our example Chebyshev design would greatly reduce Q-enhancement. This approach of designing wider-band opamps is problematic on two counts. First, it may simply not be feasible in the process technology used to design the filter due to inherent speed limitations of the transistors. Next, even if it were feasible, wider-bandwidth opamps will dissipate significantly higher power and reduce the power-efficiency of the filter.

b. *Transfer-function predistortion:* The idea behind this approach is the following. We know that finite ω_u will result in an increased Q_p. What if we deliberately aim to realize a transfer function with biquads whose Q_p's are deliberately lower, so that they result in the desired Q_p's after they become larger due to finite ω_u? In other words, if we appropriately *predistorted* the quality factors based on our knowledge of ω_u, the enhanced quality factors would be those that we desired in the first place. It is straightforward to see this should be done according to

$$\hat{Q}_p = Q_p \frac{1}{1 + \frac{2\omega_p Q_p}{\omega_u}} \tag{7.83}$$

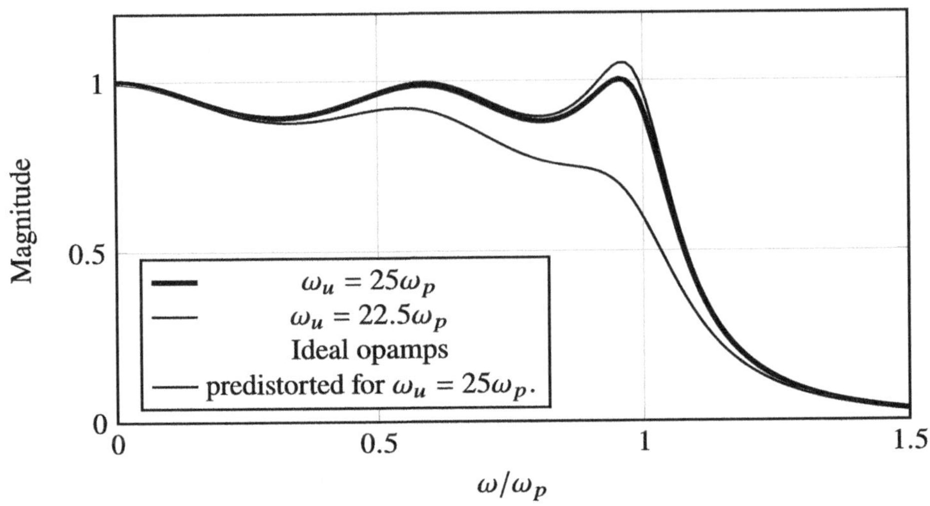

Figure 7.54 The use of predistortion to account for finite opamp gain-bandwidth product. The biquads are designed so that the band-edge response droops with ideal opamps, so that Q-enhancement restores the response to that which is desired.

where $\hat{Q}_p$ and Q_p denote the predistorted and desired quality factors, respectively. Going back to our fifth-order Chebyshev example, predistorting the quality factors to 1.29 and 3.9 should result in a close-to-ideal magnitude response. This is confirmed by simulations, as shown in Figure 7.54. The magnitude response with ideal opamps droops significantly near the band edge, and is restored to the equiripple response due to Q-enhancement caused by $\omega_u = 25\omega_p$. As seen above, predistortion allows the use of finite ω_u opamps by making the opamp a part of the filter design process. A disadvantage of predistortion, however, is that ω_u must be well controlled. Figure 7.54 also shows the response when the transfer function is predistorted assuming $\omega_u = 25\omega_p$, but ω_u reduces by 10% due to changes in the supply voltage and/or the ambient temperature. We see that there is a small residual peak in the response.

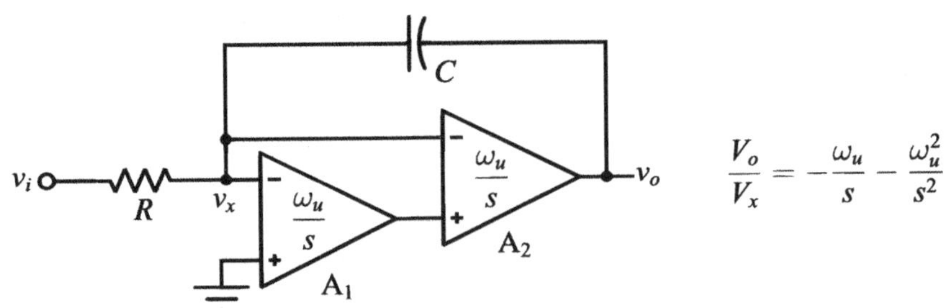

Figure 7.55 Improved integrator configuration using feed forward.

A less obvious approach is to use clever opamp configurations to design improved integrators. We illustrate one such technique—though conceived in the discrete-component era, the principle is in extensive use in today's on-chip integrators. The integrator is shown in Figure 7.55. It uses two nominally identical opamps, A_1 and A_2 in place of the single opamp in the conventional structure. The gain of this equivalent opamp is

$$A(s) = \frac{\omega_u^2}{s^2} + \frac{\omega_u}{s}. \tag{7.84}$$

Straightforward analysis shows that the integrator's transfer function is now given by

$$\frac{V_o(s)}{V_i(s)} \approx -\frac{1}{sCR}\left(\frac{1+\frac{s}{\omega_u}}{1+\frac{s}{\omega_u}+\frac{s^2}{\omega_u^2}}\right) = -\frac{\omega_o}{s}\left(\frac{1+\frac{s}{\omega_u}}{1+\frac{s}{\omega_u}+\frac{s^2}{\omega_u^2}}\right). \tag{7.85}$$

To understand how the finite ω_u of the opamp affects the filter transfer function, we recognize that the variable s has to be replaced according to

$$s \to s\left(\frac{1+\frac{s}{\omega_u}+\frac{s^2}{\omega_u^2}}{1+\frac{s}{\omega_u}}\right). \tag{7.86}$$

For $|s/\omega_u| \ll 1$, we note that

$$\left(\frac{1+\frac{s}{\omega_u}+\frac{s^2}{\omega_u^2}}{1+\frac{s}{\omega_u}}\right) \approx 1 - \frac{s^3}{\omega_u^3} \approx 1. \tag{7.87}$$

As a consequence, the transfer function $\hat{H}(s)$ of a filter built using such integrators, remains very nearly the same as the ideal transfer function $H(s)$ in spite of a finite ω_u, up to frequencies as large as a significant fraction of ω_u.

With the conventional integrator, the quality factor of a biquad with intended quality factor Q_p was enhanced by a factor $1/(1 - \frac{2\omega_p Q_p}{\omega_u})$, and was arrived at by evaluating $H(s)$ at $j\omega_p - (\omega_p^2/\omega_u)$. With the modified integrator, we need to evaluate $H(s)$ at

$$j\omega_p\left(\frac{1+\frac{j\omega_p}{\omega_u}-\frac{\omega_p^2}{\omega_u^2}}{1+\frac{j\omega_p}{\omega_u}}\right) \approx j\omega_p\left(1+j\frac{\omega_p^3}{\omega_u^3}\right) \approx j\omega_p \tag{7.88}$$

which indicates that $\hat{H}(j\omega_p) \approx H(j\omega_p)$. If $\omega_u = 25\omega_p$ for instance, $(\omega_p/\omega_u)^3 = 6.4 \times 10^{-5}$ leading us to conclude that Q-enhancement should be negligible with the improved integrator. The intuition behind the improved integrator is the following. The gain of the "composite opamp" is high at low frequencies since it benefits from the cascade of A_1 and A_2. Stability is maintained due to the feed-forward path through A_2.

7.10 Transconductance-Capacitance Integrators and Filters

We have so far explored active filters employing operational amplifiers, resistors, and capacitors. The rationale behind employing active elements is to eliminate the necessity for inductors. It turns out that employing opamps, resistors, and capacitors is not the only way to emulate an inductor. In this section, we look at an alternative method—one that employs voltage-controlled current sources (also called transconductors) and capacitors. Filters based on this approach are called transconductor-capacitor, or Gm-C filters [6, 7].

7.10.1 The First-Order Gm-C Filter

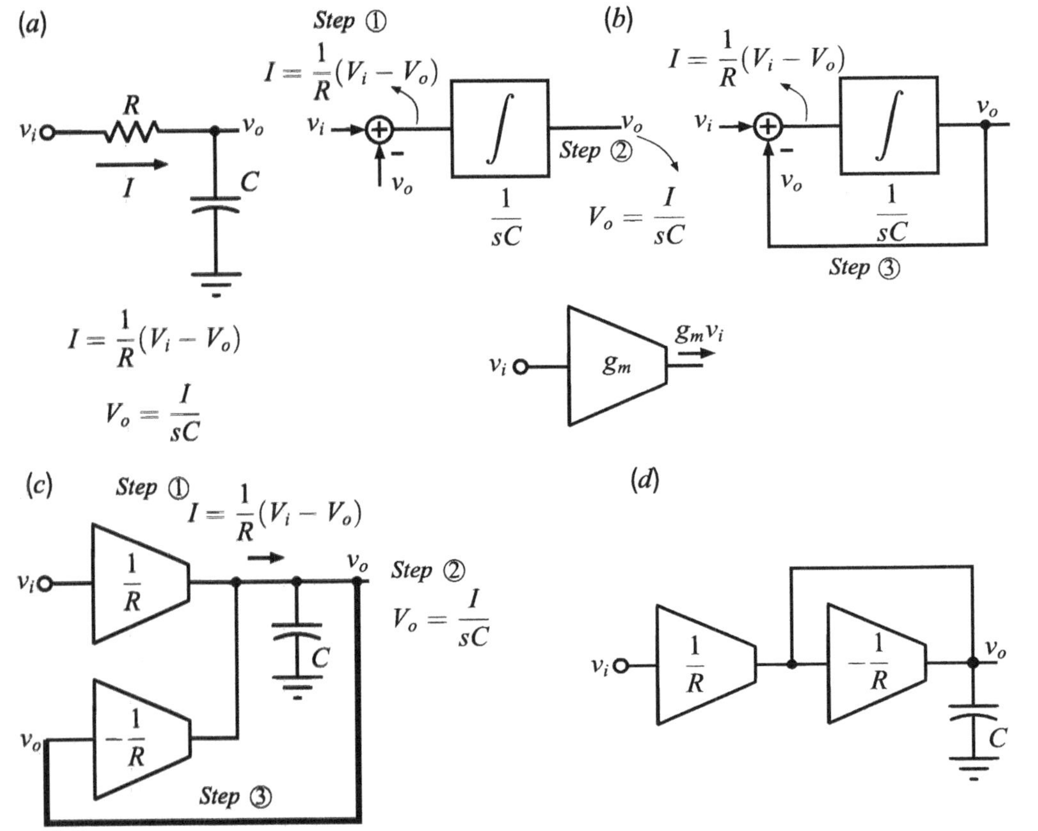

Figure 7.56 (a) First-order RC filter. (b) Signal-flow graph. (c) Implementation using transconductors. (d) A first-order gm-C filter.

Figure 7.56a shows a first-order RC low-pass filter. We see that

$$I = \frac{1}{R}(V_i - V_o), \quad V_o = \frac{I}{sC}. \tag{7.89}$$

The equations above suggest the following implementation, as shown in Figure 7.56b.

① Assuming V_o exists, generate the current I as $(V_i - V_o)/R$.
② Obtain V_o by integrating I using a capacitor.
③ Connect V_o obtained in step 2 to the V_o needed in step 1, thereby closing the feedback loop. The resulting block diagram is the first-order low-pass filter with a 3 dB bandwidth given by $1/RC$ rad/s.

To implement the strategy above, one could use voltage-controlled current sources (also called transconductors) to generate I as shown in Figure 7.56c. The output of the transconductors, which is the current I, is integrated on the capacitor C. The voltage developed across the capacitor is fed back to one of the transconductors. The resulting filter is a first-order transconductor-capacitor structure, or Gm-C filter in short, and is redrawn in compact form in Figure 7.56d.

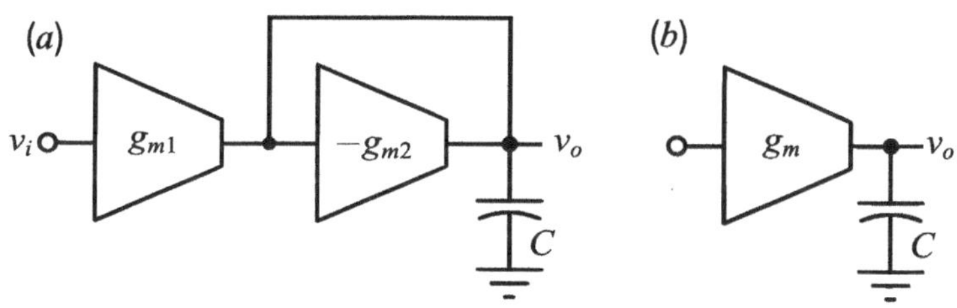

Figure 7.57 (a) First-order Gm-C filter. The dc gain is g_{m1}/g_{m2}, and the 3 dB bandwidth is g_{m2}/C. (b) The Gm-C integrator.

The general form of a first-order Gm-C filter is shown in Figure 7.57a. The input impedance is infinite, enabling it to be cascaded with another such section without loading it. The transfer function is given by

$$\frac{V_o}{V_i} = \frac{g_{m1}/g_{m2}}{1 + (sC/g_{m2})}. \tag{7.90}$$

The dc gain is g_{m1}/g_{m2}. This makes sense due to the following. At dc, the capacitor behaves like an open circuit. All the current pushed out by g_{m1} is therefore "absorbed" by g_{m2}. Consequently, the output voltage is $g_{m1}v_i/g_{m2}$, indicating that the dc gain is g_{m1}/g_{m2}. The pole frequency is g_{m2}/C and does not depend on g_{m1}. This makes sense: the time-constant is determined by the Thevenin resistance across C, which is $1/g_{m2}$.

The basic building block of the Gm-C filter is the Gm-C integrator, shown in Figure 7.57b. The unity-gain frequency is g_m/C rad/s.

7.10.2 The Gm-C Biquad

There are several ways of developing a Gm-C biquad. One way is to proceed as we did in the active-RC case (The reader is encouraged to pursue that line of thought). Another way is to begin with the parallel RLC network shown in Figure 7.58a. The transfer function from the input current to voltage across the tank is bandpass, while that to the inductor current is lowpass. Let's try to achieve the bandpass response

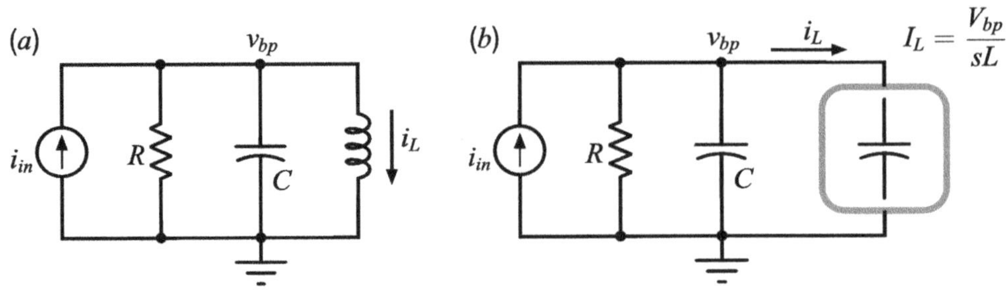

Figure 7.58 (a) RLC prototype. (b) Replacing the real inductor by a synthetic one.

without utilizing an inductor, as shown in Figure 7.58b. If we consider the gray box as a component that emulates the function of an inductor, we expect the current drawn by this box to be proportional to V_{bp}/sL, where L is the equivalent inductance it emulates. In other words, the current drawn by the box must be proportional to the integral of the voltage across it. To achieve integration without using an inductor, we must employ a capacitor since it is the only other element known for its ability to integrate. However, a capacitor integrates current to generate voltage, while what we want is the opposite—that is, we desire to integrate voltage and generate a current proportional to it. To accomplish this, we proceed in the step-by-step manner described below (Figure 7.59).

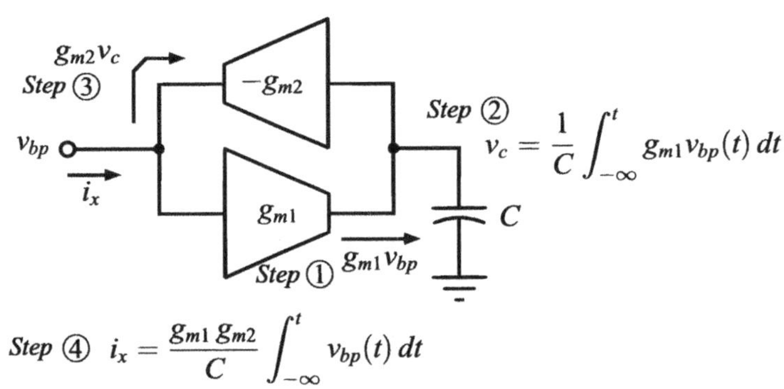

Figure 7.59 Evolution of the synthetic inductor.

① A transconductor whose transconductance is g_{m1} converts v_{bp} into a current.
② This current is integrated on a capacitor to result in a voltage v_c given by

$$v_c = \frac{1}{C}\int_{-\infty}^{t} g_{m1}v_{bp}(t)\,dt.$$

③ v_c is converted into a current using another transconductor with transconductance g_{m2}, and this current is pulled out of the input voltage source, namely v_{bp}. Note carefully that v_c is connected to the negative sign of g_{m2}.

④ The current drawn from the input voltage source, therefore, is given by

$$i_x = g_{m2}v_c(t) = \frac{g_{m1}g_{m2}}{C}\int_{-\infty}^{t} v_{bp}(t)\,dt.$$

In the Laplace domain, this means that

$$I_x(s) = \underbrace{\frac{g_{m1}g_{m2}}{sC}}_{1/sL_{eq}} V_x(s). \tag{7.91}$$

The arrangement, therefore, emulates an inductor of value

$$\boxed{L_{eq} = \frac{C}{g_{m1}g_{m2}}.} \tag{7.92}$$

In summary, the contraption we have just seen successfully mimics the functionality of an inductor by interchanging the roles of voltage and current in a capacitor. It achieves this by converting the input voltage to a current, using a capacitor to integrate this current and generate a voltage, and then converting the voltage back to a current which is drawn from the input source. The effective inductance achieved by this configuration is determined by the relationship $L_{eq} = C/(g_{m1}g_{m2})$. This approach facilitates the emulation of an inductor without the need for an actual inductor in the circuit.

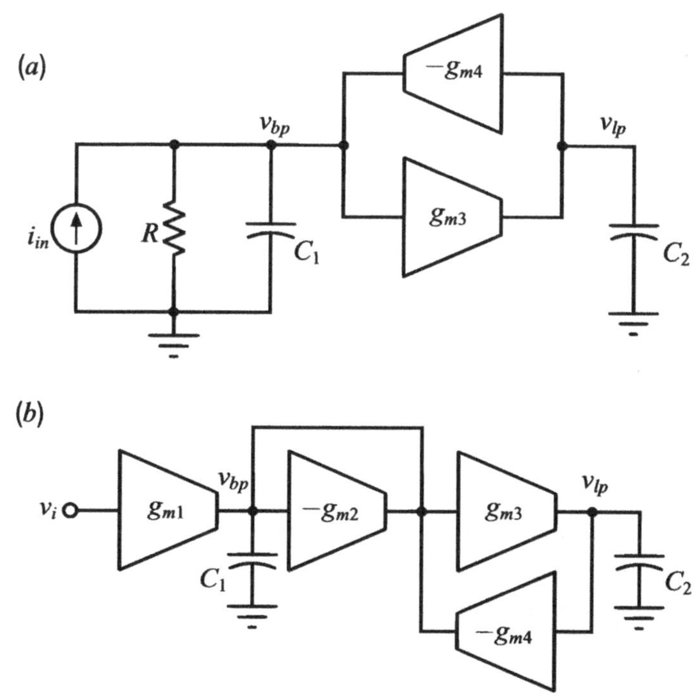

Figure 7.60 (a) RLC filter with a synthetic inductor. (b) Complete Gm-C biquad.

Replacing the inductor of Figure 7.58a with the synthetic version of Figure 7.59, we obtain the filter of Figure 7.60a. The voltage across C_1 is band pass. Since the voltage across C_2 is proportional to the inductor current, it follows that it is the low pass output. Since the synthetic inductor has a value $C_2/(g_{m3}g_{m4})$, it is apparent that

$$\omega_p = \frac{1}{\sqrt{L_{eq}C_1}} = \sqrt{\frac{g_{m3}g_{m4}}{C_1C_2}} \quad \text{and} \quad Q_p = \frac{R}{\sqrt{L_{eq}/C_1}} = R\sqrt{\frac{C_1}{C_2}}\sqrt{g_{m3}g_{m4}}. \tag{7.93}$$

The next step is to recognize that the resistor can be realized as a transconductor that is fed back on itself. Furthermore, to work with an input voltage (rather than a current as in Figure 7.60a), we can use a transconductor to convert voltage into current. The resulting biquad is shown in Figure 7.60b. g_{m2} plays the role of the resistor, and g_{m1} converts v_i into a current. The biquad now consists only of transconductors and capacitors; it is aptly termed the Gm-C biquad.

The Gyrator

The back-to-back connected transconductors play a key role in realizing the inductor, as seen in Figure 7.59. It is an important component in its own right, and is called a gyrator. It is shown in Figure 7.61 below.

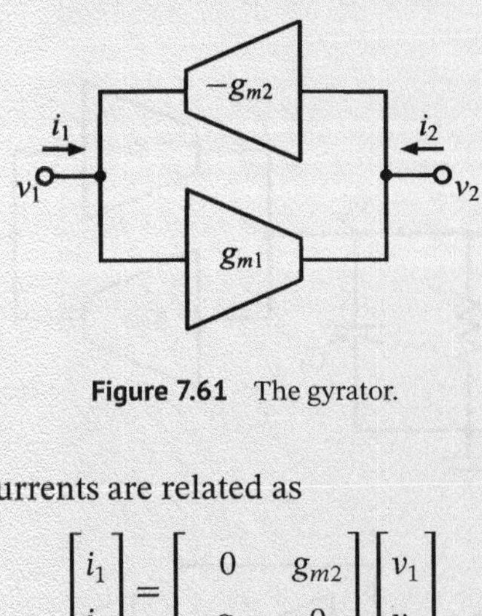

Figure 7.61 The gyrator.

The port voltages and currents are related as

$$\begin{bmatrix} i_1 \\ i_2 \end{bmatrix} = \begin{bmatrix} 0 & g_{m2} \\ -g_{m1} & 0 \end{bmatrix} \begin{bmatrix} v_1 \\ v_2 \end{bmatrix} \tag{7.94}$$

The power dissipated in the two ports, namely $v_1i_1 + v_2i_2$, is $(g_{m2} - g_{m1})v_1v_2$. If $g_{m1} = g_{m2}$, the gyrator does not dissipate power and is therefore passive. Furthermore, it is not a reciprocal two port, since $y_{12} \neq y_{21}$. This shows that a passive two port need not necessarily be reciprocal.

Straightforward analysis shows that the transfer functions from v_i to the bandpass and lowpass nodes are

$$\frac{V_{bp}}{V_i} = \frac{s}{\omega_p Q_p} \frac{(g_{m1}/g_{m2})}{\frac{s^2}{\omega_p^2} + \frac{s}{\omega_p Q_p} + 1}$$
$$\frac{V_{lp}}{V_i} = \frac{g_{m1}}{g_{m4}} \frac{1}{\frac{s^2}{\omega_p^2} + \frac{s}{\omega_p Q_p} + 1} \quad (7.95)$$

where

$$\omega_p = \sqrt{\frac{g_{m3}g_{m4}}{C_1 C_2}} \quad \text{and} \quad Q_p = \sqrt{\frac{C_1}{C_2}}\sqrt{\frac{g_{m3}g_{m4}}{g_{m2}^2}}. \quad (7.96)$$

Let us examine why the expressions for the bandpass and lowpass transfer functions above make sense. We have already discussed the intuition behind the expressions for ω_p and Q_p. The gain of the bandpass filter at $\omega = \omega_p$, which is the resonant frequency of the parallel LC tank formed by g_{m3}-g_{m4}-C_1-C_2, must be g_{m1}/g_{m2} since the tank behaves like an open circuit at its resonant frequency.

To determine the dc gain of the lowpass transfer function, we apply a dc voltage at the input. Since the output of g_{m1} is a bandpass node, the dc voltage at this point is 0. Consequently, the current through g_{m2} must be zero. All of g_{m1}'s dc output current must therefore be "absorbed" by g_{m4}. The dc gain thus has to be g_{m1}/g_{m4}.

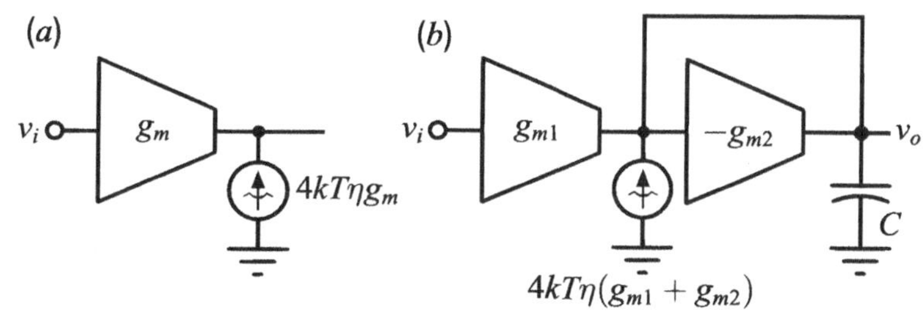

Figure 7.62 (a) Noise model of a transconductor. η is the excess-noise factor. (b) First-order Gm-C filter with noise sources.

Gm-C integrators are noisy, just like their opamp-RC counterparts. The origin of noise is the transconductor, whose noise model is shown in Figure 7.62a. The output signal current, namely $g_m v_i$ is accompanied by a current noise with spectral density $4kT\eta g_m$ A^2/Hz. η is the so-called excess-noise factor, and is dependent on the transconductor's implementation details. Figure 7.62b shows a first-order Gm-C filter with noise sources included. The noise of the negative g_m transconductor is also $4kT\eta g_{m2}$. The output voltage noise spectral density can be calculated as

$$S_o(f) = 4kT\eta \frac{1}{g_{m2}}\left(1 + \frac{g_{m1}}{g_{m2}}\right)\frac{1}{1 + 4\pi^2 f^2 C^2/g_{m2}^2}. \quad (7.97)$$

Like an active-RC integrator, where there are limits on the maximum input and output voltage swing imposed by the opamp, a Gm-C integrator has the same limitations. While the opamp is responsible for these limits in the former, the transconductor imposes swing limits in the latter. In fact, it turns out that for a given supply voltage, the peak-to-peak voltage swing that can be handled by a transconductor is significantly smaller than in a well-designed opamp-RC integrator. Dynamic-range scaling is necessary to maximize the signal handling capacity of the filter. The procedure to design a high-order Gm-C filter, therefore, is identical to that we discussed in the context of active-RC filters. It entails

a. Appropriately ordering the biquads.
b. Performing dynamic-range scaling. This maximizes the magnitude of the largest signal that the filter can handle.
c. Impedance scaling the resulting filter to achieve the desired mean-squared output noise.

We have seen two styles for realizing active filters. A natural question that arises is "which style should one choose?" The answer to this question, like in most engineering scenarios, is—it depends! Let us look at the advantages and disadvantages of the Gm-C approach. The integrator is an open-loop structure, and that is a good thing, since it enables high-speed operation. The speed advantage of an open-loop system is often concomitant with poor linearity, and the Gm-C integrator is no exception to this rule. Any attempt to improve the transconductor's linearity usually degrades noise performance. Gm-C filters are therefore restricted to high-speed applications with modest linearity and dynamic-range requirements. Furthermore, the unity-gain frequency of a Gm-C integrator is sensitive to parasitic capacitances, which can be challenging in wide-band filter designs.

An opamp-RC integrator, on the other hand, is able to achieve high linearity and parasitic insensitivity owing to strong negative feedback. The design of the opamp, therefore, is crucial. Recall that the unity-gain bandwidth of the opamp should be many times larger than the $\omega_p Q_p$ product of the highest-Q biquad employed in the filter.

Chapter Summary

- **Magnitude approximation.** We understood the principles behind the Butterworth and Chebyshev lowpass magnitude approximations.
- **Filter realization using a cascade of biquads.** The cascade-of-biquads is the most commonly used method of realizing a high-order transfer function.
- **Active-RC biquads.** A biquad can be realized in many ways. The active-RC topology, which uses an opamp-RC integrator, is a robust and parasitic-insensitive filter structure.

- **Differential operation.** We saw the practical utility of processing signals in a differential fashion; namely, disturbances common to both halves of the circuit are eliminated.
- **Noise and dynamic-range scaling.** Practical filters need to deal with two realities—noise generated by the internal noise sources, as well as saturation of integrator outputs. The "best" filter for a given output noise is that which maximizes the amplitude of the input signal that can be handled without integrator saturation. The component values that result in this "best" filter are obtained using the process of dynamic-range scaling.
- **Transconductance-capacitance filters.** High-frequency active filters benefit from realizing the integrator as a transconductance-capacitance structure, as it is an open-loop topology. The disadvantage is the loss of parasitic insensitivity inherent in an active-RC integrator.

Problems

Problem 1. *This problem explores the order needed for an anti-alias filter in an ADC signal chain. The signal bandwidth is B Hz. The sampling rate is $f_s = 2B \cdot OSR$ where OSR is called the oversampling ratio. Assuming a Butterworth magnitude response, plot the order of the filter needed as a function of OSR if a minimum attenuation of 40 dB is needed in the alias band, and if the 3 dB bandwidth is chosen as 1.1 B.*

Problem 2. *Show that the product of quality factors of all biquads in an odd-order Butterworth filter is unity.*

Problem 3. *Show that the product of quality factors of all biquads in an even-order Butterworth filter is $1/\sqrt{2}$.*

Problem 4. *Determine the expression for the group delay of a second-order transfer function*

$$H(s) = \frac{1}{\frac{s^2}{\omega_p^2} + \frac{s}{Q_p\omega_p} + 1}.$$

What is the delay at dc? At ω_p?

Problem 5. *We derived the transfer function for an nth order Butterworth filter by making $F(\omega^2)$ have $\omega = 0$ as the point of maximum flatness. We could have chosen an arbitrary frequency ω_o in the passband as the point of maximum flatness. This problem develops that line of thought. The desired $F(\omega^2)$ is shown in the figure below. As in the case of the Butterworth approximation, we constrain $F(1)$ to be unity.*

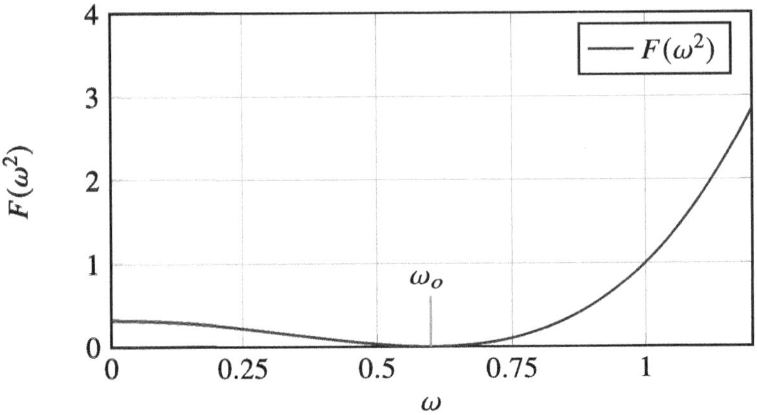

a. *Are there any restrictions on the order of $F(\omega^2)$, and thereby on the order of the filter?*
b. *What is $F(\omega^2)$ for a filter of order n?*
c. *Design a fourth-order filter of this type, with $\omega_o = 0.65$. After you find $F(\omega^2)$ proceed just as we did in the Butterworth case—that is, find the LHP roots of $D(s)D(-s)$, and the resulting $D(s)$. You might find it useful to use a computer to find the roots of the appropriate polynomial. Once you have found the denominator polynomial, normalize it so that the maximum transmission of the filter is unity. Plot the resulting magnitude response. On the same graph, plot the response of a fourth-order Butterworth filter. Which of the filters has a better stop band response ? Can you intuitively explain why ?*

Problem 6. *We saw that a Chebyshev polynomial of order n is $C_n(\omega) = \cosh(n\cosh^{-1}(\omega))$. Show that this is equivalent to writing*

$$C_n(\omega) = \frac{1}{2}[(\omega + \sqrt{\omega^2 - 1})^n + (\omega + \sqrt{\omega^2 - 1})^{-n}]$$

Derive a recurrence relation for Chebyshev polynomials by expanding $\cosh((n-1)\cosh^{-1}(\omega))$, $\cosh(n\cosh^{-1}(\omega))$, and $\cosh((n+1)\cosh^{-1}(\omega))$.

Problem 7. *A Chebyshev low-pass filter satisfying the following frequency mask is to be designed.*
a. *In the frequency range $0 < \omega < 100$ rad/s, the magnitude of the transfer function should have a maximum ripple of 0.5 dB. The maximum gain of the filter should be unity.*
b. *At $\omega = 150$ rad/s, the transfer function magnitude must be smaller than 0.02.*

Find the minimum order of the filter that will satisfy the above mask, the transfer function and the frequencies of maximum and minimum transmission in the passband. Plot the group delay $-\frac{d\phi(\omega)}{d\omega}$ of the filter.

Problem 8. *A low-pass Butterworth filter is to be designed so that it satisfies the following specifications.*

- *The dc gain is 1.*
- *The gain at 1 MHz is at least 0.95.*
- *The gain at 2 MHz is at most 10^{-1}.*

a. *Determine the minimum order needed, and the corresponding 3 dB bandwidth of the Butterworth transfer function. Plot the magnitude response expressed in dB on a scale that is linear on the x-axis.*

b. *Suppose the 3 dB bandwidth determined above varies by $\pm 5\%$. What does the order need to be so that the response still satisfies the requirements above? Show MATLAB plots to justify your answer.*

Problem 9. *It is possible to leverage off our work in deriving the Chebyshev approximation (which had an equiripple passband and monotonic stop band) to derive a transfer function which has a monotonic pass band and equiripple stop band. This problem intends to walk you through this process. Such filters are called inverse-Chebyshev filters. Proceed as follows.*

a. *Plot the function*

$$F(\omega^2) = \frac{1}{C_n^2\left(\frac{1}{\omega}\right)}$$

where $C_n(\omega)$ is the nth-order Chebyshev polynomial. Compare this with $C_n^2(\omega)$. What do you notice?

b. *Plot the function*

$$|H(j\omega)|^2 = \frac{1}{1 + \frac{1}{\epsilon^2 C_n^2\left(\frac{1}{\omega}\right)}} = \frac{\epsilon^2 C_n^2\left(\frac{1}{\omega}\right)}{1 + \epsilon^2 C_n^2\left(\frac{1}{\omega}\right)} \tag{7.98}$$

c. *Determine $H(s)$. Is it an all-pole transfer function?*

d. *Comment on the nature of the response at dc. Determine the frequencies at which the magnitude response of the filter is zero.*

Problem 10. *Determine the pole locations of an nth-order Chebyshev filter as the pass band ripple tends to zero (i.e., $\epsilon \to 0$), while the 3 dB bandwidth is held fixed at 1 rad/s.*

Problem 11. *Consider a bandpass transfer function*

$$H(s) = \frac{s/(\omega_o Q)}{(s/\omega_o)^2 + (s/\omega_o Q) + 1} \tag{7.99}$$

where $Q \gg 1$. It is excited by an input v_i, constrained so that $|v_i| \leq 1$. Determine the waveform v_i that maximizes the peak output of the bandpass filter. How does this maximum compare with what it would be if, in addition, $v_i(t)$ was also constrained to be sinusoidal?

Problem 12. *Design a sixth-order Butterworth low-pass filter as a cascade of active-RC biquadratic sections, satisfying the following.*

a. *The magnitude of the dc gain of each biquad must be 1.*
b. *The input resistance of each biquad must be 10 kΩ.*
c. *The 3 dB bandwidth of the filter must be 25 kHz.*

First, determine the Q's of the biquads. How will you order them? Then, assume ideal opamps (can be implemented in LTSPICE as a VCVS with a very large gain), and show the component values after performing dynamic range scaling. Simulate the filter with LTSPICE and show the frequency response at all the nodes over a frequency range 0–50 kHz, to verify that the peak magnitude response at all opamp outputs is 1.

Problem 13. *In the filter of the previous problem, replace the opamps used in the integrators only with a VCVS, but with a dominant-pole compensated model with a unity-gain bandwidth of 1 MHz. Calculate and verify with LTSPICE the peak magnitude response of the filter.*

Next, replace the non-ideal integrator with the contraption (Figure 7.63), where each of the opamps is a dominant-pole compensated model with a unity-gain bandwidth of 1 MHz. Using the approach used in class, estimate the peak magnitude of the filter, and verify using LTSPICE.

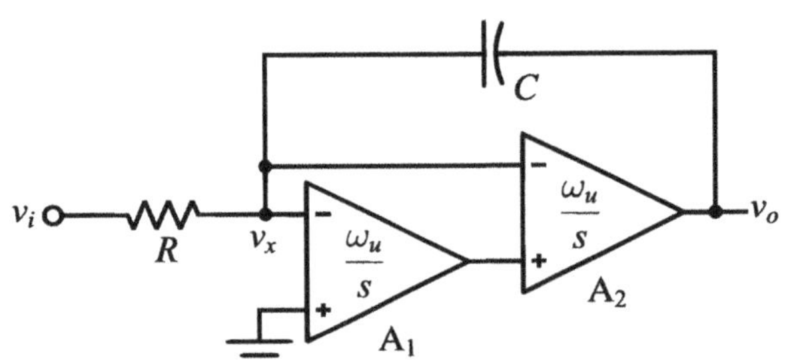

$$\frac{V_o}{V_x} = -\frac{\omega_u}{s} - \frac{\omega_u^2}{s^2}$$

Figure 7.63 Modified integrator using feed forward.

Problem 14. *Consider the bandpass filter of Figure 7.25. The opamps are ideal and noiseless. The amplifier of gain −1 is noiseless. The only noisy elements are the resistors. Assume $Q \gg 1$.*

The opamps have a peak signal swing of ±1 V at their outputs. The input $|v_i| < 1$ V. It is desired to maximize the dynamic range of the filter. The dynamic range is the maximum possible SNR at the filter output. The peak signal amplitude at the filter output is restricted to 1 V, but can be lower if one of the amplifiers saturates before the other. The noise power at the filter output is its mean square noise. The dynamic range, therefore, is given by

$$\text{Dynamic range} = \left(\frac{\text{Max. signal power at output}}{\text{Mean-square output noise}}\right) \tag{7.100}$$

The aim of this problem is to illustrate the fundamental difficulty of realizing a high-Q active bandpass filter with a high dynamic range. Choose component values so that all the following constraints are satisfied. Assume that the Q is high, so that the peaks of the lowpass and bandpass outputs occur at the same frequency.

- *Center frequency f_0 is 1 MHz.*
- $Q_p = 20$
- *Gain at f_0 of 1.*
- $C_1 + C = 100$ *pF.*

Since there are six component values to be determined but only four constraints, many component choices can realize the desired transfer function

a. *Choose component values so that the output dynamic range is maximized. Do not skip steps—show your reasoning clearly. What is the maximum dynamic range in dB? Verify your noise calculations using SPICE.*

b. *Repeat part (a) if $C_1 + C = 200$ pF.*

c. *Repeat part (a) if $Q = 40$, but all other constraints remain as in part (a).*

d. *Repeat part (a) if $f_0 = 2$ MHz, but all other constraints remain as in part (a).*

Problem 15. *Evaluate the mean-square noise at the output of the RLC network (Figure 7.64) using Bode's noise theorem to derive Eq. (7.68).*

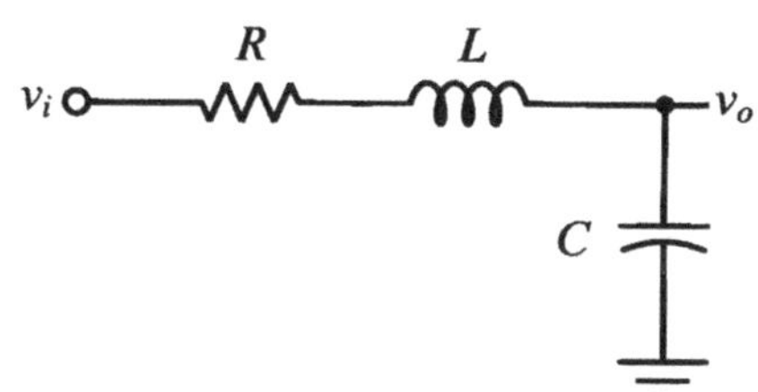

Figure 7.64 RLC network to derive Eq. (7.68).

Problem 16. *Assuming Eq. (7.68) is known, derive Eq. (7.69).*

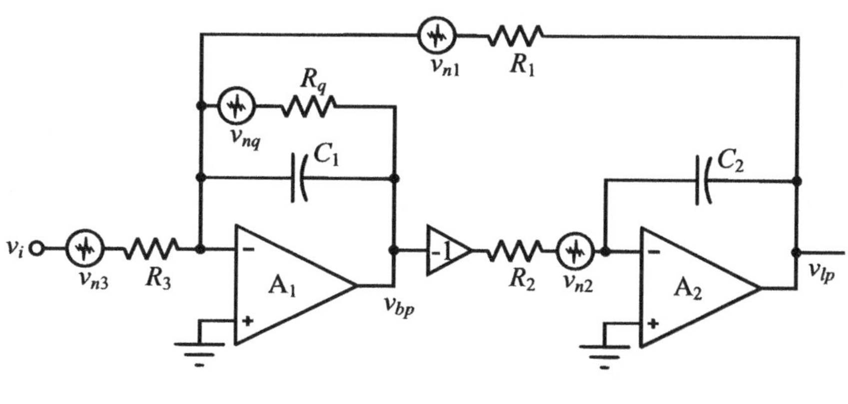

Figure 7.65 Tow-Thomas biquad with low-pass and band-pass outputs.

Problem 17. *Figure 7.65 shows a Tow-Thomas biquad.*

a. *Determine the dc gain, ω_p, and Q_p in terms of component values.*

b. *Intuitively derive the dc gain and ω_p without evaluating the entire transfer function.*

c. *Determine the transfer functions from all noise sources to the low-pass and band-pass outputs.*

d. *Determine the mean-square noise at the low-pass and band-pass outputs.*

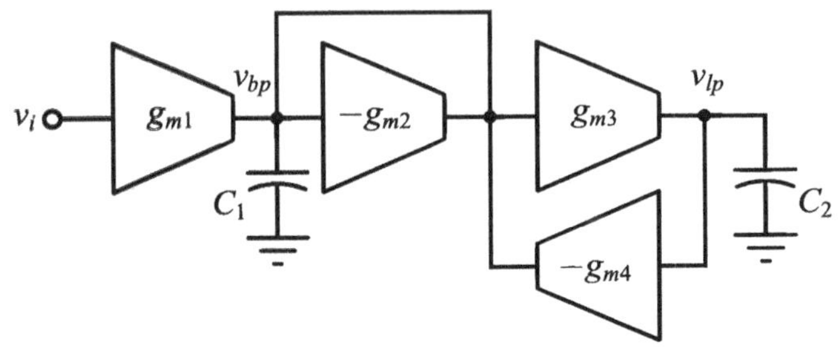

Figure 7.66 Gm-C biquad with low-pass and band-pass outputs.

Problem 18. *Figure 7.66 shows a Gm-C biquad. Every transconductor is noisy, with spectral density of the noise current at its output given by $4\,kT\eta g_m$. η is the excess noise factor.*

a. *Determine the dc gain, ω_p, and Q_p in terms of component values.*

b. *Intuitively derive the dc gain and ω_p without evaluating the entire transfer function.*

c. *Determine the transfer functions from all noise sources to the low-pass and band-pass outputs.*

d. *Determine the mean-square noise at the low-pass and band-pass outputs.*

Problem 19. *This problem is intended to compare the noise performance of Tow-Thomas and Rauch biquads. Design a Tow-Thomas Butterworth biquad with a 3 dB bandwidth of 1 MHz. Use the version with a damped second integrator, so that the noise is lower. Assume an ideal inverting stage, and ideal, noiseless opamps. The dc gain must have a magnitude of unity and input resistor must be 10 kΩ. Apply dynamic-range scaling. Plot the noise spectral density at the output, and evaluate the integrated output noise.*

Now design a Rauch biquad for the same specifications. Choose the input resistance to be the same as in the Tow-Thomas case. Plot the noise spectral density at the output, and evaluate the integrated output noise. What do you conclude?

References

1 R. W. Daniels, *Approximation Methods for Electronic Filter Design: With Applications to Passive, Active, and Digital Networks*. McGraw-Hill, 1974.

2 J. Tow, "A step-by-step active-filter design," *IEEE Spectrum*, vol. 6, no. 12, pp. 64–68, 1969.

3 L. C. Thomas, "The biquad: Part I—Some practical design considerations," *IEEE Transactions on Circuit Theory*, vol. 18, no. 3, pp. 350–357, 1971.

4 L. C. Thomas,"The biquad: Part II—A multipurpose active filtering system," *IEEE Transactions on Circuit Theory*, vol. 18, no. 3, pp. 358–361, 1971.

5 R. Schaumann, M. S. Ghausi, and K. R. Laker, *Design of Analog Filters: Passive, Active RC, and Switched Capacitor*. Prentice Hall, 1990.

6 B. Nauta, "A CMOS transconductance-C filter technique for very high frequencies," *IEEE Journal of Solid-State Circuits*, vol. 27, no. 2, pp. 142–153, 1992.

7 Y. Tsividis, "Integrated continuous-time filter design—An overview," *IEEE Journal of Solid-State Circuits*, vol. 29, no. 3, pp. 166–176, 1994.

8

Switched-Capacitor Circuits and Filters

Continuous-time active filters implement their transfer functions by creating time constants using at least two (sometimes three) different components such as capacitors, resistors, or transconductances. The accuracy of these time constants is limited by that of the components. For instance, on an integrated circuit, resistors can vary by as much as $\pm 30\%$ with process and temperature. Consequently, the bandwidth of an active-RC filter can vary by $\pm 30\%$! Tuning can reduce this variation to a few percent, but at the cost of an additional tuning stage and a potentially complicated matching protocol for all components. This makes one wonder if one can realize a filter response with a transfer function without these problems. It turns out that this is indeed possible, and the idea is one of those "out-of-the-box" approaches (which, like most elegant ideas, looks obvious after the fact).

Using the sampled-data technique described in this chapter, the accuracy of time constants, and hence of transfer functions, can be improved by several orders of magnitude [1, 2]. The idea is to relate the time constant to a periodic signal (i.e., a clock) whose frequency can be made very stable (for example, by using a crystal oscillator).

The basic concept of the new circuits is to use only capacitors as the passive elements in the circuit. To realize poles in the left-half s-plane, some form of loss (or dissipation) is necessary. In an active filter, this role is performed by resistors. How can one take a capacitor, which is a lossless element, and achieve power dissipation? Furthermore, resistors are memoryless while capacitors have memory. How can a capacitance, therefore, emulate the role of a resistance?

The answer to both these questions turns out to be the addition of switches, as explained using Figure 8.1. Part (a) shows a resistor R connected between two voltage sources V_1 and V_2, resulting in a current $(V_1 - V_2)/R$. If the resistor is replaced by an initially uncharged capacitor C_1, a spike of current with charge $C_1(V_1 - V_2)$ will flow from V_1 to V_2. To eliminate the memory of the capacitor, one needs to "remove" the voltage across it. One way to do this is shown in Figure 8.1b, where two switches are added. The series switch, controlled by p_1, (dis)connects C_1 from V_1. The switch in parallel with C_1 is controlled by p_2. To "zap-the-cap," the series switch is opened (p_1 is low), and the shunt switch is closed (p_2 is high). This act of discharging the capacitor not only removes its memory, but also creates loss since the energy stored in the

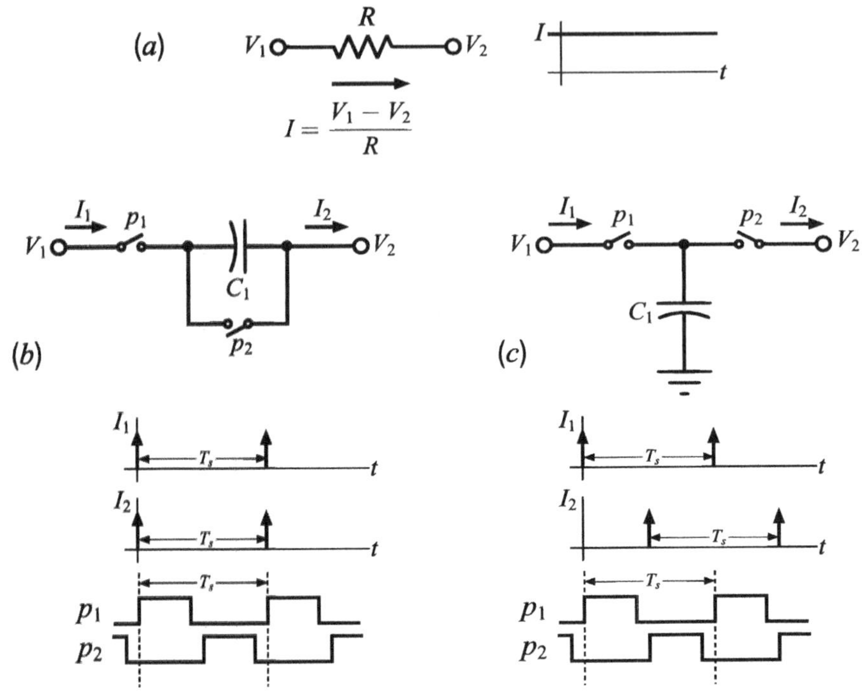

Figure 8.1 (a) A resistor connected between two dc voltage sources and the resulting current. (b) A capacitor and a switch network connected between two dc voltage sources. The capacitor's memory is destroyed during p_2, which also results in a loss of energy. The average current flowing out of V_1 (and into V_2) is $C_1(V_1 - V_2)f_s$. (c) An alternative approach to emulate a resistor using a switched capacitor.

capacitor is dissipated in the switch. In the next clock cycle, the capacitor is connected to V_1 again when p_1 goes high. Since C_1 was discharged in the previous cycle, a charge $C_1(V_1 - V_2)$ flows from V_1 to V_2. This is repeated every cycle. We thus see that unlike in a resistor, where a constant current flows from V_1 to V_2, the switched-capacitor network has current flowing in impulses, with an average given by $C_1(V_1 - V_2)f_s$. If $C_1 f_s$ is chosen to be $1/R$, the combination of switches and capacitors in Figure 8.1b emulates a resistance R in the average sense.

Another, admittedly less obvious, way of arranging the switches and capacitor is shown in Figure 8.1c. When p_1 is high, C_1 is connected to V_1, and the charge on it is C_1V_1. p_1 then goes low, and after a short dead time where both switches are open, p_2 goes high connecting the capacitor to V_2. As a result, a spike of charge $C_1(V_1 - V_2)$ flows into V_2. When C_1 is connected to V_1 during p_1 of the next clock cycle, a charge $C_1(V_1 - V_2)$ flows into C_1. In steady state, therefore, an average current $C_1(V_1 - V_2)f_s$ flows between V_1 and V_2. In this circuit, the act of connecting C_1 to either of the voltage sources serves the purpose of zapping its memory, as well as creating loss. Again, by choosing $C_1 f_s$ to equal $1/R$, the switched-capacitor emulates a resistor of value R in the average sense. Extending the arguments above, it seems as if a combination of capacitors and periodically operated switches can emulate (in an average sense) the behavior of an RC network. Such networks are called switched-capacitor circuits.

The simplest possible switched-capacitor filter is obtained by replacing the resistor in a first-order RC lowpass filter with a switched capacitor branch. Figure 8.2a shows the resulting circuit, where we chose to use the approach of Figure 8.1c to emulate the resistor. p_1 and p_2 are non-overlapping clocks with frequency $f_s = 1/T_s$. The non-overlap interval is (grossly) exaggerated for clarity. It is instructive to carefully study the properties of this seemingly simple circuit, as it brings out many important aspects of switched-capacitor signal processing.

The input v_i is sampled on C_1 at the falling edge of p_1, which is chosen to coincide with $t = 0$. During p_2, C_2, and C_1 share charge. Assuming ideal (zero-resistance) switches, v_o changes abruptly at multiples of nT_s, and is constant for the rest of the

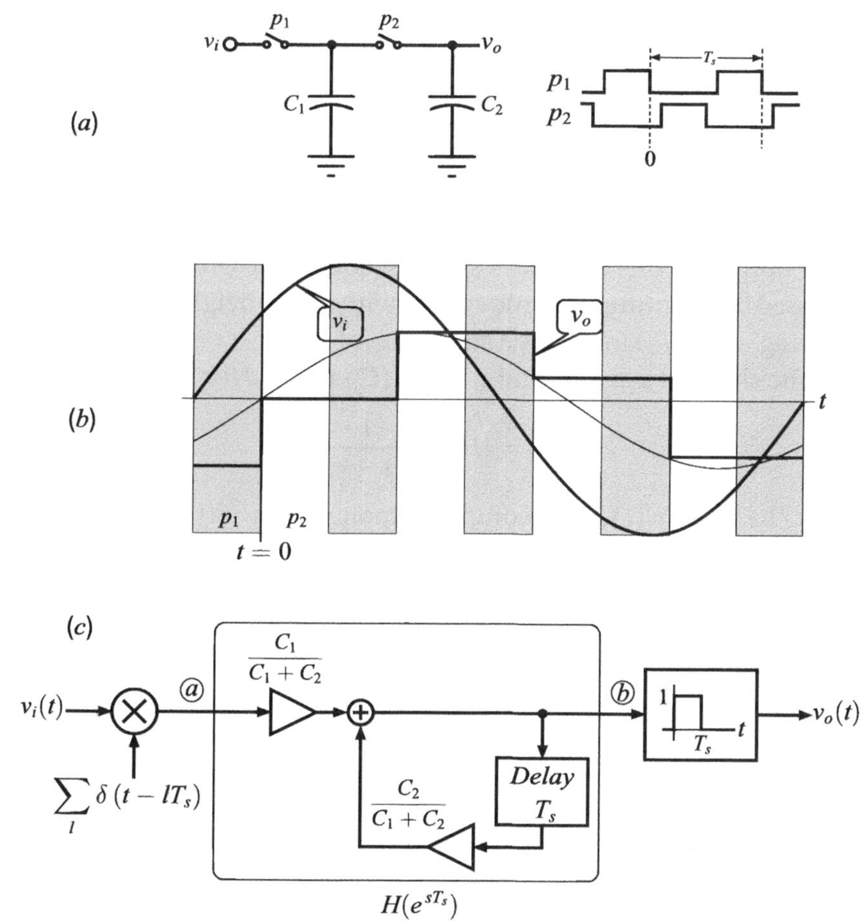

Figure 8.2 (a) A first-order switched-capacitor filter, obtained by replacing the resistor in an RC lowpass network by C_1 and the two switches. The clocks are non-overlapping. (b) Representative waveforms. (c) Signal processing equivalent of the lowpass filter.

clock cycle, as shown in Figure 8.2b. Denoting the held value of v_o in the nth clock cycle by $v_o[n]$, charge conservation yields the following

$$\underbrace{(C_1+C_2)v_o[n]}_{\substack{\text{charge on } C_1 \text{ and } C_2 \\ \text{during } p_2 \text{ of the } n^{\text{th}} \text{ cycle}}} = \underbrace{C_2 v_o[n-1]}_{\substack{\text{charge on } C_2 \text{ at the} \\ \text{end of the } (n-1)^{\text{th}} \text{ cycle}}} + \underbrace{C_1 v_i[n]}_{\substack{\text{charge on } C_1 \text{ at the} \\ \text{end of the } (n-1)^{\text{th}} \text{ cycle}}}$$

which simplifies to

$$v_o[n] = \frac{C_2}{C_1+C_2} v_o[n-1] + \frac{C_1}{C_1+C_2} v_i[n]. \tag{8.1}$$

Figure 8.2b shows representative waveforms when $v_i(t)$ is sinusoidal. The sequence sampled on C_1 is then also a discrete-time sinusoid, as is the sequence $v_o[n]$. $v_o(t)$ is the zero-order-held version of $v_o[n]$. From a signal-processing perspective, the operation of the circuit can be interpreted as follows (Figure 8.2c). $v_i(t)$ is multiplied by a Dirac impulse train with period T_s, which is then processed by a continuous-time system whose transfer function is given by

$$H(e^{sT_s}) = \frac{C_1}{C_1+C_2} \frac{1}{\left[1 - \frac{C_2}{C_1+C_2} e^{-sT_s}\right]}. \tag{8.2}$$

Note that the output of this system is still a sequence of Dirac impulses, which is further processed by a continuous-time system with a unit-height rectangular impulse response (having width T_s) to yield $v_o(t)$.

If one uses the shortcut $z \equiv e^{sT_s}$, and $\alpha = C_2/(C_1+C_2)$, $H(e^{sT_s})$ can be written as

$$H(e^{sT_s}) = H(z) = \frac{1-\alpha}{1-\alpha z^{-1}}. \tag{8.3}$$

What will $v_o(t)$ be when $v_i(t)$ is the complex exponential $e^{j2\pi ft}$? To answer this question, we first recognize that the system is linear and periodically time varying due to the multiplication of $v_i(t)$ with the periodic impulse train. Furthermore, we assume that $|f| < f_s/2$. $v_o(t)$ will therefore consist of all frequencies of the form $f + lf_s$, where l is an integer. Let us determine the zeroth-order harmonic transfer function of the circuit, i.e., the (complex) amplitude of the component of $v_o(t)$ at $e^{j2\pi ft}$. This is termed the frequency response of the switched-capacitor network. We proceed as follows.

a. Referring to Figure 8.2c, $v_i = e^{j2\pi ft}$. At ⓐ, the component at f is given by f_s. This is because all the Fourier coefficients of the delta train are $(1/T_s) = f_s$.[1]

b. The transfer function $H(e^{sT_s})$ when evaluated at the frequency f is given by $H(e^{j2\pi fT_s})$. At ⓑ, therefore, the strength of the sinusoid at f is $f_s H(e^{j2\pi fT_s})$.

c. The hold pulse is rectangular with width T_s and height of unity. Its frequency response is given by $T_s \mathrm{sinc}(fT_s)e^{-j\pi fT_s}$.

d. The strength of the component at f in v_o is $H(e^{j2\pi fT_s})\mathrm{sinc}(fT_s)e^{-j\pi fT_s}$.

[1] Recall that $\sum_l \delta(t - lT_s) = f_s \sum_l e^{j2\pi l f_s}$

The frequency response (which is also the zeroth-order harmonic transfer function of the circuit of Figure 8.2), denoted by $H_{sc}(j2\pi f)$ is given by

$$H_{sc}(j2\pi f) = \frac{C_1}{C_1 + C_2} \frac{1}{\left[1 - \frac{C_2 e^{-j2\pi f T_s}}{C_1 + C_2}\right]} \text{sinc}(fT_s) e^{-j\pi f T_s}. \tag{8.4}$$

A few remarks are in order.

1. It is evident that $H_{sc}(j2\pi f)$ depends on capacitor ratios. Consequently, the frequency response remains unchanged even if all capacitors change by the same factor. Fortunately, it turns out that it is easy to obtain very well-controlled ratios of capacitors on an integrated circuit, even though the absolute values of capacitors can change by as much as $\pm 30\%$. The robustness of the frequency response in the face of large passive component variations is a key advantage of switched-capacitor processing.
2. $H_{sc}(j2\pi f)$ depends on $fT_s = f/f_s$. This means that the shape of the frequency response is directly related to the clock frequency f_s. By changing f_s, therefore, the frequency axis can be scaled. For instance, increasing f_s by a factor of 2 while keeping C_1 and C_2 unchanged will cause the bandwidth of H_{sc} to double. As a corollary, the use of a stable sampling frequency (derived from an external crystal oscillator, for instance) results in an extremely well-controlled frequency response shape that is immune to variations in process, supply voltage and temperature.

For input frequencies much smaller than f_s so that $fT_s \ll 1$, $e^{-j2\pi f T_s} \approx (1 - j2\pi f T_s)$. Furthermore $\text{sinc}(fT_s) \approx 1$. Using this in Eq. (8.4), we obtain

$$H_{sc}(j2\pi f) \approx \frac{1}{1 + j2\pi f C_2/(C_1 f_s)}. \tag{8.5}$$

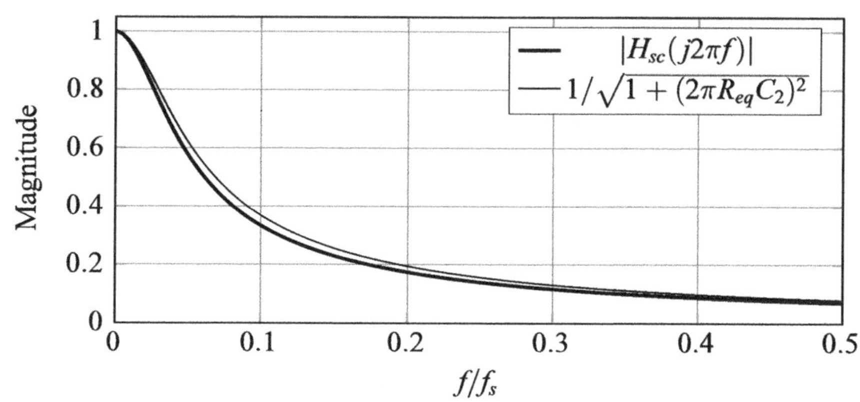

Figure 8.3 Magnitude response of the first-order switched-capacitor filter of Figure 8.2 for $C_1 = 20$ pF, $C_2 = 80$ pF, and $f_s = 10$ MHz. The response of the corresponding first-order RC lowpass filter is also shown (RC = 400 ns).

This makes sense due to the following: C_1 acts like a resistor of value $1/(f_sC_1)$. The frequency response at low frequencies must thus approximate that of a first-order RC lowpass filter with a time constant given by $C_2/(f_sC_1)$. Figure 8.3 shows the magnitude response of the first-order switched-capacitor filter of Figure 8.2 for $C_1 = 20$ pF, $C_2 = 80$ pF, and $f_s = 10$ MHz. The response of the corresponding first-order RC lowpass filter is also shown (RC = 400 ns).

To summarize, the single-pole switched-capacitor filter effectively addresses the challenge of maintaining the frequency response over large variations in the values of the capacitors (provided their ratio is maintained). However, it is fundamentally an LPTV circuit, which causes two problems. First, $v_i(t)$ is sampled, as seen in Figure 8.2c. As a result, input frequencies of the form $(f + lf_s)$ will be indistinguishable from f after sampling (this is called aliasing). This is addressed by using an anti-alias filter before the switched-capacitor circuit. One might think that the need for a continuous-time filter upfront defeats the very purpose of the switched-capacitor technique. This concern is addressed as follows. The sampling frequency f_s is chosen to be much larger than the cutoff frequency of the switched-capacitor filter. Therefore, there is adequate attenuation at frequencies beyond $f_s/2$ even if the bandwidth of the anti-alias filter changes significantly (due to RC variations). In practice, f_s is chosen to be so high that a single-pole RC filter (whose bandwidth is much higher than that of the switched-capacitor network) can be used as the anti-alias filter. The pass-band response of the overall system, i.e., the anti-alias filter + switched-capacitor filter cascade is thus defined by the latter.

The second problem with the switched-capacitor structure is that the output waveform consists of frequencies of the form $(f + lf_s)$, again a consequence of the LPTV nature of the network. This is evident in the stepped output waveform of Figure 8.2b. To remove frequencies other than the desired tone at f, a continuous-time filter with a cut-off frequency $f_s/2$ is necessary. This is termed the reconstruction (or smoothing) filter. As with the anti-alias filter upfront, its requirements are relaxed by choosing a clock rate whose frequency is much larger than the signal bandwidth.

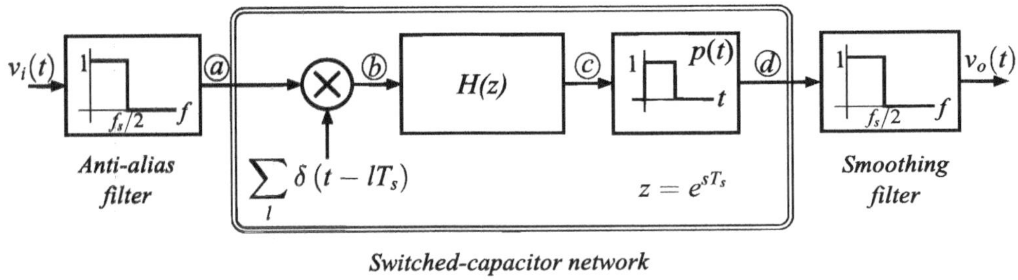

Figure 8.4 A switched-capacitor signal-processing system.

Figure 8.4 shows the block diagram of a switched-capacitor signal processing system. It operates as follows.

a. The input is first processed by a continuous-time anti-alias filter whose function is to remove all frequency components $|f| \geq f_s/2$.

b. The continuous-time signal at ⓐ, namely $v_a(t)$, consists of frequencies $|f| < f_s/2$. This is sampled by the switched-capacitor network to result in $v_b(t)$. The sampled sequence is *modeled* by multiplying $v_a(t)$ with an impulse train.

c. The continuous-time signal $v_b(t)$, which is a sequence of impulses, is processed by a linear time-invariant system whose transfer function can be expressed as $H(e^{sT_s})$. It is easy to see that $H(e^{sT_s})$ is a ratio of polynomials in e^{sT_s}. The output $v_c(t)$ is also a sequence of impulses.

d. The frequency response of $H(e^{sT_s})$, namely $H(e^{j2\pi fT_s})$ is periodic with a period f_s. We use the shortcut $z \equiv e^{sT_s}$. The coefficients of $H(z)$ depend only on capacitor ratios. Consequently, large changes in the capacitor values do not affect the frequency response, as long as ratios are maintained. The frequency response at low frequencies can be obtained by replacing $e^{j2\pi fT_s}$ with an appropriate approximation, one of which is $(1 + j2\pi fT_s)$.

e. $v_c(t)$ is held, typically for a whole clock period, to yield $v_d(t)$. This action is modeled by a linear time-invariant system with an impulse response denoted by $p(t)$. If the hold period is T_s, the frequency response of this block is $T_s\text{sinc}(fT_s)$. The attenuation due to the sinc function is zero at dc and increases with frequency. If the bandwidth of the switched-capacitor network is much smaller than f_s, which is necessary anyway, the effect of the sinc is a small droop in the magnitude response of the filter. Termed "sinc droop," it can often be ignored if the sampling rate is very high.

f. If $v_i(t)$ is a sinusoid at frequency f, $v_d(t)$ will in general consist of frequencies of the form $f + lf_s$. The role of the smoothing filter is to remove these components, thereby resulting in an output $v_o(t)$ whose frequency is f. End-to-end, therefore, the system appears like a linear time-invariant system with a transfer function given by

$$H_{sc}(j2\pi f) = H(e^{j2\pi fT_s}) \cdot (1/T_s)H_p(j2\pi f) \tag{8.6}$$

where $H_p(j2\pi f)$ denotes the Fourier transform of the hold pulse $p(t)$. The factor $(1/T_s)$ arises from the Fourier transform of the sampling operation. If one neglects the effect of the pulse shape on the frequency response of the switched-capacitor filter, $H_{sc}(j2\pi f) \approx H(e^{j2\pi fT_s})$.

Stability

For the transfer function $H(e^{sT_s})$ to be stable, its poles must lie in the left-half s-plane, with the imaginary axis $s = j2\pi f$ separating the left- and right-half planes. Since $z = e^{sT_s}$, the corresponding stability boundary in the z-plane is $z = e^{j2\pi fT_s}$, which is the unit circle. To examine stability, therefore, one can determine the roots of the denominator polynomial of $H(z)$, and check to see if they lie within the unit circle.

For instance, as seen in Eq. (8.3), the pole of the first-order switched-capacitor filter of Figure 8.2 is located at $z = \alpha$, where $\alpha = C_2/(C_1 + C_2)$. The circuit is stable, as expected.

8.1 Switched-Capacitor Filters

Suppose that we wanted to implement a switched-capacitor filter whose frequency response mimicked that of a continuous-time prototype with transfer function $H_c(s)$. How would we go about this? The discussion above suggests that we find an appropriate $H(z)$ so that $H(e^{j2\pi fT_s})$ approximates the frequency response of the prototype, namely $H_c(j2\pi f)$. We then realize $H(z)$ using switched-capacitor circuitry. For a high-order $H(z)$, this appears to be a daunting task. Fortunately, an alternative and simpler approach can be found. The thought process is the following, which exploits the rich literature on continuous-time filter design.

We recognize that an integrator is the basic building block of a continuous-time filter, and a high-order filter structure is simply a network of integrators. Suppose we created a switched-capacitor circuit whose response mimicked that of a continuous-time integrator (with transfer function ω_o/s) at low frequencies. What would the transfer function $I(z)$ of such a switched-capacitor circuit be? The integrator has a pole at $s = 0$. Since $z = e^{sT_s}$, it follows that $I(z)$ must have a pole at $z = 1$. Denoting the input and output of the switched-capacitor circuit by $x[n]$ and $y[n]$, respectively, we see that $I(z)$ is of the form

$$I(z) = \frac{Y(z)}{X(z)} = \frac{\beta}{1 - z^{-1}} \tag{8.7}$$

where β is a constant of proportionality. In the time domain, this corresponds to

$$y[n] = y[n-1] + \beta x[n]. \tag{8.8}$$

The reader will immediately recognize that $I(z)$ is the transfer function of a discrete-time accumulator, which is intuitively satisfying. To summarize, the switched-capacitor circuit that behaves like a continuous-time integrator at low frequencies is an accumulator, which we henceforth call the switched-capacitor integrator.

Once we create a switched-capacitor integrator, we replace every integrator in the continuous-time filter with an equivalent switched-capacitor structure. We should expect that the frequency response of the resulting switched-capacitor filter $H(z) = H(e^{sT_s})$ should match that of $H_c(s)$ at low frequencies. This approach allows us to leverage our knowledge of continuous-time filters, and eliminates the need to "invent" new filter topologies; the switched-capacitor structure can be obtained by simply replacing integrators in the continuous-time prototype with appropriately chosen switched-capacitor ones. The key, therefore, is to realize a switched-capacitor integrator. It turns out, as we see below, that several possibilities exist.

8.1.1 The Switched-Capacitor Integrator

Suppose that we have a voltage sequence v_i. How would we generate a sequence which is an accumulated version of v_i? In other words, we would like to generate

$$v_o[n] \propto v_i[n] + v_i[n-1] + \cdots \tag{8.9}$$

How can we accomplish this using capacitors and switches? The first idea that strikes one's mind is the following.

a. Connect a capacitor (call it C_1) to the voltage source v_i, which will result in a charge $C_1 v_i$ stored on the capacitor.

b. *Somehow* transfer this charge *completely* on to another capacitor (call it C_a). If we are successful in this endeavor, the charge on C_a will increase by $C_1 v_i$, and that on C_1 will be zero. Equivalently, the voltage across C_a will increase by $(C_1/C_a)v_i$ and that on C_1 will be zero.

c. Repeat steps (a) and (b) ad infinitum. At the nth repetition of these steps, the voltage across C_a will increase by $(C_1/C_a)v_i[n]$. The voltage across C_a will thus be

$$v_o[n] = \frac{C_1}{C_a}\{v_i[n] + v_i[n-1] + \cdots\} \tag{8.10}$$

Let us see how we can put these steps into practice. Step (a) seems easy, as shown in Figure 8.5a, where we connect C_1 to v_i during phase p_1 of a clock signal. Assuming the resistance of the switch is low enough, the voltage across C_1 at the end of p_1 is v_i. The charge on C_1 is $C_1 v_i$. How do we transfer this charge on to another capacitor C_a?

The first idea that occurs is to disconnect C_1 from v_i, and then connect it to C_a. This can be accomplished by using a switch between C_1 and C_a which is controlled by a clock p_2, as indicated in Figure 8.5a. Note that p_1 and p_2 do not overlap, i.e. there is a "dead time" between p_1 and p_2. When C_1 and C_a are connected, the voltage across the capacitors is $C_1 v_i/(C_a + C_1)$ and the charge transferred to C_a is given by

$$Q_a = C_1 v_i \frac{C_a}{C_a + C_1}. \tag{8.11}$$

We thus see that the only way to transfer *all* the charge from C_1 to C_a (i.e., $Q_a = C_1 v_i$) is to make C_a infinite as in Figure 8.5b.

How do we realize an infinite capacitor? We exploit the Miller effect, a phenomenon that is all too familiar to analog designers. The basic idea is to connect the bottom plate of C_a to an inverted and amplified version of its top plate voltage as shown in Figure 8.5c. The effective capacitance across C_1 during p_2, denoted by C_{in} is $(A+1)C_a$. Using an opamp (whose gain is infinite), C_{in} becomes infinite. As a consequence, all the charge from C_1 is transferred to C_{in} during p_2. Since C_{in} is infinite, the voltage across C_1 in p_2 is 0.

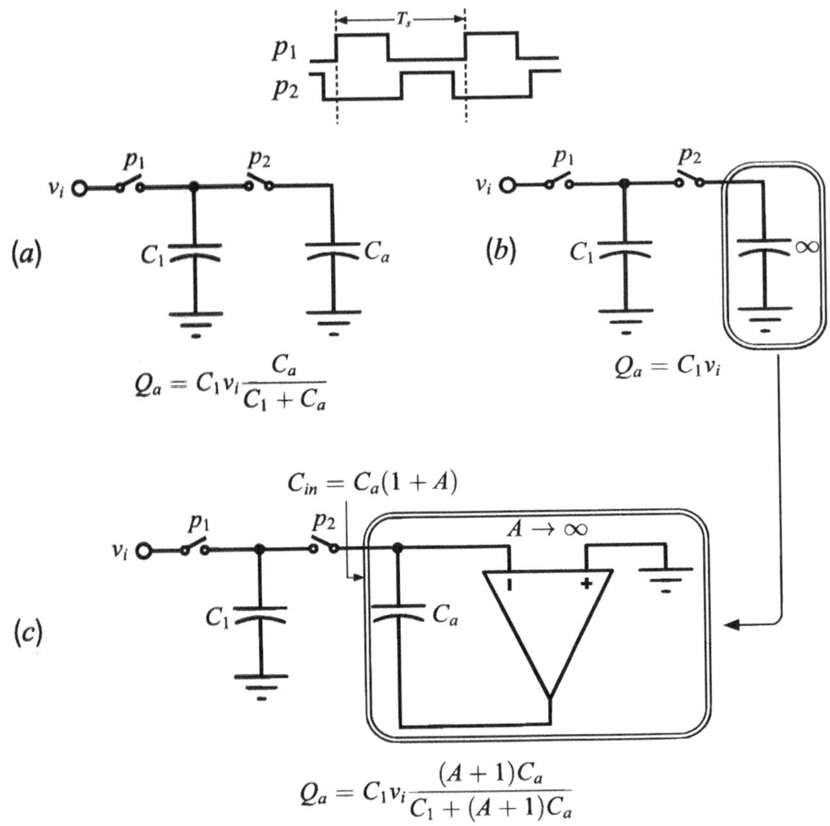

Figure 8.5 Development of a switched-capacitor integrator. (a) The input is sampled on C_1 in p_1, resulting in a charge $C_1 v_i$ stored on C_1. In p_2, some of the charge is transferred to C_a. (b) All of $C_1 v_i$ can be transferred on to C_a only if it is infinite. (c) Exploiting the Miller effect to realize the infinite capacitor.

During the next clock cycle, this whole process is repeated, and another packet of charge $C_1 v_i$ is transferred on to C_{in}, thereby causing charge to accumulate on C_{in}. But where exactly does this charge reside? That must be on the plates of C_a—there is nowhere else it can be! Every clock cycle, therefore, the charge stored on C_a increases by $C_1 v_i$, and the voltage across C_a increases by $C_1 v_i / C_a$. The top plate of C_a has a potential of zero; that of the bottom plate (i.e., the opamp output voltage) must therefore go down by $C_1 v_i / C_a$. The circuit of Figure 8.5c is redrawn to yield our preliminary version of the inverting switched-capacitor integrator as shown in Figure 8.6 (with C_a being renamed as C_2).

$$v_o[n] = -\frac{C_1}{C_2}\{v_i[n-1] + v_i[n-2] + \cdots\} \tag{8.12}$$

Note that $v_i[n]$ and $v_o[n]$ denote the values of v_i and v_o at the falling edge of p_1 in the nth clock cycle. Furthermore, the accumulated v_i in a particular clock cycle is

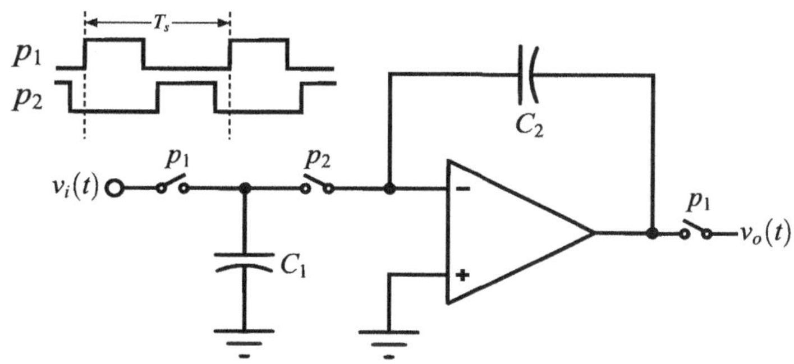

Figure 8.6 Redrawing Figure 8.5. The on periods of p_1 and p_2 are non-overlapping. The accumulated output is available in phase p_1.

available only in the next clock cycle. This integrator, therefore, is of the inverting and delaying type with a transfer function given by

$$\frac{V_o(z)}{V_i(z)} = -\frac{C_1}{C_2}\frac{z^{-1}}{1 - z^{-1}}. \tag{8.13}$$

The frequency response of the integrator is obtained by using the relation $z = e^{j2\pi fT_s}$. For low frequencies, where $f \ll f_s$, the equation above gives $H(e^{j2\pi fT_s}) \approx (C_1/C_2)/(j2\pi fT_s)$, a transfer function similar to that of an active-RC integrator. However, unlike the inaccurate time constant RC of the active-RC circuit, that of the SC integrator is $T_s(C_2/C_1)$, which can be accurately realized. This is due to the following.

a. The time constant depends on T_s, which is the period of a clock signal. This signal is usually derived from an external crystal oscillator, and is very accurate and stable over temperature and supply voltage variations.
b. C_2/C_1 is a ratio of like elements. While the absolute values of C_1 and C_2 can vary with process and temperature, their ratio remains very well controlled (often to better than 0.1%).

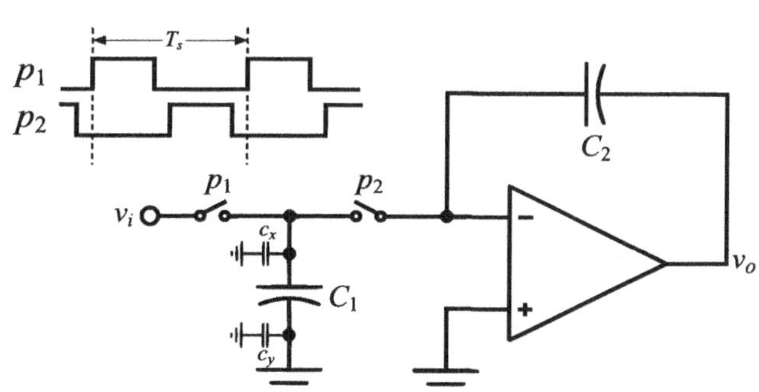

Figure 8.7 The basic switched-capacitor integrator with stray capacitances. c_x modifies the transfer function, while c_y has no effect.

As with most preliminary attempts, the circuit shown in Figure 8.6 has an important flaw, and is explained using Figure 8.7. It turns out that any on-chip circuit has parasitic capacitances loading all its nodes. For the circuit of Figure 8.6, these include stray capacitances from all capacitor plates to the substrate, as well as the capacitances of the switches connected to C_1.[2]

Referring to Figure 8.7, c_x and c_y are parasitic capacitances. During p_1, both C_1 and c_x are charged to v_i. The voltage across c_y is zero, and therefore does not have charge stored on it. During p_2, C_1 is connected to the opamp's virtual ground, and so is c_x. As a consequence, the charge stored on C_2 is $(C_1 + c_x)v_i$. Since c_x cannot be predicted accurately, the time constant (which is now $T_sC_2/(C_1 + c_x)$) is poorly controlled, nullifying in part the very motivation of using the switched-capacitor approach. The integrator of Figure 8.7 is a *stray-sensitive* circuit.

8.1.2 Stray-Insensitive Integration

How does one solve this problem? The key idea is to recognize that

a. The stray capacitor c_y has no charge stored on it (since the voltage across it is zero).
b. The charge on both plates of C_1 is equal and opposite. Therefore, apart from a sign change, it does not matter whether we move the charge from the top or bottom plates of C_1 on to C_2.

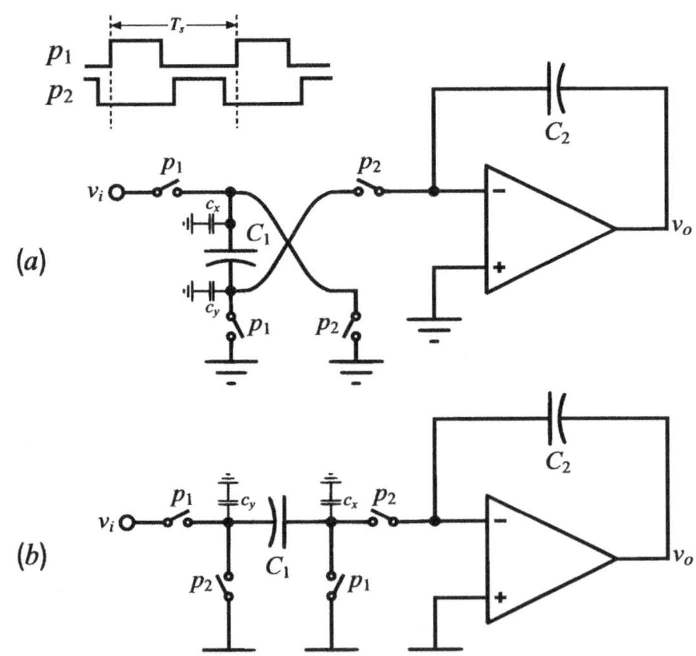

Figure 8.8 (a) Development of the stray-insensitive integrator. (b) Redrawn version: the output is delayed, and non-inverting.

[2]Incidentally, these capacitances are nonlinear as they arise from reverse-biased pn junctions of the MOS transistors used to realize the switches.

This motivates the following idea. What if we connect the bottom plate of C_1 to the virtual ground in p_2, rather than the top plate (see Figure 8.8a)? To be able to connect C_1's bottom plate to the virtual ground, one would have to disconnect it from the ground. We, therefore, need two extra switches—one that connects the bottom plate of C_1 to ground during p_1, and another that connects it to the opamp's virtual ground node in p_2. The circuit operates as follows. During phase p_1, C_1 is charged to v_i. c_x is charged to v_i while c_y does not have any voltage across it. When p_1 goes low, and before p_2 turns high, C_1 is practically floating since both its terminals are disconnected from the circuit. During p_2, the bottom plate of C_1 is connected to the opamp's virtual ground, while its top plate is grounded. Consequently, the charge on the bottom plate of C_1 (which is $-C_1 v_i$) and that on c_y (which is zero) are transferred into C_2. However, the problematic charge on c_x is discharged into ground and does not affect the charge on C_2. This integrator circuit, arrived at by adding two more switches to the simple-minded integrator of Figure 8.7, and arranging them in a clever manner[3] is *stray insensitive*. Furthermore, since a negative charge is transferred to the top plate of C_2, the output voltage of the opamp increases by $C_1 v_i / C_2$ every clock cycle. It is, therefore, of the non-inverting type.

A practical remark is in order here. By the nature of construction of high-density on-chip capacitors, the two plates of a capacitor are not equivalent in the sense that the parasitic capacitors on both plates are not identical. It turns out that the top-plate parasitic c_x is typically much smaller than on the bottom-plate, namely c_y. Since one plate of the stray capacitance connects to the substrate (which can be very noisy, for instance, due to coexistence of noisy digital circuits on the chip), it is a source through which substrate noise can couple into the filter. While the circuit of Figure 8.8a is parasitic-insensitive, we see that the stray c_y connects to the virtual ground during p_2. The virtual ground is an extremely sensitive node that must be handled with care, since the gain from that node to the output is very high. Addressing this problem is surprisingly simple, since $c_y \gg c_x$, it makes sense to interchange the top and bottom plates of C_1. The result is the stray-insensitive, non-inverting integrator of Figure 8.8b. In general, one should avoid connecting capacitor bottom plates to the virtual ground of the opamp.

The transfer function of the stray-sensitive integrator is identical to that of the one in Figure 8.6 except for a sign change. We thus have

$$\frac{V_o(z)}{V_i(z)} = \frac{C_1}{C_2} \frac{z^{-1}}{1 - z^{-1}}.$$

As discussed earlier, the integrator above is derived from the *parallel* switched-capacitor branch of Figure 8.1c. It should be possible to derive another integrator structure using the series branch of Figure 8.1b. This line of thought is pursued below.

---

[3]Such a creative, not immediately obvious, combination of capacitors and switches is not witchcraft, but switchcraft!

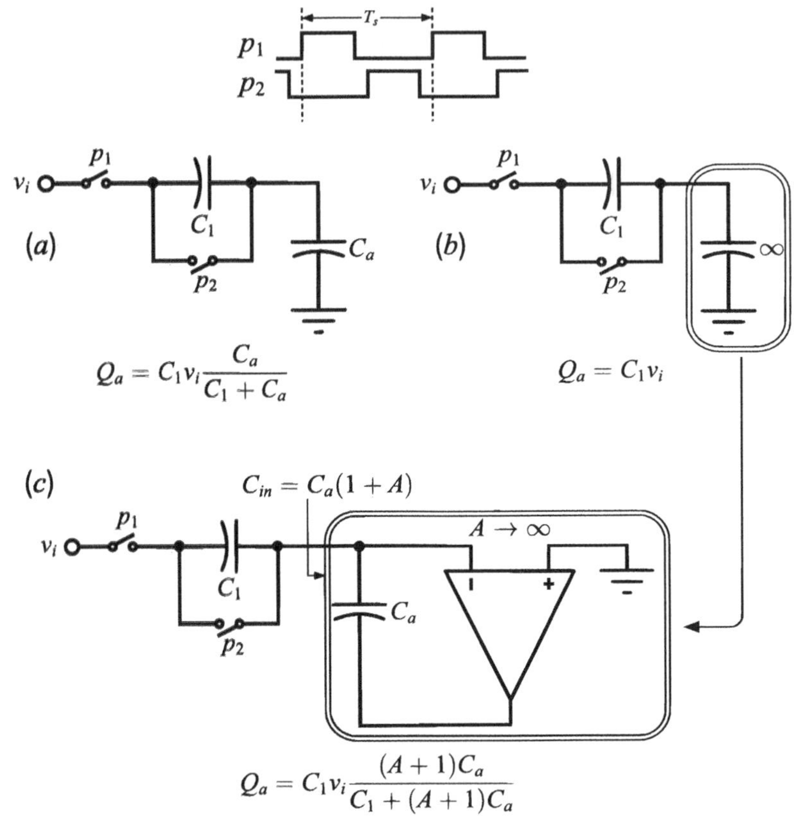

Figure 8.9 (a) Switched-capacitor integrator development using a series branch. (b) A charge $C_1 v_i$ flows into C_a only if it is infinite. (c) Realizing the infinite capacitor using an opamp ($A = \infty$).

Figure 8.9a shows an alternative attempt at delivering a charge $C_1 v_i$ to C_a. C_1 and C_a are initially uncharged. During p_1, C_1 is connected to the input voltage v_i as shown. The voltage developed across C_a is $C_1 v_i/(C_1 + C_a)$. The charge deposited into C_a is thus given by

$$Q_a = C_1 v_i \frac{C_a}{C_1 + C_a}. \tag{8.14}$$

Again, we see that the only way to achieve the transfer of $C_1 v_i$ is to make $C_a \rightarrow \infty$ as shown in Figure 8.9b. As before, the infinite capacitor is realized by exploiting the Miller effect as shown in Figure 8.9c. Redrawing the figure, we obtain the integrator of Figure 8.10a. A charge $C_1 v_i$ accumulates on the top plate of C_2 every clock cycle. Since the top plate of C_2 is at zero potential (thanks to the opamp's virtual ground), the output voltage v_o decreases by $(C_1/C_2)v_i$ every clock cycle. This is therefore an inverting integrator. Furthermore, the accumulated output is available at the end of phase p_1 of the same clock cycle. Consequently, this integrator is delay free and has a transfer function

$$\frac{V_o(z)}{V_i(z)} = -\frac{C_1}{C_2}\frac{1}{1-z^{-1}}.$$

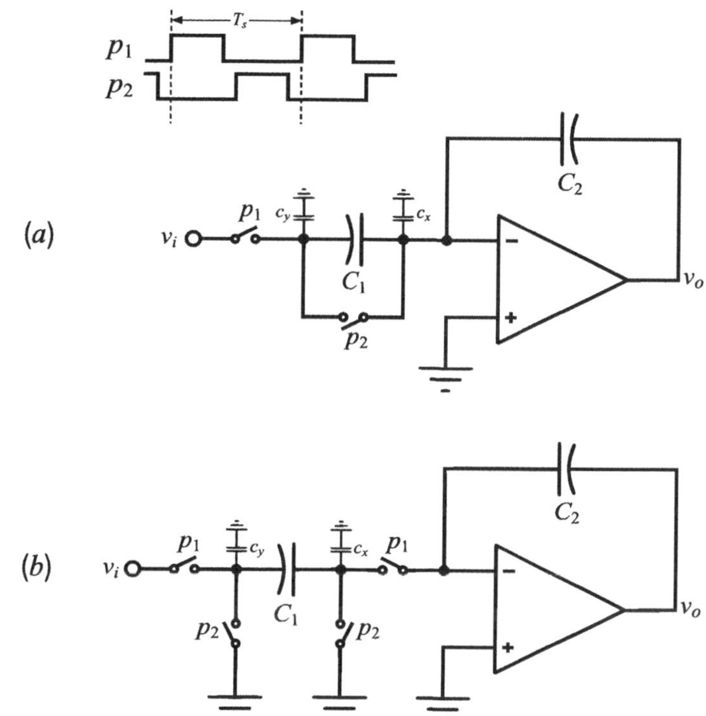

Figure 8.10 (a) The stray capacitor c_y affects the integrator transfer function. (b) Rendering the integrator insensitive to strays by adding two more switches. The resulting integrator is non-delaying and inverting.

Let us examine how stray capacitances affect the integrator's performance. During p_1, c_y is charged to a voltage v_i, while the voltage across c_x is zero. During p_2, needed to discharge C_1, c_y gets connected to the opamp's virtual ground node. As a result, a charge $c_y v_i$ is injected into C_2, resulting in a modified (inaccurate) transfer function.

Fortunately, this is easily fixed. Rather than discharging C_1 by connecting its plates together in p_2, we could instead connect both plates to ground as shown in Figure 8.10b. This way, the stray capacitance is discharged to ground (rather than the virtual ground). Consequently, the charge that accumulates on C_2's top plate every cycle is given by $C_1 v_i$, and is insensitive to parasitic capacitances. The integrator of Figure 8.10b is stray insensitive, inverting, and non-delaying.

8.2 Switched-Capacitor Filter Stages

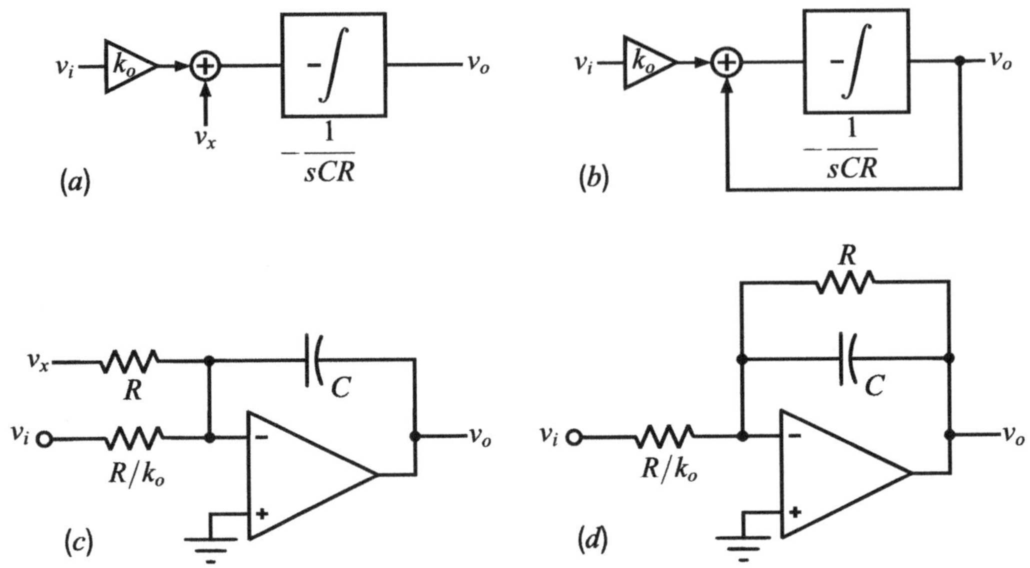

Figure 8.11 Development of a first-order continuous-time filter. (a) A summing and inverting integrator. (b) Connecting the output to one of the inputs. (c) Circuit implementation of (a). (d) First-order filter, with dc gain of $-k_o$ and a -3 dB bandwidth given by $1/(2\pi RC)$ (in Hz).

Having realized integrators, let us now attempt to realize a first-order lowpass filter. We take inspiration from the continuous-time implementation shown in Figure 8.11. Part (a) of the figure shows a summing, inverting integrator. It is easy to see that

$$V_o(s) = -\left\{k_o V_i(s) + V_x(s)\right\} \frac{1}{sCR}.$$

By feeding v_o back as v_x as in Figure 8.11b, we obtain

$$\frac{V_o(s)}{V_i(s)} = -\frac{k_o}{1 + sCR}.$$

In circuit form, Figure 8.11c shows an inverting, summing integrator. By connecting v_x to v_o, we arrive at the first-order lowpass filter of Figure 8.11d. The dc gain is $-k_o$, and the 3 dB bandwidth is $1/(2\pi RC)$ (in Hz).

To obtain a first-order switched-capacitor lowpass filter, we proceed as shown in Figure 8.12. Part (a) of the figure shows a summing, inverting integrator. Note that the top-plate switches of the C_1 and k_oC_1 capacitors are shared. During p_1, these capacitors are charged to v_x and v_i, and a charge $C_1(k_ov_i + v_x)$ is deposited on the top plate of the integrating capacitor C. In the z-domain, we have

$$V_o(z) = -\frac{C_1}{C}\left\{\frac{k_oV_i(z)}{1 - z^{-1}} + \frac{V_x(z)}{1 - z^{-1}}\right\}.$$

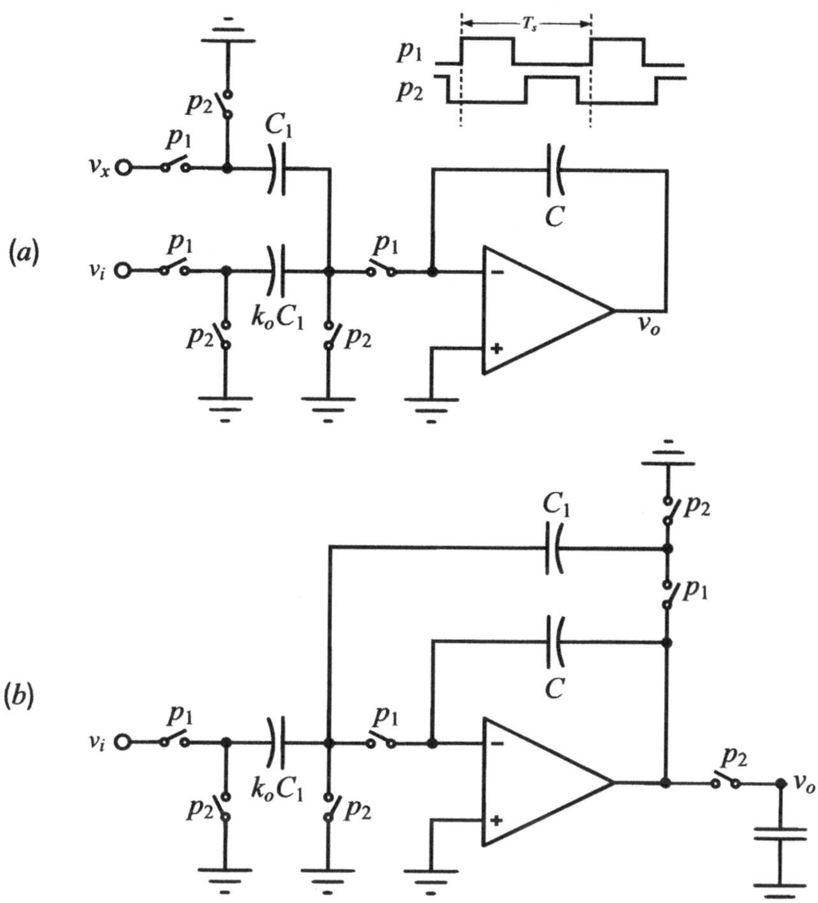

Figure 8.12 Development of a first-order switched-capacitor filter. (a) A summing and inverting accumulator. (b) Connecting the output to one of the inputs to result in a first-order filter, with dc gain of $-k_o$ and a -3 dB bandwidth given approximately by $(f_s/2\pi)(C_1/C)$ (in Hz).

By connecting v_x to the opamp output, we have $v_x = v_o$ as shown in Figure 8.12b. The equation above can be solved for $V_o(z)/V_i(z)$ to yield

$$\frac{V_o(z)}{V_i(z)} = H(z) = -\frac{k_o}{1 + (1 - z^{-1})(C/C_1)} = -\frac{k_o}{\left(1 + \frac{C}{C_1}\right) - \frac{C}{C_1}z^{-1}}.$$

The dc gain, obtained by evaluating the transfer function at $z = 1$, is seen to be $-k_o$. This makes sense due to the following. In steady state, the average current flowing through C must be zero. This means that the average current flowing into the input capacitor k_oC_1 must flow through the switched feedback capacitor, which means that the voltage across it in p_1 must be k_ov_i. Since the top plate is at zero potential (thanks to the opamp's virtual ground), the output voltage must be $-k_ov_i$. Furthermore, since $z = e^{sT_s}$, $(1 - z^{-1}) \approx sT_s$ for low input frequencies and the transfer function reduces to $-k_o/(1 + sC/f_sC_1)$. This makes sense, since the equivalent damping resistor is given by $1/f_sC_1$.

In the simple-minded single-pole example of Figure 8.2a, the output remained unchanged for a whole clock period, as seen in part (b) of the figure. The situation is different in the first-order filter of Figure 8.12b. When p_1 is high, the feedback capacitor is connected in parallel with the integrating capacitor C, and the input sinusoid injects charge into an effective feedback capacitor $(C + C_1)$. As a result, the output of the opamp has a continuously varying component during p_1. The "correct" voltage at the opamp output is that instant when p_1 goes low. This voltage, which represents the lowpass filtered sequence, is held during p_2. The output of the switched-capacitor filter, therefore, has to be "looked at" during p_2. The p_2 switch between the opamp output and v_o in Figure 8.12b indicates that the filter output is to be sampled and held during p_2. Assuming a tiny capacitor at the output node (shown in gray), v_o is held for an entire clock period beginning at the rising edge of p_2. The frequency response of the first-order filter is thus seen to be

$$H_{sc}(j2\pi f) = H(e^{j2\pi fT_s})\text{sinc}(fT_s)e^{-j\pi fT_s}. \tag{8.15}$$

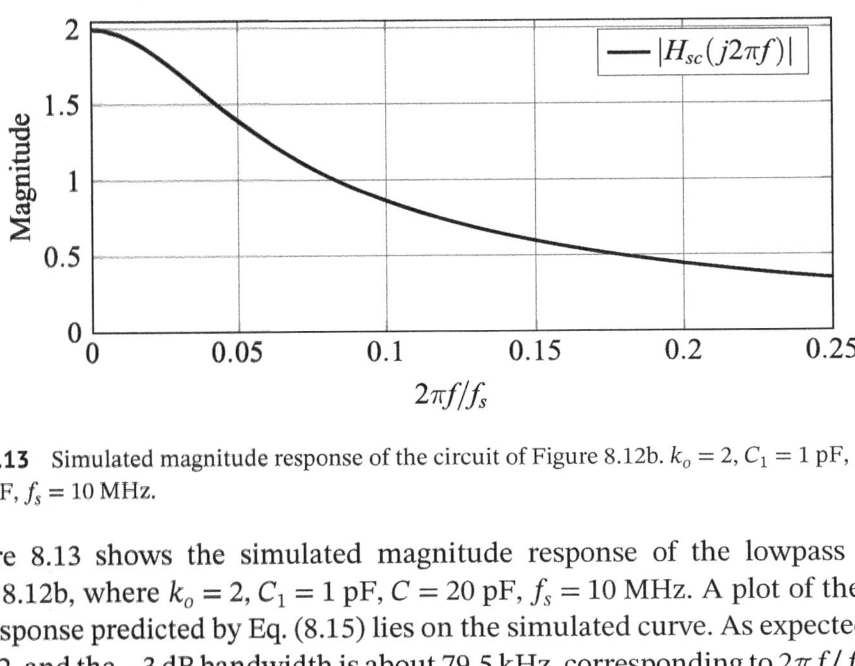

Figure 8.13 Simulated magnitude response of the circuit of Figure 8.12b. $k_o = 2$, $C_1 = 1$ pF, $C = 20$ pF, $f_s = 10$ MHz.

Figure 8.13 shows the simulated magnitude response of the lowpass filter of Figure 8.12b, where $k_o = 2$, $C_1 = 1$ pF, $C = 20$ pF, $f_s = 10$ MHz. A plot of the magnitude response predicted by Eq. (8.15) lies on the simulated curve. As expected, the dc gain is 2, and the −3 dB bandwidth is about 79.5 kHz, corresponding to $2\pi f/f_s = 0.05$.

An alternative development of the first-order switched-capacitor filter is the following. The starting point is the continuous-time prototype of Figure 8.11d. We first replace the resistors by appropriate valued capacitors C_1 and k_oC_1, as shown in Figure 8.14a. We notice that nodes ① and ② are both connected to ground during p_2 and to the opamp's virtual ground in p_1. They can therefore be shorted together, as shown by the wire in gray. The advantage of doing this is a reduced number of

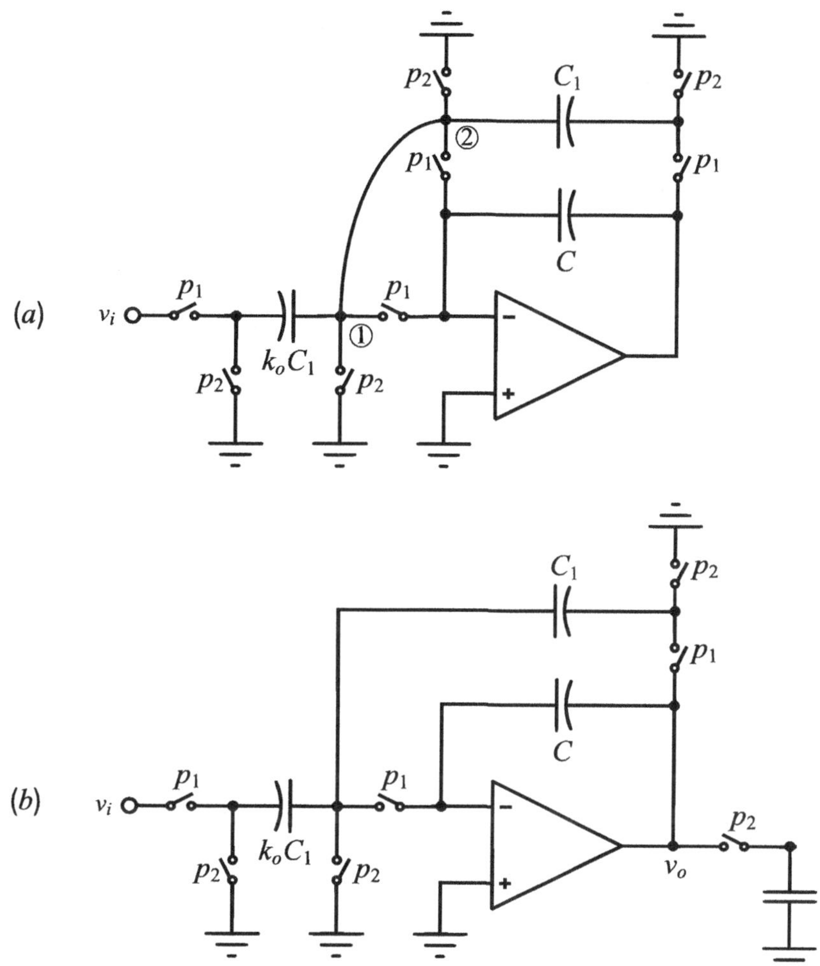

Figure 8.14 Alternative development of a first-order switched-capacitor filter. (a) Resistors in the continuous-time prototype of Figure 8.11d are replaced by switched capacitors, and switches are shared. (b) Resulting first-order filter.

switches as discussed below. The two p_1 switches connecting nodes ① and ② to the opamp's virtual ground are in parallel. One of them can be eliminated. Likewise, the two p_2 switches connecting node ① and node ② to ground are also in parallel, allowing one to be removed. This is called switch sharing. The circuit that results after eliminating the redundant switches, given in Figure 8.14b, is identical to that in Figure 8.12b.

8.2.1 Second-Order Lowpass Filters

An SC lowpass filter can be similarly constructed by replacing the resistors in an active-RC biquad by stray-insensitive SC branches, and then simplifying the circuit by employing switch sharing.

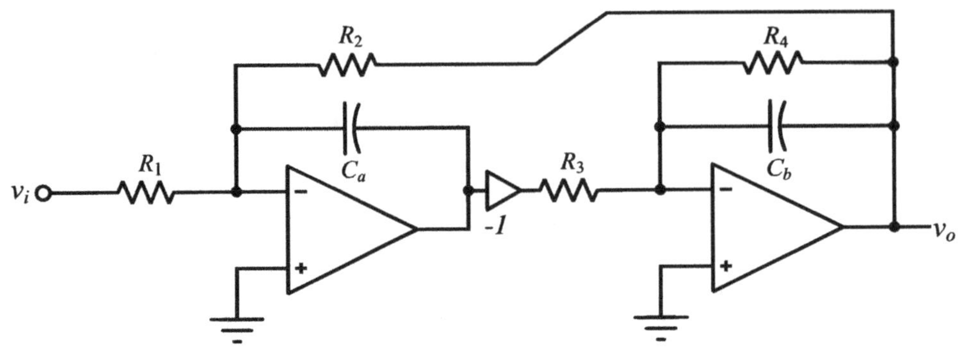

Figure 8.15 A second-order continuous-time lowpass filter prototype.

An active-RC biquad with a damped second integrator is shown in Figure 8.15. The center frequency (ω_o), Q, and dc gain of this filter are given by

$$\omega_o = \frac{1}{\sqrt{R_2 R_3 C_a C_b}}, \quad Q = \frac{R_4}{\sqrt{R_2 R_3}}\sqrt{\frac{C_b}{C_a}}, \quad \text{dc gain} = -\frac{R_2}{R_1}.$$

The SC biquad is derived as follows. First, every resistor is replaced by the equivalent four-switch parasitic-free SC branch. Then, pairs of switches performing identical roles are combined. The resulting SC biquad is shown in Figure 8.16.

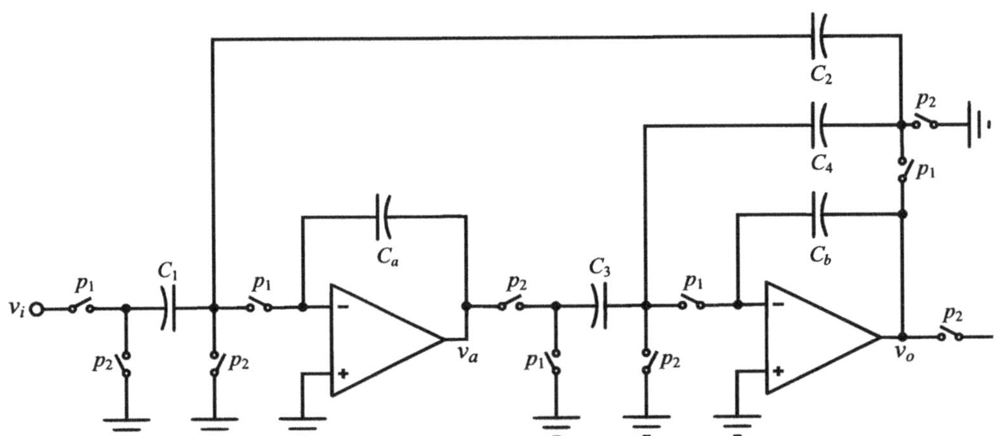

Figure 8.16 The second-order switched-capacitor filter based on the active-RC prototype of Figure 8.15. The output is available in p_2.

C_2 and its associated switches correspond to R_2, while C_1 and its switches are the replacement for R_1. The inversion and R_3 in Figure 8.15 are realized using the inverting SC branch formed by C_3 and its associated switches. The damping resistor R_4 is realized by C_4 and its switches. C_a and C_b are the integrating capacitors. To determine the transfer function $V_o(z)/V_i(z)$, we proceed as follows. v_a is the output of an inverting, delay-free, summing integrator that processes v_i and v_o. v_o can be thought of as being the sum of two parts – v_a, which is delayed, inverted and integrated, and v_o which is integrated in an inverting and non-delaying manner through C_4. We can thus write

$$V_a(z) = -\frac{C_1}{C_a}\frac{1}{1-z^{-1}}V_i(z) - \frac{C_2}{C_a}\frac{1}{1-z^{-1}}V_o(z)$$
$$V_o(z) = -\frac{C_4}{C_b}\frac{1}{1-z^{-1}}V_o(z) + \frac{C_3}{C_b}\frac{z^{-1}}{1-z^{-1}}V_a(z).$$

Solving the set of equations above for $V_o(z)/V_i(z)$ yields

$$\frac{V_o(z)}{V_i(z)} = H(z) = \frac{-(C_1/C_2)z^{-1}}{z^{-1} + \frac{C_4C_a}{C_2C_3}(1-z^{-1}) + \frac{C_aC_b}{C_2C_3}(1-z^{-1})^2}$$
$$= \frac{-C_1C_3z}{C_a(C_b+C_4)z^2 + (C_2C_3 - C_aC_4 - 2C_aC_b)z + C_aC_b}. \quad (8.16)$$

From the equations above, it is evident that the dc gain of the biquad (obtained by evaluating $H(z)$ for $z = 1$), is $-C_1/C_2$. Using $z^{-1} = e^{-j2\pi fT_s} \approx (1 - j2\pi fT_s)$, we see that

$$H(e^{j2\pi fT_s}) \approx \frac{-(C_1/C_2)}{1 + \frac{C_4C_a}{C_2C_3}j2\pi fT_s + \frac{C_aC_b}{C_2C_3}(j2\pi fT_s)^2}.$$

Comparing the denominator with that of the standard form of a second-order continuous-time lowpass filter $(1 + j2\pi f/\omega_o Q + (j2\pi f/\omega_o)^2)$, we obtain

$$\omega_o = f_s\sqrt{\frac{C_2C_3}{C_aC_b}}$$
$$Q = \sqrt{\frac{C_bC_2C_3}{C_aC_4^2}}. \quad (8.17)$$

As expected, ω_o only depends on f_s and a ratio of capacitors, while Q is only defined by capacitor ratios. Furthermore, the expression for ω_o makes sense due to the following. Referring to the continuous-time prototype of Figure 8.15, ω_o is given by $1/\sqrt{R_2R_3C_aC_b}$. In the switched-capacitor biquad, R_2 and R_3 are replaced by their switched-capacitor equivalents with values $1/f_sC_2$ and $1/f_sC_3$, respectively. Consequently, ω_o is given by $f_s\sqrt{C_2C_3/(C_aC_b)}$. Q can be derived from that of the continuous-time prototype in a similar manner.

Which phase should we "look at" for the output? After all, v_o is updated in phase p_1 and remains that way in p_2. If the opamps were ideal, in the sense that the output settles instantly after the p_1 switches are turned on, we could sample v_o at any time. In practice, however, opamps have a finite bandwidth. As a result, v_o will settle to its correct value at the end of p_1. Furthermore, the output waveform during the beginning of p_1 is not a linear function of the input due to a nonlinear phenomenon called slewing.

To summarize, therefore, the best time to sample the filter's output is the end of p_1, or equivalently, the beginning of p_2 as shown in Figure 8.16. The frequency response of the filter is given by $H(e^{j2\pi fT_s})\text{sinc}(fT_s)\exp(-j\pi fT_s)$. An example response and that of the corresponding continuous-time prototype are shown in Figure 8.17.

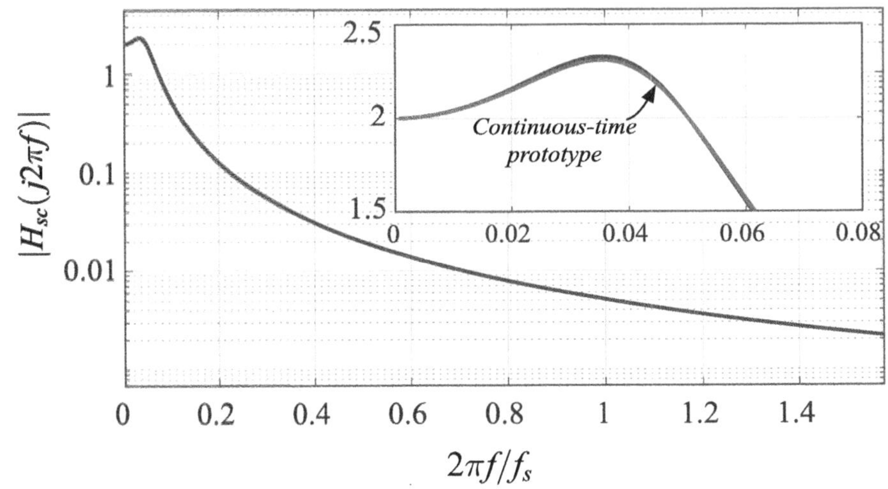

Figure 8.17 Simulated frequency response of the switched-capacitor filter, for $C_1 = 2$ pF, $C_2, C_3, C_4 = 1$ pF, and C_a, C_b =20 pF. The inset shows the passband detail, and the response of the continuous-time prototype with $Q = 1$ and $\omega_o = f_s/20$.

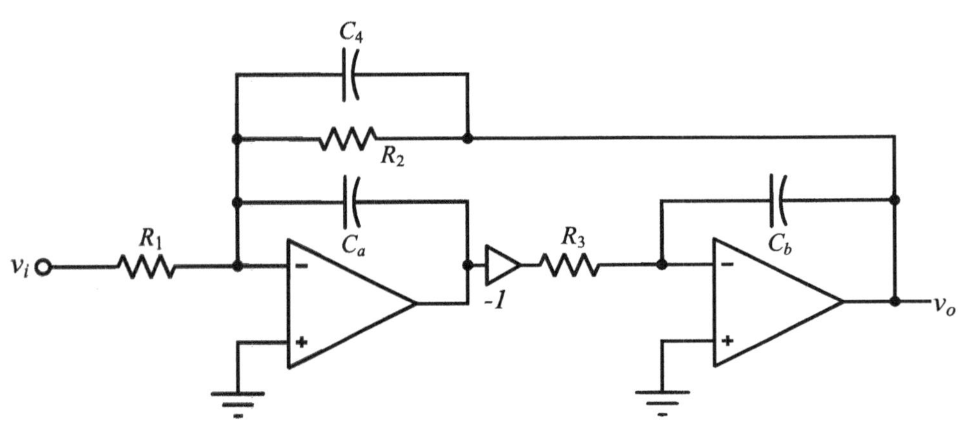

Figure 8.18 An active-RC filter with capacitive damping via C_4.

The SC lowpass filter of Figure 8.16 was derived from the active-RC prototype of Figure 8.15, where damping was introduced by R_4. When a high quality factor is needed, R_4 becomes very large. Consequently, in the SC version C_4 is very small in relation to C_3 (which in turn is small compared to C_b). The spread in capacitor values in the switched-capacitor structure thus becomes extremely large.

The problem with the large R_4 in the active-RC case can be remedied by introducing damping via the capacitor C_4 as shown in Figure 8.18. Straightforward analysis shows that the effective damping resistance around C_b is $(C_a/C_4)R_3$. A small C_4 is adequate to realize a large quality factor whereas a large resistance R_4 would be needed in the structure of Figure 8.15. In the corresponding SC filter, shown in Figure 8.19, damping is introduced by the C_4-C_a-C_3 path. When compared to the circuit of Figure 8.16, C_4 can be larger by a factor C_a/C_3. This results in a reduced capacitor spread, making the design robust.

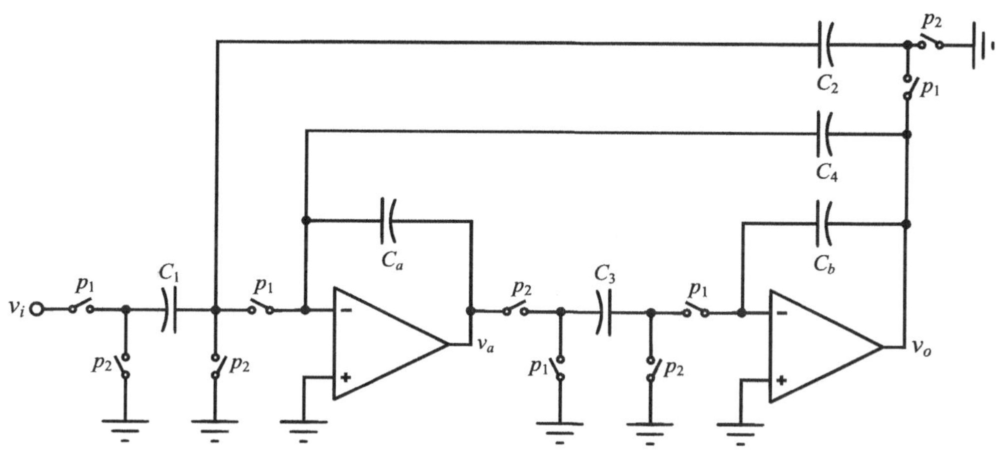

Figure 8.19 A switched-capacitor biquad suitable to implement high quality factors. Damping is introduced via C_4.

The analysis of the transfer function proceeds in a manner similar to that of the biquad of Figure 8.16. Apart from realizing a summing integrator that adds v_i and v_o, there is in addition a proportional path from v_o. The second integrator integrates v_a to yield v_o. We can thus write

$$V_a(z) = -\frac{C_1}{C_a}\frac{1}{1-z^{-1}}V_i(z) - \frac{C_2}{C_a}\frac{1}{1-z^{-1}}V_o(z) - \frac{C_4}{C_a}V_o(z)$$

$$V_o(z) = \frac{C_3}{C_b}\frac{z^{-1}}{1-z^{-1}}V_a(z).$$

Solving the set of equations above for $V_o(z)/V_i(z)$ yields

$$\frac{V_o(z)}{V_i(z)} = H(z) = \frac{-(C_1/C_2)z^{-1}}{z^{-1} + \frac{C_4}{C_2}(1-z^{-1})z^{-1} + \frac{C_aC_b}{C_2C_3}(1-z^{-1})^2}$$
$$= \frac{-C_1C_3z}{C_aC_bz^2 + (C_3(C_2+C_4) - 2C_aC_b)z + C_aC_b - C_3C_4}. \quad (8.18)$$

Using $z^{-1} = 1$ and $(1 - z^{-1}) = j2\pi f fT_s$ in $H(z)$ above yields an approximate expression for the frequency response. It turns out that ω_o is given by $f_s\sqrt{C_2C_3/(C_aC_b)}$, and $Q = (1/C_4)\sqrt{C_aC_bC_2/C_3}$. This topology, which is suited to realize filters with high quality factors, allows C_4 to be a factor C_a/C_3 larger when compared to that in the circuit of Figure 8.16. Consequently, capacitor spread is greatly reduced.

Simulating the Frequency Response of a Switched-Capacitor Filter

How does one simulate the frequency response of a switched-capacitor filter? One way is to recognize that the network is an LPTV system, and all we are interested in is its zeroth-order harmonic transfer function. If one has access to an RF circuit simulator, one can use its periodic analysis capability. Recall that the simulator inherently assumes that the network is nonlinear. All linear analyses are applied to the small-signal network that results when the circuit is linearized about its operating point. Simulation, therefore, proceeds in two steps.

a. Use the periodic steady state analysis (`.PSS`) to determine the periodic operating point of the filter. From this, the simulator derives the small-signal network, which is LPTV.

b. Run a periodic AC analysis (`.PAC`) on the filter, and specify that we only wish to evaluate the zeroth-order harmonic transfer function. Since we are interested in the complete waveform, we must choose the sample-and-held output of the filter. The resulting magnitude response, therefore, will account for the sinc droop which occurs due to the hold operation.

What does one do if one does not have access to an RF circuit simulator, i.e., one that is not able to simulate LPTV circuits? In such a situation, one can resort to a time-domain approach by recognizing that the frequency response is given by $H(z)\text{sinc}(fT_s)\exp(-j\pi fT_s)$, where $z = e^{j2\pi fT_s}$. $H(z)$ is the z-transform of the discrete-time impulse response of the switched-capacitor network, and can be obtained by running a time-domain (`.TRAN`) simulation as follows. The input is a small signal pulse whose width is adjusted so that the input capacitor samples a nonzero value only in the first clock cycle. The output sequence of the filter is monitored, $H(z)$ computed over the desired frequency range, and multiplied by $\text{sinc}(fT_s)\exp(-j\pi fT_s)$ (due to the sample-and-hold) to yield the desired frequency response.
Figure 8.20 compares the frequency responses of the biquad of Figure 8.19 obtained using the two methods described above. The center frequency is about 80 kHz, with $Q = 10$. The responses are virtually indistinguishable.

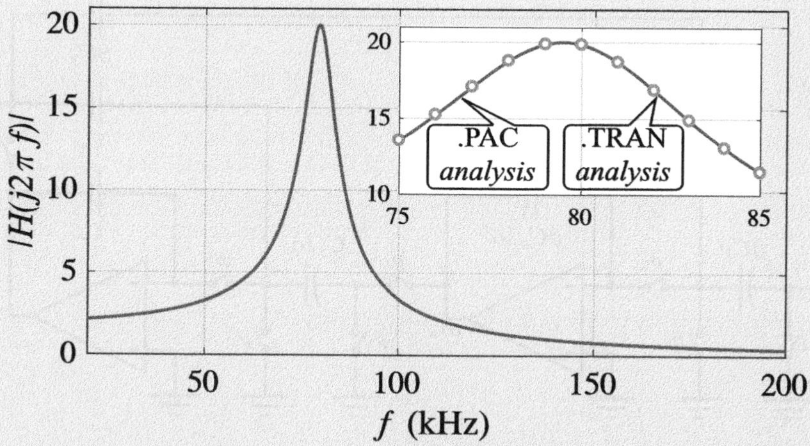

Figure 8.20 Simulated frequency response of the biquad of Figure 8.19 with $C_1 = 2$ pF, $C_2 = C_3 = 1$ pF, $C_4 = 2$ pF, C_a, $C_b = 20$ pF, and $f_s = 10$ MHz. The dc gain of the filter is 2, and the quality factor is 10. The inset shows the detail around ω_o. The markers show the results obtained using the transient analysis method, while the solid curve is that obtained using the .PAC analysis.

As was the case for active-RC filters, the switched-capacitor biquad circuit has more specifiable element values (six in the example of Figure 8.19) than necessary to achieve the desired dc gain, ω_o and Q. The additional three degrees of freedom are used for the following.

a. *Dynamic range scaling*: In general, the peak value of the magnitude responses of the transfer functions to the outputs of both opamps will not be equal. The consequence of this is that one of the opamps will prematurely limit the maximum input amplitude the filter can handle. Like in active-RC filters, therefore, the component values have to be adjusted so that the peak magnitude responses at both opamp outputs are the same. As an example, suppose that we wish to increase the peak magnitude at the output of the first opamp in Figure 8.19 by a factor α. This is accomplished by reducing C_a by a factor α, and lowering C_3 by the same factor, as shown in Figure 8.21 (assume $\beta = 1$ for the time being). This way, the transfer function from V_i to V_o remains unchanged.

b. *Noise scaling*: By increasing all capacitors in an integrator by the same factor β, the noise spectral density (NSD) due to that integrator is reduced by a factor β. When applied to the first integrator of Figure 8.19, C_1, C_2, and C_a are all increased by a factor β. Increasing all capacitors by β will not only need all switch sizes to be increased by β to maintain the same charging time constant, but an opamp that supplies β times the current (which will necessitate β times the power dissipation). The noise of the second integrator can be similarly scaled by multiplying C_3/α, C_4, C_b by a factor γ.

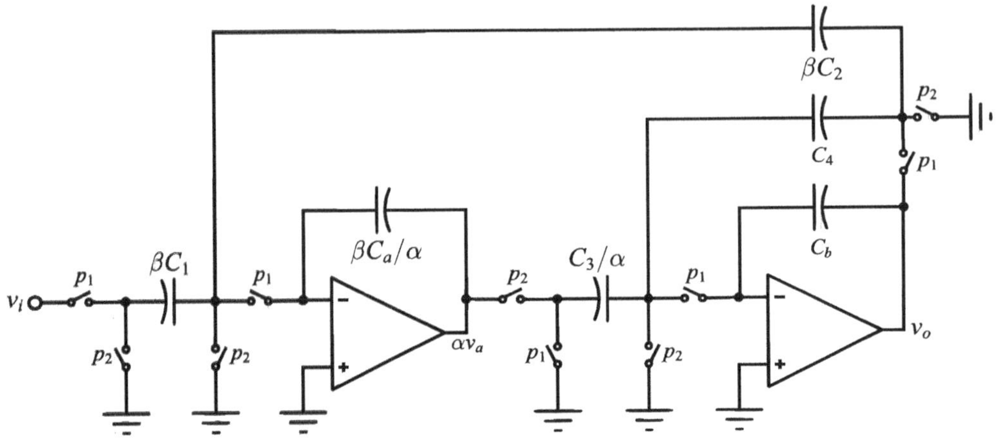

Figure 8.21 Dynamic-range and impedance scaling in a switched-capacitor biquad. The transfer function from the input to the output of the first opamp is multiplied by α, and the noise spectral density due to the first integrator is reduced by a factor β.

To summarize, the biquad of Figure 8.19 has six design variables, namely $C_1, \cdots, C_4$, C_a and C_b. These are needed to set the values of the following quantities.

1. dc gain, ω_o, and Q.
2. Peak gain at the output of the first opamp, and impedance scaling factors of the first and second integrators.

High-order switched-capacitor filters are realized by cascading biquads, much in the same way as their active-RC counterparts discussed in Chapter 7.

8.3 Switched-Capacitor Amplifiers

Amplifiers are important building blocks of many analog subsystems. Not surprisingly, the parasitic-insensitive integrators derived previously can be readily modified to result in amplifiers. The basic idea is the following. The memory of an integrator is due to the capacitance in the feedback path of the opamp. By "zapping" C_2 in Figures 8.7 and 8.10 every clock cycle, the integrators' memory is erased as shown in Figure 8.22.

Referring to Figure 8.22a, C_1 has a voltage v_i across it during p_1. Consequently, a positive charge $C_1 v_i$ flows into the top plate of C_2. The output voltage at the end of p_1, therefore, is given by $-(C_1/C_2)v_i$. In p_2, both C_1 and C_2 are discharged. Furthermore, during this phase, the opamp is in a negative feedback loop with unity feedback, and the output voltage is zero. The p_1 switch in series with $v_o(t)$ indicates that the output is available during p_1.

Figure 8.22b shows a non-inverting amplifier with a gain C_1/C_2. Here, charge from C_1 is transferred to C_2 during p_2 with v_o being available at the end of this phase. C_2

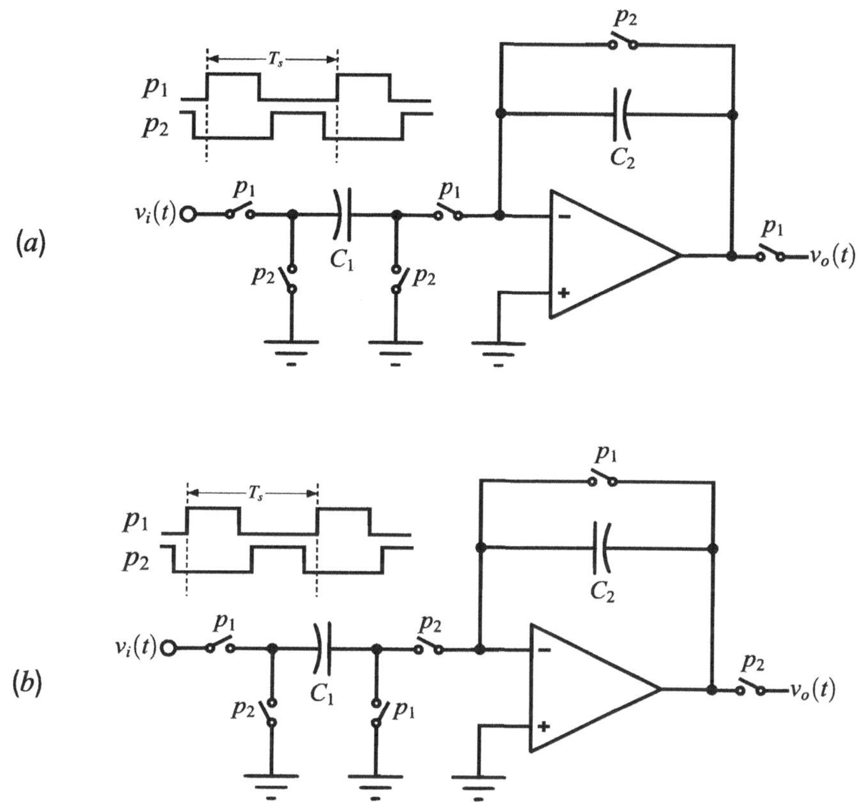

Figure 8.22 Switched-capacitor amplifiers based on the parasitic-insensitive (a) non-delaying integrator and (b) delaying integrator.

must therefore be zapped in p_1. As before, the phasing on the switch in series with $v_o(t)$ indicates that the output is available during p_2.

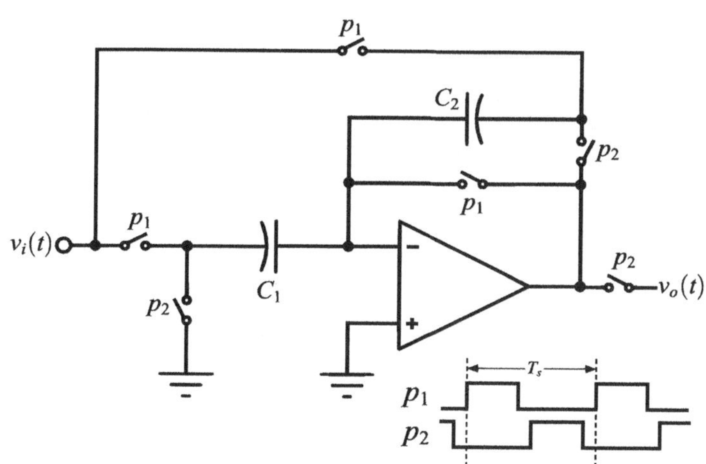

Figure 8.23 A parasitic-insensitive switched capacitor amplifier with a gain $(1 + C_1/C_2)$.

Figure 8.23 shows a non-obvious switched-capacitor amplifier. During p_1, the inverting terminal of the opamp is shorted to its output. C_1 and C_2 are in parallel, and have a voltage v_i across them. During p_2, C_1's bottom plate is connected to ground, while that of C_2 is flipped around to connect to the opamp's output. To calculate v_o during p_2, we reason as follows.

a. The charge on C_1 and C_2's top plates at the end of p_1 are $-C_1 v_i$ and $-C_2 v_i$ respectively.
b. During p_2, the charge on C_1's top plate is forced on to C_2's top plate, thereby causing the latter's charge to be $-(C_1 + C_2)v_i$.
c. The output voltage during p_2 is thus given by

$$v_o = \left(\frac{C_1 + C_2}{C_2}\right) v_i = \left(1 + \frac{C_1}{C_2}\right) v_i.$$

Necessity of DC Negative Feedback Around the Opamp

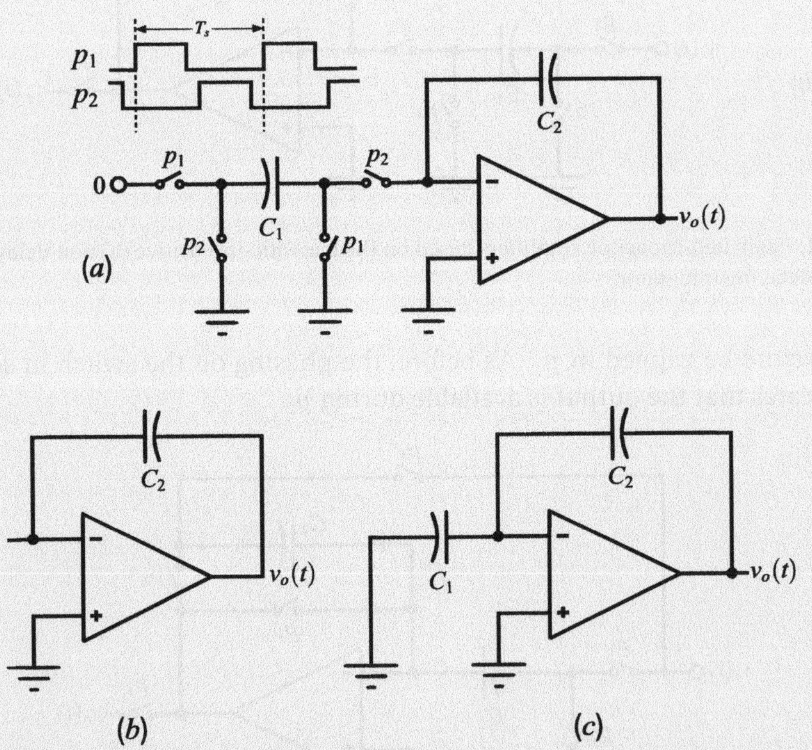

Figure 8.24 (a) The non-inverting integrator with the input set to 0. (b) DC feedback network around the opamp in p_1. (c) DC feedback network around the opamp in p_2.

As we have seen so far in this chapter, the operation of opamp-based switched capacitor circuits is based on the inverting terminal of the opamp being a virtual-ground node. This is possible only if there is dc negative feedback around the opamp. The reader may readily verify the presence of dc feedback during both phases p_1 and p_2 in all the circuits we have seen so far. Figure 8.24a shows a non-inverting integrator with the input set to 0. During p_1, the dc feedback path around the opamp is provided through C_2 which acts like a battery, as shown in Figure 8.24b. Thanks to this, the inverting terminal's potential stays at ground, and v_o does not "fly off." Figure 8.24c shows the feedback path during p_2. It consists of the capacitive divider formed by p_1 and p_2.

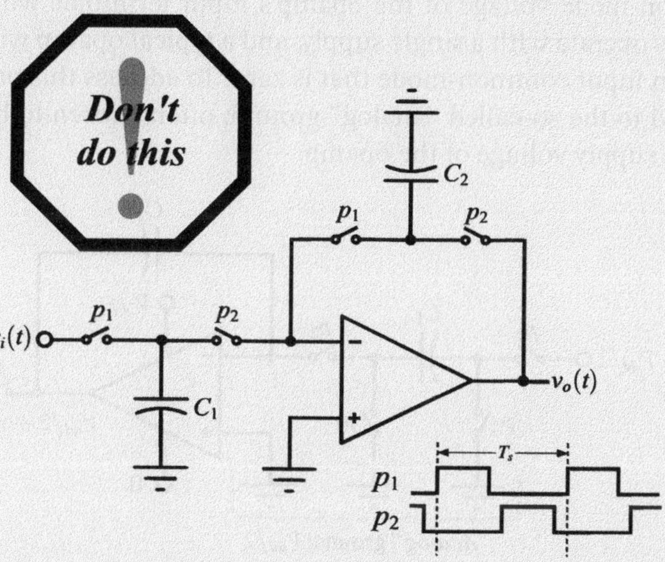

Figure 8.25 An attempt to realize an amplifier with gain C_1/C_2. It does not function as intended due to the absence of dc negative feedback around the opamp.

Figure 8.25 shows an attempt at realizing a switched capacitor amplifier by replacing the input and feedback resistors in a continuous-time inverting amplifier with switched-capacitor equivalents. It is evident that the opamp in this circuit does not have dc negative feedback around it in p_1 or p_2. Consequently, the inverting terminal is no longer a virtual ground, and the circuit does not function as intended.

8.4 Nonideal Effects in Switched-Capacitor Circuits

As discussed earlier, by using only capacitors to determine its time constants, a switched-capacitor circuit can achieve superior accuracy in realizing a prescribed transfer function. Nevertheless, all components of the implemented circuit suffer from nonidealities. In this section, the main effects will be described, and some methods for mitigating them will be discussed. Before we proceed, however, we will discuss some practical aspects of switched-capacitor circuits.

8.4.1 Single-Supply Operation

In all our discussions so far, the non-inverting terminal of the opamp was connected to ground, and the input and output voltages were referred to ground. Thanks to negative feedback, the inverting terminal of the opamp will also be at ground. This means that the common-mode voltage of the opamp's input terminals will be zero. Most practical circuits operate with a single supply, and a typical opamp will not be able to function with an input common-mode that is zero. To address this problem, all voltages are referred to the so-called "analog" ground, often chosen to be $V_{dd}/2$ where V_{dd} denotes the supply voltage of the opamp.

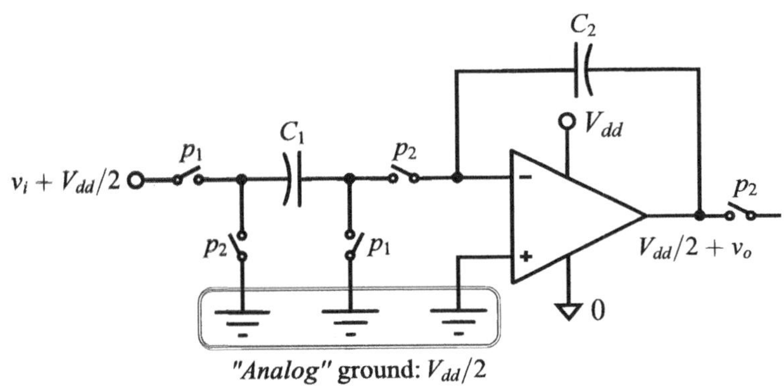

Figure 8.26 A non-inverting switched capacitor integrator with single-supply operation.

The non-inverting switched-capacitor integrator, therefore, is implemented in practice as shown in Figure 8.26. The information-carrying input voltage v_i "rides on" $V_{dd}/2$. The non-inverting terminal of the opamp is also tied to $V_{dd}/2$. Assuming that the voltage across C_2 is zero, the output of the opamp will be at a potential $V_{dd}/2$.

During p_1, the voltage across C_1 is v_i. During p_2, the bottom plate of C_1 is connected to $V_{dd}/2$. Thanks to negative feedback, the opamp's inverting terminal will remain at $V_{dd}/2$, forcing charge from C_1 to flow into C_2. The opamp's voltage will thus be $V_{dd}/2 + (C_1/C_2)v_i$. If we referred the potential of every node in the circuit to analog ground, which has a potential $V_{dd}/2$, all the analyses of the preceding section holds. When circuits are depicted, therefore, the analog ground's potential of $V_{dd}/2$, as well as the supply voltage details of the opamp are omitted.

In Chapter 7, we saw the benefits of fully differential operation in active-RC filters. The same motivation and benefit applies to switched-capacitor circuits. It is thus no surprise that differential operation is the norm. Figure 8.27 shows the schematic of a non-inverting differential integrator. Each of the inputs rides on a common-mode voltage V_{cm} (which is usually chosen to be $V_{dd}/2$). The quantity of interest is the

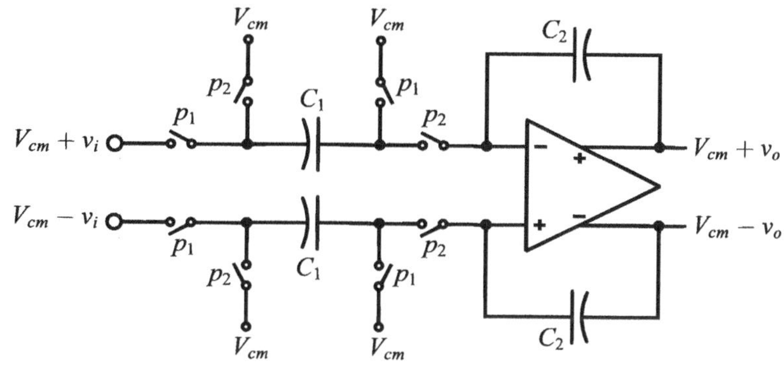

Figure 8.27 A fully-differential non-inverting switched-capacitor integrator.

difference between the two inputs. The opamp is a fully-differential structure, which produces a large output *differential* voltage in response is a small differential input. The opamp's internal circuitry maintains the common-mode component of the output nodes' potential at a fixed value V_{cm}.

During p_1, the capacitors C_1 are charged to v_i and $-v_i$. Assuming zero voltage on the C_2 capacitors and an infinite-gain opamp, the differential input and output voltages of the opamp are zero. The common-mode component at the opamp input and output nodes is V_{cm}. During p_2, charge is transferred from the C_1 capacitors to the C_2 capacitors. v_o is given by $(C_1/C_2)v_i$. While practical circuits are fully differential, single-ended schematics are often drawn for simplicity and to reduce clutter in diagrams.

8.4.2 Opamps Versus OTAs

So far, we have assumed ideal opamps, and have treated them as voltage-controlled voltage sources with infinite gain. Several practical considerations, which we will not go into here,[4] indicate that a voltage-controlled *current* source with a "large enough" transconductance is preferable to using an opamp. Such a VCCS is referred to as an operational transconductance amplifier (or OTA). An ideal OTA is a VCCS with an infinite transconductance, but is approximated in practice by a sufficiently large-valued transconductor.

Figure 8.28a shows the basic switched-capacitor integrator, but where an OTA with transconductance G_m is used in place of the opamp. During p_1, C_1 is charged to v_i. During p_2, the C_1 is connected to the virtual-ground node ⓐ. As soon as the p_2 switch is closed, v_a jumps to v_i. Thanks to negative feedback, a current $G_m v_a(t)$ is pulled out of the node, causing its potential to decay toward zero (Figure 8.28b) with a time constant C_1/G_m. When $v_a = 0$, all the charge on C_1 is transferred on to C_2.

[4]The motivation for using an OTA rather than an opamp is lower noise, and usually a superior output swing.

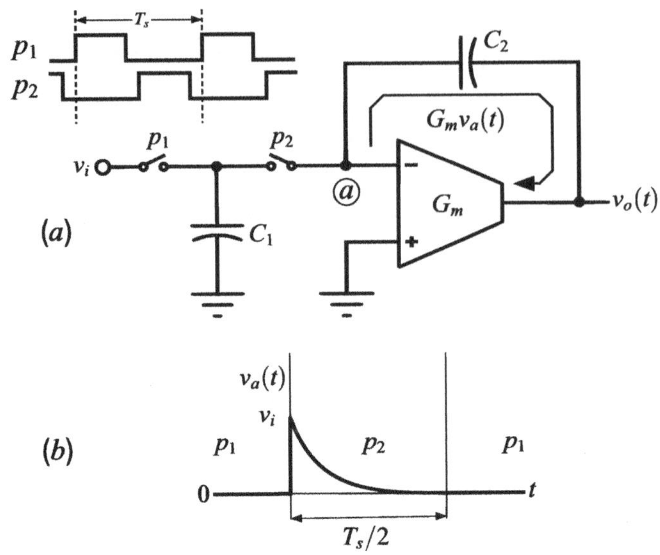

Figure 8.28 (a) The basic switched-capacitor integrator using an OTA in lieu of an opamp. (b) Voltage waveform at the virtual-ground node, which decays to zero with a time constant C_1/G_m.

G_m cannot be infinite, and is considered "large enough" if $v_a(t)$ can get sufficiently close to zero at the end of phase p_2. For instance, choosing C_1/G_m to be a tenth of the duration of p_2 results in more than 99.99% of the charge on C_1 being transferred on to C_2. In practice, the OTA has parasitic capacitances at its input and output nodes, causing the discharge time constant to increase. G_m must be appropriately increased to ensure the appropriate degree of settling. Even though an OTA is used, it has become common practice to continue to use the symbol of an opamp in schematics, and we are guilty of that same indiscretion.

8.4.3 Switches

8.4.3.1 Nonzero "on" Resistance

Figure 8.29a shows again a parasitic-insensitive SC integrator, illustrating the basic elements—switches, capacitors and opamp. As described earlier, the "ground" potential refers to $V_{dd}/2$. NMOS transistors are used to realize the switches. The gate terminal of the transistor must be connected to the supply voltage V_{dd} to turn it on, and to zero to turn it off. Ideally, each switch should become a short circuit in one of the clock phases, and an open circuit in the other. In reality, when a switch is turned on, it gradually builds up a conducting layer (the channel) between its source and drain terminals. Assuming an n-channel (NMOS) switch device, the channel will contain electrons. To generate the channel, the gate voltage V_g of the device must be higher than $V_s + V_T$, where V_s is the voltage at the source, and V_T is the NMOS threshold voltage. This condition is easy to satisfy for the three switches in the integrator which are connected to real or virtual ground but may be difficult for the switch in the input

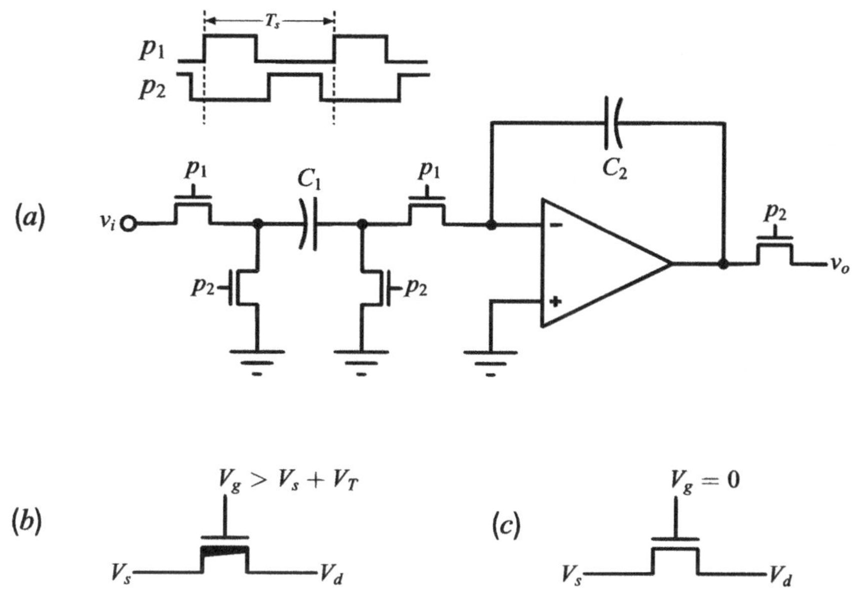

Figure 8.29 (a) A parasitic-insensitive switched-capacitor integrator, with switches realized using NMOS transistors. (b) Turning the switch on requires V_g to be greater than $V_s + V_T$ where V_T is the transistor's threshold. (c) The switch is turned off by setting $V_g = 0$.

branch, since its source voltage is the variable input voltage v_i. Thus, the switch will need a clock voltage at its gate which satisfies $V_{dd} > v_i + V_T$ for all input voltages. This restricts the allowable input voltage range significantly. State-of-art fine line technology also restricts the value of V_{dd} to prevent breakdowns in the devices.

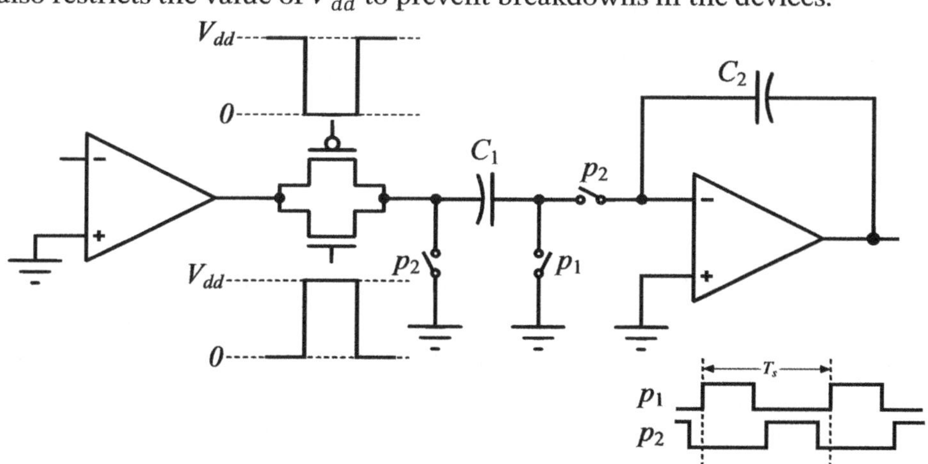

Figure 8.30 Using a transmission gate to realize the input switch.

A simple, but incomplete remedy is the use of a transmission gate, the parallel combination of an NMOS and a PMOS device, as shown in Figure 8.30. Now the NMOS switch will turn on for $v_i < V_{dd} - V_T$, and the PMOS device for $v_i > |V_T|$. If V_{dd} is greater than the sum of the threshold voltages, at least one of the devices will be able to conduct; otherwise, there will be a dead zone, usually around the middle of the range of v_i.

8.4.3.2 Charge Injection and On-Resistance Modulation

Even if there are no forbidden zones for the input voltage, there are effects by which the switch influences the signal charge flow. Two important ones are charge injection and on-resistance modulation.

When an NMOS switch is turned on with a gate voltage V_{dd}, the magnitude of charge in the (conducting) channel formed between the source and drain is given by $q_{ch} = C_{ox}WL(V_{dd} - V_S - V_T)$, where V_S denotes the potential of the source, and the other terms have their usual meanings. The resistance of the switch is given by $R_{ch} = 1/[\mu_n C_{ox}(W/L)(V_{dd} - V_S - V_T)]$. It is thus seen that the $q_{ch}R_{ch}$ product is L^2/μ_n. This indicates that a small switch resistance can only be achieved by a large channel charge. Furthermore, q_{ch} depends on V_S in a nonlinear manner due to the modulation of V_T by V_S.

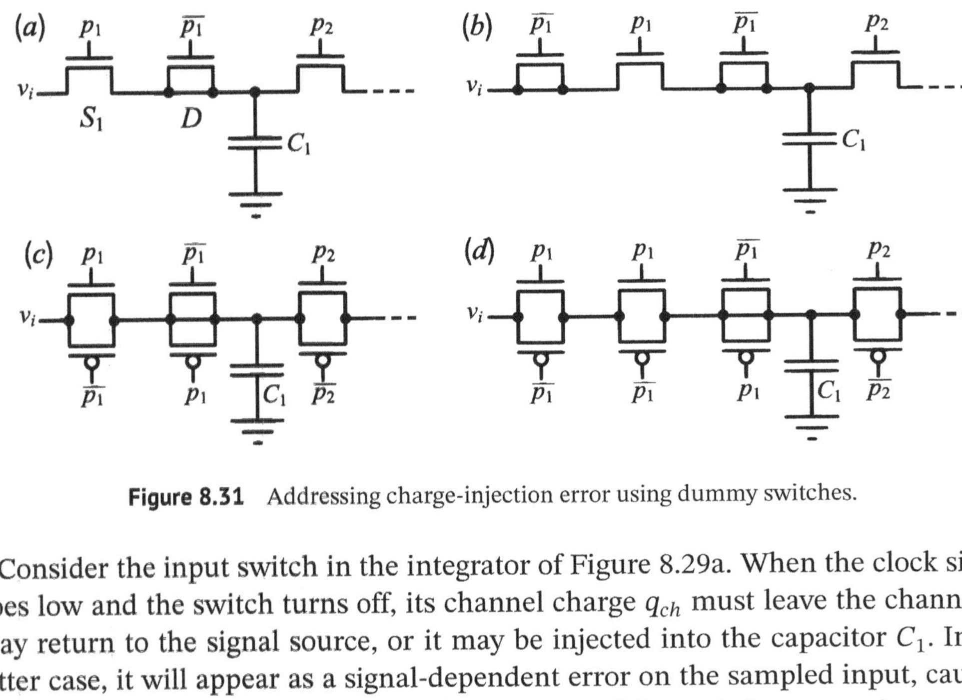

Figure 8.31 Addressing charge-injection error using dummy switches.

Consider the input switch in the integrator of Figure 8.29a. When the clock signal goes low and the switch turns off, its channel charge q_{ch} must leave the channel. It may return to the signal source, or it may be injected into the capacitor C_1. In the latter case, it will appear as a signal-dependent error on the sampled input, causing harmonic distortion in the filter. The phenomenon of channel charge causing an error on the held voltage is called charge injection.

There are several simple techniques for mitigating the effects of charge injection. One is the use of dummy switches (Figure 8.31). The dummy switch D in Figure 8.31a has half the channel area of switch S_1 and is driven by the complement of its clock signal p_1. It is intentionally short circuited. When p_1 is high, S_1 conducts, and when it cuts off it deposits an error charge approximately equal to $q_{ch}/2$ into C_1. Once this process is completed, the dummy device is turned on, and it extracts a charge $q_{ch}/2$ out of C_1, canceling the charge injection. More elaborate versions of the circuit are available. That in Figure 8.31b is aimed at making the distribution of charges from S_1 symmetric; the remaining circuits show how the principle is applied to transmission gates.

8.4.3.3 Advance-Cutoff Switching

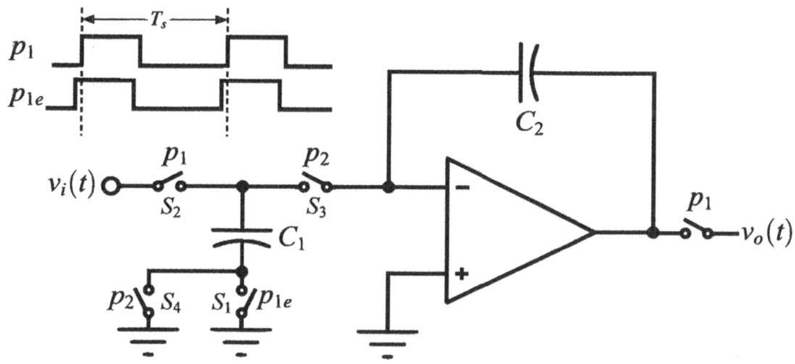

Figure 8.32 The concept of advance-cutoff switching. The bottom plate of C_1 is disconnected first.

An efficient and economical remedy for the distortion caused by charge injection is the use of advance-cutoff switching [3]. It is illustrated for the simple integrator in Figure 8.32. The clock signal p_{1e} of the added switch is similar to p_1, but it cuts off earlier. Since it is a grounded switch, when it cuts off its charge injection is signal independent. It causes a dc offset in the output of the integrator, which is usually harmless. (In fact, it appears as a common-mode jump in a differential integrator, and is of no consequence in the differential output). Since S_3 is open, most of the charge of S_1 will return into the input source, rather than into C_1.

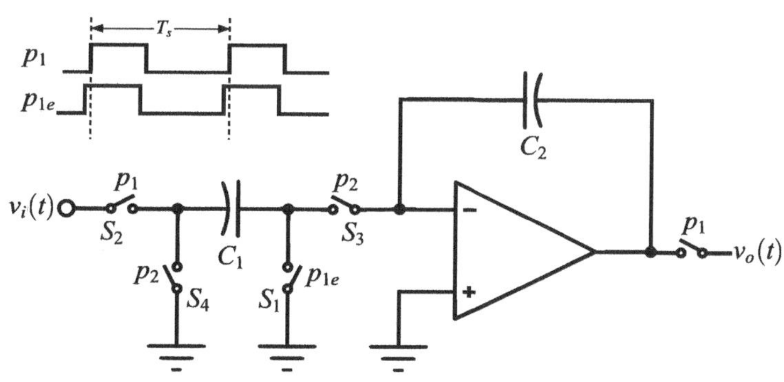

Figure 8.33 A stray-insensitive integrator with advance-cutoff switching.

In the stray insensitive integrator, no added switch is needed to implement the advance cutoff scheme, since S_1 is already present (Figure 8.33). Note that p_{1e} and p_1 must not overlap with p_2. A simple logic circuit which can generate two clock phases p_1 and p_2 and their advanced versions from a single clock signal *clk* is shown in Figure 8.34.

A more frequent problem with switches trying to sample a continuously-variable input (as opposed to sampled-and-held inputs) is the nonlinear distortion caused by

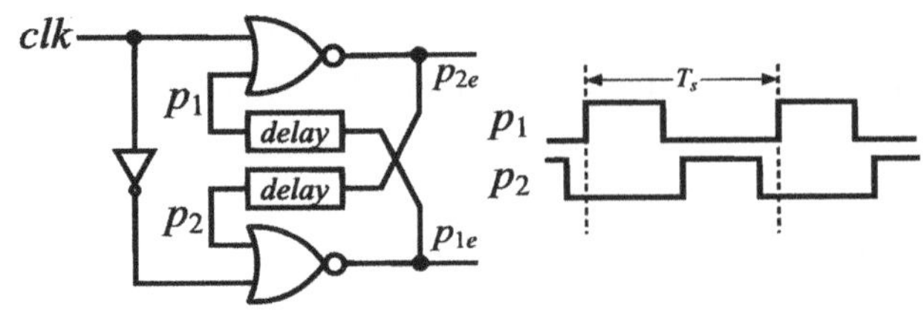

Figure 8.34 A circuit that generates non-overlapping clocks p_1 and p_2, as well as their advanced versions.

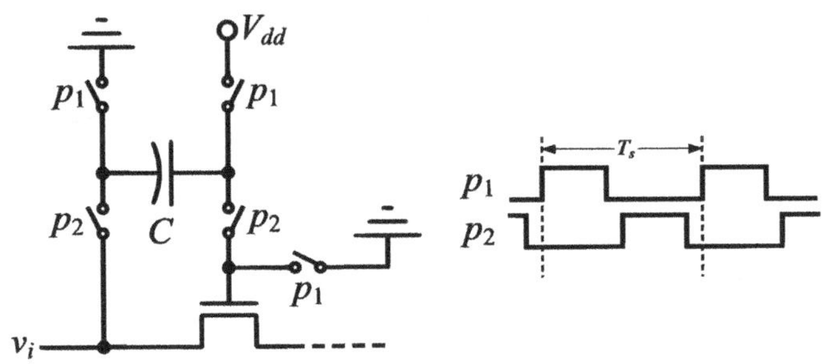

Figure 8.35 The operating principle behind the bootstrapped switch.

the signal-dependent on-resistance. Recall that the switch resistance is dependent on $(V_{dd} - v_i - V_T)$. As is evident from this expression, the root cause of this problem is that the gate-overdrive voltage of the switch varies with v_i. This suggests the following solution. Suppose the gate, rather than being connected to the fixed voltage V_{dd} to turn it on, was connected to the varying voltage $(V_{dd} + v_i)$. This way the overdrive would be $(V_{dd} - V_T)$ irrespective of v_i, resulting in linear operation.

Figure 8.35 shows the basic idea of how this could be implemented. During p_1, the gate is connected to the ground to turn it off. At the same time, the capacitor C is charged to V_{dd}. During p_2, the capacitor is connected between the gate and source of the switch, resulting in a gate voltage that is ideally given by $(V_{dd} + v_i)$. As a result, even though v_i is varying, the overdrive voltage remains $(V_{dd} - V_T)$, a value that is independent of v_i (neglecting the modulation of V_T by v_i). This technique is termed "bootstrapped sampling."

While the basic idea is straightforward, one needs to be very careful about the implementation of the extra switches to ensure that the devices used to realize them are not subject to unacceptably large voltages. At the cost of a complex added circuitry it is possible to eliminate these effects as well.

8.4.4 Capacitor Nonidealities

Under ideal conditions, the transfer function $H(e^{sT_s})$ of a switched-capacitor network depends only on the ratios of capacitances, rather than on their absolute values. Hence, their structures and layout need to achieve accurate matching. Parasitic

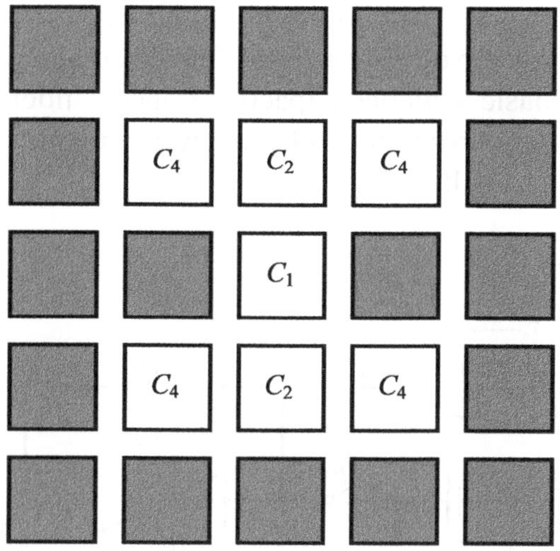

Figure 8.36 Example capacitor array to realize three capacitors with ratio 1:2:4. The shaded unit cells are dummies, so that all "active" unit capacitors see the same environment.

capacitances affect these ratios, and must be neutralized by using stray insensitive circuitry. The errors in the realized capacitance values must also be considered. They may be systematic, usually due to processing, or random, due to granularity of their plates or oxide layer.

Systematic errors may be corrected by using unit cells (equal-valued small capacitors) and applying common-centroid layout for all the capacitors requiring matching. Figure 8.36 illustrates schematically the arrangement of three capacitors $C_1, C_2 = 2C_1$, and $C_4 = 4C_1$. This arrangement makes the centers of these capacitors coincide, and compensates for gradual changes in fabrication, thus keeping the ratio of these capacitors accurate. The shaded unit cells are dummies, placed so that the environment around every active unit capacitor is nominally identical.

Systematic errors usually occur at the edges of the capacitor plates, due to etching undercut and fringing electric fields. If the matched capacitors are constructed from equal-valued unit elements, then these errors will not affect their ratio.

Random errors may be due to the granularity of the conductive plates of the capacitors, or random variation of the oxide thickness between them. They become significant for very small capacitance values, in the femtofarad range. For such values, fringe (side wall) capacitors containing metal electrodes in the top metal layers of the chip may be used.

8.4.5 Operational-Amplifier Nonidealities

In switched-capacitor circuits, the opamp is in a negative feedback configuration, and is used to create a virtual short circuit between its input terminals. In single-ended circuits, the non-inverting input node is grounded, and the inverting input becomes a virtual ground if the opamp is ideal. Furthermore, it is able to drive any load, thanks to its zero-output impedance. The above will not be true for a practical opamp, and it is thus important to develop estimates for the effects of opamp nonidealities on the actual performance of basic switched capacitor stages. Understanding these effects is a key to developing measures that can be taken to mitigate problems due to these nonidealities. These topics will be discussed next.

8.4.5.1 Finite Opamp Gain

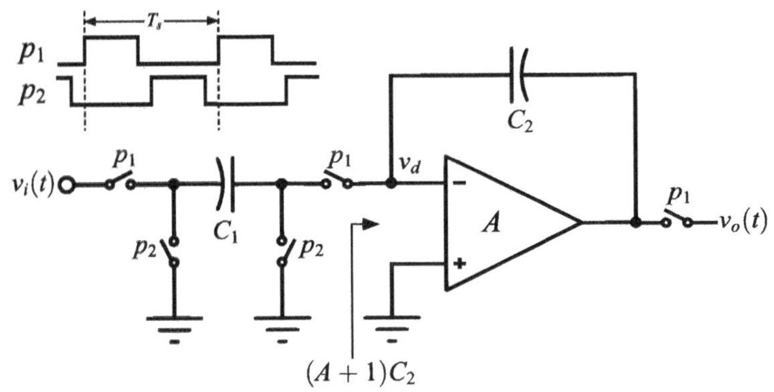

Figure 8.37 Finite gain of the opamp results in incomplete charge transfer from C_1 to C_2, as well as leakage of charge from C_2 to C_1.

We have so far assumed that the opamps have infinite gain. In reality, it is finite (surprise?) and can be traced down to the finite output resistance of transistors in the opamp. To understand the effect of finite gain on the performance of a switched-capacitor circuit, it is instructive to analyze the performance of the delay-free integrator. As shown in Figure 8.37, the effective capacitance looking into the virtual ground node is $(A + 1)C_2$. From the analysis of Section 8.1.1, and using $(A + 1)C_2$ in place of C_2, we can relate v_d to v_i according to

$$v_d[n] = (1 - \alpha)v_i[n] + \alpha v_d[n - 1] \tag{8.19}$$

where $\alpha = (1 + A)C_2/(C_1 + (A + 1)C_2)$. Intuitively, the charge transferred from the input every cycle is slightly smaller than the ideal value $C_1 v_i$ since the capacitance looking into the virtual ground is no longer infinite. More importantly, a small amount of charge is transferred from the $(A + 1)C_2$ capacitor to the input through C_1. This causes integrator leakage, and can be understood as follows. Suppose C_2 had an initial voltage across it. If $v_i = 0$, the integrator output would asymptotically reach zero, as C_2 would lose a small amount of charge in every cycle. The integrator output is given by $-Av_d$. Straightforward analysis yields

$$H(z) = \frac{V_o(z)}{V_i(z)} \approx -\frac{C_1}{C_2}\frac{1}{1 - \alpha z^{-1}}. \tag{8.20}$$

As a sanity check, $H(z)$ approaches the ideal transfer function as $A \to \infty$.

8.4.5.2 Opamp Offset

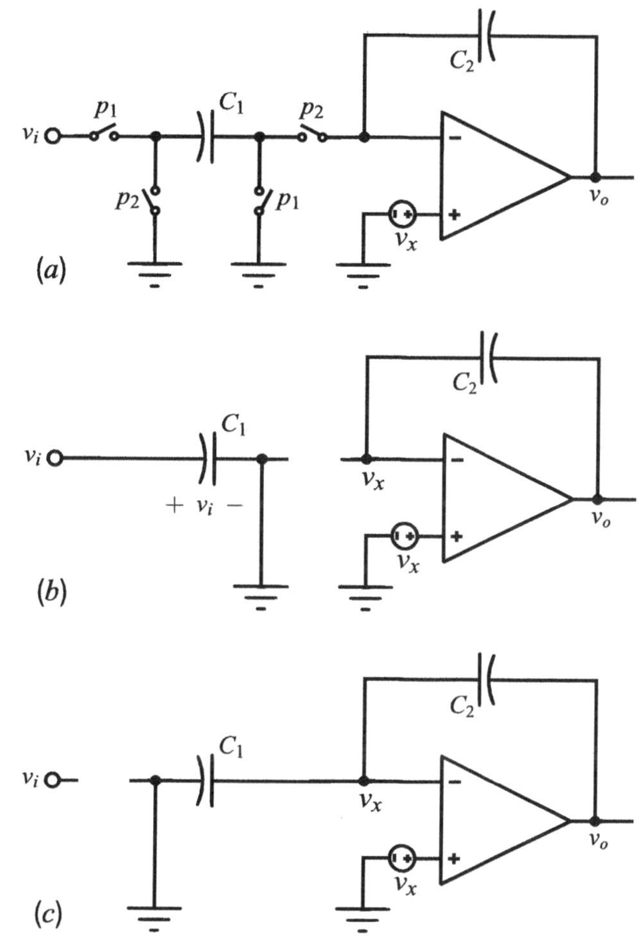

Figure 8.38 (a) Switched-capacitor integrator with opamp offset. (b) Circuit during p_1. (c) Circuit during p_2.

To understand the effect of opamp offset, it is instructive to analyze the non-inverting switched capacitor integrator of Figure 8.38a. The offset is modeled as a voltage source v_x in series with the non-inverting terminal. During p_1, C_1 is charged to v_i. Due to negative feedback, the inverting terminal of the opamp is at a potential v_x, as shown in Figure 8.38b. During p_2, the bottom plate of C_1 is grounded, and the top plate is connected to the opamp's inverting terminal (Figure 8.38c). Consequently, the charge transferred to the top plate of C_2 is given by $-C_1(v_i + v_x)$. The output voltage of the opamp, therefore, increases by $(C_1/C_2)(v_i + v_x)$. Thus, even with a zero input, the integrator output voltage increases by $(C_1/C_2)v_x$ every clock cycle.

Having understood how opamp offset affects the integrator's output, we are in a position to address this problem. The basic idea is the following. The opamp's offset manifests at its inverting terminal. During the input sampling phase, namely p_1, the opamp is "idle" and its offset is stored on a capacitor C_a, as shown in Figure 8.39a. At the end of this phase, C_1 is charged to v_i, and C_a has v_x across it.

In the charge-transfer phase p_2, when the charge on C_1 is transferred to C_2, C_a is connected in series with the opamp's inverting terminal, as shown in Figure 8.39b. Its bottom plate is the new virtual ground of the opamp. Since C_a acts like a battery of voltage v_x, the potential of its bottom plate, i.e., the virtual ground voltage, is zero. The output is offset free, since the combination of C_a and the opamp is equivalent to a zero-offset opamp. The final circuit diagram is shown in Figure 8.39c. Since the offset is sampled in one phase and canceled in the next, this technique is called correlated

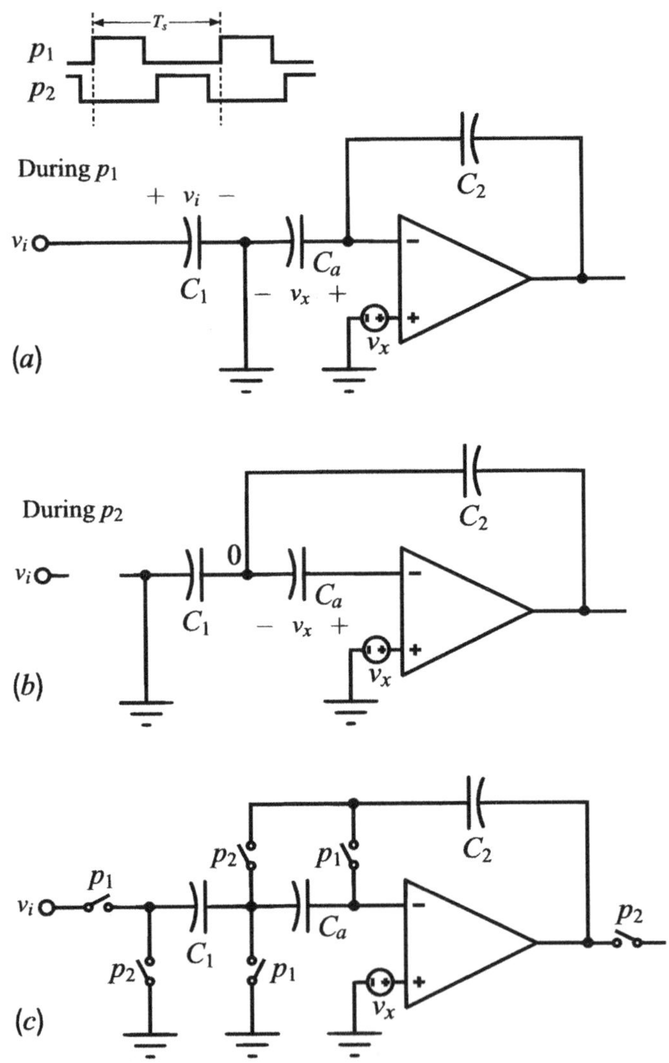

Figure 8.39 Evolution of a stray-insensitive integrator with CDS. (a) C_1 is charged to v_i, while C_a stores the opamp's offset during p_1. (b) During p_2, C_a is placed in series with the opamp, essentially canceling opamp offset. (c) Final circuit.

double sampling (CDS) [4]. CDS can also be applied to SC amplifiers and sample-and-hold stages.

A useful aspect of CDS is that it addresses any slowly-changing component of the virtual ground voltage. Flicker noise is one example of such a component. Another such instance is the following. Suppose the opamp has a finite gain, and v_i is a low-frequency signal. The virtual ground node of opamp will therefore consist of a low-frequency component during p_1. Thanks to CDS, this component will be largely eliminated during p_2. In effect, therefore, the apparent gain of the opamp appears to have increased. The analysis is left as an exercise for the reader.

8.4.5.3 Finite-Bandwidth Effects

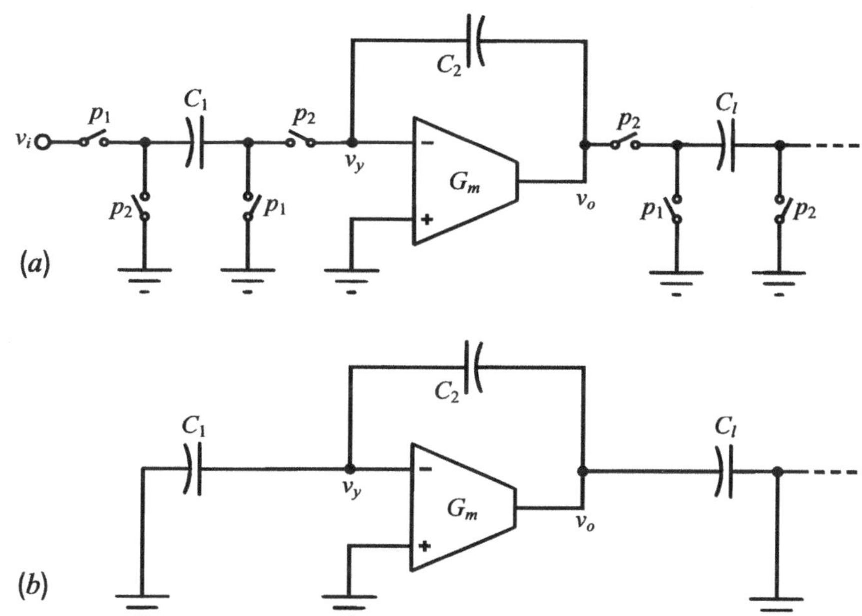

Figure 8.40 (a) Non-inverting switched-capacitor integrator driving a load C_l. (b) Equivalent circuit during phase p_2.

Another practical aspect of switched-capacitor circuits is the finite bandwidth of the opamps, or finite transconductance of the OTAs. The result is an incomplete charge transfer, and consequently, a settling error. We illustrate this for the non-inverting switched-capacitor integrator of Figure 8.40, where an OTA with a finite G_m is used. C_l denotes the load of the second stage.

The input is sampled on C_1 during p_1. In p_2, C_1 is connected to the virtual ground, while C_l is connected to the OTA's output as shown in Figure 8.40b. The resulting network is of the first order, though there are three capacitors. Analysis shows that the time constant is given by

$$\tau = \frac{C_1C_2 + C_2C_l + C_lC_1}{G_mC_2}. \tag{8.21}$$

The settled value of $v_o[n]$ is $v_o[n-1] + (C_1/C_2)v_i$. For more than 99.9% settling accuracy (0.1% settling error) at the end of p_2, $\tau < 0.5T_s/7$ (assuming a 50% duty cycle for p_2). This constraint leads to a minimum allowable value for G_m.

In addition to the finite bandwidth, the accuracy of the charge transfer in the SC integrator is also limited by the finite slew rate of the OTA. In the circuit of Figure 8.40, C_1 is charged to v_i in p_1. In p_2, when C_1 is switched to the virtual ground, the initial voltage at that node is given by $v_y = -v_i$ (assuming a nonzero resistance for the switches). If the OTA was linear, the initial current flowing out of the OTA should be $G_m v_i$. In practice, however, there is a limit on the OTA's maximum current output, which we denote by I_{max}. To a first approximation, this limit will occur for all $|v_i| \geq I_{\text{max}}/G_m$, which we term the linear range of the OTA. For inputs exceeding the linear range, the initial output of the OTA will be a ramp with a slope $I_{\text{max}}/C_{\text{tot}}$ where

$$C_{\text{tot}} = \frac{C_l C_2 + C_2 C_1 + C_1 C_l}{C_1 + C_2}.$$

Assuming v_i is positive, v_y (which initially started from a negative value) will ramp up towards zero. After some time, v_y will enter the linear range of the OTA, after which the dynamics of the circuit is linear. To summarize, slewing causes the network to operate in a nonlinear fashion for a part of the clock cycle. As a designer, one must ensure that the output settles to the desired value after taking into account both the slewing phase and linear settling phase.

8.5 Noise in Switched-Capacitor Circuits

In CMOS analog circuits, the noise sources are the resistors and opamps. The former generate thermal noise, while the latter add both thermal and flicker ($1/f$) noise. The single-sided noise voltage power spectral density (PSD) of a resistor R is given by $4\,kTR$, where k is the Boltzmann constant, and T is the absolute temperature. The input-referred thermal noise PSD of an opamp can usually be estimated from $S_a = 16\,kT/3\,g_m$, where g_m denotes the transconductance of the opamp's input devices.

8.5.1 The Noise Penalty Due to Sampling

Consider the track-and-hold circuit of Figure 8.41, which is a basic building block of switched-capacitor circuits. The switch is turned on periodically (with period $T_s = 1/f_s$) and a duty cycle given by m. As seen in Chapter 2, the mean-square noise on the capacitor is given by kT/C. How does this compare with that which would be obtained without sampling? A few practical approximations are in order here.

a. The bandwidth of the input signal is assumed to be $f_s/2$, so as to satisfy the Nyquist sampling criterion.
b. For good tracking of the input, $RC \ll mT_s$.

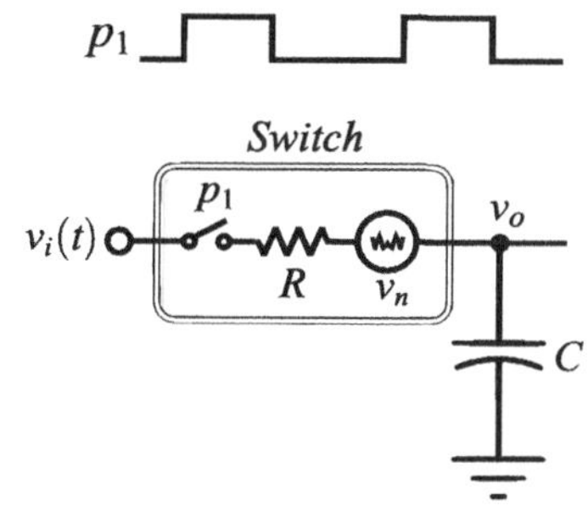

Figure 8.41 A track-and-hold circuit. For good tracking, $RC \ll mT_s$, where m and T_s denote the duty cycle and period with which the switch is operated.

Without sampling, the mean-square noise in the signal bandwidth would be $4kTR(f_s/2)$. The ratio of the mean-square sampled noise to that without sampling is given by

$$\alpha = \frac{kT/C}{4kTR(f_s/2)} = \frac{T_s}{2RC}.$$

Since $RC \ll mT_s$, it is evident that $\alpha \gg 1$, indicating a significant noise penalty associated with sampling. In the frequency domain, this can be interpreted as follows. In the tracking phase, the spectral density of the noise voltage across the resistor has a bandwidth $1/(2\pi RC) \gg f_s$. In the hold phase, where this noise is "trapped" on the capacitor, the broadband noise due to the resistor aliases, or folds into the frequency range $0-f_s/2$, thereby increasing the in-band NSD by a large factor. This increase is often referred to as the noise penalty due to sampling.

8.5.2 Noise in Passive Switched-Capacitor Circuits

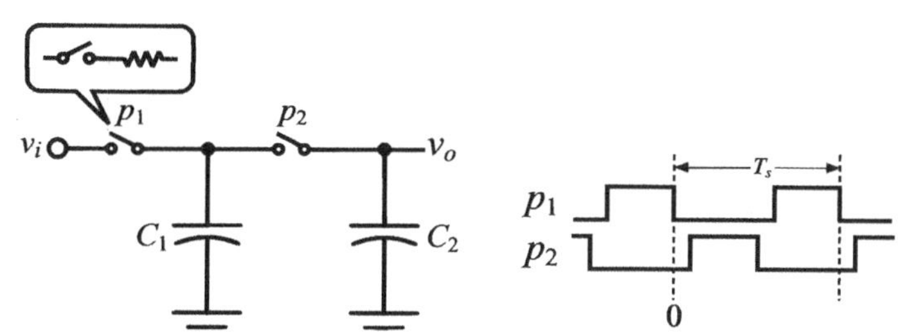

Figure 8.42 A passive switched-capacitor filter. Each switch is modeled as an ideal switch in series with a resistor that adds noise.

In networks consisting of only switches and capacitors, we can exploit passivity to determine the average NSD and total integrated noise, as seen in Chapter 6.

Consider, for instance, the first-order switched-capacitor filter shown in Figure 8.42. The total integrated noise across C_2, obtained using the energy approach of

Chapter 6, is given by kT/C_2. The average single-sided output noise voltage spectral density is given by

$$S_o(f) = 4kT\text{Re}\{Z_o(j2\pi f)\}$$

where $Z_0(j2\pi f)$ is the zeroth-order harmonic driving point impedance looking into the output node.

8.5.3 Noise in a Switched-Capacitor Integrator

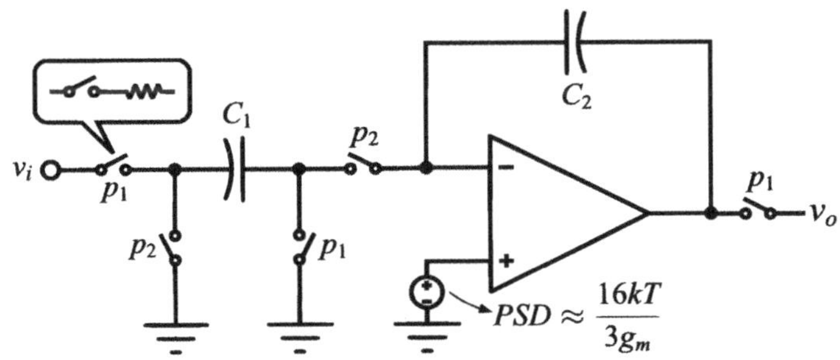

Figure 8.43 A parasitic-insensitive switched-capacitor integrator. The resistance of the switches, and the opamp add thermal noise.

We now analyze the thermal noise in the switched-capacitor integrator of Figure 8.43. Assuming an ideal opamp, the accumulated noise charges $q_{1,1}$ and $q_{1,2}$ in C_1 at the end of the two clock periods both have the mean square value kTC_1, as shown in Figure 8.44.

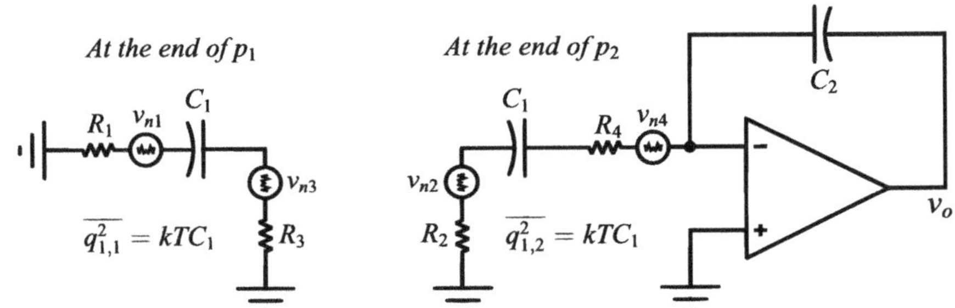

Figure 8.44 Mean-square noise charges on the capacitor C_1 at the end of phases p_1 and p_2.

At the end of phase p_2 the difference of the two charges $q_{1,1}$ and $q_{1,2}$ enters the virtual ground. Since these charges are uncorrelated random processes, the mean square value of the entering charge is going to be the sum of their powers, that is $2kTC_1$. The combined effect of the noise due to the switches can be modeled by an equivalent white noise voltage source at the input. The mean square value of this source is seen to be $2kT/C_1$ [5].

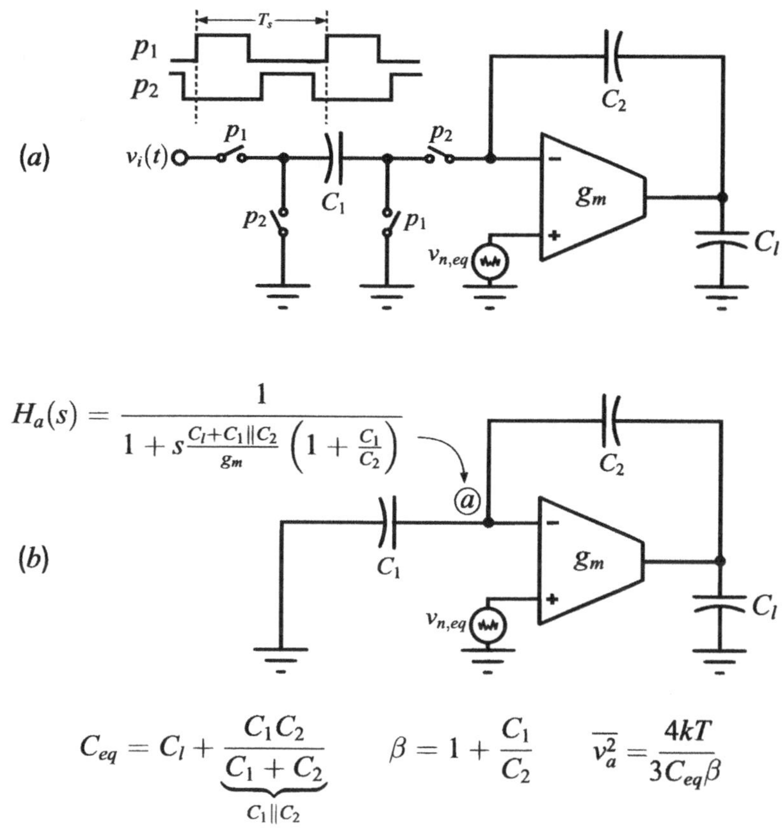

Figure 8.45 (a) Using an OTA in lieu of the opamp. (b) Noise due to the OTA.

Next, the input representation of the thermal noise of the opamp will be found. Figure 8.45a shows the circuit with the opamp and its associated noise source. The opamp, in practice, is an OTA whose transconductance is denoted by g_m. It turns out that the input-referred noise voltage spectral density of the OTA is given by $16kT/g_m$. C_l models the load capacitance driven by the integrator (which is the sum of its own output parasitic capacitance and the input-sampling capacitor of the next stage).

To analyze the effect of OTA noise, we assume that the switches are ideal. During p_1, C_1 charges to v_i. In p_2, C_1 is connected to the OTA's virtual ground, resulting in the equivalent circuit shown in Figure 8.45b. Just when C_1 is connected to the virtual ground, $v_a = -v_i$. Without OTA noise, the negative feedback loop then works in a way as to bring v_a to zero with a time constant proportional to $1/g_m$. Assuming complete settling, the virtual ground voltage would be zero and all the charge on C_1 would be transferred to C_2 during p_2.

Due to the OTA's noise, however, the virtual ground voltage is not zero. Analysis shows that the transfer function from the noise source to node ⓐ is given by

$$H_a(s) = \frac{1}{1 + s\frac{C_l + C_1 \| C_2}{g_m}\left(1 + \frac{C_1}{C_2}\right)}$$

resulting in a mean-square noise voltage

$$\overline{v_a^2} = \frac{4kT}{3C_{\text{eq}}\beta} \text{ where } C_{\text{eq}} = C_l + \frac{C_1 C_2}{C_1 + C_2} \text{ and } \beta = 1 + \frac{C_1}{C_2}.$$

As a consequence of this noise voltage, the residual noise charge with mean-square value $C_1^2\overline{v_a^2}$ remains on C_1 at the end of p_2. This is equivalent to saying that the OTA is noise free, and the input v_i is corrupted by a noise source with mean-square noise given by $\overline{v_a^2}$.

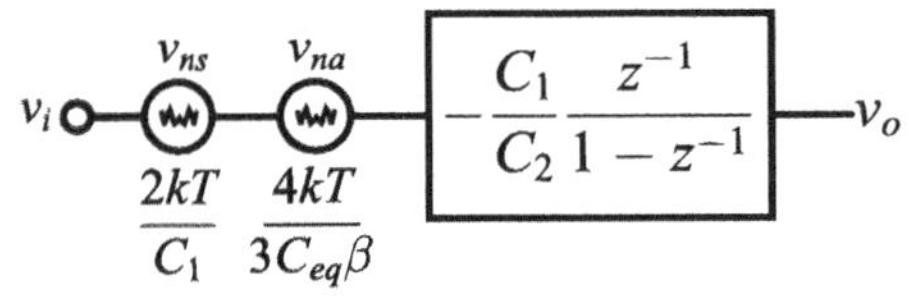

Figure 8.46 Model for the switched-capacitor integrator of Figure 8.45a including thermal noise due the switches and the OTA.

The noise model of the integrator considering the switches and the OTA is given in Figure 8.46. An observation is in order here. To reduce the input referred mean-square noise by a factor of 2 (i.e., 3 dB) while keeping the transfer function the same, C_1 (and C_2) would both have to be doubled. For the output of the integrator to settle to the same degree as before, the switch resistance would have to reduce by a factor of 2, while the OTA's transconductance would have to scale up by 2×. Consequently, the power dissipation of the OTA and the clock drivers (needed to drive the switches) would increase by a factor of 2. We thus see that to reduce noise by 3 dB, or equivalently, to increase the peak SNR by 3 dB, the power dissipation would have to double. This is hardly surprising.

Simulating Noise in Switched-Capacitor Networks

Switched-capacitor circuits are linear and periodically time-varying (LPTV) systems, varying at the clock frequency f_s. As discussed in Chapter 6, the NSD of an LPTV network at a frequency f consists of contributions from all frequencies of the form $(f + lf_s)$ where l is an integer.

The average NSD and total integrated noise in a switched capacitor network can be determined using the periodic noise analysis (referred to in most RF simulators as the `.PNOISE` analysis). The sequence of steps is the following.

a. Determine the periodic operating point using the `.PSS` analysis. This is necessary since the simulator, by default, assumes that the network is nonlinear. The first order of business, therefore, is to determine the periodic operating point about which the circuit is linearized. The result is a small-signal network which is LPTV.

b. To find the average NSD at a frequency f, the $H_{-l}(j2\pi(f + lf_s))$ must be found for *all* noise sources in the network. Inside the simulator, this is accomplished in a computationally-efficient manner using the frequency-reversal theorem as discussed in Chapter 6.

c. The spectral density of the noise sources is known from the components' noise models.

d. Assuming p uncorrelated stationary noise sources, the average NSD is given by

$$S_o(f) = \sum_{m=1}^{p} \sum_{l=-\infty}^{\infty} S_m(f + lf_s)|H_{-l,m}(j2\pi(f + lf_s))|^2 \tag{8.22}$$

where S_m denotes the spectral density of the mth noise source, and $H_{-l,m}(j2\pi(f + lf_s))$ denotes the $-l$th-order harmonic transfer function from the mth noise source to the output.

e. In the equation above, the index m sums over all the noise sources, while the index l sums over all frequencies of the form $(f + lf_s)$. Recall that noise from all these frequencies folds to the frequency f. In practice, the infinite inner summation has to be approximated as a finite one. The limit on l is thus restricted, and is referred to as the maximum number of sidebands. The limit, which is an input to be specified by the user, is not known a priori. A good strategy to achieve reliable results to keep doubling the maximum number of sidebands until the change in $S_o(f)$ is negligible.

f. Integrating the average NSD over the frequency range $0-\infty$ gives the total integrated noise.

g. In many cases, it is the sampled output of the switched capacitor network that is relevant. For example, a switched-capacitor filter might be driving an ADC. The continuous-time output of the filter is sampled by the ADC once every clock period. It is the noise at the sampling instants that is relevant. RF simulators have the option to determine the spectral density of the sampled noise, which must be integrated over the frequency range $0-f_s/2$ to yield the mean-square noise.

8.6 The History of Switched-Capacitor Filters

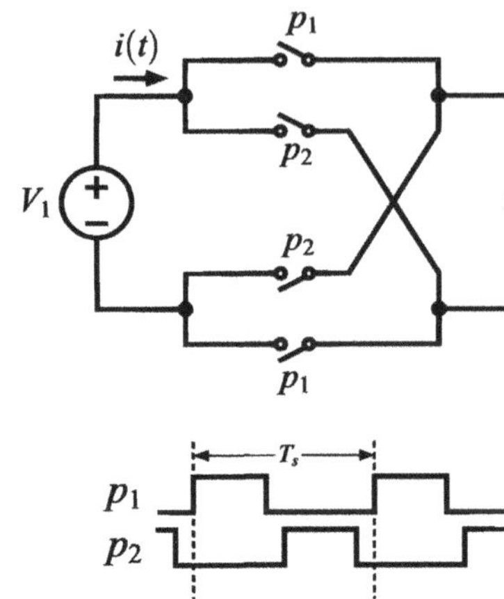

Figure 8.47 Maxwell's method of capacitance measurement. The average value of $i(t)$ is proportional to C.

The key idea of switched-capacitor (SC) circuits, which is to use switches and capacitors to realize resistors, is surprisingly old. James Clerk Maxwell published it in his groundbreaking book [6] nearly 150 years ago. He used it to show how a test set-up for measuring resistors can also measure the size of capacitors. He connected a capacitor C to a constant voltage V_1 through polarity-reversal switches as shown in Figure 8.47. Each reversal caused a charge $2CV_1$ to flow through the capacitor. If the reversals are performed with a period T_s, the average current will be $\overline{i(t)} = (4C/T_s)V_1$. This is the same value as for a resistor $R = T_s/4C$. Thus, by measuring the average current through C, its value can be found.

Nearly a hundred years later, Fried extended the concept to the design of sampled-data filters [7]. It took three more years before the idea of switched-capacitors was applied to analog-to-digital converters [8], and five years before researchers used it to design a variety of switched-capacitor (SC) filters [9]. Switched-capacitor circuits have since been (and continue to be) used in countless applications. They cover communications, consumer electronics, aerospace, and biomedical devices, containing filters, amplifiers, data converters and other analog stages.

Chapter Summary

- **Motivation and Principle**
  - Continuous-time active filters suffer from poor accuracy due to large variations in resistors and absolute component values.
  - Switched-capacitor (SC) circuits replace resistors with capacitors and switches, realizing time constants using capacitor ratios and clock frequency, both of which can be made highly accurate.
  - A switched capacitor emulates a resistor with an equivalent value $R_{eq} = 1/(f_sC)$.

- **First-Order SC Filters**
  - Replacing the resistor in an RC lowpass filter with a switched capacitor yields a first-order SC filter.
  - The filter's transfer function depends only on capacitor ratios and f_s, making it robust to process, voltage, and temperature variations.
  - At low frequencies ($f \ll f_s$), SC filters closely approximate their continuous-time RC counterparts.
- **Sampling Effects**
  - Sampling introduces aliasing, requiring an anti-alias filter before the SC network.
  - The sample-and-hold operation produces spectral images, which are removed by a reconstruction (smoothing) filter.
  - The hold operation introduces sinc droop, usually negligible when f_s is much larger than the signal bandwidth.
- **Switched-Capacitor Integrators**
  - Integrators are the fundamental building blocks for SC filters.
  - Practical integrators are realized using opamps (or OTAs) and exploit the Miller effect to achieve effective infinite capacitance.
  - **Stray-insensitive integrators** eliminate sensitivity to parasitic capacitances by careful switch placement.
  - Both delaying and non-delaying, inverting and non-inverting SC integrators are possible.
- **SC Filter Design**
  - Higher-order filters are built by replacing continuous-time integrators in RC prototypes with SC integrators.
  - First- and second-order (biquad) SC filters are derived directly from active-RC prototypes.
  - Filter parameters such as cutoff frequency, center frequency, and quality factor depend only on capacitor ratios and clock frequency.
  - High-Q designs use capacitive damping to reduce capacitor spread and improve robustness.
  - Dynamic-range scaling and noise scaling are achieved by appropriate capacitor sizing.
- **Switched-Capacitor Amplifiers**
  - By periodically resetting the integrating capacitor, SC integrators can be converted into accurate voltage amplifiers.
  - Both inverting and non-inverting SC amplifiers are realizable.
- **Nonideal Effects**
  - Switch nonidealities: finite on-resistance, charge injection, and nonlinear resistance.

(Continued)

(Continued)

- Mitigation techniques: transmission gates, dummy switches, advance-cutoff switching, and bootstrapped switches.
- Opamp/OTA nonidealities: finite gain, offset, finite bandwidth, slew rate, and noise.
- Correlated Double Sampling (CDS) effectively cancels opamp offset and reduces low-frequency noise.

- **Noise in SC Circuits**
  - Sampling introduces a significant noise penalty due to aliasing of wideband thermal noise.
  - Thermal noise arises from switch resistance and opamps/OTAs.
  - Reducing noise typically requires larger capacitors and higher transconductance, increasing power consumption.
- **Implementation Considerations**
  - Single-supply and fully differential implementations are standard in practice.
  - Accurate capacitor matching is achieved using unit capacitors, common-centroid layout, and dummy structures.
  - SC circuits trade higher power and complexity for excellent accuracy and programmability.

Problems

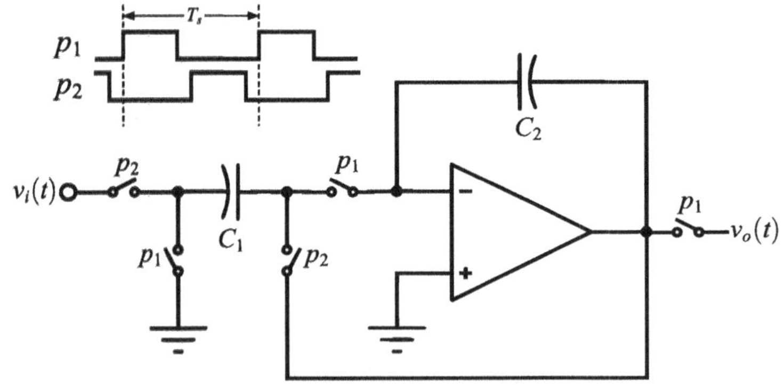

Figure 8.48

Problem 1. *For the circuit of Figure 8.48, find:*

a. *The difference equation between v_i and v_o.*

b. *The transfer function $H(z)$.*

c. *The values of C_1/C_2 for which the circuit is stable.*

What function will the circuit perform if $C_1 = C_2$? Comment on the sensitivity of this circuit to stray capacitances.

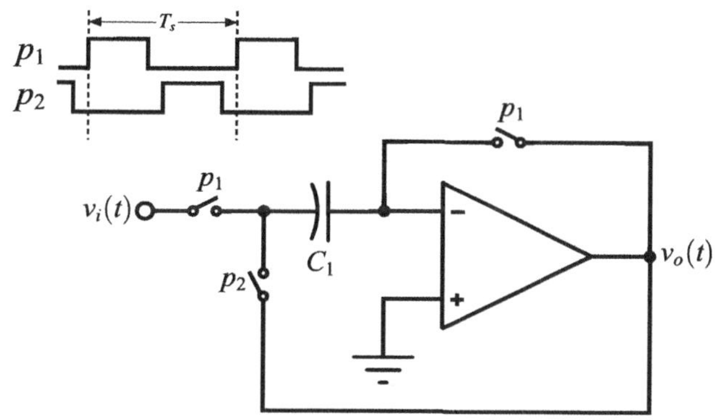

Figure 8.49

Problem 2. *The sample-and-hold stage shown in Figure 8.49 uses an opamp with a dc gain of 60 dB and an offset voltage of 3 mV. The input voltage is 1 V. Assuming ideal switches, find the output voltage when p_1 is high, and when p_2 is high.*

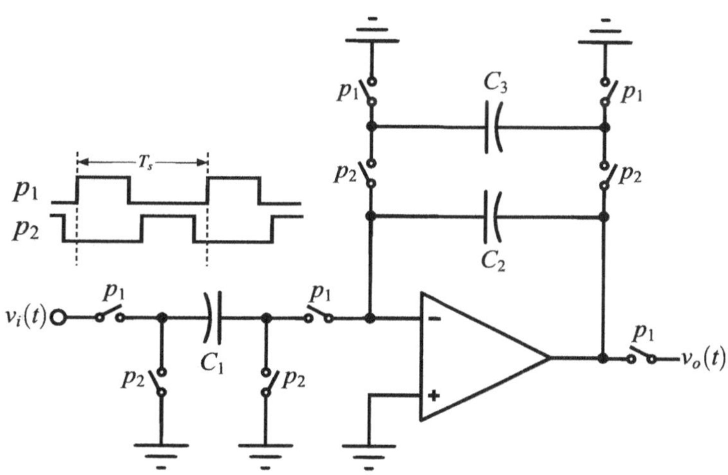

Figure 8.50

Problem 3. *In the circuit of Figure 8.50, all capacitors are of the same size. Find:*

a. *the transfer function $H(z)$ of the stage;*
b. *the pole and zero locations in the z-plane;*
c. *the dc gain of the stage;*
d. *the 3-dB frequency for $f_s = 10\,MHz$;*
e. *the output dc offset for an opamp with infinite dc gain and an input dc offset of 2 mV.*

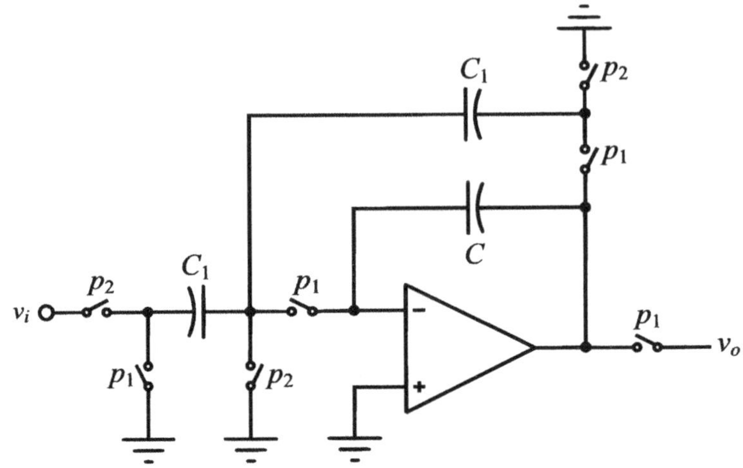

Figure 8.51

Problem 4. *In the circuit of Figure 8.51, determine the transfer function $H(z) = H(e^{sT_s})$ and the frequency response.*

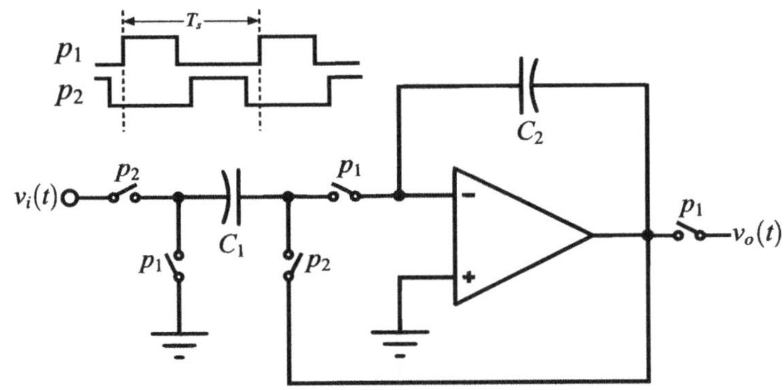

Figure 8.52

Problem 5. *In the circuit of Figure 8.52, the active element is an OTA with $G_m = 0.5$ mS. The switches have an on-resistance of 1000 Ω, and $C_1 = 0.2$ pF.*

Determine the transfer function from v_i to the voltage across C_1 during p_2. What is the time constant of the stage during sampling? What is the highest clock frequency for 0.1% and for 0.01% settling accuracy?

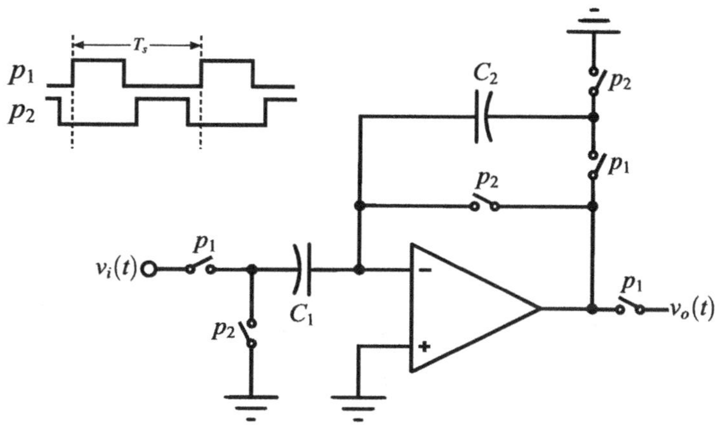

Figure 8.53

Problem 6. *In the circuit of Figure 8.53, $f_s = 10$ MHz, and 10% of each phase is allowed for slewing. Assume $C_1 = 2C_2 = 2\,pf$, and $|v_{i,\max}| = 1$.*

a. *Find the charge delivered by the amplifier during the two clock phases in the gain circuit shown below if the load is negligible.*

b. *Find the slew rate required.*

c. *How large can the output current be if all the charge is delivered during the slewing period?*

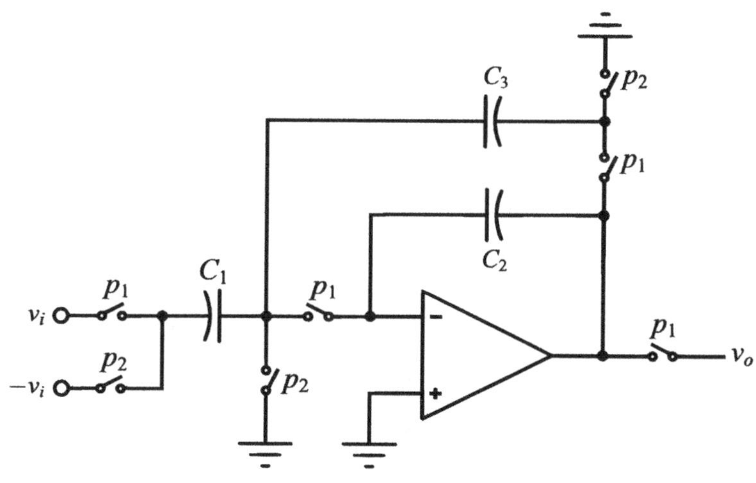

Figure 8.54

Problem 7. *In the circuit of Figure 8.54:*

a. *Find the transfer function of the filter stage shown.*

b. *Find the value of C_1 so that the dc gain is 0 dB.*

c. *Sketch the anticipated frequency response if $f_s = 100$ MHz. Calculate (or estimate) the -3 dB frequency.*

Problem 8. *Design a single-ended stray-insensitive switched-capacitor (SC) stage realizing the transfer function*

$$H(z) = \frac{-4 + 3z}{1 - z}$$

with all element values positive. The sum of all capacitances should be 10 pF.

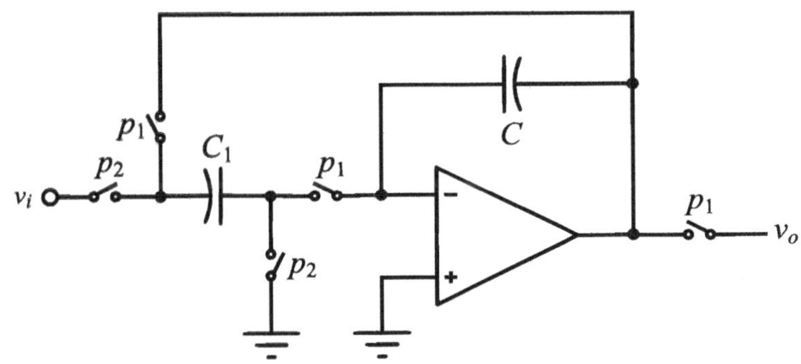

Figure 8.55

Problem 9. *For the SC stage in Figure 8.55, assume that both the input and output signals are sampled at the end of the clock phase p_1.*

a. *The transfer function H(z) under ideal conditions.*
b. *The DC gain of the stage.*
c. *The pole and zero values.*
d. *The steady-state output voltage for zero input and an input-referred offset v_{OS}.*
e. *Can capacitance inaccuracies cause the circuit to become unstable? Do they affect the DC gain?*
f. *The transfer function if a finite-gain opamp is assumed.*

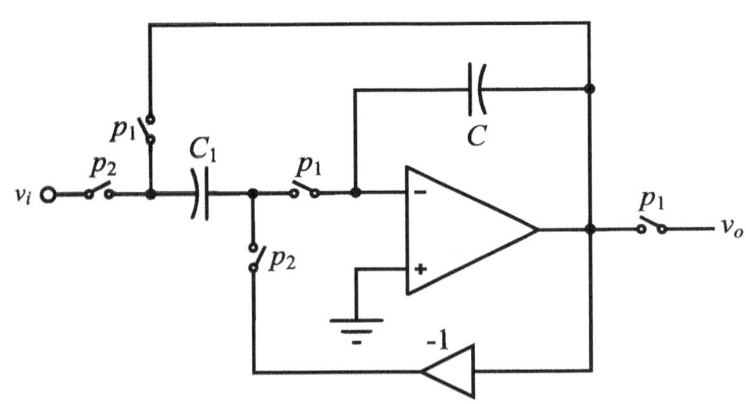

Figure 8.56

Problem 10. *Figure 8.56 shows a switched-capacitor network.*

a. *Determine its transfer function $H(z) = V_o(z)/V_i(z)$, assuming ideal operation. What function does the stage perform?*

b. *Find $H(z)$ if the opamp has a finite DC gain A.*

c. *Draw the differential implementation, without any inverters.*

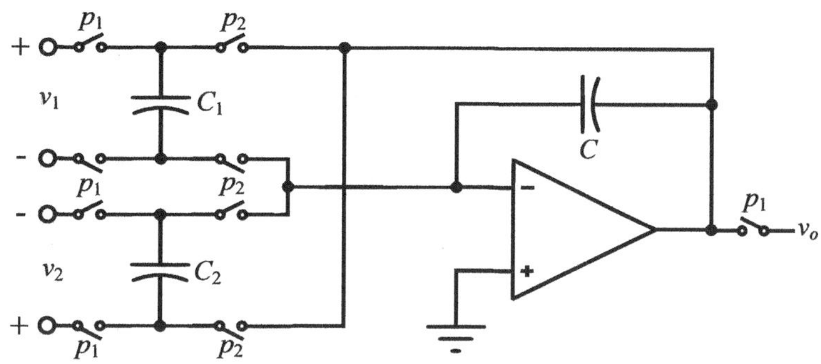

Figure 8.57

Problem 11. *For the circuit shown in Figure 8.57:*

a. *The output voltage $V_o(z)$ assuming ideal components.*

b. *The steady-state DC output offset due to an opamp input offset v_{os}.*

c. *The output voltage $V_o(z)$ if the opamp has a finite DC gain A.*

d. *Is the circuit guaranteed to be stable? Why or why not?*

e. *What are the limitations on the DC gains?*

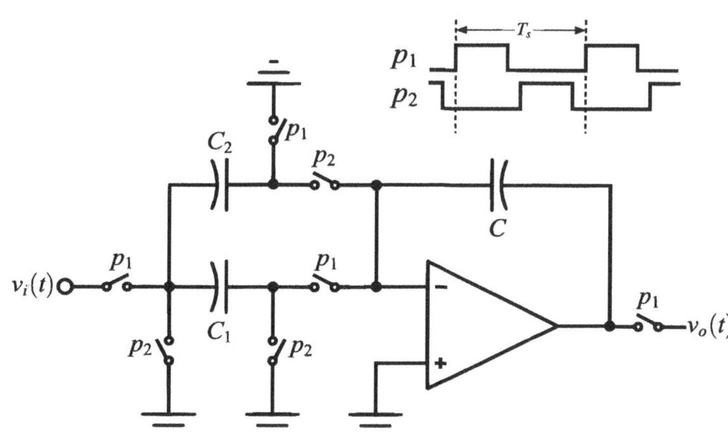

Figure 8.58

Problem 12. *Find the transfer function of the circuit of Figure 8.58. How can the circuit be simplified if $C_1 = C_2$?*

Problem 13. *What is the unity-gain frequency f_u of an SC integrator with*

$$H(z) = \frac{kz^{-1}}{1 - z^{-1}}$$

for $k = 1, 2,$ and 4? What is the phase shift at f_u for $k = 1$?

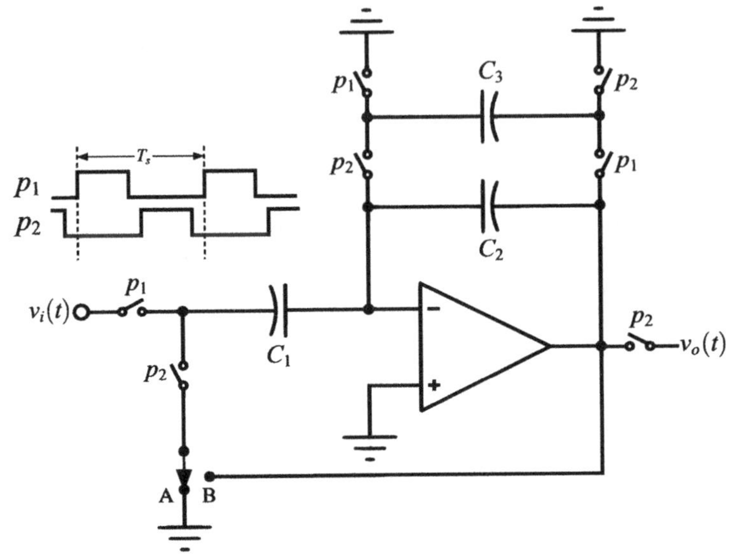

Figure 8.59

Problem 14. *Analyze the circuit shown in Figure 8.59. Is it stable? Why or why not? What if node (A) is connected to the opamp's output terminal, rather than ground?*

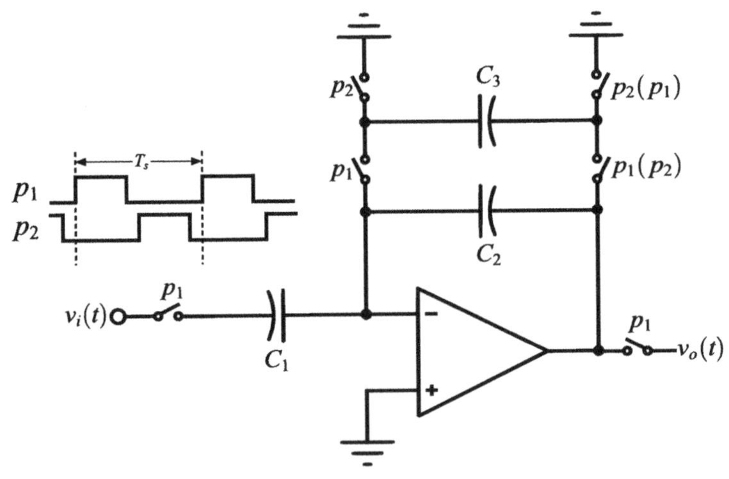

Figure 8.60

Problem 15. *The circuit of Figure 8.60 has the transfer function*

$$\frac{V_o(z)}{V_i(z)} = -\frac{z-1}{2z-1}$$

The smallest allowable capacitor value is 0.1 pF.

a. *Find all capacitor values.*
b. *Draw (qualitatively) the frequency response of the stage.*
c. *What are the lowest and highest values of the gain? At what frequencies do they occur?*
d. *What will the circuit do if two of its clock signals are accidentally interchanged, as shown in the figure? Why?*

References

1 R. Gregorian and G. C. Temes, *Analog MOS Integrated Circuits for Signal Processing*. Wiley-Interscience, 1986.

2 D. A. Johns and K. Martin, *Analog Integrated Circuit Design*, 2nd. John Wiley & Sons, 2011.

3 K.-L. Lee and R. Meyer, "Low-distortion switched-capacitor filter design techniques," *IEEE Journal of Solid-State Circuits*, vol. 20, no. 6, pp. 1103–1113, 1985.

4 C. C. Enz and G. C. Temes, "Circuit techniques for reducing the effects of op-amp imperfections: Autozeroing, correlated double sampling, and chopper stabilization," *Proceedings of the IEEE*, vol. 84, no. 11, pp. 1584–1614, 1996.

5 R. Schreier, J. Silva, J. Steensgaard, and G. Temes, "Design-oriented estimation of thermal noise in switched-capacitor circuits," *IEEE Transactions on Circuits and Systems I: Regular Papers*, vol. 52, no. 11, pp. 2358–2368, 2005.

6 J. C. Maxwell, *A Treatise on Electricity and Magnetism*, 1st. Oxford: Clarendon Press, 1873, vol. 2.

7 D. L. Fried, "Analog sample-data filters," *IEEE Journal of Solid-State Circuits*, vol. 7, no. 4, pp. 302–304, 1972.

8 J. L. McCreary and P. R. Gray, "All-MOS charge-redistribution analog-to-digital conversion techniques—Part I," *IEEE Journal of Solid-State Circuits*, vol. 10, no. 6, pp. 371–379, 1975.

9 B. J. Hosticka, R. W. Brodersen, and P. R. Gray, "MOS sampled data recursive filters using switched capacitor integrators," *IEEE Journal of Solid-State Circuits*, vol. 12, no. 6, pp. 600–608, 1977.

Index

.PAC 274, 432
.PNOISE 323, 454
.PXF 274

a

adjoint 41, 45
  biquad 145
  LPTV
    signal-flow graph 275
    time domain 283
advance-cutoff switching 443
approximation
  Butterworth 337–343
  Butterworth versus Chebyshev 352
  Chebyshev 345
  magnitude 334–337
  maximally flat 337–343
attenuator
  scattering matrix 182

b

bilinear
  extra-element 33
biquad
  differential 378
  gm-C 396
  opamp
    finite bandwidth 384
  predistortion 392
  Q enhancement 390
  Rauch 363
  Rauch versus Tow-Thomas 364
  standard form 340
  Tow-Thomas 358, 361
    differential 384, 385
    noise 369
  transfer functions 47
Bode's Noise Theorem 77
bounce diagram 165
Butterworth
  derivation 337
  highest-Q pole 343
  polynomial 342
  second order 340

c

CCVS
  LPTV 224
Chebyshev
  even order 351
  highest pole-Q 349
  odd order 346
  pole locus 348
  polynomial 346
chop
  LPTV 278
compression
  1 dB point 114
  gain 114
convolution
  linear time invariant 202
current injection
  derivation 122–149
  first-order filter 139

d

direct-conversion receiver 197
distortion
  biquad 138
  common-collector amplifier 137
  differential operation 112–113
  HD3 114
  IIP3 118
  IM3 115
  nonlinear capacitor 134–136
  second order 106
  third order 114
  Tow-Thomas biquad 140
distributed RC line 171–174
  modeling 173

e

extra-element
  example 35
  theorem 31–36

f

filter
  active-RC 356
  continuous-time
    dynamic-range scaling 374–378
    noise 367–373
  Rauch 363
  realization 353–367
frequency scaling 336
frequency-reversal
    theorem 262–266
  impedance 269

g

gyrator 399
  scattering matrix 184

h

harmonic transfer function
  definition 211
  switched-RC network 267
harmonic transfer functions
  calculation 215–234
HD2 106

i

IIP2 110
  cascade 112
IIP3
  cascaded stages 120
IM2 109, 110
integrator
  active-RC 354
  bottom-plate capacitance 366
  differential 382
  feed-forward 394
  gm-C 395
  opamp 393
    finite bandwidth 385
inter-reciprocal
  biquad 145
inter-reciprocity 39, 41, 44, 45
  opamp 47
intercept point
  second order 110
  third order 118
intermodulation
  second order 107–111
  third order 114

l

linear time-varying
  frequency response 204
linear time-varying
  impulse response 203
linear time-varying
  definition 198
lossless
  scattering matrix 183
LPTV
  adjoint 261
    signal-flow graph 275
    Zadeh expansion 277
  chop 278
  frequency response 209
  frequency reversal 262
  impulse response 208
  MNA
    equations 219
  modulated input 299–304

noise 304
offset sampling 297
Periodic Steady State 226
sampled output 287–299
first-order delta sigma 295
switched-RC network 294
time reversal 261, 285
transfer function 261
Zadeh expansion 211
LPTV circuits
admittance 234–236
Impedance 234–236

m

MNA 36–39
capacitor 28
LPTV 219, 261
CCCS 21
LPTV 223, 271
CCVS 22
LPTV 273
conductance 17
LPTV 219, 261
current source 17
equations
LPTV 219
inductor 28
opamp 22, 24
stamp
transposed 36
stamp summary 27
VCCS 19
LPTV 223, 272
VCVS 20
LPTV 222, 270
voltage source 19

n

N-path
harmonic transfer functions 243
impulse response 245, 246
noise 320
principle 238–248
noise
attenuator 82
differential 88
gm-C 400
impedance scaling 86
input referred 84
integrated
switched RLC 318
integrator
switched-capacitor 452
LPTV 304
LTI system 65
MOSFET 63
multiple sources 69
N-path 320
one port 69–72
RC circuit 66
resistor 61–68
simulation
switched-capacitor filters 454
spectral density 64
RLC 71
switched-capacitor
integrator 454
passive 451
switched-capacitor
circuits 450
total
RLC 72–79
switched-capacitor
filter 322
Tow-Thomas 369
track-and-hold 67
two port 84
two-port 79–86
noise factor
common source amplifier 92
measurement 93
two port 89–91
noise spectral density
first-order switched capacitor filter 314
switched capacitor filter
charge rotating 316
switched RLC 310

non-overlap
  clock generator 443
nonlinear
  VCVS 129
nonlinearity
  negative feedback loop 132
Norton equivalent
  LPTV networks 236–238
null double injection 35
Nyquist noise theorem 71
  switched RLC 312

p
Periodic Steady State 226
predistortion 120, 133

r
R-2R DAC 46
reciprocity 41, 44
  LPTV 257
  scattering matrix 184

s
S-parameters
  definition 178
  motivation 175–183
sampling
  bootstrapped switch 443
sensitivity 49–54
simulation
  .PNOISE 323
  noise
    switched-capacitor
      filters 454
  periodic AC 274
  periodic noise 323
  periodic transfer function 274
state-space
  descriptor 29
switch
  charge injection 442
    mitigation 442
  transmission gate 440
switched RC network
  noise 313
switched-capacitor
  amplifier 434, 436
  biquad
    high Q 431
    low Q 428
  capacitor array 444
  filter
    dynamic-range scaling 433
    first-order 424, 426
    noise scaling 433
    passive 411
    simulation 432
  integrator
    correlated double
      sampling 449
    dc negative feedback 436
    development 417
    finite bandwidth OTA 449
    fully differential 438
    offset 447
    OTA 439
    stray insensitive 421, 423
    stray sensitive 418
    transfer function 416
  resistor 410
  signal processing 414
  single supply 438
switched-RC circuit 231

t
Tellegen's Theorem 10–14
Thevenin equivalent
  LPTV networks 236–238
time reversal
  LPTV 285
time-reversal 261
Tow-Thomas 358
  noise 369
transfer function
  geometric interpretation 386
transfer function theorem 262
transmission line
  capacitance 159
  characteristic impedance 158
  circuit analysis 160–171

impedance 168
inductance 159
quarter wave 171
structure 156–160
VSWR 167
tuned-RF receiver 195
problems 196

v
vector network analyzer 185–188
calibration 186

y
Y-factor 94
Y-to-S matrix 181

z
Zadeh expansion 211
LPTV
adjoint 277
mixer-first form 213

Printed and bound by CPI Group (UK) Ltd, Croydon, CR0 4YY
17/09/2026
14964335-0007